# BOTANICAL AND MEDICINAL PLANTS

## (As Depicted in Ancient Texts, Art & Archaeology from Dawn of Civilization to the Modern Age)

## Vol. 1

AF810275

**Shantilal Nagar**

**B.R. Publishing Corporation**
[A Division of BRPC (India) Ltd.]
Delhi-110035

*Distributed by :*
**BRPC (India) Ltd.**
4222/1, Ansari Road,
Darya Ganj, New Delhi-110002
Phone : 3259196, 3259648
Fax : 3201571
E-Mail :*info@brpcltd.com*

Our Address:
425,Nimri Colony, Ashok Vihar
Phase-IV, Delhi-110052
e-mail: *brpcltd@del2.vsnl.net.in*

© 2000 Shantilal Nagar (b. 1927—)

ISBN 81-7646-140-7 (Set)
ISBN 81-7646-141-5 (Vol. 1)

*All rights including the right to translate or to reproduce this book or parts thereof except for brief quotations, are reserved.*

*Published by :*
**B. R. Publishing Corporation**
[A Division of BRPC (India) Ltd.]
3779, Ist Floor, Kanhaiya Nagar,
Tri Nagar, Delhi-110035
Phone : 7152140
E-Mail : *info@brpcltd.com*

*Typeset at :*
Amit Printers, Delhi-110006

*Printed by :*
PRAJA Offset, Delhi

**PRINTED IN INDIA**

*Dedicated to*

*Mrs. Manjula Dave*
*&*
*Dr. P. K. Dave*

4111
JNBH
C2V015

# Preface

ते र्णैश्चापि वृतो गिरीशः, स शोभते मुक्त जटाग्रभारः।
थावेन सर्ज्जकदम्बमध्ये, प्ररोहमूलोऽथ वनस्पतिर्वै।।

—*Vāmana Purāṇa*, 26.71

(Śiva surrounded by his *gaṇas* and the gods, with dishevelled locks of hair, looked like a huge banyan tree, in the forest of *Sāla* (*Vateria Indica* Linn) and *Kadamba* (*Anthocephalus indicus*).

The human association with plants and vegetation is as old as the advent of man on earth. Indeed the texts prescribe that the vegetational growth preceded the humans on earth. This seems quite logical because no human can survive on earth without vegetation, which happens to be the primary source for the supply of oxygen, so essential for human survival. Though man at the early stages of his existence on earth was quite unaware of this aspect, yet still he made use of the trees and plants as the sources for the supply of edible fruits and shelter from wild life. Gradually he became quite familiar with plant life and found them a natural resort against the hostile forces of nature like storms, floods and other natural calamities, besides the ferocious birds and beasts.

With the dawn of civilization, man found the need for covering his body and at that early stage, the tree-bark was found by him to be quite useful for the purpose; with the domestication of animals the use of vegetation as fodder for the animals was well realised by him; with the expansion of society, the caves and caverns were found to be inadequate and the need for building dwelling places dawned on him. For this purpose the wood could easily be available from trees and even the dwelling places were perched on the trees. Then the so bountiful nature provided him the fruits and tree roots to appease his hunger. As time passed he also learnt of making fire by rubbing two sticks for which the burning wood was also provided by the plant kingdom. While taking to farming, man came to know the utility of cotton which could be procured out of cotton seeds. This made a revolution in his life and he gradually replaced the bark garments with those of the woven cotton ones.

Since the early man was dwelling mostly in the forests, he gradually became conversant with even the medicinal properties of the plants and started using them for the cure of various types of ailments. The continued stay of man in the environments of nature made him observe the various qualities in the plant kingdom, which he made use of variously in his daily life. Though with the advancement of civilization man left his rural abodes, patronising urban habitats, but still the ascetics, sages and *ṛṣis* never left their forest abodes, which served as great centres of learning, knowledge and educational institutions. These recluses, by their constant research, particularly on medicinal properties of plants, served humanity at large in order to remove their human disorders. Not only that, a number of treatises on medical science were composed by these sages after conducting years of research on medicinal plants. Besides the above, the ancient Indians also developed the knowledge of the plant kingdom in general, which is evidenced from the use of the epithet *Vṛkṣāyurveda* (which means the science of the treatment of plant diseases) and *Sitādhyakṣa* (Superintendent of agriculture) by Kauṭilya in the *Arthaśāstra*.

The study of the science of botany by the ancients was quite deep rooted, and as a result of which they could even provide the classification of plant-life. These classifications were based on the botanical names of the plants; their properties (botanical) and those having food value. The earliest classification is available in the *Ṛgveda*, wherein they have been defined as—

(i) *Phalinī*—those bearing fruits.

(ii) *Aphalā*—those trees or plants which do not bear fruits.

(iii) *Apuṣpa*—those which do not blossom.

(iv) *Puspint*—those which bear flowers.

The plant life is further divided into *Vṛkṣa, i.e.*, trees; *Gulma, i.e.*, shrub; *Bhiṣaja* meaning medicinal plants; *Latā, i.e.*, creepers and *Oṣadhi*, a class of plants which could be used to cure certain ailments of human body. In the *Dharmaśāstra* two broad divisions of plant-life, *i.e., Vanaspati* and *Oṣadhi* are quite common. The *Baudhāyana Dharmasūtra* (study of which has been made by Govindasvāmin), identifies three types of plants, *viz., Oṣadhi, Vallyauṣadhi* and *Tṛṇauṣadhi;* the former class belongs to *Kuluttha*, etc. and to the latter *Vṛihi* (paddy). Pāṇini on the other hand classifies forests as—

(i) **Oṣadhi-Vana**—those producing herbs; *viz., Dūrva-Vana, Mūrva-Vana* and *Vanaspati-Vana*—those producing timber trees, *e.g., Devadāru-vana* (VIII. 4.8).

(ii) Pāṇini's use of the word *vana* had a double meaning which denoted natural and cultivated forests. The natural forests are like *Puragā-vana, Miśraka-vana* etc. (VIII.4.4.) and cultivated gardens could denote *Āmravana, Ikṣuvana*.

(iii) The ordinary forest was called *Āraṇya* (IV.I.49). The same position is testified by Kātyāyana in *Vartikā*. The dense forest was termed as *Āraṇyāṇi*.

Manu on the other hand, provides the following classification for the trees and plants—

उद्भिज्जा: स्थावरा: सर्वे बीजकाण्ड प्ररोहिण:।
ओषध्य: फलपाकान्ता बहुपुष्प फलोपगा:।।
अपुष्पा: फलवन्तो ये ते वनस्पतय: स्मृता:।
पुष्पिण: फलिनश्चैव वृक्षास्तू भयता स्मृता:।।
गुच्छ गुल्मं तु विविधं तथैव तृण जातय:।
बीजकाण्ड रुहाण्येव प्रताना वल्लय एव च।।

—*Manusmṛti*, I.46.48

Thus, according to Manu (i) *Oṣadhi*—means a class of plants which dies after the ripening of fruits; (ii) *Vanaspati*—represents those plants which bear fruits but not flowers;

(iii) *Vṛkṣa*—denotes those species of trees and plants, which bear either fruits or flowers alone; (iv) *Gulma*—denotes shrub; (v) *Tṛṇa*—the grass, (vi) *Pratana*—are the plants with tendrils; *valli*—means a creeper entwining a support. The *Sūtrasthāna* (I.36.37) or Caraka (*Upavana-vaioda, Vanauṣadhi-varga, Vaiṣyavarga, Rāmaliṅgānuśāsana*) and Śuśruta provide almost identical classification of trees and plants, with minor variations here and there.

The *Purāṇas* also provided the classification of plant-life in more than one way. As for example the *Bhāgavata Purāṇa* pronounces that there are six types of trees and plants on earth—

सप्तमोमुख्यसर्गस्तु षड्विधस्तस्थुषां च यः।।
वनस्पत्योषधिलतात्वेक्सारा वीरुधो द्रुमाः।
उत्स्रोतसप्तमः प्राथा अन्त स्पर्शा विशेषिणः।।

—(III. 10.18-19)

These species have been classified as *Vanaspati, Oṣadhi, Latā* (creeper), *Tvakasara* (those trees which have hard bark; *i.e.,* the bamboo etc.); *Virudha* (shrubs looking like clumps); *Druma* (those trees which bear flowers and fruits). The *Vāyu Purāṇa* divides the trees into two broad classes *viz., Grāmyāraṇya* and *Yajñiya Oṣadhi*—

ग्राम्यारण्याः स्मृता ह्येता ओषध्यस्तु चतुर्द्दश।
उत्पन्नाः प्रथमा ह्येता आदौत्रेता युगस्य तु।।
अफालकृष्टा ओषध्यो ग्राम्यरण्यास्तु सर्वशः।
वृक्षागुल्मलतावल्ली वीरुधस्तृण जातयः।।
मूलैः फलैश्च रोहिण्यो गृह्णन पुष्पैश्च जायते।
पृथ्वी दुग्धा तु बीजानि यानि पूर्वं स्वयम्भुवा।।
ऋतु पुष्पफलास्ता पै ओपध्यो जज्ञिरे त्विह।
यदा प्रसृष्टा ओषध्यो न प्ररोहन्ति ता पुनः।।

—*Vāyu Purāṇa*, 8.149-152

The *Matsya Purāṇa* (8.2-3; 154.304; etc.) and other Purāṇic texts, also toe the same line of the classification of plant-life, with minor deviations here and there.

The texts on the Indian system of medicines, like Caraka in *Sūtrasthāna*, however, provide classifications of vegetational growths on the basis of their purgative and astringent properties. According to Caraka, 600 species belong to the former class while about 500 to the latter class. The latter class has further been subdivided into ten *Vargas* (groups). Śuśruta on the other hand (*Sūtrasthāna*-XXXVIII) classifies the entire plant kingdom into 37 classes (*gaṇa*)—

(i) *Dhānya-Varga*—those corns which have husk.

(ii) *Samidhāna-Varga*—twelve kinds of corns, like black grams, etc.

(iii) *Śāka-Varga*—represents the vegetables.

(iv) *Phala-Varga*—represent trees bearing fruits.

(v) *Harid-Varga*—(green class) like *ārdraka* and the like which grow underground.

(v) *Āhārayogi-Varga*—seeds which produce oils.

In the course of time, trees became objects of veneration. The earliest evidence of this is available in the Stūpas of Bharhut, Sanchi and Amarāvatī, where the devotees are seen kneeling before a *Caitya-Vṛkṣa*. Such veneration of trees was quite common in those times, and this has come down even to the modern age. The veneration of plant kingdom can also be observed in the artistic representation of the subsequent period. This aspect is available in the Vedic literature too wherein Vanaspati (Lord of Forest) and Araṇyānī (the forest deity) have been mentioned. The literature of the subsequent period also carry similar traditions, and some of the trees like Pīpal, Tulasī and Udumbara are still considered to be sacred. In fact no Hindu household with oriental leanings is considered complete unless the Tulasī plant is enshrined in the centre of its courtyard.

The *Atharvaveda*, however, introduced the use of the wood of certain trees for making amulets to ward off the ill effects of evil spirits or the removal of ailments. The *Rgveda*, on the other hand, prescribes the dice to the made of *Vibhitakā* tree (*Rgveda*, VII. 86.6 and X.34.1). The same text also specifies that the juice extracted from the Soma plant was regarded as the most exhilarating drink. The wood of Udumbara has been prescribed for use in sacrifice, while the Khadira wood was used for making *Yūpas*, stakes for animal slaughter in religious rites and as the staff of a Kṣatriya Brahmacārī. The Bilva and Palāśa trees provided the staff of the Brāhmaṇa Brahmacārī.

The ancient botanists had sufficient knowledge of keeping the ground fertile and sappy. The fact that the trees and plants derived sustenance from earth was quite well known to them. The *Agni Purāṇa*, the *Bṛhatsaṁhitā* and the *Upavanavinoda* are quite vocal on the subject. To admixture of the quantities of Sesamum, goat's dung, barley powder, and water was regarded as a very effective manure in addition to the cowdung manure. The *Bṛhatsaṁhitā* contains detailed procedure for preserving seeds. Seeds, mixed with *ghee* are to be thrown into the milk; on the day following they are to be kept separately. This process has to be repeated continuously for ten days and then the seeds are required to be smeared with cowdung. Then the seeds are required to be heated in pots full of the meat of the boar or a deer. The seeds so processed when sown would be useful for the growth of flowering trees. A different method is prescribed for the treatment of seeds with a view to growing trees, rich in foliage. For the augmentation of growth of trees, particularly the mango trees, the washing of the fish is to be sprinkled on them according to the *Agni Purāṇa* (Chapter-194).

The *Agni Purāṇa* (Chapter-281) as also the *Bṛhatsaṁhitā* and the *Upavanavinoda* also deal with the treatment of various ailment of plant-life. Guṇaratna has also dealt with the subject in some detail. According to him, the excessive increase or decrease of heat and violent winds are some of the causes of the maladies of the trees. Yellow colour of leaves, drying of buds without fully blossoming, dryness of branches, exuding of juice are some of the many symptoms of their diseases. A decoction of *Kulattha* (*Dolichos biflorus*), *Māṣa*

(*Phascolus radiatus Var Roxburgi*), *Mudga* (*Ph. mungo*), *Tila* (*Sesamum indicum*) and *yava* (barley), is to be applied to the roots of trees in order to cure them of barrenness. To save the trees and plants from various diseases, mud mixed with ghee, milk mixed with water and *Vaḍaṅga* (*Embilia ribes*) mixed with flesh and fish have been required to be applied to the roots of the trees.

It is also conceived in the ancient texts that the trees and plants also have life. The contraction of leaves at night of certain trees has been described as their sleep. There are certain flowers, which bloom during the day but get closed at night; the lotus flower could be mentioned in this connection. Then there are certain plants, the flowers of which spread their fragrance either at night or during the day time. The sleep, waking up and reaction to touch have been mentioned in the Dharmottara's commentary on *Nyāyabindu* and Udayana's *Kirṇavali.* Guṇaratna in his commentary has noticed the reaction while touching creepers and shrubs like *Minosa pudica.* Some treatises equate the trees with human beings and claim that they also have infancy, childhood, youth, old age and death. The death is attributed to the lack of food and disease.

In the course of time, flowers too were considered as important item of daily life, not only by the people at large, but also by the gods. With the people no function was complete, be it social or religious one, without the personal floral adornments, especially of the ladies. The flower gardens became an essential part of the monarchy; Vātsyāyana, on the other hand, prescribes a garden for every householder. With the gods, the flowers had a different usage. They used to shower flowers or flower-petals on the victory of gods over the demons, or other auspicious occasions. The practice of planting saplings of shady fruit trees along the public pathways, besides consecration of public gardens were considered to be of special merit. The Purāṇic accounts have been quite vocal in this connection—

(i) *Agni Purāṇa*—70.1-9.

(ii) *Matsya Purāṇa*—154-511-12.

(iii) *Varāha Purāṇa*—170.34-36.

(iv) *Padma Purāṇa*—Chapter-26.

Large chunks of land in the country are hit by drought and it has been a problem to locate the underground water for human consumption as well as for plant life. *Bṛhat Saṁhitā* (Chapter-54) has dealt with the subject, and Varāhamihir finds close connection between the trees and the location of water resources, as for example—

(i) If there be an ant-hill to the east of a Jambu tree, and near it, there runs a current to the south of the ant-hill at the depth of two men.

(ii) At the distance of three cubits to the west of Udumbara and at the depth of two men and a half, there runs a current of good water.

(iii) If there be an ant-hill to the north of an Arjuna tree, there will a current at the distance of 3 cubits to the west of the hill at the depth of three men and a half.

(iv) If the plant known as Nirguṇḍi be seen to grow on an ant-hill, there will be a current at a distance of three cubits to the south of the hill and at the depth of two men and a quarter.

(v) If there be an ant-hill to the east of a Badarī tree, then to the west of it, there will be water at the depth of three men.

(vi) If an ant-hill overgrown with the Kuśa-grass be seen on the north-eastern side of Kovidara tree, there will be current at the depth of four men and a half between the tree and the ant-hill and the water will be inexhaustible.

Varāhamihir further goes on to claim in Chapter-29 of the *Bṛhat Saṁhitā*, that the blossoming of certain flowering trees results in having good harvest of certain crops of cereals. As for example, if a Sāla tree should blossom and bear fruit, Kalama rice will grow in abundance; if the red Aśoka should bear fruits and flowers, red paddy will grow; if Karika should blossom and bear fruits, white paddy grow and if black Aśoka should blossom and bear fruit, the black rice will grow. Further, according to him, the growth of the Jambu (rose-apple) tree indicates the growth of gingelly and black gram.

The growth of trees mainly depends on the soil and this happens to be the main source of production of plant life. The ancients were quite well aware of the fact that all the plants cannot grow in all types of soils. Caraka (*Kalpasthāna*-I) has accordingly mentioned the different types of soils and the set of trees that could be grown on each one of the said type—

(i) *Jaṅgala* (Arid)—The trees like Khadira (*Acacia catechu*), Sāla (*Sorea robusta*), Badarī (*Zizyphus jujuba*) which fall under Xerophytes can be grown there.

(ii) *Anupa* (marshy)—This type of land is useful for the growth of thorny shrubs, bushes and other trees like Hintala (*Phoenix peludos*) and the like. Hintala indeed is a gregarious palm, which forms impenetrable thorny thickets throughout the Sundarbans and the Andaman Islands. The trees growing on such soils are called Hydrophytes or Hygrophytes.

(iii) *Common soil*—is useful for the growth of all types of plants.

The above classification of soil compares with the one available in the *Upavanavinoda* and Śuśruta (*Sūtrasthāna*-35). The knowledge of botanical science with the ancients, indeed was sometimes so perfect that by observing certain symptoms in the trees and plants, they could predict even the rainfall, drought and other weather forecasts.

The fruits have their own importance in the country from a very remote past. The fruit trees are found mentioned in the Vedic as well as post-Vedic literature; tradition of offering fruits to the guests has been testified in the Epics and the subsequent literature, while the evidence of the use of fruits is also available in the Indus civilization of Harappa and Mohenjodaro. Indeed different varieties of fruits had been used in the country, ever since the advent of man. A vast change has occurred in the use of fruit and vegetables during the past four thousand years or so; some of them have become obsolete while some new items have been introduced. Of course, banana happens to be one of the fruits which was commonly used by the Harappan people and is still a fruit of the masses.

Besides the banana, mango enjoys an exalted position in the list of fruits in the texts. The *Padma Purāṇa* attaches great merit to the planting of the mango fruit trees—

नाकमेति स चूतस्य समारोप्यसहस्रकम्।
ततो द्वि त्रिगुणेनैव न्यूने वा प्रचुरेऽपि वा।।
आरोग्यं शौर्य सम्पन्नमारामादेव जायते।
फलानि यस्य खादन्ति जन्तवोऽथ सहस्रशः।।

—*Padma Purāṇa*, I.60.32, 34

The *Matsya Purāṇa,* on the other hand, conceives that the trees and plants cry, laugh, eject sap, drop their branches without any cause or high wind. Sometimes the plants which are only three years old start bearing fruit and trees begin to overflow with milk, oily substances, blood, honey or water and sometimes suddenly dry up without disease; sometimes the dried up begin to sprout, the fallen ones stand up, the standing ones fall down; all such symptoms forebode ill of them. There is also a mention of the weeping of the trees in the same text, which is believed to indicate the possible spread of an epidemic in future and the country on the other hand becomes annihilated by their laughing—

पुरेषु येषु दृश्यन्ते पादपादेव चोदिता।
रुदन्तो व हसन्तो वा स्रवन्तो वा रसान् बहून्।।
अरोगा व विना वातं शाखां मुञ्चत्यथ द्रुमाः।
फलं मूलं तथा कालं दर्शयन्ति त्रिहायनाः।।
पूर्ववत्स्वं दर्शयन्ति फलं पुष्पं तथान्तरे।
क्षीरं स्नेहं तथा रक्तं मधु तोयं स्रवन्ति च।।
शुष्यन्त्य रोगाः सहसा शुष्का रोहन्ति वा पुनः।
उत्तिष्ठन्तीह पतिताः पतन्ति च तथोऽत्यिताः।।
तत्र वक्ष्यामि ते ब्रह्मण विपाकं फलमेव च।
रोदने व्याधिमम्येति हसन देश विभ्रमम्।।

—*Matsya Purāṇa,* 232.1-5

The same text in another context prescribes that a devotee should procure sixteen kinds of fruit made of gold, silver and copper and give away in donation to the eligible priests. This is possibly because of the fact, that it might not be possible for anyone to procure all the prescribed fruits at one time (some of them being out of season) and as such their substitutes in metals could serve the purpose. These fruits are—

(i) **Fruits to be made in gold**

कूष्माण्डं मातुलिङ्गञ्च वार्ता कम्पन संतथा।
आम्राम्रातक पित्यानि कलिङ्गमथवा लुकम्।।
श्री फलाश्वत्यबदरञ्जम्बीरं कदलीफलम्।
काश्मरन्दाडिमं शक्त्या काल धौतानि षोडश।।

—*Matsya Purāṇa*, 96.5-6

(These sixteen fruits which are to be made in gold, according to the means of the devotee include pumpkins, citrons, egg-plants, bread fruit, mangoes, hog plums, wood-apple, Indra-grains, cucumbers, coconuts, the holy-fig, the jujubes, the lemons, the plantains, saffron and pomegranates).

(ii) **Fruits to be made in silver**

मूलकामलकं जम्बू तिन्तिड़ीकरमर्दकम्।
कङ्कोलैलाक तुण्डीर करीरं कुटजं शमी।।
औदुम्बरं नालिकेरं द्राक्षाथ बृहतीद्वयम्।
रौप्यानि कारयेच्छक्त्या फलानी मानि षोडश।।

—*Ibid.*, 96.7-8

(The fruits which are required to be made in silver according to the means of the devotee, include Muti—a kind of root vegetable, *Emblida officinalis*, or rose apple, tamarind, karaunda, kankola berry, cardamom, cotton, bamboo, kuṭaja, śami, udumbara, coconut, grapes and two kinds of egg-plants).

(iii) **Fruits recommended to be made of copper**

ताम्रं तालफलं कुर्य्यादगस्तिफलमेव च।
पिण्डारकाश्मर्य्यफलं तथा सूरणकन्दकम्।।
रक्तालुकाकन्दकञ्च कनकाह्वञ्चचिर्भिटम्।
चित्रवल्लीफलं तद्वत्कूटशाल्मलि जम्फलम्।।
आम्रनिष्पाचमधुकवटमुद्गपटोलकम्।
ताम्राणि षोड़शैतानि कारयेच्छक्तितोनरः।।

—*Ibid.*, 96.9-11

(The fruits which have been recommended to be made in copper, according to the means of a person, include—Palms, Agasti, Grandiflora, Kāyaphala, Gāmbhārī, esculent root, the red esculent root, a bulbuous root Śvarṇakhirī, cucumber, Chitravallī, Śālmalī, mango, liquorice, kidney-bean, fig-tree, patola—a species of cucumber).

Thus the texts provide interesting features of the plant kingdom, some of which might not compare favourably to the modern scientific ideology, but some of them indeed are quite relevant even during the modern times. The most remarkable aspect of the Purāṇic texts had been to inculcate a sense of dedicatioṇ towards the plant-life which is evident from the *Varāha Purāṇa*, in which it is mentioned that the one who plants a tree attains *Svarga*—heaven and does not have to go to hell—

अश्वत्थमेकं पिचतुमन्दमेकं
न्यग्रोधमेकं दश पुष्पजाति।
द्वे द्वे तथा दाडिम मातुलिङ्गे
पञ्चाम्रवापि नरकं न याति।।

—*Varāha Purāṇa*, 170.36

New Delhi **Shantilal Nagar**

# Contents

## Vol. 1

## 6. Industrial Art—172



## 7. Forests and Gardens—177



## 8. Damsel and Tree Motif—189

## 9. Medicinal Plants—194



## 10. Divine Trees and Plants—225



## 11. The Lotus—237

## 12. Agriculture, Pastures, Mountains and Irrigation in Vedic Age—245

## Appendices—279



## Bibliography—429

## Index of Indigenous Terms—437

## Index of Botanical Terms—457

## General Index—466

## Vol. 2



## Figures

## Black and White Plates

## Coloured Plates

# List of Figures

# List of Plates

# List of Colour Plates

# 1
# Introduction

अरण्यान्यरण्यान्यसौ या प्रेव नश्यसि।
कथा ग्रामं न पृच्छसि न त्वा भीखि विन्दती।।

—*Ṛgveda*, X.146.1

(Goddess of Wild and Forest who seemest to Vanish from the sight. How is it that you seekest not the village? Art thou not afraid?).

Trees and plants are of paramount importance for human life, not only in the present times but also in the remote past as well. The early man depended on them for his physical needs such as sources of food, shelter, clothing, medicines, ornaments, tools and for spiritual needs such as magic or ritualistic practices. It is quite well known that in addition to their practical and economic values, green plants are indispensable to all life on earth—through the process of photosynthesis, plants transform solar energy into chemical energy of food, which makes all life possible. A second unique and important capacity of green plants is the formation and release of oxygen as a by-product of photosynthesis. The oxygen of atmosphere is absolutely essential to all forms of life on earth.

The earliest man, like other anthropoid mammals (*e.g.*, apes and monkeys) depended totally upon the natural resources of their environments, which, until he developed methods for hunting, consisted almost completely of plants. "The behaviour of the pre-Stone-Age man can be inferred by studying the botany of aboriginal peoples in various parts of the world. Isolated tribal groups in South America, Africa and New Guinea, for example, have extensive knowledge of plants and distinguish hundred of kinds of plants according to their utility, as edible, poisonous, or otherwise important in their culture. They have developed surprisingly sophisticated systems of nomenclature and classification which approximate to the binominal system (*i.e.*, generic and specific names) found in modern biology. The urge to recognize different kinds of plants and to give them names thus seems to be as old as the human race."[1]

In due course of time early man not only collected plants but also started growing them. This domestication resulted not only in the development of agriculture but also in the greater stability of human population that had earlier been nomadic. From the settling down of agricultural people in places where they could depend upon adequate food supplies came the first villages and the earliest civilization. Because of the long preoccupation of man with plants, a large body of folklore, customs, traditions, and other scientific data has accumulated which is useful even in modern times.

The association of plants with mankind is, therefore, as old as the human race. Had there been no plants on earth, man could hardly have survived. Apparently it could be a debatable point as to who emerged on the universe first, the plant or the human race. But an answer to this riddle is available in a passage in the *Ṛgveda*, which pronounces that the plants emerged on the universe far earlier *i.e.*, three *yugas*, than the human race—

या ओषधी: पूर्वा जाता देवेभ्यस्त्रियुगं पुरा।
मनै नु ब्रभ्रूणामहं शतं धामानि सप्त च।।

—*Ṛgveda*, X.97.I

(Herbs that sprang up in old times—three ages earlier than the gods; of these whose hue is brown, will I declare the hundred power and seven).

The *Taittirīya Saṁhitā* (IV.2.6.I) and *Vājasneyi Saṁhitā* (XII.75) have also used the word *triyuga*, as in the above Ṛgvedic passage, to bring home the point that the plants (*Auṣadhi*) were born "three ages" earlier to the gods. The commentator of *Nirukta* (IX.28) thinks that the ages here meant are the *yugas* of later Indian chronology, the sense of passage being that the plants were born in the first *yuga*. The author of *Śatapatha Brāhmaṇa* (VII.2.4.26) interprets the terms triyuga in the verse as three seasons i.e., spring, rain and autumn, taking the two words *triyugam-purā* as formerly in the "three seasons", is quite adequate; the use of "three" in such case is a favourite feature in the folklore.

The testimony of the *Ṛgveda* seems to be based on factual justifications, because normally no living being could survive on earth without the plants. Not only the earliest literature, but the association of trees is amply witnessed in the prehistoric rock paintings as well as the Indus Valley civilization and other Harappan sites. Besides the above, in the Vedic and post-Vedic literature numerous trees/plants have been mentioned, a tabulation of which is available at Appendix-2 of this work. It includes the following groups—

(i) Plants with modern botanical names.

(ii) Medicinal plants.

(iii) Aquatic plants.

(iv) Sacred plants.

(v) Fruit trees

(vi) Reeds.

(vii) Grass.

(viii) Vanaspati.

(ix) Fragrant plants.

(x) Flowers.

(xi) Forests.

The above groups provide a glimpse of the nature of trees and plants which were well known to the Vedic and post-Vedic people. In this connection the point worth consideration is whether the adoration of trees was quite in vogue during the Vedic period. On this point A.B. Keith has observed as follows[2]—

"The reference to the worship of trees and plants are very scanty in the Vedic ritual and mythology alike, but they are quite adequate to show that as among all other peoples these objects were not without their share of reverence. A long hymn in the last book of the *Ṛgveda* (X.97) is devoted to the deification of the plants with special reference to their healing properties, and plants also appear in the *Atharvaveda* where they are used as spells for healing, and for driving away demons of all kinds—the plant is even besought to bestow a horse, a sheep, a garment, and the life of the patient, who doubtless was to be the instrument by which the prayer of the medicine man was to be made good if the spell succeeded in attaining its purpose. In the *Ṛgveda,* (X.97.18), Soma is already the king of plants, and they are called mothers and goddesses—the *Atharvaveda* poetically describes a plant as a goddess born of the goddess earth. On the other hand, plants have power to hinder child birth, and in that case the offering of an animal victim to them is prescribed by the *Taittirīya Saṁhitā* (ii.1.5.3) in order to procure their favour."

The cult of trees and above all the forest trees— vanaspati is well recognized in the *Ṛgveda* (VII.34.23 and X.64.8), which in a few passages invokes one or many alongwith the waters and mountains. In the later *Saṁhitās*, trees are the favourite abodes of Gandharvas and Apsarās. *Taittirīya Saṁhitā* (III.4.8.4) assigns to them, their, homes in Aśvattha, Nyagrodha, Udumbara and Plāṣka trees.

The Vedic age indeed witnessed a considerably developed form of the botanical properties. Indeed the vegetable-kingdom had been the closest, most trusted, dependable and the most useful association of man from the earliest days of his existence and as such the continued study of the subject was absolutely essential. Not only that, it also served man by providing shelter, protection from dangers, sap for life and clothings for the body. With this end in view it was natural for him to have made close observation of the pattern of the behaviour of trees and plants and equated them with his own.

The Vedic literature was followed by the Brāhmaṇas and Āraṇyakas. Of these two types of works, the latter are particularly important from the point of view of botany and horticulture, because, these highly philosophical works were created by the sages in the forests or studied there. In fact the Āraṇyakas were so called because they were composed in the environments of the forest. One could just imagine the competence of the composers of Āraṇyakas, who could feel quite homely in the inhabitable dense forests, which would be beyond the reach of an ordinary man, and compose literature which could be useful to the society. However, the adoration of the trees/plants in one or the other form, continued during the period of Āraṇyakas.

But during the Upaniṣad period, a basic change in the approach in dealing with the trees and plants is discernible when the trees were equated with human beings. The *Yājñavalkya* and *Bṛhadāraṇyaka Upaniṣads* are quite vocal on the subject—

ऊवध्यः सिकताः सिन्धवोगुदा यकृच्च कलोमानश्च
पर्वता ओषधयश्च वनस्पतयश्च लोमानि।

—*Bṛhadāraṇyaka Upaniṣad*, I.I.I

(The half digested food is sand, the rivers the bowls, the liver and mountains the lungs, the hair the herbs and trees).

The Upaniṣads further go on to testify that both humans and trees are real and have a common pattern of life. A man has hair, so the trees have leaves; as a man has skin, so the tree has bark. If there are some scratches on man's skin, it results in bleeding. Similarly when the bark of a tree is scratched somewhere, it results in oozing out of some liquid. Just as a man has flesh below the skin, so the tree has a soft content beneath its hard bark. If a man has tissues inside the fleshy part of his body, tree has the same inside its soft content. If man has bones inside the flesh, trees have hard wood. Finally if man has nervous system inside his body, tree has similar system in its own. Each one of them is born of its peculiar seed—

यथा वृक्षो वनस्पतिस्तथैवपुरुषोऽमृषा।
तस्य लोमानि पर्णानि त्वगस्योत्पाटिका बर्हिः।।
त्वचएवास्य रुधिरं प्रस्यन्छिद त्वच उत्पटः।
तस्मात् तदातृणणत् प्रैतिरसो वृक्षादिवाहतात्।।
माः सान्यस्य शकराणि किनाटः स्नाव तत् स्थिरम्।
अस्थीन्यन्तरतो दारूणि मज्जा मज्जोपमा कृता।।

—*Bṛhadāraṇyaka Upaniṣad*, III.9.28

(As a mighty tree in the forest, so in truth is the man, his hair are the leaves, his outer skin is the bark. From his skin flows forth blood, sap from the skin of the tree; and thus from the wounded man comes forth blood as from a tree that is struck. The humps of his flesh are in the tree the layers of wood, the fibre is strong like the tendons. The bones are the hard wood within, the marrow is made like the marrow of the tree).

Besides all these similarities, there are some basic differences between the structures of both, the trees and the humans, regarding their survival and death. If a tree is cut at its trunk, it sends up shoots from within its hewen part, which, in course of time resume the form of the original tree. But curiously enough, the same does not happen with man. This poses a problem. If the two organic properties are so akin to each other as stated above, it was not expected of them to differ so radically here. It is with this question that Yājñavalkya baffles his contestants in the famous debate held in the court of Janaka. When they all failed to answer, he himself rounded up the debate by suggesting that *Jīvātman* was the basis of man's birth. Just as the hewen tree grows out of its roots, so the man is reborn on the axis of his *Jīvātman*. Obviously both of them are invisible products. But while the root, inspite of its apparent invisibility, is a thoroughly tangible object, and gives rise to the tree quite understandably. The same is lacking in the case of *Jīvātman*. It is not only intangible, but as a cause of individual, it is also ordinarily incomprehensive. That is why it requires to be symbolized by something bearing fair equivalence to it and at the same time understandable in its functions. The *Bṛhadāraṇyaka Upaniṣad* (III.9.28) enlightens on the subject thus—

यद्वक्षोवृक्णो रोहित मलानवतरः पुनः।
मर्त्यः स्विन्मृत्युना वृक्णः कस्मान्मूलात् प्ररोहति
रेतस इति मा वोचत जीवतस्त् प्रजायते।
धानारुह इव वै वृक्षोऽञ्जसा प्रेत्य सम्भवः।।
यत्समूलमावृहेयुर्वृक्षं न पुनराभवेत्।
मर्त्यः स्विन्मृत्युना वृक्णः कस्मानमूलात् प्ररोहति।।
जात एव न जायते को न्वेनं जनयेत् पुनः।
विज्ञानमानन्दं ब्रह्म रातिदातुः परायणं तिष्ठमानस्य तद्विद इति।।

(But the tree, when felled, grows up again more young from the root, from what root, tell me, does a mortal grow up after he has been felled by death. Do not say "from seed", for the seed is produced from the living, but the tree springing from a grain, clearly rises again after death. If the tree is pulled up with the root, it will not grow again; from "what root" tell me, does a mortal grow up after he has been felled by the death? Once born, he is not born again, for who should create him again? Brahman who is knowledge and bliss, he is the principle both to him, who gives the gift, and also to him who stands firm and knows).

The above discussion of the composer of the Upaniṣad, also suggests another idea which is the real motive behind all the Upaniṣadic deliberations. It is the principle of liberation. Instead of merely hewing the trunk, if the whole root of the tree is taken out, asks Yājñavalkya, what would happen to the tree? The answer is obvious. The tree would not grow any more. Likewise, says the sage, if the cause of the rebirth is eliminated, the individual would be liberated and would not be reborn.

In the *Chāndogya Upaniṣad*, Sage Aruṇi, while explaining the mysteries of *Atman* to his son Śvetaketu cites the example of a tree and its vitality. If the tree is struck at its root, a kind of sap oozes out of it. If it is struck in the middle or at the top, the same kind of sap comes out. It suggests therefore that the tree is pervaded from top to bottom. If the sap of vitality withdraws itself from any one of the branches, it gets dried up. If it withdraws from the whole of it the entire tree dries up. Similar is the case with the human body. All the pleasures, all joys, all the vitality that it feels and possesses are due to *Ātman* which pervades it. When the *Ātman* forsakes it, it meets death. The *Ātman* according to the sage, is wholly separable from the body. The association of the *Ātman* with the body is the pleasure of the latter. Indeed separation of the same from the body amounts to death. Moreover, just as in the case of the tree, it is the gross body which dies and by no means, the *Ātman*, which only withdraws from the body. Thus the vitality of the tree comes to form an analogue and hence a potential symbol of *Ātman*. The testimony of the said Upaniṣad is provided hereunder—

अस्यसोम्य महतो वृक्षस्य यो मूलेऽभ्याहन्यज्जीवनम्
स्रवेद्यामध्येऽभ्याहन्याज्जीवन् स्रवेद्योऽग्रे
ऽभ्याहत्याज्जीवन् स्रवेत् । स एष जीवेनाऽऽत्मना
ऽनुप्रभूत: पेपीयमानो मोदमान तिष्ठति।
अस्य यदेकाꣳशाखां जीवोजहात्यथ सा
शुष्यति। द्वितीयां जहात्यथ सा शुष्यति।
तृतीयां जहात्यथ सा शुष्यति। सर्वं जहाति सर्व: शुष्यति।
एवमेव खलु सोम्य विद्धीति हो वाच।
जीवापेतं वाव किलेदं म्रियते न जीवोम्रियत
इति। स य एषोऽणि मैतदात्म्यमिदꣳ
सर्वम्। तत्सत्यम्। स आत्मा। तत्त्वमसि
श्वेतकेतो इति। भूय एव मा भगवान्
विज्ञापयत्विति। तथा सोम्येति होवाच।

—*Chāndogya Upaniṣad*, VI.II.I-3

(If some one were to strike at the root of this large tree here, it would bleed, but live. If he were to strike it at its stem, it would bleed but live. If he were to strike its top, it would bleed, but live. Pervaded by the living self that tree stands firm, drinking in its nourishment and rejoicing. But if the life, the living self, leaves one of the branches, that branch withers; if it leaves a second, that branch withers; if it leaves a third that branch withers. If it leaves the whole tree, the whole tree withers. In exactly the same manner, my son, know this. Thus he spoke. This body indeed withers and dies not. That which is the subtle essence, in it all that exists has its self. It is the True. It is the self, and thou Śvetaketu art it).

According, however, to Buddhist conception and the followers of Kaṇāda, the trees are held to be unconscious. They, according to them, live, because one sees how their sap runs and how it dries up, just as one sees the sap in the living body, which, as we see, is produced by food and water. Therefore, the simile holds good. The life, or , more correctly, the liver, the living self, pervades the tree, as it pervades man, when it has entered the organism which produces breath, mind and speech. If any accident happens to a branch, the living self draws himself away from that branch, and then the branch withers. The sap, which caused the living self to remain, goes, and the living self goes away with it. The same applies to the whole tree. The tree dies when the living self leaves it, but the living self does not die; it only leaves an abode, which it had earlier occupied.

Thus the vitality of the tree comes to form an analogue and hence can be considered to be a potential symbol of *Ātman* as conceived in the Upaniṣadic literature. Though it would be a debatable point as to who emerged on the universe, the seed or the tree, but the composers of the Vedic hymns, did not ignore this essential or rather the basic ingredient of the vegetable kingdom. *Bīja* which denotes the seed, which is connected with the sowing seed (*vap*) is referred to in the *Ṛgveda* several times.

तद्विदन्त्यद्रयो विमोचने यामन्नञ्जस्पा इव घेदुपब्दिभिः।
वपन्तो बीजयिव धान्याकृतः पृञ्चन्ति सोमं न मिनन्ति बप्सतः।।

—*Ṛgveda*, X.94.13

(This, the stones proclaim, what time, they are disjoined and when with ringing sounds they move and drink the balm. Like the tillers of the ground when they are sowing seed, they mix the Soma nor devouring, minish it).

In a metaphysical sense, the term is used in the Upaniṣads of classes of beings according to the origin, of which *Chāndogya Upaniṣad* (VI.3.1) enumerates three and the *Aitareya Upaniṣad*, four. The former list includes—*aṇḍaja*, egg-born; *jīvaja*, born alive and *udbhijja* produced from sprouts—

तेषां खल्वेषां भूतानां त्रीण्येव बीजानि
भवन्त्याण्डजं जीव जमुद्भिज्जमिति।।

—*Chāndogya Upaniṣad*, VI.3.1

(Of all living beings, there are indeed three origins only, that which spring from an egg (oviparous) and that which spring from a germ).

But the *Aitareya Upaniṣad* (3.4) adds another type of origin *i.e., Svedaja*, which means born out of sweat or perspiration.

Just as the sap, the seed of the tree, also is used by Aruṇi as a potential symbol to explain to his son, the mystery of *Ātman* and its manifestations. He asks his son to bring a fruit of the Banyan tree and break it. When the son does so, the father asks him to observe the inside contents. The son then reports to the father that the same was invisible. The sage then impresses upon the son that it was out of a similar seed of invisible content that the Banyan tree of such a gigantic proportions had grown and that the seed in question has itself the potentiality of a similar tree concealed in it. Thus with the aid of the potential symbol, the father brings home to the son, the fact that it is very well possible for *Ātman* to manifest in human form or in any other form whatsoever, though itself it is intangible. The *Chāndogya Upaniṣad* (VI.12.1-3) enlightens us on the subject as follows—

न्यग्रोध फलयत आहरेति। इदं भव इति। भिन्द्धीति।
भिन्नं भगव इति। किमत्र पश्यसीति अण्व्य, इवेमा
धाना भगव इति। आसामङ्गैकां भिन्द्धीति।
भिन्ना भगव इति। किमत्र पश्यसीति। न किंचन भगव इति।।१।।
तꣳहोवाच—यं वै सोम्यैतमणिमानं न निर्भालयस
एतस्य वै सौम्यैषोऽणिम्न एवं महान्यग्रोधस्तिष्ठति।
श्रद्धत्स्व सोम्येति।।२।।
स य एषोऽणिमैतदात्म्यमिदꣳ सर्वम्। तत्सत्यम्।
स आत्मा तत्त्वमसि श्वेतकेतो इति। भूय एव
मा भगवान् विज्ञापयत्विति। तथा सौम्येति होवाच।।

("Fetch me from thence a fruit of Nyagrodha tree." "Here is one, Sir." "Break it". "It is broken Sir". "What do you see there?" "These seeds, almost infinitesimal". "Break them". "It is broken, Sir". "What do you see there?" "Not anything sir." The father said, "My son, that subtle essence which you do not perceive there, of that very essence this great Nyagrodha tree exists. Believe it my son, That which is subtle essence, in all that exists has itself. It is the true. It is the self, and thou Śvetaketū, act it." "Please Sir, inform me still more", said the son. "Be it so, my child" said the father.)

The falling of the fruit from the tree has also been conceived by Yājñavalkya in a highly philosophical manner. He visualises the event of falling of a fruit from the tree as potentially symbolic of regeneration. When a mango, a fig or a pīpala fruit becomes ripe, it gets loosened from its twig and falls on the ground. But its falling down does not in any way results in its loss. On the contrary, after being separated from the tree, the fruit is taken hither and thither as destined for it and then it sprouts and grows in the form of another tree. The very fall of the ripe fruit from the tree is quite important, because unless it falls from the tree it cannot grow into a huge tree which gives shelter and abode to many. Similar is the case with the human body. When weakened by disease or old age, it falls automatically like a fruit from the twig. Just as the fallen fruit grows into another tree, so the deceased person must be born in the form of another person. Thus the analogy of fruit suggests and supports the possibility of rebirth as propounded by the Upaniṣadic literature. This aspect has been amply highlighted in the *Bṛhadāraṇyaka Upaniṣad* (IV.3.36) as under—

स यत्राय मणि मानं न्येति जरया वोपतपता वाऽणिमानं
निर्गच्छति। तद्यथाऽऽम्रं वोदुम्बरं व पिप्पलं व बन्धनात्
प्रमुच्यते। एवमेवायं पुरुषएभ्योऽङ्गेभ्यः सम्प्रमुच्य
पुनः प्रति न्यायं प्रतियोन्याद्रवति प्राणायैव।।

(And when the body grows weak through old age, or becomes weak through illness, at that time that person, after separating himself from his members as an *āmra*—mango, or fig, udumbara or pīpala fruit is separated from the stalk, hastens back again as he came, to the place from which he started to new life).

The *Chāndogya Upaniṣad*, on the other hand, finds in a lotus leaf the similarity of a realised soul. When dipped into the water other things get wet or soaked. The only noted exception is the lotus leaf or lotus flower. Though always dwelling in water the lotus leaf or flower remains absolutely unaffected from the water. Similar is the case with the humans. Before being fully enlightened, he gets attached to every thing he comes in contact with. But once awakened in the supreme consciousness, he though still living amidst allurements of the world, remains absolutely unaffected—

यथा पुष्कर पलाश आपो न श्लिष्यन्त
एवमेवंविदि पापं कर्म न श्लिष्यन्त इति।।

—*Chāndogya Upaniṣad*, IV.14.3

(As the water does not cling to a lotus leaf, so no evil deed clings to one who knows it).

On account of its unique aptness, the lotus leaf has been highly significant in the development of the idea of detachment in the Indian spiritual thought. Indeed, where logical statements fail to make any appeal or carry only a week appeal, symbols do make an appeal and that also quite permanently. In fact, they clarify, strengthen and perpetuate what they are made to bring out.

Besides Upaniṣadic symbolism, with reference to the trees, the Vedic literature also does not ignore the fact the trees and plants have life. This could be surmised from a passage from the *Ṛgveda* (III.8) wherein the faith in the life of the tree is clearly discernible. This passage highlights the treatment of the tree from which the sacrificial post is to be taken. A.B. Keith[3] has observed in this connection as follows—

"The necessary cutting is performed after a blade of grass has been put over the place where the blow is to be inflicted, and the blade is hidden to protect the tree, while the axe is expressly ordered not to harm the tree; the prescription is precisely the same as that adopted in the domestic ritual when the hair is cut, and in the ordinary cult when the victim is slain; the aim is clearly to avoid injury to the life in the tree by pretending that it is not being injured. The tree is also, when made into the form of a post, adorned with a band placed around it and is anointed, and this dobutlessly prove its living character, while it is possible that verse in the *April litany* of the animal sacrifice, addressed to the forest tree, is really an adjuration to the sacrificial post. The evidence must not be unduly pressed, as in this case, owing to the presence of the god at place of sacrifice, it is beyond doubt that the surroundings of the sacrifice are filled with his presence, and that, therefore, they may attain a sanctity not their own; but the conception of life of the tree is clear, even if its divinity is not so certain in this special instance."

The medicinal properties of the vegetable world were quite well known to the Vedic people and such plants are termed as *Auṣadhis* in the Vedic literature. In one of the Ṛgvedic verses, the use of herbs by the physicians has been amply highlighted—

यत्रौषधी समग्मत राजानः समिताविव।
विप्रा य उच्यते भिषग्रक्षोहामीवचातनः।। —*Ṛgveda*, X.97.6

(He who hath store of herbs at hand like kings amid a crowd of men; physician is that sage's name, fiend slayer, chaser of disease).

Roughly speaking, the plant kingdom is divided in the Vedic literature between *Oṣadhi* or *Virudha* plants and *vana* or *Vṛkṣa*—trees; *Auṣadhi* is employed in opposition to *Virudha* to denote plants as possessing power or some other qualities, useful for human beings, while *Virudha* is rather a generic term for minor vegetable growths, but sometimes, when occurring besides *Auṣadhi* signifies those plants which do not possess medicinal properties.

A list of minor parts of which a plant is made up is provided in the *Taittirīya Saṁhitā*[4] and *Vājasneyī Saṁhitā*[5], which comprises of—

(i) *Mūla* or the root,

(ii) *Tula* or panicle,

(iii) *Kanda* or the stem,

(iv) *Valsa* or the twig,

(v) *Puṣpa* or the flower,

(vi) *Phala* or the fruit

(vii) *Skandha* or the corona

(viii) *Śākhās* or the branches,

(ix) *Parṇa* or the leaves.

The *Atharvaveda* provides an elaborate division of plants as under—

प्रस्तृणती स्तम्बिनीरेकशुङ्गा: प्रतन्वतीरोषधीरा वदामि।
अंशुमती: काण्डिनीर्या विशाखा ह्वयामि ते वीरुधो
वैश्वदेवीरुग्रा: पुरुषजीवनी:।।

—*Atharvaveda*, VIII. 7.4

The classification of the plants by the above text, includes—

(i) *Pra-stṛṇatīḥ*—those which expand,

(ii) *Stambinīḥ*—those which are bushy,

(iii) *Ekaśuṅgaḥ*— those which have only one sheath

(iv) *Pra-tanvatīḥ*—those which are creepers or climbing plants

(v) *Aṁśumatīḥ*—those having many stalks

(vi) *Kāṇḍinīḥ*—those which are jointed,

(vii) *Viśākhāḥ*—those which have spreading branches,

The *Ṛgveda*,[6] however, provides further classification of the trees/plants as follows—

(i) *Puṣpavatīḥ*—Blossoming,

(ii) *Prasūvarīḥ*—having flowers

(iii) *Phalinīḥ*—Laden with fruit or yielding fruit.

While dealing with plants and vegetation, the insects and worms are perhaps an inseparable part of the same. The Vedic texts have also identified a number of such insects, some of which are listed hereunder—

(i) *Upa-kvasa* (*Atharvaveda*, VI.50.2)

(ii) *Kapana* (*Ṛgveda*, V.54.6)

(iii) *Kaṣkaṣa*, (*Atharvaveda*, V.23.7)

(iv) *Kīṭa* (*Atharvaveda*, IX.4.16)

(v) *Kṛmi*, (*Ṛgveda*, I.191; *Atharvaveda*, II.31.32)

These worms have been extensively dealt with in the *Atharvaveda*, some of which have been identified to be poisonous and are said to be found in the mountains, forests, in water, in plants and even in human beings. In accordance with the widespread primitive ideas, these worms are considered to be the cause of diseases in men, animals and plants.

These Vedic texts did not lag behind in spelling out ways and means to serve as insecticide. The *Atharvaveda* contains three hymns as charms directed against them. The first of these hymns is of general character and is meant to destroy worms in cattle and the third is intended to cure children of these worms. When found in men, worms are said to have their place in head or ribs, eyes, nose and teeth. These germs are also said to adversely affect and destroy the vegetational growth of trees and plants—

ये क्रिमयः पर्वतेषु वनेष्वोषधीषु पशुष्वप्स्वान्तः।
ये अस्माकं तन्वऽमाविविशुः सर्व तद्धन्मि जनिम क्रिमीणाम्।।

—*Atharvaveda*, II.31.5

(The germs which are in mountains, forests, trees and plants, animals, water and in human body, I destroy them all.)

The *Atharvaveda* also describes other insects like locusts (*Śalabha*, IX.5.9). *Vṛkṣa—sarapi* (IX.2.22) and others which are responsible for the destruction of crops.

The belief in the presence of spirits in the trees is, of course, ethnic and the Buddhist literature mentions it variously showing the different sets of conceptions which are easily formed; the tree is primarily the spirit, then the spirit with or without its children lives in the trees, and if it perishes, has to go away to seek a new abode, while at a further stage the tree becomes more and more connected with a deity. The literature is also instructive, in fact that it mentions other than tree spirit, spirits of terrible form, besides beings of uncanny and hostile nature, who may be of double origin, arising from a direct or animalistic conception of the darkness and hostility of woods, and also from the conception of unfriendly spirits, which develops from the belief in hostile spirit of the dead. The *Ṛgveda* in a hymn in the last book, besides describing the goddess Araṇyānī, also mentions the uncanny sounds heard in the solitudes, the many beasts, which abound, and the food which is raised without tillage.[7] It is easy from this hymn alone to realise how spirits of dangerous character could, independently of any connection with the spirits of the dead, be deemed to abound in the forest, as per the ancient belief.

The earth too is an essential part of the vegetable kingdom and without earth nothing can grow. The earth has been conceived as a goddess in the *Ṛgveda* who is invoked together with Dyaus. She has but one short hymn of her own (V.84) and a long and an interesting hymn (XII.1) is ascribed to her in the *Atharvaveda*, which is conspicuous rather for its accuracy of enumeration of the sights of the earth than for religious fervour. She is described as rich in heights, bears the burden of mountains and supports on the ground the forest trees. She is said to be great, shining and firm, and quickens the earth, by scattering rain from the clouds, a fact which shows that she is not rigidly confined to her element, doubtless because she has borrowed an attribute of Dyaus herself. The *Taittirīya Brāhmaṇa* (i.I.3.5) follows the *Ṛgveda* in referring her name as broad to the fact of her having been extended.

Besides the earth, the cultivable fields are also not lost sight of in the Vedic texts. The fields are controlled according to the Vedic texts by a deity called Kṣetra-pati, the "lord of fields." Here *Kṣetra* stands for a field. Kṣetra-pati is invoked in the *Ṛgveda* (IV.57) to grant horses, cattle, and with sweetness to fill heaven and earth with the plants and the waters. Indeed it could be argued that no non-carnivora type of animals like the horses, cows and goats etc., could survive without vegetation. In another hymn,[8] he is asked along with Savitr, the Dawns and Parjanya to best of prosperity and elsewhere it is said that worshippers are fain to have him as a neighbour. The full nature of the deity is made clear in Gṛhya[9] rituals which prescribe offerings to him when the field is ploughed, and he is, it is clear, no more than a deity believed to be in the ploughed land. Similarly the *Ṛgveda*[10] contains an invocation of Sītā, the furrow, to grant rich blessings of crop. The figure of this goddess naturally has more life in the Sūtras[11] which deal with the operations of agriculture; she appears to be the wife of Indra, which could be due to the fact that in the *Ṛgveda*[12] the god is once called Urvarāpati, *i.e.* the "lord of plough field", and her existence is manifested in the deities like Sītā, Āśā, Ārdā and Anaghā, who are to be worshipped at the furrow sacrifice, the sowing sacrifice, the threshing-floor sacrifice, the reaping and putting the corn into the barn.[13] In the epithet of Urvarāpati the word *Urvarā* denotes a piece of plough land, fertile field[14] as well as waste fields.

Intensive cultivation by means of irrigation is also testified in the *Rgveda*; thus—

य आपो दिव्या उत वा स्रवन्ति खनित्रमा उत वा या: स्वयंजा:
संमुद्रार्था या शुचय: पावका स्ता आपोदेवीरिह मामवन्तु।।

—*Ṛgveda*, VII.49

Similar spirit is available in the passage of *Atharvaveda*—

श न: खनित्रिमा आप: शमु या: कुम्भ आभृता
शिवा न: सन्तु वार्षिकी।।

In both the above cases, different sources of water, viz; drawn from well, river-water, rain-water, water found in the desert and the water found in the areas having enough of water storage have been mentioned.

Besides, the natural water sources, like the rain water, river water, etc. we find in the Ṛgvedic texts, the mention of artificial sources as well. In this connection, the word *Avaṭa* used in the *Ṛgveda* (I.55.8; I.85.10-11) is quite important. It denotes a well artificially made. This is in contrast with a spring (*Utsa*), though the latter expression is also applied to an artificial well. Such wells were covered by the makers (*Ṛgveda*, I.55.8) and are described as unfailing (*Akṣita*) and full of water (*Ṛgveda*, X.101.6). The water was raised by a wheel (*Cakra*) of stone to which was fastened a strip (*Varatra*) with a pail attached to it. (*Ṛgveda*, X.101.7) and when raised the water was poured into the buckets (*Ahava*) of wood. Sometimes these wells were used for irrigation purposes, the water being led off by broad channels (*Surmī Suśira*) (*Ṛgveda*, VIII.69.12).

The plants and fields would require manure and the use of manure has also been clearly suggested in a passage of the *Atharvaveda* (III.14. 3-4), wherein the cow has been described as करीषिणी or the one with good quality of cowdung which is used as manure. Actually the word *Karīṣa* denotes cowdung and the use of cowdung manure is testified by the Vedic literature.

For the purpose of agriculture as well planting and upkeep of trees and plants, the use of garden implements cannot be once emphasise. There is no very essential distinction between the worship of natural objects conceived as living and the worship of objects made by human hands, but it is obvious that the worship of such objects tends to be restricted in effect and importance. This is borne out by the evidence of Vedic literature where such religion is of very minor importance, despite the fact that it receives further attention in the *Ṛgveda* than might *a priori* have been expected. Two types of implements of the ploughman, called *Suna* and Sīra which could be the plough share and the plough are invoked and that in the *Śatapatha Brāhmaṇa* a cake is assigned to them in sacrifice. The warrior invokes his arrow as divine[15] and begs it to protect him and assail his does; his armour, his bow and quiver, are also celebrated, and the drum is invoked to drive away the demons and the danger. In the *Atharvaveda* (V.20) an entire hymn is devoted to its praises. But the priest naturally is more prominent in the *Ṛgveda*, where not only the sacrificial post is deified, perhaps as the remanent of tree worship, is called divine, and is asked to allow the sacrifice to go to the gods, but the sacrificial grass[16] and divine doors are celebrated.[17] Adverting however to the issue of the implements used by the ploughman or the gardener, it may be pointed out that the texts speak of various types of implements used for the purpose some of which are mentioned here—

(i) *Abhri* (Spade)—This is a Vedic word of frequent occurrence. Various possible material and forms of this type of implement are enumerated in the *Śatapatha Brāhmaṇa* (VI.3.1.30). It could be made of a bamboo or the wood of Vikankara or Udumbara Trees. It could be a pan or a cubit in size. It is hollow and each one of the edges are sharp. From this it would appear that the handle was made of wood and the head of some metal.

(ii) *Astra*—This is the plougher's goad and the badge of agriculture. It is mentioned several times in the *Ṛgveda*.[18]

(iii) *Kinasa*—This is the term used for a ploughman or a cultivator of the soil as per the *Ṛgveda* and other texts.[19]

By the time of epics, the secrets of Botanical kingdom were better known to man. The forests became the usual abodes of the sages, who carried out their austerities there, which were occassionally disturbed by the *Rākṣasas*. The earliest reference to the dwelling the sages in the forest possibly is that of Vālmīki, whom Nārada briefs on the story of Rāma, in the opening chapter of the Bālakāṇḍa of the *Vālmīki Rāmāyaṇa*. Then the sage while bathing on the outskirts of the forest saw a pair of healthy *krauñca* birds dallying in amorous sport and singing in melodious notes. At that moment suddenly a fowler appeared and killed the male bird. Then the female bird finding her mate thus slain and rolling in dust besmeared with blood, raised piteous cries of despair under the pangs of separation from her copper-crested, amorous companion of fluttering wings. The very sight of this scene, made the sage overwhelmed with grief and he pronounced a curse on the hunter thus—

मा निषादप्रतिष्ठा त्वमगमः शाश्वतीः समा।
यत्क्रौञ्चमिथुनादेकमवधी काममोहितम्।।

—*Vālmīki Rāmāyaṇa*, I.2.15

(O Fowler ! Since thou hast slain one of the pair of *krauñcas*, while engaged in love, thou shalt never attain any fame).

Then there is the story of Ṛsyaśṛṅga, son of Vibhāṇḍaka, who dwelt in the thick forest. Their complete isolation from the world in the secluded cottage in the forest is evidenced from the fact, that the young Ṛsyaśṛṅga had been brought up in the woods and was ignorant of women and the sensual pleasures—

ऋष्यश्रृंग इति ख्यातस्तस्यपुत्रो भविष्यति।
स वने नित्य संवृद्धो मनिर्वनचरः सदा।।

—*Vālmīki Rāmāyaṇa*, I.8.4

This young sage was enticed by the courtesans of the king of Anaṅga, who ultimately married his daughter Śāntā to him. This sage performed the *putreṣṭhi yajña* of Daśaratha, as a result of which four sons were born to the king.

Indeed, the birth of Sītā is attributed to the king Janaka's efforts to take to the ploughing the dry and parched land due to draught. Two things are clear from this episode—firstly, that the society of that time depended on the agricultural produce, and that when the exigencies of the situation so demanded, even the king would not hesitate to till the land.

Then there is the emergence of Viśvāmitra in the Court of Daśaratha seeking the help of Rāma and Lakṣmaṇa for the destruction of the demons in the forest hermitages. There is an extensive tour of the Ayodhyā princes in the company of Viśvāmitra in the forest. Thereafter almost the entire episodes of *Rāmāyaṇa*, consequent to the banishment of Rāma with Lakṣmaṇa and Sītā to the forests for fourteen years were enacted in the environments of the forests. Their exile was spent in Pañcavaṭī, Daṇḍakāraṇya, Kiṣkindhā and the forests near Lankā.

Besides the forests, numerous trees, plants, creepers, groves and pleasure gardens find elaborate mention in the epic.

The *Mahābhārata,* the other great epic of the country, extensively deals with the vegetable kingdom. Here too the great Pāṇḍavas were exiled to the forests for twelve years. In the case of these princes too, their living in the forest was quite eventful and numerous episodes of the epic were enacted there.

In the Vanaparvan of this epic, there is a conversation between Bhīma, the Pāṇḍava hero and Hanumān (Chapters 147-150). In this conversation Hanumān has narrated the story of the *Rāmāyaṇa* to Bhīma, introducing himself as "Hanumān son Vāyu" from the wife of Kesarī and the friend of Sugrīva, the son of Āditya—

अहं कसरिणः क्षेत्रे वायुना जगदायुषा।
जातः कमलपत्राक्ष हनूमान्नाम वानरः।। —(3.147.24)

Hanumān, while narrating the story of Rāma, further informs Bhīma that Sugrīva had been driven out by his brother, and he dwelt for a long time with Hanumān at *Ṛṣyamūka* mountain when Dāśarathī Rāma, who was Viṣṇu in human shape, with his wife and younger brother resided in Daṇḍaka forest and Rākṣasa Rāvaṇa carried away Sītā, Rāma's queen from Janasthana, assisted by Rākṣasa Mārīca in the form of a deer marked with gems and gold—

तस्य भार्या जनस्थानाद्रावणेन हृता बलात्।
वाञ्चयित्वा महाबुद्धिं मृगरूपेण राघवम्।। —(3.147.30)

Rāma slew Bāli and installed Sugrīva in the kingdom; Sugrīva sent monkeys to inquire of Sītā and according to the communications of the vulture Sampāti, Hanumān crossed the ocean over 100 *yojana*, discovered Sītā in the abode of Rāvaṇa, burnt Laṅkā and returned to Rāma. Rāma constructed the bridge across the deep sea, and followed by myriads of monkeys, he slew Rāvaṇa, his brother, sons, and kindred with all the Rākṣasas and installed the pious Rākṣasa Chief Vibhīṣaṇa in the kingdom of Laṅkā. Rāma then returned with Sītā to his city of Ayodhyā and was established in the kingdom. Rāma granted a boon to Hanumān that he should live as long as the history of Rāma's deeds is recited on earth

गावद्रागकथा वीर भवेलोकेशु शत्रुहन्।
तावज्जीवेयमित्येवं तथास्त्विति च सोब्ररीत्।। —(3.147.37)

According to the epic, Rāma went to his own abode after having reigned for 11,000 years. From that time all excellent objects of entertainment were, through the grace of Sītā supplied to Hanumān, who constantly stayed in the forest listening to the virtuous deeds of Rāma, sung by Apsarās and Gandharvas.

Hanumata was then prevailed upon to grant the request of Bhīma, and showed himself as great as the Vindhyas or the Maināka mountain, and said that he could become as tall as he wished; he would have himself slayed Rāvaṇa but he would not obscure the glory of Rāma. He pointed out to Bhīma the way to Saugandhikavana and the garden of

Kubera, which was guarded by Yakṣas and Rākṣasas, and warned him against plucking the flower himself.

In the Rāmopākhyāna-parva of Vanaparvan (Ch.273) Mārkaṇḍeya begins to narrate the history of Rāma. The birth of Rāma and his brothers is described along with the performance of austerities by Rāvaṇa and his brothers in order to obtain boons from Brahmā. The celestial Dundubhi is stated to have been born as Mantharā. Daśaratha intends to install Rāma as Prince-regent, Kaikeyī insists on the king to banish Rāma, who sets out for forest along with Lakṣmaṇa and Sītā. The mutilation of Sūrpaṇakhā is narrated; she goes back to Rāvaṇa with a complaint. Rāvaṇa desires Mārīca to entice Rāma away from his forest abode; Mārīca does so and imitates the voice of Rāma. Rāvaṇa carries away Sītā; Rāvaṇa tries to seduce Sītā but gets a rebuff. Hanumān on the other hand crosses the 100 *yojana* wide sea, reaches Laṅkā, traces Sītā and returns to Rāma with tidings from Sītā. The journey of Hanumān through the sea is described. Then the monkeys flock to Sugrīva and Nala and Nīla construct the bridge over the vast sea. Then Rāvaṇa is killed, Sītā is restored to Rāma and all of them go to Ayodhyā, where Rāma is installed as the king.

A significant aspect of both the epics has been the equation of trees with the great heroes, in the *Vālmīki Rāmāyaṇa*, Rāma has been compared to a huge Banyan tree, whose flower and fruit is Sītā, and the monkey chiefs like Sugrīva, Jāmbvān, Kunuda, Nala, Nīla Dvida, Aṅgada, Mainda, Hanumān and Suṣeṇa, besides the other commanders of the Rāma's army were the branches of the said tree—

राम वृक्षं रणे हन्मि सीता पुष्प फलप्रदम्।
प्रशाखा यस्य सुग्रीवो जाम्बवान् कुमुदो नलः।।
मैन्दश्च द्विविदश्चैव ह्यङ्गदो गन्धमादनः।
हनूमांश्च सुषेणश्च सर्वे च हरि यूथपाः।।

—*Vālmīki Rāmāyaṇa*, 6.100.4-5

Even Rāvaṇa has been compared to a huge tree, which has patience as his leaves, his adament nature as the beautiful flowers, his *tapasya* as his strength, and his stubborn nature as its trunk—

धृतिप्रवालः प्रसहाग्र्यपुष्पस्तपोबलः शौर्यनिबद्धमूलः।
रणेमहान राक्षसराजवृक्षः संमर्दितो राघवमारुतेन।।

—*Vālmīki Rāmāyaṇa*, 6.112.10

(The great storm of Rāma, uprooted the gigantic tree of Rāvaṇa, in the battlefield, whose patience was its leaves, his adament nature as beautiful flowers, his tapas as its strength and his stubborn nature as its trunk).

Interestingly the trees were freely used by the monkey army as their weapons in their battle against the army of Rāvaṇa.

Commenting on the importance of the composition of the *Mahābhārata*, Sūta in chapter 1 of the Ādiparvan, has remarked as under—

"As the full-moon with its mild light opens the buds of the water-lily, so does this Purāṇa (*Mahābhārata*) with the light of *Śruti* expand the human intellect. The whole house of the womb of nature is properly and completely lighted by the lamp of history, which destroys the darkness of ignorance. This work (*Mahābhārata*) is a tree, the chapter of contents is its seed, the divisions *Paulama* and *Astika* are its roots, the *Sambhava* is its trunk, the books (*parvas*), *Sava*, *Araṇya* are the roosting perches, *Arani* is the knitting knot, *Virāṭa* and *Udyoga*, the *pitha*, Bhīṣma, the main branch, Droṇa, the leaves, Karṇa its beautiful flowers, *Salya* their fragrance, *Stri* and *Aisika* are its cooling shades and Śānti its great fruit. Aśvamedha is its immortal sap, *Āśramavasika* the place, where it grows and *Mansula* is the Vedas. This tree will be highly respected by all the virtuous Brāhmaṇas. This tree of *Bharata* will be inexhaustible as the clouds and be the means of livelihood of many illustrious poets."

—*Mahābhārata*, I.1.85-92

Even Duryodhana and Yudhiṣṭhira have been compared with the great trees as under—

दुर्योधनो मन्यु मयो महाद्रुमः स्कन्धकर्णः शकुनिस्तस्य शाखा।
दुःशासन फलपुष्पै समृद्धमूल राजाधृतराष्ट्रौऽमनीषी।।
युधिष्ठिरो धर्ममयो महाद्रुमः स्कन्धोऽर्जुनोभीमसेनोऽस्य शाखा।
माद्री सुतौ पुष्पफलं समृद्धे मूलं कृष्णोवहन च ब्राह्मणश्च।।

—*Mahābhārata*, Ādiparvan, I.108.109

(Duryodhana is the great tree created out of passion, Karṇa is its trunk, Śakuni serves as branches, Duśāsana is its fruit and flowers, and weak Dhṛtarāṣtra is its root. Yudhiṣṭhira is the great tree, created out of virtue and religion, Arjuna is its trunk, Bhīma serves as branches, two sons of Mādrī are its flowers and fruit, Kṛṣṇa, Brahmā and Brāhmaṇas are its roots.

The different types of trees and plants detailed in both the epics have been tabulated in Appendix-3 but some of the forests, gardens, groves and pleasure parks mentioned in the *Mahābhārata* are given hereunder—

(i) *Dvaitavana* (III.24.928)

(ii) *Kāmyakavana* (1.2.432, 464, 472 etc; II. 52. 1877 etc.)—As Yudhiṣṭhira declared that Bhūriśrava, Śalva, Jarāsandha etc. and Karṇa (master of all weapons) and Dhṛtarāṣṭra together with the kings who had been injured by the Pāṇḍavas, could not now be vanquished, that all the Kaurava soldiers and their families had been honoured with wealth and luxuries by Duryodhana and would sacrifice their lives for him, that, although Bhīṣma, Droṇa and Kṛpa all knowing celestial weapons and being invincible, behaved equally towards both parties, they would certainly, to pay off the royal favours they enjoyed, lay down their lives in battle, Bhīma remained silent. While they were thus conversing, Vyāsa appeared, and taking Yudhiṣṭhira into a corner, taught him, after he had purified himself (*Suddhaye*) the science of *Pratismṛti* by means of which Arjuna should proceed to Mahendra, Varuṇa, Kubera and Yama, to receive weapons from them. He also advised him to go to some other forest in order not to cause discomfort to sages living there and to save the destruction of plants, flora and fauna. After the advice of Vyāsa, Pāṇḍavas went to the Kāmyaka forest on the bank of the River Sarasvatī and lived there for quite some time.

(iii) *Khāṇḍavavana*—(1.1.150; 1.222-34)—At the instance of Dhṛtarāṣṭra and Bhiṣma, the Pāṇḍava at Indraprastha began to bring other kings under their sway. All their subjects lived happily. Once in the summer Arjuna and Kṛṣṇa sported on the banks of the Yamunā with Draupadī and Subhadrā. They were then approached by a Brāhmaṇa whose complexion was like molten gold. It was Agni. He asked them to help him to consume the Khaṇḍava forest and the creatures therein, because the same forest was always prevented by Indra from being consumed by Agni for serpent Takṣaka dwelt there. Arjuna said that he needed a stronger bow and a quiver of inexhaustible arrows and a resplendent chariot to ride in, and Kṛṣṇa needed a weapon with which he could slay the Nāgas and Piśācas. Then Agni thought of Varuṇa, who presented Arjuna with the *Gāṇḍīva* bow, two inexhaustible quivers, a chariot with Yoked horses; Varuṇa then presented Kṛṣṇa with *Kaumadī gadā*. Both Kṛṣṇa and Arjuna destroyed the forest with its inmates. However, the Dānava Maya was rescued by Arjuna.

(iv) *Kadalī Vana*—This forest is associated with Hanumān in the epic. Bhīma, one of the Pāṇḍava heroes, met Hanumān here and the Rāma Kathā was narrated by Hanumān to Bhīma. Hanumān also displayed his gigantic form to Bhīma here.

(v) *Saugandhikāvana* (III.84.7082; III.150.11286)—The place is connected with Hanumān and Bhīma in the epic. Bhīma went to this forest to collect the celestial flower demanded by Draupadī.

Other prominent forests mentioned in the Mahābhārata are as follows—

(vi) *Lodhravana*

(vii) *Bamboo forests.*

(viii) *Gardens of Dvārkā*

(ix) *Terrible forests*

(x) *Forests of Kailāśa*

(xi) *Forests of Himālayas*

(xii) *Śveta Parvata grove.*

(xiii) *Gandhamādana*

(xiv) *Malaya Parvata groves*

(xv) *Various kinds of trees and foliage found in the forests.*

During the epic period, the human society evidently became more conversant with the botanical wealth of the country, which was further popularised during the Purāṇic period. The composers of the Purāṇic texts were fully conversent with the different properties of the Vegetable kingdom. A statement showing the plants, trees and forests mentioned in some selected Purāṇas is available at Appendix 4.

The early Buddhist literature also did not lag behind in the knowledge of vegetable kingdom. Many of the episodes relating to the life of the Buddha, starting from his birth are closely connected with groves and forests. Besides most of the Jātaka tales abundantly describe the flora and fauna, the forests, groves and tree spirits.

In the Jaina texts the trees and plants have their own importance, inasmuch as all the Jaina Tīrthaṅkaras, attaind the Supreme light under one tree or the other. A statement showing the trees associated with each one of twenty-four Tīrthaṅkaras is available in the relevant chapter of this work.

The presence of trees, plants and gardens etc. found a place of considerable importance in classical Sanskrit literature, right from a few centuries before the Christian era to the late medieval times.

The Indian Scientists have traced the presence of botanical species in the Palaeolithic times, followed by the archaeological evidence from the Harappan and post-Harappan cultures witnessed from Harappa, Mohenjodaro, Kalibangan, Rupar, Lothal and other sites in India, which provide a continued awareness of the contemporary society about the botanical wealth of the country. The trees are commonly found on the seals from Indus sites. At Mohenjodaro,[20] the trees are found on seals 509 and 522, in which a man is seated on the latter, appears to present some kind of acacia; upon one side of a round thin seal No. 604, a very conventionalised tree is carved, which cannot be identified, and exactly the same kind is seen on another seal.[21] The tree carved on the two seals could be regarded as sacred, since so much space was devoted to it. The Pīpala and Nīma trees have been identified on some of the seals from Mohenjodaro. Both these trees are considered even today as sacred throughout India. The leaves of the latter are supposed to repel snakes and they are eaten to cure various ailments. It is also the dwelling place of deities, especially the godlings of diseases; and the twig of the Nīma tree are much favoured as tooth brushes in the country side. Indeed the Harappan seals portray trees like Arjuna, Banana, Banyan, *Dalbergia Sisu*, besides the motifs like portraying the tree deities on seals and amulets, marriage of trees, offerings to trees, and planting of trees, besides their adoration. In addition to the sealings and seals, trees are also found portrayed over the early specimens of pottery starting from Harappan sites. Besides those depicted on the seals and sealings, trees are also found on the early Indian coinage as well.

The trees, plants, flora and founa are abundantly found in the sculptural art starting from the Aśokan period followed by the gateways of the Stūpas at Bharhut, Sanchi and

Amarāvatī, the early cave temples and the fresco paintings of Ajanta, Ellora, Bagh, Elephanta and other sites. This tradition was followed in the art of miniature paintings.

ध्रुवा एव वः पितरो: युगे युगे क्षेमकापासः सदसो न युञ्जते।
अजुर्यासो हरिषाचो हरिद्रव आ द्यां रवेण पृथिवीमशु श्रवुः।।

—*Ṛgveda*, X.94.12

(Your fathers, verily stand firm from age to age—they, loving rest, are not dissevered from their seat. Untouched by time, never lacking green plants and green trees, they with their voice have caused the heavens and earth to hear).

## Notes and References


1. *The New Encyclopaedia Britannica*, Vol. 3, pp. 65-66.
2. Keith, A.B., *The Religion and Philosophy of the Vedas and Upaniṣads*; (Harvard Oriental Series, Vol. 31), p. 184, (Reprint Delhi, 1970).
3. *Ibid.*, p. 185.
4. *Taittirīya Sāṁhitā*, II.5.3.2; II 7.3.19.
5. *Vājasneyi Saṁhitā*, XXII.28.
6. *Ṛgveda*, X.97.3; X. 97.15.
7. *Ibid.*, X. 146, 1-6.
8. *Ibid.*, VII.35.10.
9. *Āśvālayana Gṛhya Sūtra*, II.10.4; *Sāṁkhyāyana Gṛhya Sūtra*, IV.13.5; *Bhāradvāja Gṛhya Sūtra*, II.10.
10. *Ṛgveda*, IV.57.6.
11. *Pāraskara Gṛhya Sūtra*, II.17.19.
12. *Ṛgveda*. VIII.21.3.
13. *Gobhila Gṛhya Sūtra*, IV.4.27.
14. *Ṛgveda*, 1.127.6.
15. *Ibid.*, VI.75. II. 15, 16.
16. *Ibid.*, II. 3.4; X.70.4.
17. *Ibid.*, 1.142.6.
18. *Ibid.*, IV 57.4; VI.53.9; VI.58.2.
19. *Ibid.*, IV. 57.8; *Atharvaveda*, V.II.10; VI.30.I; *Vājasneyi Sāmhitā*, XXX.II; *Taittirīya Brāhmaṇa*, II.4.8.7.
20. Mackay, E.J.H., *Further Excavations at Mohenjodaro* I.p. 341 (N.D, 1930)
21. Marshall John, *Mohanjodaro and Indus Civilization*, p. 527, Pl. CXIV.

# 2
# Literary Evidences

The story of horticulture, botany or forests in India is of considerable antiquity. No other culture bears such a deep impress of the forest, vegetation and flowers as that of the Aryans in India. Their religious books, the *Vedas* and the *Upaniṣads* which portray their religions, moral and intellectual ideas were inspired by beauty and majesty of forests of India. In the forest retreats of their sages, pervaded beauty and mysticism which had a profound influence on their pupils who sat with them under natural surroundings in the *āśramas* under the shades of trees discussing fundamental problems of philosophy and life. These Aryans appeared on the Indian scene about 1600 B.C. They are believed to have migrated from the regions below the Caspian and passing through Iran and Afghanistan reached the plains of Punjab. They were Bronze Age people who kept large herds of cattle and had chariots drawn by horses. They settled down in villages in forest clearings and cultivated barley, wheat and pulses. They, it is believed, were familiar with copper and gold but not iron. Their houses were constructed of wood with mud walls and thatched roofs. Their sacred books, the Vedas, frequently mention the forest and vegetation.

While tracing the genesis and evolution of the vegetable kingdom, in the ancient literature, one has naturally to peep into the Vedic literature, which are considered to be the oldest of the Indo-Aryans, written in Sanskrit and happen to be the fountainhead of the Indian literature. According to the orthodox faith, the Vedas are *apauruseya*, and not of human composition, being supposed to have been directly revealed by the Supreme Being, Brahman and are *Śruti, i.e.* "what is heard or revealed" as distinguished from *Smṛti*, and the several sages, to whom the hymns of the Vedas are ascribed, are, therefore called seers and not the composers. In the light of the above position it may not be surprising if the trees and plants in a variety of forms appear in Vedic and post-Vedic literature. In general the following aspects of the plant and vegetable kingdom have been highlighted in these texts—

(i) Trees as symbols

(ii) Trees of general usage

(iii) Medicinal plants

(iv) Aquatic plants

(v) Sacred trees and plants

(vi) Grafting of trees

(vii) Fruit trees

(viii) Reeds

(ix) Grass

(x) Vanaspati

(xi) Auṣadhi

(xii) Fragrant trees

(xiii) Flowers

(xiv) Forests

(xv) Seeds

A comparative study of the ancient Indian literature would reveal that in the earliest literature like the Vedas the number of trees and plants is somewhat of a restricted nature, and this number went on increasing during the period that followed the Vedic and post-Vedic times, and finally in the Purāṇas the number is swelled considerably, introducing many more trees, plants and many such species which are not found in the earlier literary works. The account that follows, provides chronological details of the presence of plants, trees, herbs, medicinal plants, etc. as enshrined in the literary works in India, from the earliest times—

## 1. Vedic and Post-Vedic Literature

### (i) *Ṛgveda*

#### (a) *The Symbolism*

In the *Ṛgveda*, the Brahman has been equated with trees or a forest. In a hymn to Viśvakarman, the primal father of the world is said to have acted as a *hotṛ* priest offering the creation to fire and then concealed himself at the first instance and subsequently to have entered into the lower creatures. At last, the question is put as to what served as the resort and what as the material when he did all this. It is by way of reformulating this question that the images of forest and trees have been introduced in the *mantra*. Forest is to symbolize the place of creation and trees are the material. The answer to the question is implied in the account of one god who, having eyes all around, mouths all around, arms all around, feet all around, is said to blow the heaven and the earth by means of his hands and wings, while creating them. This is the creative movement of this omnipresent, omnipotent, omniscient Creator who assumes the form of the creation comprising the heaven and earth. He serves both, as the efficient and the material cause of the world besides providing the background of the creative act—

विश्वतश्चक्षुरुत विश्वतामुखो विश्वतो बाहुरुतविश्वतस्यात्।
सं बाहुभ्यां धमति सं पतत्रैर्धवा भूमीजनयन्देव एकः।।
किं स्विद्धनं क उ स वृक्ष आस यतोद्यावापृथिवीनिष्टतक्षुः।
मनीषिणो मनसापृच्छतेदु तद्यदध्यतिष्ठद् भुवनानिधारयन।।

—*Ṛgveda* X. 81. 3-4

Indeed the tree has been recognised in the *Ṛgveda* as a sort of living being, source of fruits; wood and something of constant and wavy nature (1.64.20,22). In view of these qualities it was well on the way of being conceived as the symbol of the universe as well as the human body. This has been achieved in the *mantra* in which two companion birds, with beautiful wings, are described as perched on the same tree; one of them eating sweet fruits while the other observing it without eating. When this tree is traced back to its deepest root, it comes close to being conceived, root and branch as Brahman himself—

द्वा सुपर्णा सयुजा सखाया समानं वृक्षं परिषस्वजाते।
तपौहन्यः पिप्पलं स्वाद्वत्त्यनश्नन्न्यो अभिचाकरीति।।

—*Ṛgveda*, 1.164.20

Another passage from the same text goes on to announce about the person eligible for obtaining the fruits of the said tree—

यस्मिन्वृक्षे मध्वदः सुपर्णा निविशन्ते सुबते चाधिविश्वे।
तरयदाहुः पिप्पलं स्वाद्वग्रे तन्नोन्नशद्यः पितरं न वेद।।

—*Ṛgveda*, 1.164.22

(On which the meat-eating beautiful birds live and procreate, it is to the top of this tree that the sweet fruit is said to belong to. He cannot obtain it who does not know the father).

Taking into consideration the inner spirit of these passages it could be safely stated that in this account the bird is identified with the guardian, the guardian with the father and the father with immortality. What apparently appears to be the associate of the fruit-eating bird and is thus expected to share in the act of eating, is in reality the guardian of fruit. As a guardian he sees to it that the fruit is enjoyed properly. In the next turn the guardian is viewed as associated with the tree of world not fortuitously but as its father. And when the universe is conceived as a *mantra*, the father becomes the supreme principle of immortality. Thus, whatever is realized by the seers as the principle of immortality in the innermost of his being, proves to be the Creator of the Universe. It is only due to ignorance that the Creator is taken to be a partner of the individual in life. The bird can be aware of the guardian only when it stops eating the fruit.

The same idea is propounded in the *Muṇḍaka Upaniṣad*, (III.1.3) according to which when the fruit-eating bird frees itself from eating and looks around, it finds a person who is golden in colour and is the creator, ruler and the source of creation; it gets rid of the sin as well as of the virtue and attains to the supreme harmony. Obviously *Puruṣa* in this Upaniṣad is the same who in the Ṛgvedic account is envisaged as the guardian, father and immortality.

### (b) *Medicinal Herbs*

Several passages in the *Ṛgveda*, bring out the medicinal properties of the plant kingdom, as would be evident from some of the numerous examples shown here—

(i) या ओषधी: पूर्वा जाता देवेभ्यस्त्रियुगं पुरा।
मनै नु बभ्रूणामहं शतं धामानि सप्त च।।

—*Rgveda*, X.97.1

(The herbs that sprang up in the olden times, three ages earlier than the gods. Of these whose hair is brown, will I declare the hundred powers and seven).

This is the most important passage in as much as it not only propounds the theory that the vegetable growth came into existance prior to the emergence of humans on earth. The difference of the period between the birth of plants and the humans on the earth has been mentioned here to be three *yugas* or ages. Now this *Yuga* has been variously interpreted to mean—

(a) A period or astronomical cycle of five or six years;

(b) A lustrum (especially in the cycle of Jupitar);

(c) An age of the world, long mundane period of years (of which there are four, *Kṛta* or *Satya, Treta, Dvāpara* and *Kali*, of which the first three have already elapsed; while the *Kali* began at mid-night between 17th and 18th of February, 3102 B.C.; is that in which we live).

The exact period, to which the three *yugas*, mentioned in the above Vedic hymn, refers to, would be difficult to state with authenticity, but one thing is quite clear that the emergence of human and other population on earth was subsequent to the growth of vegetation.

(ii) शतं वो अम्ब धामानि सहस्रभुत वो रुहः।
अधा शतक्रत्वो यूयमिमं अगदं कृत।।

—*Rgveda*, X.97.2

(Ye Mothers, have a hundred homes, yea, and a thousand are your growths; Do ye who have a thousand power free this my patient from disease).

यत्रोषधी समग्मत राजानः समिताविव।
विप्रः स उच्यते भिषग्रक्षोहामीवचतनः।।
अश्वावतीं सोमावतीगूजयन्ती मुदौजसम्।
आवित्सि सर्वा ओषधींरस्मा अरिष्टतातये।।
उच्छुष्मा ओषधीनां गावो गोष्ठादिवेरते।
धनं सनिष्यन्तीनामात्मानं तवपूरुष।।
इष्कृतिनाम वो भाताथो यूयंस्थ निष्कृतीः।
सीराः पतत्रिणीः स्थन यदामयति निष्कृथ।।

—*Ṛgveda*, X. 97. 6-9

(He who hath store of herbs at hand like kings amid a crowd of men; Physician is that sage's name, find-slayer, chaser of disease;

Herbs rich in *Soma*, rich in steed, in nourishment in strengthening power; All these have, I provided here, that this man may be whole again;

The healing virtues of plants stream forth like cattle from the stall; Plants that shall win my store of wealth, and save thy vital breath, O men;

Reliever is they mother's name and hence Restorer are ye called; Rivers are ye with wings that fly, keep for whatever brings disease).

The above pessages point to the institution of physician, who could apply the medicinal herbs not only for the removal of the diseases, or recovery of the man from ailment but also store energy besides the healing properties of the herbs.

**(c) *Life Saving Drugs***

Some of the passages from the *Ṛgveda* point to the life saving properties of the herbs—

मुञ्चन्तु मा शपथ्या३ दथो वरुण्यादुत।
अयो यमस्य पड्वीशात्सर्वस्माद्देवकिल्विषात्।।

—*Ṛgveda* X.97.16

(Release me from curse's plague and woe that comes from Varuṇa—Free me from Yama's fetter, from sin and offence against the gods).

**(d) *Fruit Trees, Their Blossoming and Trees without Fruits***

There is also a mention of plants/trees, which blossom and bear fruit and are useful for mankind—

ओषधीः प्रतिमोदध्वं पुष्पवती प्रसूवरी।
अश्वाइव सजित्वरीर्वीरुधः पारयिष्ण्वः।।

—*Ṛgveda* X.97.3

(Be glad and joyful in plants, both blossoming and fruit bearing; Plant that will lead us to success like the mares who conquer in the race).

Another passage from the same text, besides pointing to the blossming of trees and their bearing fruit, also mentions the fruitless trees—

या फलानीर्या अफला अपुष्पा याश्च पुष्पिणीः।
बृहस्पतिप्रसूतास्ता नो मुञ्चन्त्वंहसः।। —*Ibid.*, X.97.15

(Let fruitful plants and fruitless, those which blossom and blossomless; urged onwards by Bṛhaspati relieve us from our pain and grief).

**(e) *Regaining of Lost Strength by Application of Herbs***

यदिमा वाजयन्नहमोषधीर्हस्त आदधे।
आत्मा यक्ष्मस्य नश्यति पुरा जीवगृभो यथा।।

—*Ibid.*, X.97.II

(When bringing back the vanished strength, I hold these herbs within my hands; the spirit of the disease departs, he can seize upon the life).

**(f) *Tree Worship***

While discussing the subject, A.B. Keith has observed that the "reference to the worship of tree and plants are very scanty in the Vedic rituals and the mythology alike, but they are quite adequate to show that as among all other peoples, these objects were not without their share of reverence." It is, however, a fact that the tree was conceived as a goddess in the *Ṛgveda*—

ओषधीरीति मातरस्तद्वो देवीरूप ब्रुवे।
सनेयमश्वं गां वास आत्मानं तव पूरुष।।

—*Ibid.*, X.97.4

(Plant, by this name, I speak to you, Mothers to you the goodess; steed, cow and garment may I win, win back thy very self, O man).

Not only that in Chapter 146 of the tenth *Maṇḍala* of the *Ṛgveda* the tree-goddess has been conceived as Araṇyāṇī as would be evident from the following passages—

अरण्यान्यरण्यान्यसौ या प्रेव नश्यसि।
कथाग्रामं न पृच्छसि न त्वा भीरिव विन्दतीः।।
वृषारवाय वदते यदुपावति चिच्चिकः।
आघाटिभिहिव धावयन्नरण्यानिर्महीयते।।
उत गावइवादन्त्युत वेश्मेव दृश्यते।
उतो अरण्यानिः सायं शकटीरिव सर्जति।।
गामङ्गैष आह्वयति दार्वङ्गैषो अपावधीत्।
वसन्नरण्यान्या सायमक्रुक्षदिति मन्यते।।
न वा अरण्यानिर्हन्त्यन्यश्चेन्नामिगच्छति।
स्वादो फलस्य जग्ध्वाय यथा कामं निपद्यते।।
आञ्जनगन्धि सुरभिबह्वन्नामकृषीवलाम्।
प्राहं मृगााणां मातरमरण्यानिमशंसिषम्।।

—*Ibid.*, X.146. 1-6

(Goddess of wild and forest who seemest to vanish from the right. How is it that thou seekest not the village? Art thou not afraid.

What time the grass-hopper replies and swells the shrill cicala's voice.

Seeming to sound with tinkling bells, the Lady of the wood exults.

And; yonder, cattle seem to graze, what seems a dwelling place appears—

or else at eve the lady of the forest seems to free the wains.

Here one is calling to his cows, another hath felled a tree—At eve the dweller in the wood fancies that somebody hath screamed.

The goddess never slays, unless some murderous enemy approach.

Man eats of savoury fruits and then takes, even as he will, his rest.

Now have I praised the Forest Queen, sweet-scented, redolent of balm.

The mother of all sylvan things, who tills not but hath stores of food).

### (g) *The Tree-god*

Besides the tree-goddess or Aranyāṇī, the Tree-god or Vanaspati has also been invoked in the *Ṛgveda* on a number of occasions. The very term vanaspati denotes the one who could be the lord of the forest or vegetation. A number of hymns have been devoted to the wood-god—

(i) अव सृजा वनस्पतेदेवदेवेभ्यो हविः। प्रदातुरस्तु चेतनम्।।

—*Ibid.*, I.13.11

(God Savran of the wood, present his oblation to the gods. And let the giver be renowned).

God Savran of the wood, Vanaspati, Lord of wood is usually a large tree; here said to be an Agni, as if the fuel and the burning of it were identified).

(ii) उप त्मन्या वनस्पते पाथोदेवेभ्यः सृज अग्निर्हव्यानिसिष्वदत्।

—*Ibid.*, I.188.10

(Send to gods, Vanaspati, thyself, the sacrificial drought, Let Agni make the oblation).

(iii) वनस्पतेऽव सृजोपदेवानग्निर्हविः शमिता सूदयति।
सेदु होता सत्यतरो यजातियथादेवानां जनिमानि वेद।।

—*Ibid.*, III.4.10

(Send to the gods the oblation, Lord of Forests; and let the Immolator Agni dress it; He as the truer priest shall offer worship, for the god's generations well he knoweth).

Here too Vanaspati is a large tree.

(iv) वनस्पतिरवसृजन्नुपस्थादग्निर्हविः सूदयाति प्रधीभिः।
त्रिधा समक्तं नयतु प्रजानदेवेभ्यो दैव्यः शमितोप हव्यम्।

—*Ibid.*, II.3.10

(Vanaspati shall stand near and start us and Agni with his arts prepare oblation; Let the Skilled heavenly Immolator forward on to gods the offerings thrice oriented).

(v) अञ्जन्ति त्वामध्वरे देवयन्तो वनस्पते मधुना दैव्येन।
यदूध्वतिष्ठा द्रविणेह धत्ताद्यद्वा क्षयो मातुरस्या उपस्थे।।

—*Ibid.*, III.8.I

(God serving men, O Sovran of the forest, with heavenly meath at sacrifice anoint thee; Grant wealth to us when thou art standing upright as when reposing on his mother's bosom).

(vi) उच्छ्रयस्व वनस्पते वर्ष्मन्पृथिव्या अधि।
सुमिति मीयमानो वर्चो धा यज्ञवामहे।। —*Ibid.*, III.8.3

(Lord of the forest, raise thyself up on the loftiest spot of earth; Give splendour, fixity, and measured well, to him who brings the sacrifice).

(vii) यत्र वेत्थ वनस्पते देवानां गुह्या नामानि।
तत्र हव्यानि गामय।। —*Ibid.*, V.5.10

(Vanaspati, wherever thou knowest the god's mysterious names, send your oblation thither with).

(viii) उपाव सृजत्मन्यासमञ्जदेवानां पाथ ऋतुथा हवीषि।
वनस्पतिः शमिता देवो अग्निः स्वदन्तु हव्यंमधुना घृतेन।।
—*Ibid.*, X.110.10

(Send our offerings which thyself thou blamest the companies of gods in ordered seasons; Agni, Vanaspati, the Immolator sweeten our offered gifts with meath and butter).

**(h) *The Grass***

The text speaks of a varied type of grass, used for a variety of purposes, including animal fodder or making of ropes. But the most important aspect available concerning the grass is its trimming. An indication of the same is available in *Rgveda*, (I.142.5)—

(i) इला सरस्वती मही तिस्रो देवीर्मयो भुवः।
बर्हिः सीदन्त्वस्रिधः।। —*Ibid.*, I.13.19

(Ilā, Sarasvati, Mahi, three goddesses bring delight; be seated peacefully on the grass).

(ii) ईलते त्वामवस्यवः कण्वासो बृक्त बर्हिषः।
हबिष्मन्तो अरंकृतः। —*Ibid.*, I.14.5

(The sons of Kaṇva fain for help adore thee, having strewn the grass; with offerings and all things prepared).

(iii) स्तृणानासो यत स्रुचो बहिर्यज्ञे स्वध्वरे।
वृञ्जेदेवव्यचस्तममिन्द्रायशर्मसप्रथः।। —*Ibid.*, I.142.5

(The ladle-holders stream trimmed grass at this well adored sacrifice.)

(iv) सरस्वतीसाधयन्तीधियं न इलादेवीभारती विश्वतूर्तिः।
तिस्रो देवीः स्वधया बर्हि रेदमच्छिद्रं पान्तु शरणं निषद्य।।
—*Ibid.*, II.3.8

(Sarasvatī, who perfects on devotion, Ilā divine, Bhāratī all surpassing; Three great goddesses with power inherent, seated, protect the holy grass, our flawless refuge).

(v) शतंवेणूञ्छतं शुनः शतं चमाणि म्लातानि।
शतं मे बल्वजस्तुका अरुषीणां चतु:शतम्।। —*Ibid.*, 8.55.3

(A hundred bamboos, a hundred dogs, a hundred skins of beasts, well tanned;

A hundred tufts of *Balbaja* grass (*Abutilon indicum*, Linn.) four hundred red hued mares are mine).

The foregoing account provides only a glimpse of the vast coverage provided in the Ṛgvedic texts, which goes on to testify about the existence of hundred varities of plants and trees—

या ओषधीः सोमराज्ञी बह्वीः शतविचक्षणा।
तासां त्वमस्युत्तमारं कामाय शं हृदे।। —*Ibid.*, X.97.18

(Of all the many plants whose king is Soma; Plants of hundred species; Thou art the plant most excellent, prompt to the wishes and sweet to the heart).

## (ii) *Yajurveda*

The text contains several references to agriculture, trees, plants, and medicinal plants, forests and agriculture, a few of which are brought out here—

### (a) *Agriculture*

भूताय त्वा नरातये स्वरभि विख्येषन्दृश्ठंहन्तां
दुर्य्याः पृथिव्यामुवन्तरिक्षमन्वेमि। —(I.11)

(I resort to agriculture and craft for removing poverty and ministering happiness to all. May I feel happiness in my domestic life.

### (b) *The Forests*

अद्रिरसि बानस्पत्थो ग्रावासि पृथुबुध्नः प्रति त्वादित्यास्त्वग्वेत्तु।।
—(I.14)

(Clouds receive moisture from forests, reside in the vast space and receive water from air. May the learned understand the building of houses, which serve as the skin of the earth.

### (c) *Medicinal Herbs*

पृथिवि देवयजन्योषध्यास्ते मूलम्माहिश्ठंसिषं
वृजगुङ्छगोष्ठानं वर्षतुते द्यौर्वधानदेव सवित:
परमस्यां पृथिव्यां शतेन पार्शर्य्योस्मान्द्वेष्टि
यंच वयं द्विष्मस्तमतो माकौक।। —(I.25)

(O God, the creator of the solar bright regions we implore. Thee that through Thy grace, may we not destroy the medicinal herbs on earth, on which the learned perform the *yajñas*. May the *yajña* reach the clouds. Let the sun pour rain or earth through its rays. O heroic persons, bind by various fetters, a wicked man in this world, who is opposed to us and is opposed by us, and release him not).

## (iii) *Sāmaveda*

### Sun as the Giver of Life to Plants and Trees

The rays of sun are said to have been helpful not only to the plant world but also to the mobile creatures, which were created, by providing them with a shelter which means that matter, which formed the shelter, was derived from the rays of the sun. The fact is that both energy or the various types of radiations and the matter have been derived from their parents *i.e.* ideas. Both are forms of ideas differing in the density of ideas with which they are packed. The ideas are much more densely packed in matter than in other forms like energies. The fact is that ideas constitute the basic material for everything in the Universe. All these ideas are objective counterparts of the universal and intelligent subject, the creator.

In the *Sāmaveda* the sun has been projected as the one who puts life into the plants and vegetables thus—

शाक्मना शाको अरुणः सुपर्ण आ महः शूर सनाश्नीडः।
यच्चिकेत सत्यमित्तन्न मोघं बसु स्पार्हमुत जेतोत दाता।।

—*Sāmaveda*, 9.1.1783

(The ruddy rays of the sun are helpful to vegetation, they have pleasing effect on these valiant creatures (being manifested) as from olden times with nest. The sense activities perceived their desires and installed them in existence, Indra did not strike them down, they were agreeable, he won for them and bestowed them also).

## (iv) *Atharvaveda*

In the *Atharvaveda*, the plant-kingdom has been defined varioulsy with numerous usages like—

**(a) *As Healing Agents***

*VII.7.1*— "The plants that are brown, and those that are white; the red ones and the speckled ones; the sable and the black plants, all these we invoke."

*VII.7.2*— "May they protect this man from the disease, sent by the gods, the herbs whose father is the sky, whose mother is the earth, whose root is the ocean."

*VII.7.3*— "The water and the heavenly plants are foremost; they have driven out from every limb thy disease, consequent upon sin."

*VII.7.4*— "The plants that spread forth, those are bushy, those that have single sheath, those that creep along, do I address; I call in thy behalf the plants that have shoots, those that have stalks, those that divide their branches, those that are derived from all the gods, the strong plants that furnish life to man."

*VII.7.5*— "With the might that is Yoms, ye mighty ones, with the power and strength, that is yours, with that do ye, O plants, rescue this man from this disease, I now prepare a remedy."

*VII.7.6*— "The plants *jivata* (quickening) *na-gha-risa* (for sooth no harm), *jivanti* (living) and the *arundhati*, which removes disease, is full of blossoms, and rich in honey, do I call them to protect from injury."

### (b) *Plants Used to Allay Jeolousy*

भगमस्या वर्च आदिष्यधि वृक्षादिव स्रजम्।
महाबुघ्न इवपर्वतो ज्योक् पितृष्वास्ताम्।।
एषाते राजन्कन्याऽवधूनां धूयतां यम्।
सामातुर्बध्यतां गृहेऽथो भ्रातुरथो पितुः।।
एषा ते कुलपा राजन्तामु ते परि दद्मसि।
ज्योक पितृष्वासाता आ शीर्ष्णः समोप्यात्।।
असितस्य ते ब्रह्मणा कश्यपस्य गयस्य च।
अन्तः कोशमिव जामयोऽपि नह्यामि ते मगम्।। —(I.14.1-4)

(I have taken unto myself her fortune and her glory, as a wreath of a tree. Like a mountain with broad foundations may she sit a long time with her parents. This woman shall be subjected to thee as thy wife, O king Yama—(till then) let her be fixed to the house of her mother, or her brother, or her father. This woman shall be the keeper of thy house, O king (Yama), and her do we make over to thee. May she long sit with her relatives, until her hair drops from her head. With the incantation of Asita, of Kaśyapa and of Gayā do I cover up thy fortune, as woman cover something within a chest).

### (c) *Plants Used to Deprive One of the Virility*

In verse VI.138 (1-2) of this Veda the plants are invoked to deprive one of his virility saying, that "as the best of plants thou art reputed, O herb; turn this man for me today into an eunuch that wears his hairdress, and into one that wears a hood. Then Indra with a pair of stones shall break his testicles both."

### (d) *Use of Plants as Charm by a Woman Against a Rival or a Co-wife*

In a passage III.18. (I-4) the above aspect has been projected thus—

"I dig up this plant of herbs the most potent, by whose power, rival woman are overcome and husbands are obtained. O thon (plant) with erect leaves, lovely, do thou, urge on by the gods, full of might, drive away my rival, make my husband mine alone. He did not for sooth, call thy name, and thou shalt not delight in this husband. To the very farthest distance do we drive our rival. Superior am I, O Superior (plant) superior truly, to superior women. Now shall my rival be inferior to those that are inferior."

**(e) *Plants Used for Arousing Love or Passion***

The above aspect is reflected in a passage VI.139.(1-3) as under—

"Clinging to the ground thou didst grow (O plant), that producest bliss for me; a hundred branches extend from thee, three and thirty grow down from thee, with this plant of a thousand leaves, thy heart do I porch. Thy heart shall porch with love for me and thy mouth shall porch with love for me. Languish, moreover with love for me with parched pass thy days. Thou that causet affection, kindlest love, brown, lovely plant, draw us together draw together yonder woman and myself, out heart make the same."

**(f) *Father and Mother of Plants***

III.9.I; III.23.6; VIII.7.2—These passages from *Atharvaveda* prescribe Dhyaus, the heaven and Pṛthivī, the earth as the father and mother of plants.

**(g) *Plant Used as Charm Against Ophthalmia***

The use of the aforesaid aspect of plants is reflected in a passage (VI.16.1) as under—

"O Abayu (and even if) thou art not Abayu, strong is thy juice, O Abayu, we eat a gruel compounded of thee. Vakalha is thy father's name, Madavati thy mother's name. Thou art verily not such as to have consumed thy ownself. Tauvilika, do be quiet. This howling one has become quiet. O brown one, and brown-eared one, go away. Go out."

**(h) *Charm with Apāmārgā Plant Against Demons, Sorcery and Enemies***

The Apāmārga plant is conceived for the destruction of demons and evil spirits, (IV.17.1-3) thus—

"We take hold, O victorious one, of thee, the mistress of remedies. I have made thee a thing of thousand fold strength for every one, O plant. Her the unfailingly victorious one, that ward off curses, that is powerful and defensive (her and) all the plants have I assembled, intending that she shall save us from this trouble. The woman who has cursed us with a curse, who has arranged fire misfortune for us, who has taken hold of our children, to rob them of their strength—may she eat her own offspring. The magic spell which they have put into the unburned vessel, that which they have put into raw flesh, with these slay thou those that have prepared the spell. Evil dreams, troubled life, *Rākṣas*, gruesomeness, and grudging demons (*arayi*) all the evil named, evil speaking powers, these do we drive out from us."

The tabulation that follows, would provide a bird's eyeview of the use of plants in the *Atharvaveda* for medicinal and other purposes. This is an illustrative account and not an exhaustive one, and serves as a window to peep into the indepth view of the botanical properties as known to the contemporary intellectuals who did conceive the use of plant kingdom for a veriety of purposes like the life saving drugs to the charms of different kinds.

**(i) *Trees and Plants Used for the Cure of Leprosy, Ailments and Other Purposes***

(i) *Those which Cure Leprosy*

| | | |
|---|---|---|
| (1) | "Even colour" epithet of a plant which cures leprosy.... | 1.23.1; 1.24.1 |
| (2) | Magic and medicinal plants used as universal remedy | VII.7.1-28 |

(ii) *Other Ailments*

| | | |
|---|---|---|
| (3) | Trees used as the charm against fever | 1.25.1-4<br>VI.20.1-3<br>VII.116.1-2<br>V.4.1-10<br>XIX,39.1-10 |
| (4) | Charm with the plant Arundhati.... (lākṣa) for cure of fracture | VI.12.1-7. |
| (5) | Charm with the plant Sitaci (lākśa) for cure of wounds | VI.5 1-9 |
| (6) | Use of pepper corn for cure of wounds | V.109.1-3 |
| (7) | Use of plants as charm against ophthalmia | VI.16.1-4 |
| (8) | Use of plants for the growth of hair | VI.21.1-3<br>VI.136.1-3<br>VI.137.1-3 |
| (9) | Use of trees for charm to promote virility | IV. 4.1-8 |
| (10) | Charm with the plant *Ajaśṛṅgī* to drive out Rākṣasas, Apasarās and Gandharvas | IV.37.1-82 |
| (11) | Possession of demons of disease cured by an amulet of ten kinds of wood | II. 9. 1-5 |
| (12) | Charm with plant Pṛṣparṇī against the demon of disease, called Kaṇva | 11.25.1-5 |
| (13) | Charm with an amulet derived from the Jāngiḍa tree against diseases and demons | 11.4.1-6<br>XIX.34.1-10<br>XIX.35.1-5 |

| | |
|---|---|
| (14) Exorcism of disease by means of an amulet from the Varṇa tree | VI.85.1-3 |
| (15) Use of Cipadru tree as panacea | VI.127.1-3<br>VI. 96.1-3 |
| (16) Use of barley and water as universal remedies | VI.91. 1-3 |
| (17) Charm with Apāmārgā plant against sorcery, demons and enemies | IV.17.1-8<br>IV.18.1-8<br>IV.19.1-8<br>VII.65.1-3 |
| (18) Prayer for protection addressed to an amulet made from the wood of the Śrakṭya tree | VIII.5.1-22 |
| (19) Praise of the virtues of an amulet derived from the Varaṇa tree | X.3.1-25 |
| (20) Praise of the virtue of the amulet of Khadira wood in the shape of a plough-share | X.6.1-35 |
| (21) Use of Aśvattha tree for obtaining a son | VI.II.1-3 |
| (22) Use of plants as charm to secure the love of a man | VII.38.1-5 |
| (23) Use of plants to serve as charm against a rival or co-wife | III.18.1-6 |
| (24) Use of plants to serve as charm for depriving a man of his virility | VI.138.1-5 |

**(j) *List of Trees and Plants in the Atharvaveda***

| | |
|---|---|
| (1) Abayu (Mustard) | VI.16.1 |
| (2) Ajaśṛṅgī (*Odina wodiar*) | IV.37.2 |
| (3) Ahava | VIII.8.4 |
| (4) Aśvattha (*Ficus religiosa*, Linn.) | VIII.8.3 |
| (5) Khadira (*Abutilon catechu*, Willd.) | VIII.8.3 |
| (6) Tajalbhaṅgā (*Ricinus communis*) | VIII.8.3 |
| (7) Vadhaka | VII.8.3 |
| (8) Alāśilā (*Green creeper*) | VI.16.4 |
| (9) Apāmārgā (*Achyranthes aspera*, Linn.) | IV.17.7 |
| (10) Arātakī (*Accacia suma*) | IV.37.6 |
| (11) Arka tree (*Calotropis gigantea*, R. Br.) | 1.12.1 |
| (12) Arundhatī (*Sida cordifolia*, Linn.) | IV.12.1 |

| | |
|---|---|
| (13) Avaka tree, A Reed (*Blyxa octandra,* Rich) | IV.37.8 |
| (14) Banyan (Nyagrodha) (*Ficus bengalensis*) | V.5.5 |
| (15) Plāska (*Ficus infectoria*) | V.5.5 |
| (16) Dhava (*Anogeissus latifolia,* Wall.) | V.5.5 |
| (17) Parva (*Butea frondosa*) | V.5.5 |
| (18) Darbha-grass (*Desmostachya bipinata,* Staff.) | VI.43.1 |
| (19) Barley (yawa) | VII.7.20 |
| (20) Rice | VII.7.20 |
| (21) Dūrva plant, millet (*Desmostachya bipinnata,* Staff.) | VI.106.1 |
| (22) Jāngiḍa tree (*Terminalia arjuna*) | XII.4.1 |
| (23) Jayantī | V.5.3 |
| (24) Ghṛitācī (*Large cardamoms*) | X.4.24 |
| (25) Jivala | XIX.39.3 |
| (26) Jīvanta (A medicinal plant) | XIX.39.3 |
| (27) Jīvantī (*Desmotrichum fimbriatum,* Blume) | VII.7.6 |
| (28) Kipadru | VI.127.2 |
| (29) Kudi plant | V.19.12 |
| (30) Kuṣṭha plant (*Costus speciosus*) | V.4.1-10 |
| (31) Lākṣa (Lac) | IV.12.1-7 |
| (32) Madhavatī or Mādhavī (*Hiptage madablota,* Gaertn) | IV.7.4 |
| (33) Muñjā-grass (*Saccharum munja*) | I.2.4 |
| (34) Nitani | VI.136.1 |
| (35) Pata plant (*Buchanania latifolia*) | II.27.1 |
| (36) Papper | VI.109.1 |
| (37) Pṛṣṇiparṇī (*Uraria logopioides,* L.) (A medicinal plant) | II.25.1 |
| (38) Putadru tree (*Pinus longifolia*) | VIII.2.28 |
| (39) Vetasā Reed (*Calamus rotang*) | II.29.1-7 |
| (40) Sadampuṣpā (Ever flowering plant) | VI.20.1 |
| (41) Śāmi (*Prosopis spicigera Minosa suma*) | IV.37.9 |
| (42) Silangala (Green creeper) | VI.16.4 |
| (43) Soma plant (Intoxicating plant) (*Asclepias acida,* Roxb.) | VI.91.20 |
| (44) Śrakṭya tree (*Clerodendrum phlomoides*) | VII.5.3 |

(45) Tajalbhaṅgā tree (*Ricinis communis*) VIII.8.3

(46) Udumbara (*Ficus glomerata*) II.14.10

(47) Kuśā (*Poa cyonosurloides*) II.14.4.16

(48) Vadboka II.14.4

(49) Varana (*Cratoosva nurvalla*) VI.85.1

(50) Jāngiḍa (*Terminalia arjuna*) XIX.35.1-5

(51) Vibhindati IV.19.5

(52) Vilhalha VI.16.2

(53) Viṣaṇakā (*Gymnema sylvestra*) VI.44.3

## 2. *Saṁhitā* Literature

### *Taittirīya Saṁhitā*

The text speaks about the origin of plants, thus (IV.2.6);

(a) "The plants born,
Three generations before the gods,
Of the brown ones I celebrate
The seven and a hundred abodes,

(b) A hundred, O mother, are your abodes,
A thousand too your shoots,
Therefore do ye, with a hundred powers,
Make him whole for me.

(c) With flowers, with shoots,
Fruit bearing and without fruit,
Like steed victorious,
The plants are strong for help.

(d) Plants, O ye mothers,
I hail you, O goddesses,
Go bearing away defilement,
Defilement Destroying.

(e) In the Aśvattha is your seat,
In the *parṇa* is your dwelling made.

(f) These plants in my hand,
The Soul of disease perisheth
As before one that taketh alive.

(g) When the plants come together,
Like princes in the assembly,

Sage is the physician called,
Slayer of Rākṣasas and slayer of disease.

## 3. The *Brāhmaṇas*

The Vedas and Saṁhitās were followed by *Brāhmaṇa* literature which happen to be the explanations and utterances of the learned priests upon any point of ritual and upon the science of *yajña*. While tracing the presence of vegetable kingdom in the *Brāhmaṇas*, it may be pointed out that in these works, sometimes Brahman has explicitly been put in the form of a tree. In the *Śatapatha Brāhmaṇa* (VI.6.3.7) the *Palāśa* tree has been equated with Brahman—

अथैता उत्तरा पालाश्यो भवन्ति। ब्रह्म वै पलाशो ब्रह्मणैवे-
नमेतत्सन्धिे यद्धेव पालाश्यः। सोमो वै पलाश एषो ह
परमाहुतिर्यत्सोमाहुतिस्तामस्मिन्नेतज्जुहोति तयैनमेतत्प्रीणाति।।

(Rest of the *Samidhās* are of Palāśa (*Butea frondosa*) Palāśa tree is Brahman. So Brahman is lighted from the Palāśa wood. And, again, why they are Palāśa ones—because *Soma* is Palāśa tree. This is supreme offering since *Soma* is invoked. This is offered to fire and with this he gratifies Agni—Fire).

The above passage projects the enkindling of sacrificial fire with Palāśa wood. Since Palāśa is Brahman, it is argued, that enkindling of fire by means of its wood is to be regarded as equivalent to enkindling it by means of Brahman. Palāśa indeed is very much prone to ignition. Brahman, is also regarded as luminous. Its luminosity becomes manifest as soon as the mind is devoted to it, in the same way as the Palāśa wood is ignited as soon as it is brought into contact with fire. On account of this fundamental resemblance, Palāśa has been equated with Brahman.

### (i) The *Śatapatha Brāhmaṇa*

The *Śatapatha Brāhmaṇa* (I.2.4.16), while praying to the earth goddess declares that the *Auṣadhis* (Vegetable growth) will not be harmed. (पृथिवि देवयजन्योषध्यास्ते मूलं मा.....) Not only that, but the passage under reference further goes on to testify, though indirectly, that the plants have life and the roots are to be protected. The English rendering of the same is provided here—

"O' earth, that affordest the place for making offerings to gods may I not injure the roots of thy plants. He thereby makes her, as it were, with roots remaining in her. Whilst he takes up (the earth dug up by a pick-axe); he thus addresses her, "May I not injure the roots of thy plants, and in further saying, 'Go the fold, the abode of the Cows, when he is about to throw it away (on the heap of rubbish), he causes it not to forsake him; for that which is within the fold does not forsake him', for that reason, he says, 'Go to the fold, the abode of the cows.' He further says (whilst looking at the hole in the ground), 'May the sky rain on thee,' wherever, in digging into her, they wound and injure her—water being (a means of) expiation that he thereby expiates by the water which is (a means of) expitiation; that he thereby makes good by means of water—that is the reason why they say, 'May the sky rain on thee'..."

The above passage makes a few things concerning the plants clear, though obliquely, that (i) the plants possess life; (ii) their roots are in the earth, and cannot be exposed out of it; (iii) serving of cowdung manure to plants, and finally (iv) water them by rain water or otherwise. Other aspects of botany and horticulture have been highlighted in the *Śatapatha Brāhmaṇa* thus—

**(a) *Grass and Sprinkling of Water Thereon***

In the passage (at I.3.3.2) the following procedure has been prescribed for the purpose—

"He then sprinkles the altar with the text (*Vājasneyī Saṁhitā*-II.I). Thou art the altar; I sprinkle thee, agreeable to the *barhis* (Sacrificial grass covering). He thereby makes it sacrificially pure for the grass."

(I.3.3.3) "He then hands the sacrificial grass to him (the *Adhvaryu*). The latter puts it down on the altar with the knot turned to the east, and sprinkles it, with the text (*Vājasneyī Saṁhitā*, II.I).

Indeed in the beginning of a sacrifice the *Adhvaryu* makes of the load of Darbha grass or sacred grass, which has been brought to the sacrificial ground, seven *muṣṭhis* or bunches, each one of which is tied together with a stalk of grass just as the Baresona (Barsom) of Parsis. These are unloosened and spread all over the *vedī*.

Another passage speaks about the watering of grass (1.3.3.4) as under—

"Thereupon he pours the sprinkling water which is left on the roots of the (grass) plants with the text (*Vājasneyi Saṁhitā*, II. 2). A moistening art thou for Aditi. Aditi, indeed, is this earth; hence it is for the latter that he, thus moisten the roots of these plants; thereby these plants become root-moistened; and even if their tops are dry, their roots at least remain moist."

**(b) *Fruits***

The *Śatapatha Brāhmaṇa* (II.5.2.11) speaks of the Karīra fruit thus—

"Upon both (the dishes of curds), he scatters Karīra fruits; for with Karīra fruits, Prajāpati bestows happiness on the creatures, and so does he (the sacrificer) thereby bestow happiness on the creatures."

Karīra is a tree identified as *Capparis aphylla*, which has fruits which have several usages. According to Sāyaṇa, (Om *Taittirīya Brāhmaṇa*, I.8.3) it is Karīra shoots, which resemble *Soma* creepers, which are so used; but he also mentions that some authorities take Karīra to mean fruit. He further quotes, that over a hundred Śamī-leaves and over a thousand Kariras should be strewn over the two dishes of curds. (*Taittiriya* Brāhmaṇa, 1.6.5.5)

Karkaṇḍa is another fruit mentioned in the *Śatapatha Brāhmaṇa* (XII.9.1.5) thus—

"The two cups of Aśvins are his (Brahmaṇ's) eyes, and the ground wheat and *kuvalas* (*jujubes*) his eye lashes; two cups of Sarasvatī are nostrils and the ground Indra-grain and *badara (jujubes)* the hair in his nostrils; the two cups of Indra are his ears and the ground barley and Karkaṇḍa (*jujubes*) the hair of his ears and his eyebrows."

**(c) *The Sacred Trees***

The *Śatapatha Brāhmaṇa* (XII. 9.1.3) mentions Aśvattha, Nyagrodha, Udumbara as the sacred trees, while Asmi-gandhā, Bhūmipaśā, Pṛṣṇiparṇī (*Uraria logopioides*, Linn.), Aśvattha (*Ficus religiosa*), or a Vibhitakā (*Terminalia bellerica*), Tilvaka (*Symplocos racemosus*) or Sphurgaka (*Diospyros perrigeri*) or a Haridru (*Pinus deodora*) are also mentioned as sacred elsewhere (XIII. 8.1.16).

**(d) *As a Symbol of Brahman***

The use of tree as a symbol of Brahman is found in the *Taittirīya Brāhmaṇa* (II. 8.9), wherein Brahmaṇ is compared to a forest and the tree out of which both the heaven and earth were manufactured. It describes Brahmaṇ as the resort of the act of manufacturing as well. This expression comes as an answer to the following question in a passage from the same text (*Taittirīya Brahmāṇa*, II.8.8)—

"What probably was the forest and what, indeed, the tree out of which they manufactured heaven and earth? O wise men, ask it also through the mind as to what he took to be his resort, while supporting the creations."

This discussion did not originate in the Brāhmaṇa literature, but the genesis of the ideology is found planted in the *Ṛgveda* itself, which has been suitably dealt with while discussing the issue in theVedic literature in this work.

**(e) *The Forests***

An Udumbara forest has been mentioned in the *Pañcaviṁśa Brāhmaṇa* (XXVI.6.4).

## 4. *Āraṇyakas*

### *Aitareya Āraṇyaka*

The *Āraṇyakas* come next to the *Brāhmaṇas* and are considered to be the forest texts, so called because of their secret character, which could be learnt in the forests alone. The *Āraṇyakas* are the concluding portions of the *Brāhmaṇas* but they deal with mysticism and symbolism of the *Yajña* and are partly speculative and partly ritualistic. They form a transition between the Brāhmaṇas and Upaniṣads. The text mentions several types of trees and plants, besides their usage, refers to the swing, which is made of wood and used by the ladies. While dealing with the making of a swing, the text prescribes (1.2.3) that "there should be three planks, for three fold are these worlds and (the swing should) resemble them. But this is not to be accepted. There should be two planks, for these two worlds seem most real and the other between them is the sky. So let there be two planks. Let them be of Udumbara tree. The Udumbara is sap and proper food, and plants of it serve to win sap and proper food. Let them be raised in the middle. For in the middle food delights man, and so he places the sacrificer in the middle of proper food. There are two kinds of ropes, the right and the left. The right serves for some animals, the left for others. When there are both kinds, they serve to win both kinds of animals. The rope should be of Darbha grass. For, of all the plants, Darbha is free from evil, and so they should be of Darbha grass—

> एकं फलकं स्यादित्याहुरेकधा ह्येवायं वायु पवतेऽस्य रूपेणेति। तत्रन्नादृत्यम्।
> त्रीणिफलकानि स्युरित्याहुस्त्रयो वा इमे चिवृतो लोका एवां रूपेणेति। तत्रन्नादृत्यम्।
> द्वे एव स्यातां द्वौ वा श्मौ लोकवद्धातमाविव दृश्यते
> य उ एने अन्तरेणा काशः सोऽन्तरिक्षलोकस्तस्माद्वे एव स्याताम्।
> औदम्बरे स्यातामूर्वा अन्नाद्यमुदुम्बर अर्जोऽन्नाद्यस्यावरुद्ध्यै।
> मध्यत उद्भृते स्यातां मध्यतो वै प्रजा अन्नं धिनोति मध्यत
> एव तदन्नाद्यस्य यजमानं दधाति। उभ्य्यो राज्जवो भवन्ति
> दक्षिणाश्च सव्याश्च दक्षिणा वा एकेषां पशूनां रज्जवः सव्या
> एकेषां तद्यदुभय्यो रज्जवो भुवन्त्युभयेषां पशूनामाप्तौ।
> दार्भ्यः स्युर्दर्भो वा ओषधीमापयहतमाप्मा तस्माद्दार्भ्यः स्युः।
>
> —*Aitareya Āraṇyaka*, I.2.3

In the next passage (I.2.4) the mention of trees, plants and the bearing of fruit is made thus—

> "The swing is masculine, the seat feminine, and they are united. The union is made at the beginning of the hymn for the sake of offspring. Children and cattle are his who knows this. Now the swing is food and the seat prosperity. Thus they mount to food and fortune. The Hoṭraka with the Brāhmaṇa sit down on seats of grass. Plants and trees having grown up bear fruit."

The seed on the other hand, which is the cause of the vegetable kingdom, has been interpreted (II.1.3) thus—

> "The seed of mind is speech. The seed of speech is action. The act done is this man, the abode of Brahman...". It is further stated (II.3.2) that there "are plants, trees and animals and he knows the self more clearly in them. For in plants and tree's sap only is seen, in animals consciousness...".

## 5. *Upaniṣads*

Going one more step towards *Jñāna-mārga* or speculation, the Upaniṣads, which on account of their mysterious character, were only communicated within a narrow circle of privileged persons took, however, various forms. Gradually, these Upaniṣads became the recognised name of the philosophical treatises contained in the Vedic literature. As to the exact meaning of the word *upaniṣad* it may be stated that the different schools of thought have interpreted it variously. Some of the ancient Indian philosophers derive it either from the root *sad*, in the sense of approaching, because a knowledge of Brahman comes near us by means of Upaniṣads, or because we appraoch Brahman with their help. Another explanation proposed by Śaṅkara in his commentary on the *Taittirīya Upaniṣad* (II.9) is that the highest bliss is contained in the *Upaniṣads*). Still some of the philosophers, wherever the word *Upaniṣad*, occurs, take it to be meaning of a doctrine, secret doctrine or is simply used as the title of the philosophical treaties which constitute the treasure-house or *Jñānakāṇḍa* of the knowledge, as opposed to the *karmakāṇda*, the work or the Ceremonial portion of the Vedas. In the *Trikaṇḍaśeṣa Upaniṣad* the word has been taken to mean as *Samīpasādhana* or sitting down near a person. Most of the European scholars are also agreed in deriving *upa-ni-ṣad* from the root *sad* to sit down preceded by the two prepositions *ni*-down, and *upa*-near, so that it would express the idea of session, or assembly of pupils sitting down near their teacher to listen to his instructions.

Though the theme of these *Upaniṣads* is highly philosophical in nature, still, the seeds, trees and plants are frequently mentioned therein. Sometimes they portray the highly philosophical doctrines in a variety of forms dubbed in the highly imaginative symbolism, a few examples of which are detailed hereunder—

## (i) *Symbolising the Tree with the Universe or Om*

(a) According to the *Chāndogya Upaniṣad* (I.1.2) the essence of all beings is the earth, the essence of earth is water, the essence of water is the plants, the essence of plants man, the essence of man is speech, the essence of speech *Ṛgveda*, the essence of the *Ṛgveda*, the *Sāmaveda*, the essence of *Sāmaveda* the *udgītha* (which is *Om*)—

एषां भूतानां पृथिवी रसः। पृथिव्या आपोरसः।
आपमोषधयोरसः ओषधीनांपुरुषो रसः।
पुरुषस्य वाग्रसः। वाच ऋग्रसः। ऋचः साम रसः।
साम्न उद्गीथो रसः।।

Thus this passage testifies that *udgītha* (*Om*) is the best of all the essence, the highest, deserving the highest place, the eighth. In these eight places *Oṣadhis* (plants) enjoy the third place.

(b) The *Maitrāyaṇa-Brāhmaṇa Upaniṣad*, testifies that the three-footed Brahman has its root upward; the branches are ether, wind, fire, water, earth etcetra. This one Aśvattha by name, the world is Brahman, and of it that is the light is called the Sun, and it is also the light of that Syllable *Om*. Therefore let him for ever worship that (breath, Sun, as manifestation of Brahman) with the syllable *Om*—

ऊर्ध्वमूलं त्रिपाद्ब्रह्म। शाखा आकाश वाय्वग्न्युदकभूम्यादया
एको अश्वत्थनामा। एतद्ब्रह्म। एतस्यैतत् तेजो यदसा आदित्यः।
ओमित्येतदक्षरस्य चैतत्। तस्मादोमित्यनेनैतदुपासीताजस्रमिति।
एकोऽस्य संबोधयितेत्येवःस्वाहा।

(c) The *Aitareya Āraṇyaka Upaniṣad* (II.3.6.9.10) goes on to symbolize the *Om* as the flower and fruit of speech. The one who is able to become celebrated and of good report, speaks for the true *Om*, which is the flower and fruit of speech. The composer of the Upaniṣad further goes on to say that the untrue is the root of speech and as a tree whose root is exposed dries up and perishes; thus a man who says what is untrue exposes his root, dries up and perishes. Thus one should not say what is untrue, but guard oneself from it.

The *Chāndogya Upaniṣad* (II.23.3) goes on to symbolise the trees with *oṁkāra* or Brahman, saying that Prajāpati, brooded on the worlds and from them, thus brooded on, issued the *Om*. All the leaves are attached to a stalk, so is the speech (all words) attached to the *Oṁ* (Brahman). *Om* is all this, yea, *Om* is all this.

तान्यभ्यतपत्। तिभ्योऽभितप्तेभ्य ओकाहः
संप्रास्नवत्। तदरथा शंकुना सर्वाणि पर्णानि
संतृण्णान्येवमोकारेण सर्वा वाक्य संतृण्णा।
ओंकार एवेदँ। सर्वमोंकार एवेदः सर्वम्।।

In the foregoing account the tree has been equated with Brahman. It may be stated here that, if the physical symbols represent Brahman in its unlimited pervasiveness, luminosity and power of assuming all varieties of form, organic symbols bring it home to us as an organic being, self-propelled, self growing, self procreating, self-sustaining and self-enjoying. Tree is one of such symbols. In the concluding portion of *Kaṭha Upaniṣad*, which has the account of a fig tree, having its roots upwards and branches downwards. It is characterised as eternal, bright and immortal and is regarded as Brahman. All the worlds are supposed to rest in it and none is taken to be capable of transcending it. Finally it is said to be the same as the Supreme Being and is referred to as neuter *tat*, that.

ऊर्ध्वमूलोऽवाकशाख एषोऽश्वत्थः सनातनः।
तदेव शुक्रं तद् ब्रह्म तदेवामृतमुच्यते।
तस्मिल्लोकाः श्रिताः सर्वे तदु नात्येति कश्चन। एतेद्वेतत्।।

—*Kaṭha Upaniṣad*, VI.I

(There is the ancient tree, whose roots grow upward and whose branches grow downward—that indeed is called the bright, that is called Brahman, that alone is called immortal. All the worlds are contained in it and no one goes beyond. This is that).

The commentator says that the tree is the world, and its root Brahman, but there is nothing to support this view in the original, where tree, roots, and branches are taken together as representing the Brahman in its various manifestations.

One of the most important position about this symbolic tree is its inverted position. Obviously the inversion is intended for bringing out Brahman's exalted position. Brahman is admitted as the cause of everything. It was, therefore, natural for it to lie beneath all its effects which constitutes the world. It could, for instance, be conceived as *Prakṛti* of the *Sāṁkhyās* or the Nature of Aristotle. In this case, however, it would have been difficult to account for the spiritual values and upward tendency of the world. To meet this difficulty, while the Sankhyas brought in the Puruṣa, Aristotle introduced God. But any duality was not acceptable to the Upaniṣadic sages, though they accepted Brahman as the sole cause of the universse. This involves considerable difficulties. These may be put briefly as obvious contradictions of eternity and flux, unity and multiplicity, delightfulness and wretchedness, etc. obtaining between cause and effect. The symbol of inverted tree resolves these contradictions to a considerable extent.

In the first place, placing the root of the tree upward and its branches downward accounts well for the origin of the universe from such a lofty being as Brahman and its tending downward. If a tree is turned upside down, it would naturally grow downward, though otherwise, it must have soared upward. This suggests that there is something in the very scheme of creation which, despite the fact that it has its own origin in Brahman, makes it debased. The scheme may be taken to have its root in the creation joy of Brahman. The joy probably lies in the sportive desire to see oneself in a position one ordinarily is not in. In the present case, it may be supposed to consist in moving from being to non-being, from consciousness to inconscience and from bliss to destitution.

In the second place, the symbol of tree is well suited for explaining the idea of unity in diversity which characterizes the Upaniṣadic view of Reality. Brahman is non-dual, while the universe is a system of staggering diversities. Then how could non-dual Brahman be the sole cause of diversities? Any example from the inorganic world does not help us much in resolving the paradox, except the tree, a single seed of which results in the growth of a tree of gigantic proportions, with innumerable branches, the leaves and the fruit.

Finally the tree also stands for constancy in change. Its fruits, leaves, twigs and even branches come and go, but the root and trunk remain constant beyond individual's memory or, indeed, span of life. This duality of the tree's life is well reflected in the description of the Brahman tree as eternal on the one hand and changing on the other. These two strands, however, depart from each other later on and adapted in the *Svetāsvatāra Upaniṣad* (III.9)—

गरगात् परं नापरमस्ति किंञ्चिद्यस्मान्नाणीयो न ज्यायोऽस्ति किंचित्।
वृक्ष इव स्तब्धो दिवि तिष्ठत्येकस्तेनेदं पूर्णं पुरुषेण सर्वम्।।

(Than whom nothing is higher; nothing lower; than whom nothing is smaller; nothing greater. The one stands like a tree grown in the heaven. By that *puruṣa* is filled all this.)

## (ii) *The Seed*

The natural interpretation of the word *Bīja* or seed is, from which a tree or a plant is grown. It has also been taken to mean semen of men or animals or any germ element, primary cause or principle, besides being the source of origin. But in the Upaniṣads, the word *Bīja* has been dealt with in a variety of forms, besides rendering it a symbolical interpretation saying that a single seed sprouts and grows into a tree branching and leafing around abundantly, all the leaves and branches being steeped in the original root and yet doing little to impair its wholeness. None of the leaves or branches can afford either to disassociate itself from the tree's properties. Similar is the case with Brahman and the universe. The universe has emanated from Brahman without impairing the latter even in the least and has to keep itself constantly centred round the nucleus. The account that follows, provides a glimpse of the varied types of interpretations rendered for *Bīja* in the Upaniṣads—

**(a) *Bṛhadāraṇyaka Upaniṣad***

(1) *III.7.23*

यो रेतसितिष्ठन रेतसोइन्तरो यःरेतो न वेद यस्य रेतः
शरीहंयो रेतोऽन्तरो यमं यत्येषतः आत्माऽन्तर्याम्यमृतः।
?? द्रष्टाऽश्रुतः श्रोताऽमतोमन्ताऽविज्ञातो विज्ञाता।
नान्योऽतोऽस्तिद्रष्टा नान्योऽतेऽस्ति श्रोता नान्योऽहित
मन्ता नान्योऽतोऽस्ति विज्ञाता। एव त अत्माऽन्तर्याम्य मृतः।
अतोइन्यदार्तम। ततो होद्दालक आरुणि रुपरराम।।

(He who dwells in the seed, and within the seed, whom the seed does not know, whose body the seed is, and who rules the seed within, he is thyself, the ruler within the immortal; unseen, but seeing; unheard, but hearing; unperceived but perceiving; unknown but knowing. There is no other seer but he; there is no other hearer, but he; there is no other knower, but he. This is thy Self, the ruler within, the immortal. Everything else is of evil. After that Uddālaka Aruṇi held his peace).

(2) *VI.1.12*

एतो होच्चक्राम। तत्संवत्सरं प्रोष्यागत्योवाच कथमशबत मदृते जीवितुमिति।
ते होचुः यथा क्लीवा अप्रजाममाना रेतसा प्राणन्तः प्राणेन वदन्तो
वाचा पश्यन्तश्चक्षुषा शृण्वन्तः श्रोत्रेण विद्वाःसोमनसैवमजीविष्मेति।
प्रविवेश ह रेतः।।

(The seed departed, and having been absent for a year, it came back and said—'How have you been able to live without me?' They replied—'Like impotent people, not generating with seed, but breathing with the breath, seeing with the eye, hearing with the ear, knowing with the mind. Thus we have lived'. Then the seed entered in.)

**(b) *Chāndogya Upaniṣad***

(1) *III.14.3*

एव म आत्माऽन्तर्हृदयेऽणीयान्त्रीहेर्वा यवाद्वा सर्वपाद्वा
श्यामकाद्वा श्यामक तण्डुलाद्वा। एष म आत्माऽन्तर्हृदये

ऽयायान्पृथिव्या ज्यायानन्तरिक्षाज्ज्यायान्दिवो ज्यायानेभ्यो
लोकेभ्यो लोकेभ्यः।।

(He is myself within the heart, smaller than a corn of rice, smaller than a corn of barley, smaller than the mustard seed, smaller than a canary seed or the kernel of a canary seed. He also is myself within the heart, greater than the earth, greater than the sky, greater than the heaven and greater than all these worlds).

(2) *V.8.2*

तस्मिन्नेतस्मिन्नग्नौ देवारेतो जुह्वति।
तस्या आहुतेर्गर्भः संभवति।।

(On the altar the Devas or *prāṇas* offer seed. From that oblation rises the germ).

(3) *VI.12.1-2*

न्यग्रोधफलमतं आहरेति। इदं भगव इति।
भिन्द्धीति। भिन्नंभगव इति। किमत्र पश्यसीति।
अण्व्य इवेमा धाना भगव इति।
असामङ्गैकां भिन्द्धीति भिन्ना भगव इति।
किमत्र पश्यसीति। न किंचन भगव इति।।
तꣳहोवाच—यं वै सोम्यैतमणिमानं न निमालयस एतस्य
वै सौम्यैषोऽणिम्न एवं महान्यग्रोधस्तिष्ठति। श्रद्धत्स्व सोम्येति।

(Fetch me from thence a fruit of Nyagrodha tree; "Here is one Sir". "Break it". "It is broken sir". "What do you see there?" "These seeds, almost infinitesimal". "Break one of them". "It is broken Sir". "What do you see there?" "Not anything Sir". The father said, "My Son, that subtle essence which you do not perceive there, of that very essence this great Nyagrodha tree exists).

**(c) *Aitareya Āraṇyaka Upaniṣad***

(1) *II.1.1.1.*

This Upaniṣad describes the origin of the seed thus—

"Next follows the origin of the seed. The seed of Prajāpati are the *Devas* (gods). The seed of the *Devas* is the rain. The seed of rain are herbs; the seed of herbs is food. The seed of food is seed. The seed of seed are creatures. The seed of creatures is the heart. The seed of heart is the mind. The seed of mind is speech (Veda). The seed of speech is action. The action done (in a former state) is this man, the abode of Brahman."

(2) *II.3.7.3*

The human seed has been symbolised in this text, thus—

"Here the blood of the woman is the form of Agni (fire); therefore no one should despise it. And the seed of man is a form of Āditya (sun); therefore no one should despise it. This self, (the woman) gives herself (skin, blood and flesh) to that self (fat, bone and marrow), and that self (man) gives his self (fat, bone and marrow) to this self (skin blood and flesh). Thus these two grow together. In this form (belonging to the woman and the fire) he goes to that world (belonging to the man and the sun), and in that form (belonging to the man and the sun) he goes to this world (belonging to the woman and to fire."

Indeed the body consists of six elements, and is thus called *Ṣatakauṣika*. Of these three having a white appearance (fat, bone and marrow) come from the sun and from man; three having red appearance come from fire and from women.

(3) *II.5.1.1-2*

"Verily from the beginning he, (the self) is in man as a germ, which is called the seed. This seed, which is the strength gathered from the limbs of the body, he (the man) bears as self in his self (body). While he commits the seed to woman, then he (the father) causes it to be born. This is his first birth."

(4) *III.2.2.5*

"Marrow is the real breath (life), for marrow is seed and without breath (life) seed is not sown. Or when it is sown without breath (life), it will decay and will not grow."

## (iii) *Birth of the Tree*

### *Chāndogya Upaniṣad*

This text testifies about the birth of a tree as follows—

अभ्रं भूत्वा मेघो भवति। मेघो भूत्वा प्रवर्षति।
त इह व्रीहियवा ओषधि वनस्पतयस्तिलमाषा इति
जायन्ते अतो वै खलु दुर्निष्प्रपतरम्। यो यो ह्यन्नमत्ति,
यो रेतः सिञ्चति, तद्भूयं एव भवति।।

(Having become the mist, he becomes the cloud, having become a cloud, he rains down. Then he is born as rice and corn, herb and trees, sesamum and beans. From thence the escape is beset with most difficulties. For whoever the person may be that eat the food, and beget offspring, he henceforth becomes like on to them).

Similar ideology is propounded in the *Taittirīya Upaniṣad* (2.1)

## (iv) *Tree as Human Food and Animal Fodder*

The trees are also conceived to serve as animal fodder and as the human food in the *Aitareya Āraṇyaka Upaniṣad* (11.3.1.3) thus—

"To whom who knows the food and the feeder in that in *Uktha*, a strong son is born, and food is never wanting. Herbs and trees are food, animals the feeder, for animals eat herbs and trees."

### (v) *Tree as the Resting Place for Birds*

The trees, in the *Praśna Upaniṣad* (IV.7) are described as the resting place of the birds—

स यथा सोम्य वयांसि वासोवृक्षं संप्रतिष्ठन्ते।
एवं ह वै तत्सर्वं पर आत्मनि संप्रतिष्ठते।।

(As birds go to a tree to roost, thus all this rests in the Highest Ātman.)

The same theme is expressed in the *Aitareya Āraṇyaka Upaniṣad* (1.2.4.7-8) as under—

"They say—Let him touch the swing with his chin (*Khubuka*). The parrot (*Śuka*) thus mounts on a tree and he is of all birds the one who eats most food. Therefore let him touch it with his chin. Let him mount the swing with his arms. The hawk thus swoops on birds perched in the trees and he is of all the birds, the strongest. Therefore let him mount with his arms."

### (vi) *Fruit Trees and Ripening of Fruits*

The *Aitareya Āraṇyaka Upaniṣad* (II.1.7.2) goes on to testify that from *Puruṣa*'s speech earth and fire were created. "Herbs are produced on earth and Agni (fire) make them ripe and sweet."

### (vii) *Sap in the Trees*

The *Aitareya Āraṇyaka Upaniṣad* (11.3.2.2) describes that there are herbs and trees and all that is animated and he knows the self (Brahman) gradually developing in them. For in the herbs and trees, sap only is seen, but thought (*Citta*) in animated beings.

### (viii) *The Grass*

Grass is an important part of the vegetable kingdom and it has not been lost sight of in the Upaniṣads, which account for a variety of them—

#### (a) *Aitareya Āraṇyaka Upaniṣad (1.2.3. 9-10)*

This text propounds that there are two kinds of rope, twisted towards the right and twisted towards the left. The right ropes serve for some animals, the left ropes for others. If there are both kinds of rope, they serve for the attainment of both kinds of cattle. Let them be made of *Darbha* (*Kuśa* grass), for among plants, *Darbha* is free from evil, therefore, they should be made of *Darbha* grass.

#### (b) *Kauśītakī Upaniṣad (11.3)*

This Upaniṣad prescribes the spreading of grass and watering the surface at the time of *homa* ritual—

अथात एकधनावरोधनम्। यदेकधनमभिध्यात् पौर्णमस्यां
वाऽमावस्यां व शुक्लपक्षे व पुण्ये नक्षत्रे ऽग्निमुपसमाधाय
परिसमूह्य परिस्तीर्थ पर्यक्ष्योत्पूर्य दक्षिणं जान्वाच्य स्रुवेण
वा चमसेन व कंसेन वेता आज्याहुतीर्जुहोति।।

(Now follows the attainment of the highest treasure. If a man meditates on that highest treasure, let him on a full moon or a new moon, or in the bright fortnight under an auspicious Nakṣatra, at one of these proper times, bending his right knee, offer oblation of *ghee* with a ladle *Sruva*, after having placed the fire, swept the ground, strewn the sacred grass, and sprinkled water...)

The strewing of grass around the house is also mentioned elsewhere in the same Upaniṣad (11.15).

The same idea of spreading the grass during the time of sacrifice and watering it is contained in the *Bṛhadāraṇyaka Upaniṣad* (VI.3.1) which goes as follows—

"If a man wishes to attain greatness (wealth or performing sacrifice), he observes the *upasad* rule during twelve days (*i.e.* he lives on small quantity of milk), beginning on an auspicious day of the light half of the moon during the northern progress of the sun, collecting at the same time in a cup or a dish made of Udumbara wood all sorts of herbs including fruits. He sweeps the floor (near the house altar), sprinkles it, lays the fire, spreads the grass round it according to rule...".

## (ix) *Juice from Trees*

The *Chāndogya Upaniṣad* (III.9.1-2) testifies about the extraction of juice from the tree and to reduce it into one form—

यथा सोम्य मधु मधुकृतो निस्तिष्ठन्ति नानात्ययानां
वृक्षाणाः रसान् समवहारमेकताः रसं गमयन्ति।।
तेयथा तत्र न विवेकं लभन्तेऽमष्याहं वृक्षस्य रसोऽस्म्य
मुष्याहं वृक्षस्य रसोस्मीति। एवमेव खलु सोम्येमा: सर्वा
प्रजा सति संपद्य न विदु: सति सपंद्यामह इति।।

(As the bees, my son, make honey by collecting the juice of distant trees, and reduce the juice into one form. And as these juices have no discrimination, so that they might say, I am the juice of this tree or that, in the same manner, my son, all these creatures when they have become merged in the tree, know not that they are merged in the tree.)

## (x) *Worms, Insects or Germs*

Wherever there is vegetation, the presence of worms or insects becomes a natural phenomena. The Upaniṣad seers were well aware of this aspect, which they have projected in their works in one or the other form, a few instances of which are given hereunder—

| | |
|---|---|
| *Chāndogya Upaniṣad* (VII.2.1) | All beasts down to worns, midgets and ants. |
| *Ibid.*, (VII. 7.1 and VI.10.2) | -do- |
| *Bṛhadāraṇyaka Upaniṣad* (VI.1.14) | Worms, insects and birds. |

## 6. The *Gṛhya Sūtras*

The vegetable kingdom is also mentioned in the *Gṛhya Sūtras* which form part of the Vedāṅga and deal primarily with the sacraments (*Saṁskāras*) and other rites or ceremonies prescribed for the householders. They enshrine rules concerning forty sacraments, of which the first eighteen (from conception to marriage) are called bodily sacraments. The *Gṛhya Sūtras* also treat of the fivefold sacrifices, (*Pañca-mahāyajña*) of the householder, duties of student life (*Brahmacarya*), of the forest life (*Vānaprastha*) and the renunciation, (*Saṅnyāsa*). Since the domestic rituals are not discussed in the Brāhmaṇa literature, these have been enshrined in the *Gṛhya Sūtras* which specify the use of a variety of trees, plants and grass for the performance of various rituals or ceremonies—

### (i) *Sāṁkhyāyana Gṛhya Sūtra*

(i) In the ceremonies connected with the full-moon days, it is prescribed that "In the morning, when the sun shines on the top of the great trees, that is the most auspicious time for all kinds of sacrifices." (I.3.2).

(ii) Fruits and flowers are prescribed in the engagement ceremony of a boy and the girl, thus—

(a) "When they (the girl's parents) arrive (at boy's place) they take flowers, fruits, barley and a pot of water (I.6.2).

(b) "When the matter pleases both sides, let them touch a full vessel into which have been put flowers, fruits, fried grain, barley and gold and let them recite a formula,—" (I.6.5)

(iii) After the engagement is finalised, a ceremony is to be performed which requires the strewing of grass around the fire and the performer—

(a) Strews eastward-pointed *kuśa* grass around the fire, in three or five layers, covering the roots of the grass blades with the points; adorn with five flowers; (I.7.1-7).

(b) Takes up with the left hand, the *kuśa* blades and arranges them on the ground with his right hand; (I.7.9).

(c) Measures off with the span of his hand, two *kuśa* blades, which are not unequal, with unbroken points, bearing no young shoots in them, and severs them from their roots, with a *kuśa* blade." (I.8.14)

(d) If the oblation consists in *Ajya*, let him seize the *kuśa* blades in the left hand, with his right hand at their points and with the left hand at their roots and let him wet their points (with *Ajya*) in the *Sruva*, the middle and the roots in the *Ajya* pot. (I.9.13).

(iv) At the marriage ceremony, the bridegroom ties to the bride, reciting a verse (*Ṛgveda* IV.57.3), "Paul of honey the herbs" Madhūka flowers (I.12.9). Now this Madhūka has been interpreted to mean *Bassia latifolia,* from the blossoms and seeds of which arrac is distilled and oil extracted.

(v) While performing *homa* at the time of marriage the groom throws into the fire tree branches with milky sap and leaves, of a tree the name of which is masculine, together with *kuśa* grass (I.13.6)

### (ii) *Āśvālayana Gṛhya Sūtra*

(i) For performing a sacrifice, one is required to sprinkle that place with water, lit a sacred fire and put two or three pieces of fuel on it, wipe the ground round (the fire), strew grass round it (1.3.1). Thereafter two *kuśa* blades are used as strainers, the purifying of the *Ajya* is done (1.3.2)

(ii) In the fourth month of pregnancy the *sīmāntonnayana* ceremony (parting of the hair) is performed (1.14.1). "He then three times parts her hair upwards (*i.e.,* beginning from the front) with a bunch, containing an even number of unripe fruits, and with a porcupine's quill having three white spots and with three bunches of *kuśa* grass, he gives orders to two lute players to sing." (1.14.4).

### (iii) *Pārāśara Gṛhya Sūtra*

In the ceremony relating to the performance of domestic sacrifice the use of *kuśa* grass has been prescribed (1.1.4). Again at the time of the marriage ceremony, the girl's brother is required to pour out of his two hands into her joined bands, fried grains mixed with *Śāmi* leaves (1.6.1). (This *Śāmi* tree has been identified as *Prosopis spicigera* or *Mimosa Suma,* possessing a very hard wood supposed to contain fire. It was used to kindle the sacred fire. A legend relates that Pururavās generated primeval fire by the friction of two branches of *Śami* and *Aśvattha* trees.

During the *Puṁsavana* ceremony aimed at securing a male child the wife is required to bathe, fast and put on new garments which we have not yet been washed, and after having in the night time crushed in water descending roots and shoots of Nyagrodha tree, he inserts (that into her right nostril) as above, the two verses—"The Gold-Child" (1.15.3)

### (iv) *Khadīra Gṛhya Sūtra*

According to this Sūtra, the sacrifical cord has to be made of a string or of *kuśa* grass, (1.1.4)—and Darbha grass has to be strewn to the west of the fire (1.2.9). At the marriage ceremony, the bride's mother is required to put fried grain mixed with *Śāmi* leaves into the basket (1.3.18). While commenting on the offering of food in charity by a householder it is prescribed (1.5.37) that—

> "From the rice harvest, till the barley harvest, or from the barley harvest till the rice harvest, he should offer the *balis* himself."

### (v) *Gobhila Gṛhya Sūtra*

This *Sūtra* prescribes that in the *Śrāddha* ceremony, the householder "takes his *yajñopavīta* (then Sacrificial Cord) a string or a garment, or simply a rope of *kuśa* grass"

(1.2.1). Then the *Asasya-Bali* is required to be offered, starting from the barley-harvest till the rice-harvest and from rice-harvest to the barley-harvest (1.4.29).

During the ceremonies to be performed by a householder, during the full moon days, it has been desired that—

(i) He then gets pieces of wood ready, which are to be put on fire, of Khadīra or Palāśa trees (1.5.14).

(ii) If Khadira or Palāśa wood cannot be obtained, it may be wood—as it serves the purpose—of any tree, with the exception of Vibhidaka, Tilvaka, Niva, Nimba, Rajavṛkṣa, Śālmali, Aralu, Dadhittha, Karidara, Slesamataka wood (1.5.15).

(iii) The *Barhis* consist of *kuśa* grass cut off at the points at which the blades diverge from the main stalk (1.5.16).

(iv) The blades of *kuśa* grass should be cut off near the roots in the ceremonies directed to the fathers (1.5.17).

(v) If the *kuśa*-grass cannot be obtained, he may take any kind of grass with the expection of *śuka*-grass of Saccharum reed, of such grass as is opt to break, of *Balbaja* grass, of Mutava of Amphidonax reed of *Suṇṭha* (II.5.18).

(vi) He than washes the mortar, the pestle and the winnowing basket, strews to the west of the fire east-ward-pointed Darbha-grass and puts the mortar on that grass (1.7.1)

## (vi) *Hiraṇyakeśin Gṛhya Sūtra*

The use of Darbha grass has been prescribed in this Gṛhya Sūtra, in the Upanayana Ceremony (I.II.II-14). The use of Palāśa-wood has been prescribed during the *homa* ceremony (1.3.9.4). Then there is a mention of burying a lump of dung in a cow-stable or near an Udumbara tree or in the clump of Darbha-grass (1.3.9.18). Interestingly however there is also the mention of cleansing the teeth with a stick of Udumbara wood. (1.3.9.19). The medicinal properties of the herbs have also been recognized and it is stated that the herbs is protecting, overcoming and powerful (1.3.10.6). The use of Darbha-grass and the udumbara wood has also been considered essential during the *Śrāddha* ceremony (II.5.15.9).

## (vii) *Āpastambha Gṛhya Sūtra*

The text provides for the use of Darbha grass and pieces of wood for *homa* during the Upanayana and other ceremonies (I.2.1; and IV.10.6). Then there is the provision of using *Mūñja*-grass as a girdle (IV-10.II). While the Palāśa-wood is required to be used as offering in the homa-fire (V.12.3), the *Brahmacārī* is required to hide the girdle at the root of an Udumbara tree or in the tuft of Darbha-grass (V.12.5).

At the *Puṁsavana* (ceremony to secure the birth of a male child) the use of Nyagrodha tree has been stated as under—

(i) From a branch of a Nyagrodha tree, which points east-ward or north-ward, he takes a shoot with two fruits which look like testicles. The putting of wood on the fire and other ceremonies are performed (VI.14.11).

(ii) He causes a girl who has not yet attained maturity to pound (the Nyagrodha shoot), on an upper mill-stone with another mill-stone and to pour water on it (VI.14.11).

The use of wood or trees in the house-building ceremonies has been ordained as under—

(i) The ground for building is required to be initially swept with the broom of a balsa wood or of Śāmi-wood (VII.17.I).

(ii) During *homa* ceremony, the pieces of Palāśa Wood are to be offered in fire (VII.17.7).

(iii) Darbha grass is to be strewn so that its points are turned in every direction. Then rice is poured together with barley grains over the grass (VII.17.9).

(iv) Then there is the provision of the offering of the Kiṁśuka flowers (VII.18.6).

## 7. The Epics

The Epics, *Rāmāyaṇa* and *Mahābhārata*, constitute a large bulk of Sanskrit literature. The nucleus of the works is pretty ancient, as these used to be sung from place to place till at last the storices were written down. Both the Epics project the tree and plant kingdom, since most of the episodes contained therein were enacted in the forest. Not only that Rāma and Lakṣmaṇa, as also the Pāṇḍavas were exiled to stay in the forests for a long period of 12-13 years. Even otherwise in both the cases the heroes had to visit the forests quite frequently. The account that follows, throws considerable light on the subject—

### (i) *Vālmīki Rāmāyaṇa*

Forestry and plant kingdom dominate the epic, because most of the scenes of the epic were enacted in the forests. The very start of the poetry of *Rāmāyaṇa* is attributed to a scene in the forest, where two Krauñca-birds, who were busy in their love-sport, and one of them, the male, was killed by an arrow by a hunter, a *Vyādha*. The outcry of the birds moved the tender heart of the sage Vālmīki, who had witnessed the scene. The pathetic sight enraged the sage, who showered, instantaneously, a curse upon the hunter, thus—

मा निषाद प्रतिष्ठात्वंमगमा शाश्वती समा।
यत्क्रौंच मिथुनादेकमवधी काम मोहितम्।। —*Vālimīki Rāmāyaṇa*

According to the Bālākāṇḍam of this epic, the sage Sṛṅgi was residing in the forest, where no woman folk had ever gone. The same sage was seduced by the courtesans, to marry him with Śāntā, the daughter of the king Romapada. The same sage is summoned by Daśaratha for performing the *putreṣṭhi-yajña* of the king of Ayodhyā.

The sage Viśvāmitra calls on the king Daśaratha to ask for the help of Rāma and Lakṣmaṇa, for destroying the demons in the forest.

In the end of Ayodhyākāṇḍam, Rāma is exiled to forests for twelve years and the rest of the happenings in the epic were enacted in the forests. There are numerous *āśramas*, forests and groves listed in the epic, some of which are listed below—

| ***Forests/groves*** | ***Āśramas*** |
|---|---|
| (1) Daṇḍakāraṇya | (1) Siddhāśrama |
| (2) Citrakūṭa | (2) Bhāradvāja *āśrama* |
| (3) Pañcavaṭī | (3) Sutikṣana's *āśrama* |
| (4) Godāvarī Forests | (4) Agastya's *āśrama* |
| (5) Seashore groves | (5) Ṛṣyaśṛṅga *āśrama* |
| (6) Mālyavan forests | (6) Prasavanagiri *āśrama* |
| (7) Tungabhadrā groves | |
| (8) Kiṣkindhā forests | |
| (9) Madhuvana | |
| (10) Vindhya Kānana | |
| (11) Ariṣṭagiri forest | |
| (12) Sahya Parvata forests | |
| (13) Parks and groves of Laṅkā | |

Besides the above, the epic also mentions some forests which are *sugama* (easily accessible) and *durgama* (inaccessible forests). In the Yuddhakāṇḍam of the epic, some *divya* forests and cursed forests are mentioned. In the Kiṣkindhākāṇḍam, the Ṛṣyamūka forest is said to be inaccessible for Bāli, because of a curse pronounced on him.

But of all the groves and parks mentioned in the epic, the Aśoka grove or popularly called Aśokavāṭikā in Laṅkā, where Sītā was lodged after abduction, has its own importance.

In order, however, to have a glimpse of the botanical aspects of Aśokavāṭikā, when Sita was lodged in Laṅka, after her abduction one has to peep through the window provided by the sage Vālmīki in the fourteenth *sarga* of Sundarakāṇḍam of the *Vālmīki Rāmāyaṇa*. It provides various aspects of the horticulture, floriculture and Botany simultaneously. It is evident therefrom that the sage had not only an extensive knowledge of the subject, but also was aware of the need for proper land seaping of the garden besides providing natural and artificial lakes, stepwells, water channels and streams, etc.

The said Vaṭika was named as Aśoka-Vāṭikā which ostensibly means a grove of Aśoka trees. This could mean that the Vāṭikā had a large number of Aśoka trees. But this is not all, there were other trees as well like Sāla, Aśoka, Champā, Nāgakesara, Mango and others, blossoming with beautiful flowers.

सालानशोकान् भव्यांश्च चम्पकांश्च सुपुष्पितान्।
उद्दालकान् नागवृक्षांश्च चूतान् कपिमुखानपि।।
तथाऽऽम्रवण सम्पन्नाँल्लताशत समन्वितान्।।
ज्यामुन्क्त इव नाराचः पुप्लुवे वृक्षवाटिकाम्।।

—*Rāmāyaṇa*, Sundarakāṇḍam, 14.3-4

This indeed was the type of a garden which immensely pleased the human as well as the fauna and birds of various species ( प्रहृष्टमनुजां काले मृग पक्षी मदाकुलम्). The plenty of flowers in the said gardens adornig the blossomed trees, have been conceived as resembling the mountain of flowers—

पुष्पावकीर्णः शुशुभे हनूमान् मारुतात्मजः।
अशोक वनिकामध्ये यथा पष्पमयो गिरि।।

—*Rāmāyaṇa*, Sundarakāṇḍam, 14.11

Continued and sufficient water supply is an essential part of a good garden. This aspect was also well taken care of by the Sage Vālmīki while describing Aśokavāṭikā. According to his description, there were several step-wells of various dimensions, which were full of clean water and various types of flowers. These step-wells had a number of trees adorning their borders.

वापी च विविधाकाराः पूर्णा परमवारिणा।
महाहैमणिसोपानै रुपपन्नास्ततस्ततः।।

—*Rāmāyaṇa*, Sundarakāṇḍam, 14.22

These step wells had a constant and regular supply of water by means of several channels and streams, which surrounded them—

दीर्घाभिद्रुमयुक्ताभिः सरिद्भिश्च समन्ततः।
अमृतोपमतोदाभिः शिवाभिरूपसंस्कृताः।। —*Ibid.*, 14.25

Besides these step-wells, Hanumān saw a number of natural streams which were also filled with clean water and various types of trees were decorating their banks. Close by the place, where Sītā was lodged, the monkey-god also saw an artifical lake which was also full of clean and cold water—

कृतिमां दीर्घिकां चापि पूर्णां शीतेन वारिणा।
मणिप्रवरसोपानं मुक्तासिकतशोभिताम्।। —*Ibid.*, 14.33

Landscaping is also an essential part of a modern garden and this aspect was not lost sight of by the sage, when he described the presence of a black hillock with strange type of peaks and having several stone caves. This hillock too was surrounded by numerous trees. This has apparently been studded in the garden in order to break the monotony—

ततोऽम्बुधरसंकाशं प्रवृद्धशिखरं गिरिम्।
विचित्रकूटं कूटैश्च सर्वतः परिवारितम्।।
शिलागृहेवततं नानावृक्षं समावृतम्।
ददर्श कपि शार्दूलो रम्यंजगति पर्वतम्।। *Ibid.*, 14.27-28

It would thus be evident that the grand garden of Aśokavāṭikā was a huge botanical complex, surrounded by a boundary wall ( सतु संह. सर्वाङ्गः प्राकारस्थो महामधिः). It had thousands of Aśoka trees ( किंचित् तत्राशोकाः सहस्रशः) besides other innumerable flower and fruit-trees, plants creepers, shrubs, etc. There were natural as well as artificial lakes and streams to ensure the constant supply of water. The entire layout of the garden could compare with any of the well-organised and well-developed modern garden where landscaping and dwelling places (like modern farm houses) were well looked after.

In order, however, to ensure the natural environmental moorings, the sage has pointed out that the several species of birds had nested in the trees of Aśokavāṭikā and fauna like stag and deer were present in flocks everywhere in the garden. Though it could not be assumed that the black hillock mentioned in the above account was artifical or natural, but the very presence of the same amidst the botanical surroundings does serve as a beautiful landscape, which could be found in any modern park or a garden.

The use of medicinal plants is also found mentioned in *sarga* 74 of the Yuddhakāṇḍam of the epic when all the monkey army was wounded by the *Brahmāstra* issued by Meghanāda, Jāmbvān asked Hanumān to bring four types of medicinal plants from Kailāśa. These plants were named—*Mṛtasanjīvanī, Viśalyakarṇī, Savarṇyakarṇī* and *Sandhānakarṇī* which were said to illumine the ten regions—

तयो: शिखरमध्ये प्रदीप्तमतुल प्रभम्।
सर्वौषधियुतं वीर द्रक्ष्यस्योषधि पवतम्।।
तस्य वानरशार्दूल चतस्रो मूर्घ्नि सम्भवा:।
द्रक्ष्यस्योषधयो दीप्ता दीपयन्त्यो दिशो दश।।
मृतसंजीवनी चैव विशल्यकरणीमपि।
सावर्ण्यकरणीं चैव संधानकरणीं तथा।।

—*Vālmīki Rāmāyaṇa*, 6.74.31-33

The three herbs were possibly used for the following purposes—

(i) *Mṛtasanjīvanī*— To save a dying person who could be in coma to heavy loss of blood.

(ii) *Viśalyakaraṇī*— This herb could heal an arrow wound. This has been identified with *Timospora Cordifolia* or *Coculus Cordifolius*. The stem, roots, leaves and watery extract of this plant have long been valued in Hindu medicine. The entire plant is regarded as a valuable alternative and tonic, and is used in general debility, fever, jaundice and skin diseases, rheumatism, urinary diseases and dyspepsia.

(iii) *Savarṇyakaraṇī*— This could possibly revive the speech of a person.

(iv) *Sandhānakaraṇī*— This could perhaps join the broken limbs, because the word *Sandhāna* has been interpreted to mean, joining, uniting, healing, or a small egg plant, or the act of placing or joining or uniting together.

## (ii) *Ādhyātma Rāmāyaṇa*

In the Bālakāṇḍam of this epic, Viśvāmitra, with the permission of Daśaratha, took Rāma and Lakṣmaṇa with him and entered the region known as the forest of Tāṭakā—

तत उत्तीर्य गंगा ते ताटका वनमागमन।। —(1.4.26)

After spending a night in the forest *āśrama* of Viśvāmitra, they proceeded further and reached Siddhāśrama, the abode of the Siddha. Viśvāmitra then performed the sacrifice and the great demons Subāhu and Mārīca were killed by Rāma and Lakṣmaṇa (1.5.4-9). Both the princes stayed in the *āśrama* of Viśvāmitra for four days and were fed with ripe fruits and other edibles (1.5.10-11). From there they all moved to the *āśrama* of Gautama, which was full of trees, laden with celestial flowers and fruits, but without any living creatures, (1.5.17-18). In the same forest, Ahalyā, who was turned into a stone, was liberated by Rāma by the touch of his foot. Then while on journey to Mithilā, the party reached over the banks of the Gaṅgā, which they had to cross. But the boatsman refused to carry Rāma into the boat saying—

क्षालयामि तव पादपंकजं नाथ दारु दृषदोः किमन्तरम।
मानुषीकरणचूर्णमस्ति ते पादयोरिति कथा प्रथीयसी।।
पादाम्बुजं ते विमलं हि कृत्या पश्चात्परं तीरमहं नयामि।
नोचेत्तरी सद्युवती मलेन स्याच्चेद्विभो विद्धि कुटुम्बहा।। —(I.6.3-4)

("O Lord, I would like to cleanse your feet first. What is the difference between a stone and a wooden object? The news has spread that the dust of your feet can transform anything into a human being. Therefore I shall clean your feet of all the dust on it and then take you to the other shore. If this is not done, there is a danger of my boat being converted into a young woman. In that case, O Master, my family will be endangered.")

In the Ayodhyākāṇḍam, of this epic, Rāma is exiled to forest. Rāma enters the forest, reaches the Bhāradvāja *āśrama* (II.6.28-41) stays there for the night and next day he alongwith Sītā and Lakṣmaṇa reaches the *āśrama* of the sage Vālmīki (ii.6.42-64). The hermitage was full of animals and birds and the trees were laden with flowers and fruits. On the advice of the sage Vālmīki, Rāma with his wife and brother built a hermitage at Citrakūṭa and started staying there.

In the Araṇyakāṇḍam of this epic, Rāma, Lakṣmaṇa and Sītā entered the Daṇḍakāraṇya which has been described as a notorious place, infested with Rakṣasas—

विद्यते दण्डकारण्ये श्रुतपूवमरिन्दम।। —(III.1.4)

Virādha a great *Rakṣasa* was killed in this forest by Rāma. After this Rāma with his party visited the *āśramas* of the sages Sarbbanga and Sutīkṣaṇa. (II.2.1-40). In this connection Rāma has been conceived as blue like the *Dūrva*-grass, Lotus eyed, and dressed in tree-bark cloth and wearing matted locks. (III.2.7). From here Rāma's party goes to the *āśrama* of Agastya (III.3.1-16) and then to Pañcavaṭī, where Lakṣmaṇa built a specious cottage on the bank of the river Godāvarī. The place was surrounded by a thick growth of fruit bearing trees like Kadamba, Panāsa, and Āmra. Everyday Lakṣmaṇa went in the forest to collect roots and fruits of the forest and fed all (III.4.9-14).

After the abduction of Sītā, both the brothers left for Kiṣkindhā, meeting Śabarī on the way. Hanumān meets the Ayodhyā princes in the Kiṣkindhā forest. For the alliance of Sugrīva with Rāma, makes the latter to undergo a test of strength, at the instant of the former—

यदित्वमेक बाणेन विद्धवा छिद्रं करोषि चेत्।
हतस्त त्वया तदा बाली विश्वासो मे प्रजायते।। —(IV.I.73)

(If you can pass an arrow through all these seven trees then I can believe that Bālī's death at your hand is like an accomplished fact).

Besides the above, the text mentions several types of trees in Laṅkā, the Aśokavana and other places.

## (iii) *Mahābhārata*

Like the *Rāmāyaṇa* of Vālmīki, the *Mahābhārata* has extensively dealt with the plant kingdom in a variety of forms. Like Rāma, the Pāṇḍavas too were exiled into forests for twelve years and they kept on wandering in the forests during this period. An exclusive *Parvan i.e.* Vanaparvan is devoted to the exploits of the Pāṇḍavas in the forests. Though a tabulated statement projecting the trees, plants and creepers etc. mentioned in the *Mahābhārata* appears at Appendix-3(a) of this work, the account that follows, highlights some of the prominent forests appearing in the epic.

### (a) *Dvaitavana*

The Pāṇḍavas with Kṛṣṇa and Dhaumya, mounted on the costly chariots, yoked with excellent steeds, went into this forest. The citizens walked round Yudhiṣṭhira and the Brāhmaṇas and principal men of *Kurujangulu* of all castes and saluted him blaming the Dhṛtarāṣtras. Arjuna told them that of their exile (Pāṇḍava's) they would take away the good name of their enemies and asked them to beseech the asceties for their good. At last they returned to their respective abodes (III.23). After a conference between Yudhiṣṭhira and Arjuna, they resolved upon dwelling for these twelve years at the sacred lake of Dvaitavana in the forest of the same name, on the bank of the river Bhogavatī (Sarasvatī). The king saw the huge forest covered at the end of the summer, with Sālas, palms, mangoes, Madhūkas, Nipas, Kadambas, Sarajas, Arjunas, Karnikaras, clothed with flowers. (III.24.17). Approaching the picturesque Bhogavatī, he saw many asceties of accomplished piety in that forest in the hermitage of the pious Ṛṣis, of purified souls, and wearing bark and matted locks (III.24.20). It is further stated (III. 24.26) that huge trees were bent down with the weight of the creepers, with those five illustrious bowmen sitting under it for rest, appeared like a mountain with fine gigantic elephants resting at its side.

**(b) *Gandhamādana***

The Pāṇḍavas accompanied by Brāhmaṇas entered into the Gandhamādana and at the same time a terrible wind arose, followed by heavy rains, lightning and torrents, during which they took shelter under trees. When the tempest was over they continued their journey (III.143). When they had proceeded only a *krasa* Draupadī sank down. Nakula ran forward and supported her, and the others hastily came and Yudhiṣṭhira took her in his lap and lamented. Dhaumya and other Brāhmaṇas comforted and blessed him and she regained consciousness. Then Bhīma offered them that "Ghaṭotkaca will carry us all." Then Ghaṭotkaca, alongwith other *Rakṣasas* carried the entire party across the snowy mountains (III.145).

**(c) *Kāmyaka Vana***

The Pāṇḍavas had been residing in this forest for quite some time, but then one day the sage Vyāsa appeared there and taking Yudhiṣṭhira into a corner, taught him, after he had purified himself, the science of *Pratismṛti* by means of which Arjuna should proceed to Mahendra, Rudra, Varuṇa, Kubera and Yama to receive weapons from them. The sage also advised him as follows (III.36.35-37)—

> "O Son of Kunti, O ruler of the world, think of going from this forest to another forest, that may be fit for your abode. To reside at one place for a length of time is never pleasant. It might create anxieties to the asceties. As you support numerous Brāhmaṇas, learned in the Vedas, your continued residence here in the forest may exhaust the deer of the dorests and may be destructive to the creepers and plants."

The above passage seems to be veiled concern expressed by the sage over the disturbing of ecological balance or the environmental pollution of the forest due to the continued presence of a large number of humans there.

**(d) *Khāṇḍava***

At the instance of Dhṛtarāṣṭra and Bhiṣma, the Pāṇḍavas at Indraprastha began to bring other kings under their sway. All the subjects lived happily. Once in the summer Arjuna and Kṛṣṇa sported on the banks of the Yamunā with Draupadī and Subhadrā. They were then approached by a Brāhmaṇa whose complexion was like the molten gold (I.222). It was Agni. He asked them to help him to consume the forest of Khāṇḍava and the creatures therein; for Indra always prevented him from consuming it, because he was a friend of Nāga Takṣaka who dwelt there with his family and followers. Arjuna said that he wanted a stronger bow and an inexhaustible quiver full of arrows and a resplendent chariot to ride in and that Kṛṣṇa needed a weapon with which he could slay Nāgas and Piśācas. Then Agni thought of Varuṇa, who presented Arjuna with Gāṇḍiva bow, two inexhaustible quivers, a chariot yoked with horses. Then Agni gave to Kṛṣṇa, the *Sudarśanacakra* and mace *Kaumudī*. They were ready to fight and Agni began to consume the forest, Indra poured in showers, which were first dried up in the air by the flames of the fire. Then more rain was poured down. Arjuna dispelled it with the shower of his arrows. Takṣaka was not present there, having gone to Kurukṣetra. Thus the entire Khāṇḍava forest with its inmates, was destroyed. However Maya—a *dānava* was saved from destruction along-with Aśvasena and four Sārangakas (a kind of birds).

### (e) *Malaya Forest*

The Malaya mountain is known for its Sandal trees. The Cola and Pāṇḍya kings brought as tribute sandal juice from Malaya forest, sandal aloe tree from Dardura. This forest was seen by Mārkaṇḍeya in the stomach of Nārāyaṇa (III.118).

### (f) *Malyavat*

The mountain with forests measures according to the epic as 11,000 *yojanas*. The men there are of the complexion of gold, who had fallen from Brahmaloka. The Pāṇḍavas passed through this mountain on their way to Gandhamādana (III.280). Rāma is believed to have stayed on this mountain for four months. (III.282).

### (g) *Saugandhikavana*

This is the forest where Bhīma had gone to bring celestial lotus flowers and on the way he met Hanumān. When Bhīma was embraced by Hanumānt his fatigue vanished. Hanumat asked Bhīma not to tell anyone that he lived there and vanished. Going in search of the place where grew the lotuses, Bhīma saw romantic forests, groves, rivers, lakes, trees adorned with blossoms and silvan woodlands beautified by various flowers (III.152.3-4). When proceeding quickly he saw in the forest deer with quick glances with grass in their mouth standing along their mates. He passed by lakes, each having romantic descents and woods, adorned with lotuses and lillies which were swarmed by maddened black bees. On account of the presence of lotus-buds, they appeared as if they had joined hands (to greet the Pāṇḍava hero). (III.152.7-9). He also came across a great river in the forest full of fresh grown lotuses. Then he saw the great Sougandhika forests which was quite delightful and effulgent as the sun. Then he reached the lake where the Saugandhika lotuses were grown. It was filled with beautiful, variegated, charming and fragrant golden lotuses with beautiful stalks made of *vaidūrya* gems. The very sight of the lake with fragrant flowers deligted Bhīma immensely (III.154).

### (h) *Kadalī Vana*

While living in the forest, the wind brought a beautiful celestial lotus flower, and Bhīma was asked to bring more flowers of the same species. The great Bhīma went through the forest crossing many hurdles on the way and reached a charming forest of plantain trees (III.146.51) extending to many *yojanas*. The most powerful hero, disturbing the forest, went with great speed towards it. Like a large elephant breaking many trees and uprooting innumerable plantain trees. (III.146.52-53). There Bhīma met Hanumān. With the desire to do good to Bhīma, and to ensure his safety, Hanumān obstructed the path of Bhīma. Thinking thus he lay across the narrow path adorned with plantain trees, with the object that the Pāṇḍava hero might not invite a curse or defeat by entering the plantian forest, the gigantic Hanumān lay down there as if to overcome drowsiness. In the mean time Bhīma arrived there and asked Hanumān to leave passage for him, which the latter refused saying that he was too old to move about. Ultimately when Bhīma was unable to move aside even the tail of Hanumān, the latter disclosed his true identity.

Besides the above, a large number of mountains, trees, plants, creepers and foliage, etc. find frequent mention in this epic. There are also some divine forests, Himālayan forests, bamboo forests which have been interestingly mentioned therein.

## 8. The *Purāṇas*

Purāṇa means anything belonging to the ancient or olden times or a thing or event of the past, an ancient tale or legend or old traditional history. In the ancient religious texts, this represents a class of sacred works, supposed to have been complied by the sage Vyāsa dealing with five topics. The chief Purāṇas are known as eighteen and grouped in three divisions—

**(i) Rajas, exalting Brahma, include—**

(1) *Brahma Purāṇa.*
(2) *Brahmāṇḍa Purāṇa.*
(3) *Brahmavaivarta Purāṇa.*
(4) *Bhaviṣya Purāṇa.*
(5) *Mārkaṇḍeya* and
(6) *Vāmana Purāṇa.*

**(ii) Satvika, exalting Viṣṇu, include—**

(1) *Viṣṇu Purāṇa.*
(2) *Bhāgavata Purāṇa.*
(3) *Nāradiya Purāṇa.*
(4) *Garuḍa Purāṇa.*
(5) *Padma Purāṇa*, and
(6) *Varāha Purāṇa.*

**(iii) Tamasa, exalting Śiva, include—**

(1) *Śiva Purāṇa.*
(2) *Liṅga Purāṇa.*
(3) *Skanda Purāṇa*
(4) *Agni or Vāyu Purāṇa*
(5) *Matsya Purāṇa*, and
(6) *Kūrma Purāṇa*

Besides the above Purāṇas, there are *Upa-Purāṇas* as well which deal with a variety of subjects. But the description of Plant-Kingdom, the forests, flowers, creepers, fruit trees, are described in many forms. The study of the Purāṇas would reveal that the sages who had made the forests as their abodes, had obtained extensive knowledge not only of the plant-kingdom in general, but were also become conscious of the environmental pollution due to the senseless felling of trees, resulting in making the land barren. The account that follows, highlights different aspects of the plant kingdom as described in the various Purāṇas—

## (i) *Agni Purāṇa*

### (a) Consecrating or Dedicating a Tree and the Merits that Accrue

The Brāhmaṇical abound in describing the rites prescribed for the consecrating a tree and the merits that accrue from such a sacred act. Chapter-282 of the *Agni Purāṇa* describes the method of consecrating a tree as follows—

> "The consecration of trees imparts enjoyment in this world and salvation in the next. Water containing solution of the drugs (*Oṣadhis*) known as *Sarvoṣadhis* should be poured over the heads of trees from whose boughs garlands should be hung down, and the tree should be plastered with the rice paste and covered over with pieces of cloth. The ceremony of *Karṇabheda* (perforation of the ball of the ears) should be performed on to them, with needles made of gold (1-2), and on the masonry platform around them, the *Adhibasa* rite of the pitcher should be celebrated. Offering should be made in honour of the god Indra and his companion deities, and the *homa* ceremony should be performed on to vanaspati. The Earth goddess should be invoked out of the hearts of the trees by uttering *Abhiśeka mantra*. The renowned Brāhmaṇas should pour water over the head of the consecrator by uttering *Ṛk*, *yaju* and *Sāma mantras* and the *mantras* which are sacred to the sun and the sea-god Varuṇa, together with those which are called the *maṅgala-mantras*. The consecrator wearing all the ornaments should make gifts to the Brāhmaṇas of clothes, cows, ornaments and proprietory rights in land. For four consecutive days the Brāhmaṇas should be feasted with the thickened milk, among other things. The *Homa* ceremony should be performed with Plākṣa tree and handfuls of *sesamum orientale*. The sacrificial priest should have remuneration double in value to what should be given to other Brāhmaṇas and the mystic sacrificial diagram should be drawn." (3-8)

The same text testified further that the consecration of trees and gardens imparts the highest merit to the consecrator and absolves him of all the sins and demerits.

### (b) Planting of Trees near Dwellings

Chapter 282 of the *Agni Purāṇa* speaks about the actual plantation of the trees in the vicinity of the places of dwellings as also the type of trees to be planted—

"The Tree known as the *Plākṣa* should be planted to the east of the dwelling house, the *Vaṭa* on the east, the mango on the south and *Aśvattha* on the west. Thorny shrubs should be so planted as to edge the southern boundary of the ground of the house. The flower garden should be laid out adjoining a dwelling house or blooming plant of *Sesamum orientale* should be planted after having worshipped the Brāhmaṇas and this sovereign, the moon-god and under the asupices of the asterism presided over by the wind god, Prajāpati and god Viṣṇu respectively, or those known as the Mūla, and the Drubhas. (1-4)

Tanks should be excavated in the garden and arms of rivers should be made to run into them. The excavation of tanks, or of artificial reservoirs of water, should first be commenced under the auspices of the asterisms known as the *Hasta*, the *Magha*, the *Puṣya* or under those presided over by Maitras the Varuṇas and three *Uttaras*. The gods such as Varuṇa, Viṣṇu, Parianya should be first worshipped and propitiated before starting the excavation. Trees such as Ariṣtha, Aśoka, Plantain, Punnaga, Śirīśa, Jamber, Bākula, Pomegranate, or any other trees carefully planted in a garden should be watered every morning and evening, during the summer months, on each alternate day during winter, and at the break of dawn during the rains, when the earth would be dry and require watering. (5-8)

"Trees planted in rows, twenty cubits apart, should be deemed as the best planted, while those having a space of sixteen or ten cubits left intervening between them should be deemed as ranking second best in respect of fruitfulness. Trees should be transplanted twelve times, nor should they be planted too close or adjoining each other, in which case the branches that touch one another, should be lopped off with a chopper as otherwise they would bear no fruits. In case where barrenness would be apprehended, the leaves and branches of a tree should be sprinkled with a solution of cold water saturated with *Vidaṅga*, clarified butter, pulverised *Mudga, Māsa* and Kulattha pulse. Similarly a tree sprinkled with a solution of cold water and clarified butter, becomes laden with abundant fruits and blossoms. A similar result is obtained by manuring the soil with powdered barley, sesamum and offal matter of goat, mixed together and soaked in washing of beef, for seven consecutive nights. And good growth of tree is secured by sprinkling them with the washing of fish. Fish and meat powdered and mixed with a quantity of pulverised *Vidaṅga* should be sprinkled over a tree for pregnancy-present or for the purpose of curing of all distempers. (9-13)

**(c) Medicinal Plants**

According to *Agni Purāṇa* (283. 1-16) "The paste of Śṛṅgi and Ativiṣa mixed with honey or pulverised Ativiṣa alone, should be applied on the tongue of a child suffering from Cough, Catarrh or from inflamatory fever incidental thereto. Similarly, extracts of Yaṣṭikā and Śatpuṣpi administered to a child through the vehicle of milk, oil, clarified butter or thickened milk, tends to improve its complexion, speech and memory and to increase the longevity of life. A decoction of drug known as Ksna treated with the cake of *Mṛdvika* should be known as a strong vermifuge, which proves especially efficacious in case of infantile worm. The expressed juice of *Triphala* and Bṛṅga may be added to the recipe above, enumerated and may be administered to a child, suffering from worms in intestines, through the medium of honey and clarified butter. The express juice of *Dūrva* grass used as snuff is best specific for the bleeding of the nose." Similarly, the expressed juice of garlic should be administered in all cases of inflamation of the ear. A decoction of Tulam, Ginger, Jatya, acts as a specific for intestinal colic and as an absorbent in cases of inflammation of the lower lip. An oil prepared from the decoction of Jatipatra, Phalam, Abhaya should be deemed to be the best cure for all diseases of the teeth. A person suffering from tonsilitis, or an enlargement of Vulva, should gargle his mouth everyday with a decoction of paddy, cocoanut, urine of cow and Visva. A person suffering from a scorfulous inflammation of glands about the neck, or from glandular swelling in general, should snuff everyday, a quantity of oil prepared with the cake of Languli and the expressed juice of Nirgunthika. A person suffering from skin disease should rub the affected part with a solution composed of cow's urine, and saturated with the juice of Arka, Pallava, Sinhi and Putika. A continuous exhibition of the medicine composed of Vakuchi and Sesamum for a year, may prove a successful cure for leprosy. Similarly the oil of Balātakā as well as the oiled Guda-piṇḍī is a safe cure for that dreadful disease. A man afflicated with condylomatous growths about the region of the anus, should use a potion composed of whey saturated with the essence of Jutika, Vanhi, Rajani, Triphala and Vyosa or should take the treacle and Abhaya everyday untill the polypus would fall off, or become dead or inert. A man suffering from any sort of urethral discharges, should drink the decoction of Phala, Darvi and Visa, or the expressed juice of Dhatri fruit mixed with the boiled essence of Rajani. The essence of Vasa taken through the medium of castor oil is a very efficacious remedy for the form of disease known as Tata-Rakta, while the drug Pippalī should be known as possessed of the virtue of curing enlarged spleens. (2-16)

A man suffering from any sort of abdominal dropsy, should drink goat's milk, boiled with the essence of *Chavya*, Danti, Agni, *Vidaṅga* and *Vyosa*. Similarly medicated clarified butter prepared with whey, essence of meat, *Granthika, Ugra, Abhaya, Kṛṣṇa* and *Vidaṅga*, should be prescribed in case of diarrhoea, piles, jaundice, worms and abdominal tumours in general. A decoction of the drugs known as *Triphala*, Vasa, Amṛta, *Cireta*, taken with honey as an anxiliary potion should be deemed as a very reliable remedy in case of jaundice with oedema. A man suffering from Haemoptysis, should take with honey a potion composed of the essence of *Vasa, Suṇṭhi, Vari*, grapes, *Vala* and *Surasa* or the essence of each of the abovesaid drugs separately pressed out and administered (17-20).

A man suffering from Phthisis, should take with honey and clarified butter, a pulverised compound of drugs, known as *Vari*, the *Vidari*, the *Valas*, the *Vasakam*, and the *Svadanstra*. A man suffering from erysiplatous ulcer should take *Sigu, Arka* with *Sindhuvara* (*Vitex negundo*) and honey. In case of fistula in ano, the ulcerated part should be covered over with a paste of *Trivata, Jivati, Danti, Manjistha*, Turmeric, *Daruharidra, Taraksja* and pulverised leaves of the *Nīmba* tree. A limp saturated with paste of *Budhata, Rajani,* pulverised shallac, honey and Vasa should be inserted in an incised wound or ulcer for antiseptic purposes, as well as for the purpose of arresting its spread. Oil, boiled and distilled with *Śyāma, Yaṣṭhi, Nisa.* Shellac, *Padma, Utpala,* sandal-wood and pepper mixed with thickened milk should be applied over a boil, or an ulcer, to bring it into a difinite head and accelerate a process of healthy suppression (21-24).

## (ii) *Garuḍa Purāṇa*

The Dhanvantari Saṁhitā of the *Garuḍa Purāṇa* (Chapters 146-226) speaks of various drugs, recipes, etc. for the cure of a number of diseases, which are made from medicinal plants. Chapter 226 of this Purāṇa, however, lists a number of medicinal herbs used in the cure of diseases, a few of which as detailed below—

| | | |
|---|---|---|
| (1) | Sālparṇi | *Desmodium gangeticum* |
| (2) | Guha | *Hembonitis cardifolia* |
| (3) | Varsabhu or Punarnava | Spreading hogneed or *Boerhaavia diffusa,* Linn. |
| (4) | Karavallah or Katillaka | *Momordica charantia* |
| (5) | Eraṇḍa | Castor oil plant or *Ricinus Communis,* Linn. |
| (6) | Nagavala or Jhasa | *Uraria lagopodioides* |
| (7) | Svadamṣṭrā or Gokṣura | *Tribulus alatus,* Delile |
| (8) | Śatāvarī or Vara | *Asparagus racemosus,* Willd. |
| (9) | Vyāghrī or Vṛhati | *Solanum jacquini* |
| (10) | Vṛścikalī, Amṛta, Kalī, Sarpadaṁṣṭikā | *Eulophia campestris,* Wall. |
| (11) | Kṣudra-saha or Mudgaparṇī | *Phaseolus trilolus* |
| (12) | Mahā Saha | *Grangea maderaspatra* |
| (13) | Nyagrodha (Vaṭa) | Banyan tree |
| (14) | Plākṣa | *Ficus infactoria* |
| (15) | Arjuna | *Terminalia arjuna* |
| (16) | Nandī Vṛkṣa | Species of a fig tree. |
| (17) | Vañjula/Vetasā | Species of a fig tree |
| (18) | Bhallataka or Aruskara | *Semecarpus anacardium* |
| (19) | Lodhara/Sarvaka | *Symplocas racemosa* |
| (20) | Valaphala or Mahajambu | *Engenia jambulana* |
| (21) | Pippalī | *Piper longinum* |

| | |
|---|---|
| (22) Maricam | *Piper nigrum*, Linn. |
| (23) Suṇṭhi | Dry ginger |
| (24) Trikatu—Suṇṭhi, Pippalī, Maricam | Tryshunam |
| (25) Gajapippalī | *Pothis officinalis* |
| (26) Trayanti | *Thalictrum foliolosum* |
| (27) Citrakam | *Racenus communis* |
| (28) Sadgrantha/Ugra | Sweet flag |
| (29) Kuṭuja | *Wrightia antidysenterica* |
| (30) Muṣṭaka | *Maricus cyprus* |
| (31) Elā | Cardamom |
| (32) *Brāhmaṇa Yaṣṭika* | *Celerodendron Siphonanthus* |
| (33) Murvā | *Sanseviera roxburghiana* |
| (34) Jamani | *Eugenia kurzu*, Duthic |
| (35) Vidāṅga | Seeds of *Erycibe paniculata* |
| (36) Jirakan | Cummin seeds |
| (37) Upakañcikī | *Nigella Sativa* |
| (38) Kaṭurohiṇī | Black Hellibore |
| (39) Varaṅgakam | *Cinnamon Bark* |
| (40) Hṛveram | *Pavonia odorata* |
| (41) Nāgakesara | *Mesua ferrea* |
| (42) Kumkum | Saffron |
| (43) Kadalī | Plantain |
| (44) Kataphala | *Solanum melongena* |
| (45) Sallakī | *Artocarpus locuca* |
| (46) Āmalakam | *Emblic myrobalans* |
| (47) Vibhītaka | *Terminalia bellerica* |
| (48) Haritakī | Indian Myobalans |
| (49) Madhūkam | *Liquoritee-root* |
| (50) Dhātakī | *Woodfordia fruitcosa*, Kurz. |
| (51) Karkatasṛṅgī or Mahāghoṣa | *Rhus succedanea* |
| (52) Vaṅśalocana | Manna of the Bamboo |
| (53) Drākṣa (Grapes) | *Vitis vinifera*, Linn. |

| | |
|---|---|
| (54) Mṛṇālam | *Andropogon muricatus* |
| (55) Bhadra | *Tithyamalus antiquorum*, Linn. |
| (56) Dāru-haridra | *Berberis aristate*, D.C. |
| (57) Vira Vṛkṣa | *Timospara cordifolia* |
| (58) Sūryabhakta | *Cleome viscosa* or *Palanisia icosundara* |
| (59) Pasanabheda | *Coleus amboinius* |
| (60) Kusumbha | *Carthamus tinctorius*, Linn. |
| (61) Pītaśilā (Sarasa or Bījaka) | *Pentaptera tomentosa* |
| (62) Snuha | *Euphorbia nerifolia*, L. |
| (63) Sāla | *Artocarpus locuca* |
| (64) Tinisa | *Dalbergia ujjeinensis* |
| (65) Tulasī (Sarasa) | *Holy basil* |
| (66) Tulasī (Sītā) | *Ocimum sanctum*, L. |
| (67) Tulasī (Arjunaka) | *Ocimum basilicum* |
| (68) Sindhuvara | *Vitex negundo* |
| (69) Kaliyakamr | Yellow sandal wood |
| (70) Khadira (Gāyatrī) | *Abutilon catechu*, Willd. |
| (71) Indīvaram, Kuvalyam, Sangandhikam | *Nelum bicim speciosum* |
| (72) Nīlotpalam | *Nymphaea stellata* |
| (73) Saraja | *Vatica robusta* |
| (74) Sulesmataka | *Cordia myxa* |
| (75) Chatra | *Andropogon schoenanthus*, L. |
| (76) Draviṇam | *Hyperanthera moriaga* |
| (77) Amlavetasā | *Emblica officinalis* |
| (78) Aśvattha | *Ficus religiosa*, Linn. |
| (79) Rajadanam | *Mimusops hexanbra* |
| (80) Dāḍimvam | Pomegranate |
| (81) Nikumbha | *Ipomoea turpethum* |
| (82) Carmakṣa | *Mimosa abstergens* |
| (83) Gavakṣī | *Colocynth-citrullus* |
| (84) Kampillaka | *Kamila-melloctas-phillipinesis* |

(85) Nagavala — *Sida spinosa*

(86) Śālmali — *Salmalia malabarica*, Schott

(87) Apāmārga — *Achyranthes aspera*

(88) Sati — *Curcuma zedoria*

(89) Sugandhika — *Myrica sopida*

(90) Durlabha — *Alhagi Maurorum Fogonia-arbarica*

(91) Bhṛṅgarāja — *Eclipta alba*, Linn.

(92) Ikṣāku — Bottle gourd

## (iii) *Padma Purāṇa*

सर्वान् कामानवाप्नोति पदंचानन्तमश्नुते।
यश्चैवमपिराजेन्द्र वृक्षं संस्था पयेब्दुधः।।
सोपि स्वर्गेवसेद्राजन्यावदिंद्रायुत्रयम्।
भूतान्भव्यांश्चमनुजांस्तारयेद्रोम समिंतान्।।
परमां सिद्धिमाप्नोति पुनरावृत्ति दुर्लभाम्।
यइदं श्रुणुयान्नित्यं श्रावयेद्वापिमानवः।।
सोपिसंपूज्यतेदेवैर्ब्रह्म लोके महीयते।
अपुत्रस्यचपुत्रित्वं पादपा एव कुर्वते।।
तीर्थेषु पितुदानादीन् रोपकाण ददंतिते।
यत्नेनापिन्त्र राजेन्द्र अश्वत्थारोपणं कुरु।।
सते पुत्र सहस्त्रस्य कृसमेकः करिष्यति।
भनी नाश्वत्थवृक्षेण अशोकः शोकनाशनः।।
ल्पक्षोयज्ञप्रदः प्रोक्तः क्षीरीचायु प्रदः स्मृतः।
जम्बुकी कन्यका दात्रीभार्यादा दाडिमीतथा।।
अश्वत्थ रोग नाशाय पलाशो ब्रह्मदस्तथा।
प्रेतत्वं जायते पुंसोरोपयेद्योवभीतकम्।।
अंकोले कुलवृद्धिस्तु रत्नादिरेणाप्यरोगिता।
निंब प्ररोह काणांतु नित्यंतुष्येद्दिवाकरः।।
श्रीवृक्षेशंकरोदेवः पाटलायांतु पार्वती।
शिशुपायामप्सरसः कुन्देगंधर्व सत्तमाः।।
तिंतिडीकेदासर्वगावंजुलेदस्यवस्तथा।
पुण्यप्रदः श्रीपदश्चन्दनः पनस्तथा।।
सौभाग्यदश्चंपकश्चंपकश्च करीरः पारदारिकः।
अपत्यनाश्कस्तालो वकुलः कुल वर्द्धनः।।
बहु भार्यानारिकेलाद्राक्षासर्वांग सुन्दरी।

इतिप्रदा तथा कोली केतकी शत्रु नाशनी।।
एवमादिनगाश्चान्ये येनोक्तास्तेपि दायका:।
प्रतिष्ठांते गमिष्यंति यैस्तु वृक्षा प्ररोपिता:।।

(O Lord of Kings, he who thus installs (the plants) a tree, would also live in heaven as long as three myriads of Indras (Rule there) and would save (from falling into the hell), past and future men (*i.e.* his relatives) equal to the number of hair on the body. He attains great prosperity (making) rebirth difficult. Even that man who listens to this, or makes others to listen to it (*i.e.* reads it to others), is honoured by gods and in the world of Brahman. The tree alone makes a sonless person have a son. They offer libations, etc., to plants at sacred places, O lord of kings, plant an *Aśvattha* tree even with great effort. It alone will give you a thousand sons. And by planting an *Aśvattha* tree a man becomes wealthy; the *Aśoka* tree destroys grief. The *Plākṣa* tree is said to give (the fruit of) a sacrifice; the *Kṣira* tree is said to cause longlife. *Jambuki* daughters; *Dadini* gives a wife. *Aśvattha* leads to the destruction of diseases; *Plākṣa* takes one to Brahman. A man who plants a *Viblutaka* tree becomes a ghost. The planting of an *Ankola* tree leads to the expansion of family. Health is caused by the planting of the *Khadīra* tree. The sun is ever pleased with *Nimba* shoots. Lord Śiva is pleased when a tree is planted and Pārvatī is pleased with the red *Lodhara* when planted. The celestial nymphs are pleased with the planting of *Simsapa* and the best Gandharvas with *Kunda* tree. One would get a group of servants if one plants *Tintidika* tree. Similarly, the planting of *Vanjula* leads to the rise of robbers. *Candana* and *Panāsa* also give religious merit and love. Planting of *Campaka* gives good fortune and that of *Karīra* makes one adulterer. Planting of *Tala* destroys one's progeny; while planting of *Vākula* (Bākula) expands the family. The planting of *Coconut* tree brings many wives, while planting of a grape-creeper makes one have all beautiful wife. Similarly planting of *Kali* causes sexual enjoyment. Planting of *Ketaki* destroys one's enemies. These trees and others which are not mentioned here are givers of various objects. Those who have planted trees will attain high position.)

Besides the above, the following methodology has been prescribed for the ceremonies connected with the planting of the tree—

"After having accomplished everything in the case of rites with reference to the lakes, he (the householder) should worship the priests, parts of the pavilion and also the preceptor in the same way. He should similarly honour Brāhmaṇas with gold, garments and anointments. He should cover the tree with garments after having sprinkled them with water containing all herbs and with curds, and (after having) decorated them with curds and sacred grains. He should also apply Collyrium to them with a small golden stick. He should also get fashioned seven or eight silver fruits; and should put each one of them on the altar (in the form) of the tree. Here the incense of *guggulu* is the best. O king having filled pitchers with seven kinds of grains, having covered them with cloth, (having applied) sandal and anointed them and having placed them in copper vessels, he should put them on the trees. After worshipping them in the evening, and after properly making offerings to the regents of the quarters (and gods) like Indra, the Brāhmaṇas should thus consecrate the tree." (1.38.1-10).

## (iv) *Matsya Purāṇa*

### (a) Botanical Beauty

The *Matsya Purāṇa* has described the botanical beauty in the episode relating to Satyavāna and Sāvitrī. When the time of the death of Nala comes nearer Satyavāna went to fetch flowers and fruits from the forest with his father's permission, followed by Sāvitrī. There in the forest Sāvitrī tried to engage the attention of her husband by enquiring about the qualities of various kinds of flowers, forest trees, besides flora and fauna. Satyavāna then spells out the beauty of the forest wealth thus—

वनेऽस्मिन् शाद्वलाकीर्णं सहकारं मनोहरम्।
नेत्र घ्राणसुखं पश्य वसन्तं रति वर्धनम्।।
वनेऽप्यशोकं दृ्वैनं राग वन्तं सुपुष्पितम्।
वसन्तो हस्ती वायं मामे वयत लोचने।।
दक्षिणे दक्षिणैतां पश्य रम्यां वनस्थलीम्।
पुष्पिते किशुकैर्युक्तां ज्वलितानलसप्रभैः।।
सुगन्धि कुसुमामोदो वनराजिविनिर्गतः।
करोति वायुदाक्षिण्यमायवोः कलमनाशनम्।।
पश्चिमेन विशालाक्षी कणिकारैः सुपुष्पितैः।
काञ्चनेन विभात्येषा वनराजीमनोरमा।।
अतिमुक्त लता जाल रुद्धमार्गा वनस्थली।
रम्या सा चारु सर्वाङ्गी कुसुमोत्कर भूषणा।।

—(*Matsya Purāṇa*, 109.1-6)

"Dear! look at the forest smiling with verdure and nice trees. The entire atmosphere is so pleasing to the eyes and the nose. It indeed fills the mind with amorous feelings. Look at the Aśoka tree laden with flowers. O, one with beautiful eyes! the spring is really smiling on us. Look at the *Kiṁśuka* blossoms to the south of this charming forest. The *Kiṁśuka* flowers look like a blaze of fire and are fragrant. In this forest flows the wind laden with the sweet aroma of flowers of the golden hue. Most of the thoroughfares of this forest are choked with luxuriant blossoms. Indeed the whole place abounds with flowers and looks charming."

**(b) Dedicating or Consecrating a Tree**

The *Matsya Purāṇa* (Chapter-59) prescribes rites and ceremonies connected with the dedication or consecration of trees for the benefit of one and all. In this connection one has to keep in mind the importance the ancient people attached for the planting of trees which are considered so important to contain pollution and eradication of diseases, besides other benefits of improved environments. These rites and ceremonies are almost the same as those prescribed for the installation of images of various deities which are as follows—

> "I shall explain to you, the rites of dedicating as well as those prescribed for laying of groves. One should follow the rites as dictated for consecrating the tank and should then plant trees; the Ṛtvijas, the pavilion and the Ācāryas should be arranged in the same way, the Brāhmaṇa should be adorned with sandal paste, etc. Afterwards the trees should be bathed with water, mingled with all prescribed medicinal herbs, and then, garlanding them they should be wrapped with cloth. Then with the needle of gold, the *Karṇavedha* ceremony should be performed on the tree and similarly with the same golden stick, the eyes of the tree should be painted with black pigment. Thereafter seven or eight fruits made of gold should be placed on the altars. The incense of resin (*guggulu*) is considered best on such an occasion. Copper pitchers, full of water should be placed and worshipped with cloth and incence, etc. after placing corn on them. The pitchers containing gold inside them should be placed close to the tree, and sacrifice should be made after which offerings should be made to Indra, Vanaspati, Lokapāla and others according to the prescribed rites.
>
> "Thereafter a milch cow, covered with a piece of white cloth and wearing gold ornaments and having his horn covered with gold should be let loose between the trees planted with her face turned towards the north. A milking vase of bell-metal should also be included to be given away with the cow. Then the tree should be watered with the water contained in the copper pitchers, after reciting the *mantras* of the *Ṛk*, *Yaju* and *Sāma* addressed to Varuṇa, accompanied by music and song. The worshipper should then bathe with water taken from the above pitchers and wear white garments and then with a peaceful mind should confer on the officiating priest, the cows, the riches, bed sheets, sacred thread, rings of gold, clothes, sandals, alongwith other things, and feed them on milk for four days. Libations of mustard, barley and black sesamum and wood of palāśa (*Butea frondosa*), should be offered in the fire and on the fourth day, festivities should be observed and *dakṣiṇa* should be given to the Brāhmaṇas according to the means of the devotee."

The Purāṇa goes on further to attest that, one who observes the rties in dedicating the plantation of trees, obtains all his desires and attains everlasting felicity—

अनेन विधिना यस्तु कुर्यात वृक्षोत्सवं बुधः।
सर्वान् कामानवाप्नोति फलञ्चानन्त्यमश्नुते।।

—*Matsya Purāṇa*, 59.16

### (c) Merits from Planting Trees and Gardens

The *Varāha Purāṇa* also describes the merits accured from planting of a tree or gardens, as follows—

भूदानेन ये लोका गोदानेन च कीर्तिता:।
ते लोका प्राप्यते पुंभि: पादपानां प्ररोपणे।।
अश्वत्थमेकं पिचुमन्दमेकं न्यग्रोधमेकं दशपुष्पजाती:।
द्वे द्वे तथा दाड़िममातुलिङ्गे पञ्चाम्रवापी नरकं न याति।।
यथा सुपुत्र: कुलमुद्धरेद्धि यथाति कृत्स्नान्नियम प्रयत्नात्।
तथा प्रारोपा: फल पुष्प मूला स्वं स्वमिनं नरकादुद्धरन्ति।।

—(170.35-37)

(On who implants the trees, gets the merit comparable to the donating of land or a cow. The one who constructs a lake, and implants a Pīpala tree, a Nīma tree, Vaṭa-Vṛkṣa, ten flower trees, two trees of pomegranate, two lemon trees and five mango trees never goes to hell. As the son liberates the entire dynasty, similarly the fruit and flower trees planted with strict discipline and effort liberate their master from the hell).

## (v) *Śiva Purāṇa*

### Merits of Planting a Tree

*Śiva Purāṇa* is quite vocal about the merit one can derive from the planting of a tree. This Purāṇa (Umā Saṁhitā, Chapter-12) declares that if a person plants a tree in a far flung and inaccessible place, he relieves his past and future generations, of all the sins. Therefore one should plant the trees—

अतीतानागतान् सर्वान् पितृवंशास्तु तारयेत्।
कान्तारे वृक्षरोपी यस्तास्माद् वृक्षांस्तु रोपयेत्।।

—*Śiva Purāṇa*, Umā Saṁhitā, 11.7

The same Purāṇa further goes on to say that the one who plants a tree, enjoys eternal bliss of the heaven, from where he is never discarded.

## (vi) *Harivaṁśa Purāṇa*

In this Purāṇa the plants, trees, flowers, greenery as well as vegetation has been dealt with extensively like several other Purāṇas, but there is a conspicuous mention of the treatment of the forest of Braja, by the local populace, as a result of which it has become uninhabitable not only to the humans, but also to the birds and animals. The people had made use of all the vegetation, wood, fruits and even the firewood of the said forest. Because of the careless felling of the trees, the land had become barren. Earlier the said forest was so dense that nothing was visible due to the shade of the trees. The trees of cattleshed had also been burnt out, and even for collecting firewood one had to traverse a long distance. No water, grass or even shade for shelter was available. All the trees had become useless, as a result of which even the birds had left these trees. Indeed this account provides a glimpse of the environmental disaster of a sort, due to thoughtless felling of trees—

अवगीतमिदं सर्वमावाभ्यां मुक्तकाननम्।
प्रक्षीणतृणकाष्ठं च गौपैर्मथितपादपम्।।
घनीभूतानि यान्यासन्काननानि वनानि च।
तान्याकाशनिकाशानि दृश्यन्तेऽद्य यथासुखम्।।
गोवाटेष्वपि ये वृक्षाः परिवृत्तार्गलेषु च।
सर्वेगोष्ठाग्निषु गताः क्षयमक्ष यवचसः।।
संनिकृष्टानि यान्यासन्काष्ठानि च तृणानि च।
तानिदूरावकृष्टासु मार्गितव्यानि भूमिषु।।
अरण्यामिदमल्पोदमल्पकक्षं निराश्रयम्।
अन्वेवितव्यं विश्रामं दारुणं विरलद्रुमम्।।

—*Harivaṁśa Purāṇa*, Pūrva Viṣṇu Parva, IX. 9.13

Thereafter the entire populace of Braja migrated to the forest known as Bṛindavān. Besides, Chapter-8 of Viṣṇu Parva contains the descriptive account of the uprooting of the twin Arjuna trees (Yamalārjuna) by the child Kṛṣṇa when he was tied to them by his mother.

In the second Viṣṇu Parva of this Purāṇa (Chapter-41) there is a mention of Gomantaka vana, with a beautiful landscape and a number of trees including—

(i) *Trees and Plants*

| | |
|---|---|
| Kathala | Punnaga |
| Mango | Kadru |
| Cane | Jamin |
| Tinisa | Aśoka |
| Sandal | Plantain |
| Tamala | Bākula |
| Ingudi | Bilva |
| Rala | Tinduka |
| Sāla | |
| Nīma | |
| Arjuna | |
| Padara | |
| Hitala | |

(ii) *Flower Plants*

Pippalī
Kewaḍa
Campā
Nāgakesara

(iii) *Creepers*

Cardamom
Black pepper

(iv) *Water Plants*

Lotus
Lilly

## (vii) *Devī Bhāgavata Purāṇa*

The plant kingdom or trees and plants find extensive coverage in this Purāṇa in variety of forms. But the most interesting aspect of the subject is that in several cases the names of the countries are said to have been derived from the trees, a few instances of which are given hereunder—

(i) *Jambūdvīpa*— The eighth skandha testifies that this country derived its name because of the presence of a gigantic rose-apple (or a kind of black plum) tree there.

प्लाक्षाख्यश्च स्वयं जंबुप्रमाणो द्वीपरूधृक।।

—(8.12.3)

(In Jambūdvīpa, there is a gigantic tree of rose-apple and a Plakṣa tree of the same size).,

(ii) *Plākṣadvīpa*— The same text (8.12.3) also testifies the presence of Plākṣadvīpa, named after a gigantic Fig tree (*ficus religiosa*) standing on its soil.

(iii) *Śālmalidvīpa*— In this island, it is stated, that there is a gigantic silk-cotton tree, which derives its name from the said tree. Garuḍa, the kingof birds is said to reside there—

यत्र वै शाल्मली वृक्षः प्लक्षायामः प्रकीर्तितः।।
स्थानं तत्पक्षीराजस्य गरुड़स्य महात्मनः।

—(8.12.18-19)

(iv) *Kuśadvīpa*— There is a thick bush in the Kuśadvīpa, and the name of the country is derived from the same—

घृतोदेनावृतः सोऽयं कुशद्वीप प्रकाशते।
यस्मिन्नास्ते कुशस्तम्बो द्वीपारव्यकरणो ज्वलम्।।

—(8.12.28)

In the ninth Skandha of this Purāṇa, there is mention of the adoration of holy-basil (*Tulasī* plant). In the Twelfth Skandha, it is stated that the Maṇidvīpa had a large number of trees, some of which are mentioned here—

गोपुर द्वारसहितो बहुवृक्ष समन्वितः।
या वृक्षजातयः सन्ति सर्वास्तत्र सन्ति च।।
निरन्तरं पुष्पयुताः सदाफलसमन्विताः।
नवपल्लव संयुक्ता परसौरभ संकुला।।
पनसा बकुला लाघ्रा कर्णिकाश्च शिंशुपा।
देवदारु काञ्चनारा आम्रश्चैव सुमेरवः।।
लिकुचा हिंगुलाश्चैला लवंगाः कट्फलास्तथा।
पाटला मुचुकुन्दाश्च फलिन्यो जघनेफला।।

ताला तमाला: सालाश्च कंकोला नाग भद्रका:।
पुन्नागा: पीलव: साल्व का वै कर्पूर शाखिन:।।
अश्वकर्णा हस्तिकर्णास्ताल पर्णाश्च दाड़िमा:।
गणिका बन्धुबीजाश्च जम्बीराश्च कुरंडका:।।
चांपेया बन्धुजीवाश्च तथा वै कनकद्रुमा:।
कालागुरुद्रुमाश्चैव तथा चन्दन पादपा:।।
खर्जूरायूथिकास्ताल पर्ण्यश्चैव तथेक्षव:।।
क्षीर वृक्षाश्च खदिराश्चिंचाभल्लातकास्तथा।।
रुचका कुटजा वृक्षा बिल्व वृक्षास्तथैव च।
तुलसीनां वनान्येवं मल्लिकानां तथैव च।। —(12.10.19-27)

(The *gopuras* are decorated with various trees. All the species of trees are available here. These trees always remain laden with fruits, flowers and tender shoots and leaves. Jack-tree, Bākula (*Manilkara hescandra,* Roxb.), Lodhara (*Symplocos racemosa*), Karṇikar (a flowing plant), rose-wood (*Dalbergia Sisu*), Devadāru (Cedar). Kachanāra (*Bauhinia tomentosa,* Linn.), mango and Sumeruka trees were there. Besides Badhala (*Artocarpus laucha*), Hingula (a thorny egg plant) cardamom, clove, Jāyaphala (nutmeg), Mucukunda (a kind of big tree), Pāṭalā (a tree of trumpet flowers), Phalini and Kathamara (a wild fig tree) were also found there. The Palmyra, Tamāla (*Crataeva roxburgii*) Sāla, Sākhu, Kankola, Nāgbhadra, Pīlu, Aśvakarṇa, Hastikarṇa, Ekaṅgi, Mura, Pomegranate, Jambhīrī, Lemon, Katasariya, Campā, red flower plant, Dhattūrā, Kālāguru, Sandal, Palm, Lilly, Talaparṇi, Sugarcane, Kṣira-vṛkṣas, Jujube, Tamarind, Metava Benda, Kudakeria. Tulasī, Jasmine and various other types of fruit trees were also there.)

### (viii) *Brahmavaivarta Purāṇa*

This Purāṇa brings out the importance of the forests which were primarily used by monarchy for merry making purposes or the visits to these forests were made primarily for the purpose of seclusion. In the context of the pleasure trip of Candra and the guru's wife Tārā, in the forests, various forests are mentioned which include Mādhawivana, Candanavana, Campakavana, Malayavana, Svarṇarekhatatavana, Puṣpabhadravana and Puṣkaravana (Prakṛti khaṇḍa, 58.25, 28, 37, 38). In addition to the above forests the names of Visramakavana, Suvasana and Śrīvana are also mentioned in connection with the pleasure trips of Upabarhana and his wife Mālavati. In the *Śrimadbhāgvata Purāṇa,* it is mentioned that at the time of the air journey by the Sage Kardama with Devahuti, they visited Viśrāmakavana, Surasana, Nandana, Puṣpabhadra and Manasavanas. (3.3.40-41).

Besides the forests, names of several trees are mentioned in the Brahma-Khaṇḍa of this Purāṇa, which include Kalpavṛkṣa, Mandāra, Puṣkara, Vanigarṇa, Tāmbūla, Svetatala, Patola, Kusmāṇḍā, Tumbi, Kalambika. Simbi, Putika and Varta (Brahma Khaṇḍa, 27.20,26,29,57). In the Prakṛti Khaṇḍa, of the same Purāṇa, however, the trees and plants like the Taṇḍula, Tila, Ikṣu, Śukladhānya, Rāmbhāphala, Nārikela and Kesara are also mentioned. In the Kṛṣṇa-Janma-Khaṇḍa, while discussing on the building of the city of Dvārakā, the good and evil qualitites of the various species of trees are also mentioned. Accordingly, Āmra, Panāsa, Jambira, Nārikela, Jambū, Kadamba, Kadalī, Rose, Kharjūra, Karkari, Alābu, Kūṣmāṇḍa, Kiṁśuka. Cereals like Canaka, Ikṣu, Aśoka, Śirīṣa and Haridra are included in the best variety of trees, while those like Vaṭa, Śālmali, Titindi, Dhatūra, Bimba and Badarī, come under lesser auspicious trees.

The composer of the Purāṇa also was aware of a variety of creepers including Mādhavī, Mālatī, Yuthika, Kunda-Ketaki, Nāgesvara, Mallikā, Bākula and Aparājitā. The use of different types of *vanaspatis* in different months have also been brought out in this Purāṇa.

A list of the trees and plants mentioned in this Purāṇa is available at Appendix-3(a).

The Purāṇa also contains an interesting episode of the felling of Yamalārjuna, the twin Arjuna trees. According to this episode the naughty Kṛṣṇa alongwith other children of Gokul, had been breaking the pots of *Gopīs* containing butter, milk and butter milk. Yasodā, the mother of Kṛṣṇa wanted to punish him in order to prevent him from committing the mischief, but Kṛṣṇa runs away. With considerable efforts, Yasodā caught hold of Kṛṣṇa and tied him up with a piece of cloth to the twin Arjuna trees. But the very touch of the child Kṛṣṇa makes the trees to fall down uprooted; out of these trees Nalakubera and Maṇigrīva emerged in human form. He happended to be the son of Kubera, but was cursed to be a twin tree. (Khaṇḍa-14,1-14 and 29-41). This episode is beautifully depicted.

## (ix) *Bhāgavata Purāṇa*

The same story is repeated in the *Bhāgavata Purāṇa,* according to which Kṛṣṇa after stealing the butter, broke the vessels of curds and butter-milk. Snapping the churning stick and taking a pot full of butter, Kṛṣṇa ran away to play with the cowherd boys. He found a mortar placed upside down, over which he mounted and sat, causing his companions to sit down around him. He bagan to chuckle and share with them the butter. The monkeys and cowherd boys had a hearty feast over the butter and the spilt butter-milk. On discovering what had happened, Yasodā chased Kṛṣṇa and threatened to beat him. She tied him with a rope to the mortar. However, the rope was not long enough and was found to be two inches (*aṅgulas*) short. She joined another rope to it, and then another, but every time it was found to be a little short. Taking pity at the exertion of the mother, who was perspiring, Kṛṣṇa allowed himself to be tied to the mortar. After tying him Yasodā returned to her domestic duties.

The above scene is depicted in a Kāngra painting from the *Bhāgavata Purāṇa* in which, in the courtyard of the house, are shown two Arjuna trees. Kṛṣṇa managed to crawl between the trees alongwith the mortar to which he was tied and gave a jerk uprooting the trees. Out of the trees emerged Nalakubera and Maṇigrīva sons of Kubera, who on account of their haughtiness were cursed to become trees by the curse of Nārada. Hearing the crashing of the trees, the cows broke their ropes in panic, and Nanda, Upananda and Yasodā as well as the neighbours all rushed to the place. The two *Siddhas* rid of their sins, circumambulated and bowed to Kṛṣṇa tied to the mortar, and taking their leave, left for Alakā in the Himālayas.

The panic created by the crashing of the trees is shown in a dramatic manner. Balarāma is pointing towards Kṛṣṇa tied to the mortar and their foster parents are expressing a sense of relief that he has not been injured. Tied to the mortar, Kṛṣṇa is showing his resentment to Nanda for the manner in which he has been treated. To the right are Nalakubera and Maṇigrīva standing with folded hands expressing their gratitude to Kṛṣṇa. The two trees connected with the main story have been beautifully drawn, with green leaves and branches. The roots of the trees are shown in the most naturalistic manner.

The *Bhāgavata Purāṇa* also contains an episode relating to Kṛṣṇa's bringing the Pārijāta tree from Indra's heaven which is also repeated in the *Brahmavaivarta Purāṇa* (Kṛṣṇa Khaṇḍa—Ch—114-120). According to these Purāṇas, the Pārijāta tree was one of the products of the churning of the ocean of Milk, *Kṣīrasāgara* and was planted by Indra in his gardens. "Its barks were of gold and it was embellished with young sprouting leaves of copper colour and fruit stalks bearing numerous cluster of fragrant fruits." Nārada who delighted in sowing discord, brought a flower of this tree to Dvārakā and presented it to Kṛṣṇa. He waited to see to which of his wives Kṛṣṇa would give the flower. Kṛṣṇa gave it to Rukmiṇī, where upon Nārada wearing a sad look on his face went to Satyabhāmā, Kṛṣṇa's other consort. On her enquiring the cause of his sorrow, the sage replied that he had presented Kṛṣṇa with a flower of Pārijāta tree thinking that he would present it to her, his favourite wife, but he was grieved to find that he had given it to Rukmaṇī instead. Satyabhāmā's jealousy was aroused, and she asked Nārada as to what could be done to spite Rukmaṇī. The sage advised her to ask Kṛṣṇa to bring the Pārijāta tree itself to Dvārakā and plant it in her abode. After giving this advice, he went back to celestial region, and informed Indra that thieves were about and what he should guard Pārijāta tree with care.

Satyabhāmā repaired to the anger chamber and when Kṛṣṇa came to her, she said, "You say that I am your favourite wife, but you treat me as Rukmaṇī's handmaid." She asked him as to what made him present the Pārijāta flower to Rukmaṇī. Kṛṣṇa admitted his fault and asked her what could he do to expiate it, Satyabhāmā replied that she would not be satisfied with anything less than the Pārijāta tree itself. Kṛṣṇa proceeded to Amarāvatī, stole into Indra's gardens and uprooted Pārijāta tree. Mounted on Garuḍa, he escaped with the tree, but Indra, warned by Nārada, followed him. A battle followed and Indra was defeated and Kṛṣṇa brought the tree to Dvārakā. Now he had to face the problem of fulfilling his promise to Satyabhāmā without offending Rukmaṇī. The problem was solved by planting the tree in such a position that while the base and the trunk lay within Satyabhāmā's garden, its branches extended over the adjoining palace of Rukmaṇī, scattering flowers in the early morning. The Pārijāta tree has been identified with Harasingār-*Nyctanthes arbortristis*, which shed its flowers at night time when they spread exquisite fragrance.

In a Basohli painting of A.D. 1780, Kṛṣṇa is shown riding on Garuḍa, warding off Indra's attack. Behind Kṛṣṇa is Nārada holding a *Vīṇā* and the Parijāta tree. Indra is riding on his white elephant. The marks on Indra's body are to be noted. In the ancient texts Indra is known as Sahasrākṣa or a thousand-eyed-one.

The *Bhāgavata Purāṇa* also conceives a tree the entire universe in the form of cosmic manifestation (3.9.16).

> यो वा अहं च गिरिशश्च बिभुः स्वयं च
> स्थित्युद्भवप्रलय हेतव आत्ममूलम्।
> भित्वा त्रिपादूवृध एक उरुप्रोह
> स्तस्मै नमो भगवते भुवनद्रुमाय।।

(Your Lordship is the prime root of the tree of the planetary systems. This tree has grown by first penetrating the material nature in these trunks and myself. Śiva and yourself, the Almighty for creation, maintenance and dissolution, and we three have grown with many branches. Therefore I offer my obeisances unto you, the tree of cosmic manifestation).

The cosmic manifestation is grossly divided into three worlds, the upper, lower and middle planetary system, and there it broadens into the Cosmos of fourteen planetary systems with the manifestation of the Supreme Personality of Godhead as the supreme root. Material nature which appears to be the cause of the Cosmic manifestation, is only the agency or the energy of the Lord. This is confirmed in the *Bhagvadgītā* (9.10) मयाध्यक्षेण प्रकृतिः सूयते स चराचरम्।। "Only under the superintendence of the Supreme Lord does the material nature appear to be the cause of all creations, maintenance and dissolution." The lord expands himself into three, Brahmā, Viṣṇu and Śiva, for creation, maintenance and destruction respectively. Of the three principal agents controlling the three modes of material nature, Viṣṇu is the Almighty; even though he is within the material nature for the purpose of maintenance, He is not controlled by the laws of material nature. The other two, Brahmā and Śiva although as greatly powerful as Viṣṇu, are within the control of the material energy of the Supreme Lord. The conception of many gods controlling the many departments of material nature is ill-conceived by the panthiest. God is one without a second and He is the primal cause of all causes. As there are many departmental heads of governmental affairs, so there are many heads of management of the universal affairs.

Due to the poor fund of knowledge, the impersonalist does not believe in the personal management of things as they are. But in this verse, it is clearly explained that everything is personal and nothing is impersonal. The tree of material manifestation is described in the Fifth Chapter of the *Bhagavadgītā* as an *aśvattha* tree whose root is upwords (अश्वत्थसर्व वृक्षाणाम्). One can have the actual experience of such a tree to site viewing the shadow of a tree on the bank of a reservoir of water. The reflection of the tree on the water appears to hangdown from its upward roots. The tree of creation which is described here is only a shadow of the reality, which is Paramabrahman, Viṣṇu. In the internal potential manifestation of the Vaikuṇṭhalokas, the actual tree exists and the tree reflected in the material nature is only the shadow of the actual tree. The impersonalists' theory that Brahman is void of all variegatedness is false because the shadow tree described in the *Bhagavadgītā* cannot exist without being the reflection of the real tree. The real tree is situated in the eternal existence of spiritual nature full of transcendental varieties, and lord Viṣṇu is the root of that tree as well. The root is the same—the Lord—both for the real tree and the false, but the false tree is only the perverted reflection of the real tree. The lord being the real tree, is here offered obeisances by Brahmā on his own behalf and also on behalf of Śiva.

The remembrance of life in the forest and the cowpens of Braja came to Balarāma at Dvārakā. Alongwith Kṛṣṇa he visited Nanda and Yaśodā at Gokul. He saw cows wandering about in the woods, lowing and panting, their mouths open and not grazing grass, in their distrees on account of separation from Kṛṣṇa. The young women of Braja with emaciated bodies and dishevelled hair were in deep despondency. On seeing Balarāma and getting the news of Kṛṣṇa, their spirits revived, and they danced with him. One night, while dancing with milkmaids, Balarāma under the influence of wine, commanded the Yamunā to flow to his feet so that he might bathe conveniently. The Yamunā in her conceit paid no attention to what Balarāma said, and he angrily drew her towards himself with his plough, and bathed. The bend in the river Yamunā, at Gokula remains to this day.

The above scene is portrayed in a Basohli painting with the following inscription on the back—

वहसि वपुषि विशदे बसनं जलदाभं
हलहतिभीतिमिलित यमुनाभं।
केशवधृत हलधर रूप
जयजगदीश हरे।।

"Keśava, Salutation to thee in Thy Haladhara from with Thy fair person resplendent in its azure mantle. It looks as though the Yamunā, fearful of Thy plough, has vested Thee with her treasured Charm."

The said painting dates back to A.D. 1730-40. The river Yamunā is shown as a small girl draped in red, holding a golden vase following the plough of Balarāma. Kṛṣṇa is embracing a cowherd in the right corner, and cows are grazing or seated over the lush green grass or pasture land. A few trees are also beautifully depicted.

## (x) *Mārkaṇḍeya Purāṇa*

Besides numerous references to the plant kingdom and forests, etc. the Purāṇa makes a pointed reference to some territories which were named after the presence of huge trees or groves, grown on their soils.

जम्बुद्वीपे तथाग्नीध्रं राजानं कृतवान् पिता।
प्लक्षद्वपेश्वरश्चापि तेन मेधा तिथि: कृत:।।

शल्मलेस्तु वपुष्मन्तं ज्यतिष्मन्तं कुशाह्वये।
क्रौञ्चद्वीपे द्युतिमन्तं भव्यं शाकाह्वयेश्वरम्।।
पुष्कादिपतिञ्चापि सवनं कृतवान् सुतम्।
महावीतो धातकिश्च पुष्कराधिपते सुतौ।। —(53.17)

(The father (Priyavrata), gave *Jambūdvīpa* to Agnighra, *Plakṣa-dvīpa* to Medhatithi, *Śālmalidvīpa* to Vapusman, *Kuśadvīpa* to Jyotisman, *Kraunchdvīpa* to Dyutimān, *Śākadvīpa* to Bhavya, *Puṣkaradvīpa* to Savana).

Out of the above six *dvīpas* or countries names of five of them viz—*Jambūdvīpa, Puṣkaradvīpa, Śālmalidvīpa, Kuśadvīpa* and *Śākadvīpa* are named after the plant kingdom.

## (xi) *Brahma Purāṇa*

The *Brahma Purāṇa* also testifies that the names of these countries were based on the plant kingdom as follows—

(i) जम्बूवृक्ष प्रामाणान्तु तन्मध्ये सुमहातरु:।
प्लक्षस्तन्नामसंज्ञोऽयं प्लक्षद्वीपो द्विजोत्तम:।।

—*Brahma Purāṇa*, 20.18

(ii) शाल्मलिश्च महावृक्षो नाम निवृत्तिकारक:। —*Ibid.*, 20.33

(iii) कुशद्वीपो कुशस्तम्ब: संज्ञा तस्य तत्स्मृतम्।। —*Ibid.*, 20.45

(iv) शाकश्चात्र महावृक्ष: सिद्धगन्धर्व सेवित:। —*Ibid.*, 20.64

## (xii) *Varāha Purāṇa*

A somewhat identifical information of these countries is provided in the *Varāha Purāṇa* 85.45.

## (xiii) *Viṣṇu Purāṇa*

In Chapter IV, Book-II of this Purāṇa, the above statement of the *Bhāgavata* and the other Purāṇas is duly supported as follows

"In this *Dvīpa*, is a large fig-tree of similar size as the Jambū-tree of *Jambūdvipa* and this Dvīpa is called Plākṣa, after the name of the tree...*Plākṣadvipa* is surrounded by a disc by the sea of molasses, of the same extent as the land....A large Śālmalī (silk-cotton) tree grows in this *Dvīpa* and gives it its name...Sur Sea is entirely encircled by *Kuśa Dvīpa* which is so named from a clump of Kuśa grass growing there....There grows a large Śāka (teak) tree in the *Śākadvīpa* from which the name of the Dvīpa is derived. *Puṣkaradvīpa* is said to be double of the size of *Śākadvīpa*. There is a Nyagrodha-tree (*Ficus bengalensis*) grown on this *Dvīpa*."

Wilson has opined on the above Dvīpas mentioned in the Purāṇas, as follows—

"The description of Dvīpas in the *Agni, Brahma, Kūrma* and *Vāyu Purāṇa* agrees with that of our texts. The *Mārkaṇḍeya, Liṅga* and *Matsya* contain no details. The

*Bhāgavata* and *Padma* follow the same order as Viṣṇu, etc. but alter all the names and the measurements. The account of the *Mahābhārata* is very irregular and confused. The variations throw no additional light upon the geographical systems of the Purāṇas. Some traces of this appear discoverable in the west; and the seven Dvīpas with the surrounding seas may have some connection with the notion of seven climates as Wilford has supposed. The learned but fanciful writer bestowed great pains upon the verification of these fictions and imagined the different *Dvīpas* to represent actual divisions of the globe. *Jambūdvīpa* being India; Kuśa, the Kush of scripture or the countries between Mesopotamia and India; Plākṣa being Asia Minor; Śālmali eastern Europe; Krauncha, Germany; Śāka the British Isles; and the Puṣkara, Iceland...." (Wilson, H.H., *Viṣṇu Purāṇa*, p.166).

### (xiv) *Vāmana Purāṇa*

This text has interpreted the growth of various trees and plants from entirely a different angle, inasmuch as they are conceived to have emerged from the body of one deity or the other—

मासि चाश्वयुजे ब्रह्मन यदा पद्म जगत्पते:।
नाभ्या निर्याति हि तदा देवेष्वेतान्यथोऽभवन्।।
कन्दर्पस्य कराग्रेतु कदम्बश्चारु दर्शन:।
तेन तस्य परा प्रीति: कदम्बेन विर्वद्धते।।
यक्षाणामधिपस्यापि मणिभद्रस्य नारद:।
वटवृक्ष समभवत् तस्मिंस्त तस्मिंस्तस्य रति: सदा।।
महेश्वरस्य हृदये धत्तूर विटप: शुभ:।
संजात: स च शर्वस्य रति कृत तस्य नित्यश:।।
ब्रह्मणो मध्यतो देहाञ्जातो मरकराप्रभ:।
ख दर: कण्टकी श्रेयानभवद्विश्वकर्मण:।। —(18.1-5)

(In the month of Āśvina, lotus was born out of the navel of Viṣṇu; Kadamba (*Anthocephalus cadamba*, Miq.) was born out of the forearm of Kāmadeva; Vaṭavṛkṣa, Indian fig tree, was born out of Maṇibhadra, the king of Yakṣas; Dhattūra, the Torn-apple, was born out of the heart of Śiva; Khadīra, *Mimosa catechu* was born out of the middle of Brahmā's body; Kautaki or a throny bush was born out of Viśvakarmā).

The emergence of other trees from other deities as testified (in the verses 6-10) of the same (Chapter-18) are noted below—

| | | |
|---|---|---|
| (i) | Palm of Girinandini | Kanda-gulma |
| (ii) | Gaṇapati | Sindhuvaraha |
| (iii) | Yama | Right side—Palāśa Left side—Udumbara |

| | |
|---|---|
| (iv) Rudra | Udvejaka Vṛsa |
| (v) Skanda | Bandiyava |
| (vi) Sun | Aśvattha |
| (vii) Kātyāyanī | Śāmi |
| (viii) Lakṣmī | Belā |
| (ix) Śeṣanāga | Sarastamba or Sarepata |
| (x) Vāsuki (back) | White Dūrva |
| (tail) | Black Dūrva |
| (xi) Heart of Sadhyas | Harita Sandal |

## (xv) *Vāyu Purāṇa*

The *Jambūdvīpa* is reported to have a Jambū tree in the *Vāyu Purāṇa* thus (35.26-32)—

"There is a Jambū tree (Rose Apple) on the top of the Southern mountain. It is resorted to by the Devas. It ever blooms and bears fruits and branches; the sweet and soft fruits of large size falling on the peak of the mountain are nectar-like in taste. From the table land on the top of this excellent mountain, flows the river Jambū with currents of honey. Gold known as Jambūnala with the lustre of burning fire is formed here. It is the incomparable ornament of Devas and is the destroyer of sins. Devas, Dānavas, Gandharvas, Yakṣas, Rākṣasas and Paunāgas (serpents) drink the honey, the juice exuding from the Jambū tree, which is as sweet as nectar. The eternal Jambū tree is well known in the world and is the landmark in the Southern Continent. The continent is called *Jambūdvīpa* after its name."

Besides the Jambu tree, this Puraṇa also associates some other continents with the trees as would be evident from the following examples—

***Tree***

**(1)** ***Aśvattha***—It grows on the peak of the great Vipula mountain (35.33). It has hanging garlands, and golden platforms set with gems. Its trunk and branches are quite lofty. It is the abode of many living beings. It abounds in many good qualities (35.34). Symbolised with the two names (Ketu and Māla) joined together, the continent is known as Ketumāla and lies in the western region. It is very extensive and is well known in the world and the heaven (35.41).

**(2)** ***Fig-tree***—On the northern peak of the mountain Supārśva a great fig tree grows up with extensive branches. It covers an area of many *Yojanas*. It is resorted to by the Siddhas and the Cāraṇas. From its branches clusters of garlands of sweet fragrance hang loose. It shines with them (35.43). This tree is the landmark of the northern Kuru country. It shines with fruits full of honey. The fruits resemble the coral cup.(35.44). Seven mind-born sons of Brahmā, the blessed younger brothers of Sanatakumāra, stay here. They are known as Kurus (35.45).

Besides the *Jambūdvīpa*, this Purāṇa, deals with other Dvīpas as well and the bird's-eye view of the same is provided here—

| | Name | Size | Other aspects |
|---|---|---|---|
| (1) | *Plākṣadvīpa* | Twice the size of *Jambūdvīpa* (41.2). | *Plākṣadvīpa* is large and glorious. It abounds in reaches and foodgrains. It is endowed with divine medicinal herbs and fruits. It has many kinds of plants and herbs (49.36). |
| (2) | *Śālmali-dvīpa* | Twice the size of *Plākṣadvīpa* (49.29). | There is a *Śālmali* or Silk cotton tree in the middle of this *Dvīpa* which is named after it. (49.44). |
| (3) | *Kuśadvīpa* | Twice the size of *Śālmalidvipa* | |
| (4) | *Śākadvīpa* | Twice the size of *Kraunchadvīpa*, | This has been identified with the grasslands of Iran, Iraq and the hot deserts. |

The chapter-37 of this Purāṇa contains a beautiful description of forests in various valleys and lakes and the forest beauty in general.

## (xvi) *Bhaviṣya Purāṇa*

This Purāṇa in its Madhyaparva (Chapter-3) prescribes in detail the ceremonies connected with the formal planting of the trees and the gardens. It is also stated therein that if any tree after having been planted in the garden is uprooted and dies or its branches are broken, the consequent sin for such a mishap does not befall on the person who plants such tree or garden—

वृक्षाग्रात् पतितस्यापि आरोहात् पतितस्य च।
मरणे वास्ति भङ्गे वा कर्ता: पापैर्न लिप्यते।।

—*Madhyaparva*, 3.1:31

The second and third Chapters of the Madhyaparva, not only define the setting up of small gardens but also speak of the pasture lands around villages. It may be recalled here that till recently, there was a practice of leaving pasture lands in each village in the country, which was exclusively used for the grazing of the cattle. (This practice however is vanishing fast). But the sanctity of the pasture land has been detailed in this Purāṇa, equating it with form of Śiva, being sacred and dedicated to the cause and welfare of the cattle—

शिवलोकस्तथा गावः सर्वदेव सुपूजिता:।
गोभ्यो एषा मया भूमि: सम्प्रदत्ता शुभार्थिना।।

Besides the above, the merits for dedicating pasture land have also been described.

The Chapters 4-8, of the third Madhyaparva of this Purāṇa is devoted to the planting of Aśvattha tree besides construction of stepwells and water reservoirs. Chapter-9 of the same Madhyaparva prescribes the rites and ceremonies connected with the setting up of gardens of *Vaṭa, Bilva, Pugiphala* and other trees. Similarly Chapters 14-17 of the same Madhyaparva, speak of the planting of flower gardens and the Tulasī plants, and while doing so, even the auspicious *Nakṣatras* in which the Tulasī plants have to be planted have been pointed out. Chapter-17 of the same Madhyaparva, prohibits the formal ceremonies in planting of the Jayantī, Samavṛkṣe, Panāsa (Kathala), Kadamba, Nimba, Kanakapatala, Śālamali, Nimbaka, Bimba and Aśoka trees. It is also enjoined that the trees like Bhadraka, Samikona, Candataka, Baka and Khadira should be planted with due rites, but the ear-piercing ceremony should not be performed for them.

Chapter-60 of Uttaraparva of this Purāṇa, describes *Śrīvṛkṣa Navami-vrata* to be performed for the adoration of *Bilva* tree also known as *Śrīvṛkṣa*. The Purāṇa further goes on to say that at the time of the churning of the ocean, Lakṣmī emerged out of the sea and took refuge under a *Bilva* tree, as a result of which it came to be known as *Śrīvṛkṣa* as well. Then the worship of this tree with the following verse has been prescribed—

श्रीनिवास नमस्तेऽस्तु श्रीवृक्ष शिववल्लभ।
ममाभिलषितं कृत्वा सर्वं विघ्नहरो भव।। —*Uttaraparva*

Chapter-66 of Uttaraparva, defines the merits of the Araṇya-*Dvādaśī-Vrata*, which, it is stated, was advised by Rāma to be performed by Sītā, during her stay in the forests.

## 9. Early Indian Polity and Classical Sanskrit Literature

Besides the Vedic, post-Vedic literature, followed by the Epics, the Purāṇas, India has the proud possession of rich cultural heritage in the form of works on Indian polity and other classical Sanskrit Works. While the Arthśāstra of Kauṭilya and the Code of Manu are unique works composed in the Sanskrit language, other works of the famous poets have enriched the Sanskrit language in many ways. These Sanskrit works emerged on the Indian literary horizon by about the fourth century B.C. and were continued for centuries that followed. The works of Kālidāsa are held in high esteem by scholars. Though in the pre-Kālidāsa period poetry, prose and drama works were indeed composed, but much of these texts have been lost. Some of the works belonging to pre-Kālidāsa period were composed by Aśvaghoṣa and Bhāṣa. The former composed works entitled *Buddhacarita* and *Saundrānanda* and Bhāṣa composed works like *Abhiṣekanāṭaka* and others. These works are attributed to 1st century A.D. In all these works, the botanical science found prominent place, as would be evident from the account that follows—

### (i) 4th Century B.C.

#### (a) *Arthaśāstra* by Kauṭilya

According to the *Arthaśāstra*[1] of Kauṭilya, a Horticulturist/Agriculturist should possess the knowledge of the science of agriculture dealing with the plantation of bushes, trees, or he should be assisted by those who are trained in such sciences; such an official shall in time collect the seeds of all kinds of grains, flowers, fruits, vegetables, bulbous roots, fibre producing plants, and cotton—

सीताध्यक्षः कृषितन्त्र शुल्ब वृक्षायुर्वेदज्ञस्तज्ज्ञसखो वा
सर्वधान्य पुष्पफल शाक कन्दमूलवाल्लिक्यक्षौमकार्पास-
बीजानि यथा कालं गृह्णीत। —*Arthaśāstra*, 2.24.1

The said officer shall employ slaves, labourers, and prisoners to sow the seeds on crown lands which have been often and satisfactorily ploughed. (वहुहलपरि कृष्टायां स्वभूमौ दास कर्मकर-दण्ड प्रति कर्तृभिर्वापयेत्, *Arthaśāstra*,2.24.2). The work[2] of the above men should not suffer on account of any deficiency in plough (कर्षण यन्त्र) and other necessary instruments or of the bullocks.[3] Nor shall there be any delay in procuring to them the assistance of balcksmiths, carpenters, borers, ropemakers as well as those who catch snakes and similar persons. Any loss due to the above person's negligence was punishable with a fine, equal to the amount of loss.[4]

Further, while commenting on the rainfall, so essential for the crops of various kinds, Kauṭilya points out that the "forecast of rainfall can be made by observing the movement, position and pregnancy (*garbhādhāna*) of Jupiter (Bṛhaspati), the rise, setting and motion of Venus, and the natural and unnatural aspect of the sun—

तस्योपलब्धिबृहस्पतेः स्थानगमनगर्भधानेभ्यः
शुक्रोदयास्तमयचारेभ्यः सूर्यस्य प्रकृति वैकृताच्च।
—*Arthaśāstra*, 2.24.9

***Sprouting of the Seeds***

According to the *Arthaśāstra* of Kauṭilya, the sprouting of the seeds could be inferred from the sun; from the position of Jupiter, the formation of grains can be inferred, and from the movement of Venus, rainfall can be inferred—

सूर्याद्बीज सिद्धि। बृहस्पतेः सस्यानां स्तम्बकरिता।
शुक्राद्वृष्टिरिति।। —*Arthaśāstra*, 2.24.10-12

***The Sowing of the Seeds***

According to Kauṭilya, the sowing of the seeds should depend upon the rainfall and the kinds of seeds sown. Because for some seeds much water is not required while for others excessive rain would be required.[5] Thus Sali—a kind a rice, *Vṛhi* (rice) Kodrana (*Paspalum scrobiculatum*), Tila (*Sesamum indicum*, DC.), Priyaṅgu (ponic seeds), Daraka and Varaka (*Phaseolus trilolus*) are to be sown at the commencement[6] of the rainy season. Mudga (*Phaseolus mungo*, L.), Māsa (*Phaseolus radiatus*), and Sailuja are to be sown in the middle of the rainy season.[7] Kusumbha (Safflower), Masūra (*Lens esculenta*), Kulthā (*Dolichos biflarus* Linn.), Yava (barley), Godhūma (wheat), Kataya (leguminus seeds), Atasi (linseed) and Sarsapā (mustard) are to be sown[8] last.

In addition to the above, Chapter-27 of the same work is exclusively devoted to the forest produce, which includes lists of plants, trees, creepers, fruit trees, etc.

**(b) *Pratijñayaugandharāyaṇa* by Bhāṣa**

In the first Chapter of the play, Yogandharāyaṇa, the minister declares the king was about to proceed to the thick forest called Nayavana, near Veṇuvana, next day, therefore he aught to meet him immediately.

***Punishment for Cutting of Trees***

Kauṭilya provides for stringent punishment for the cutting of tender sprouts of fruit trees, flower trees or shady trees in the parks near the city. A fine of five *paṇas* is to be imposed for the offence. For the cutting of minor branches of the same trees, the fine prescribed is double the above amount, *i.e.* 12 *paṇas;* and for the cutting of big branches 24 *paṇas*. Cutting of trunks of these trees was punishable with the first amercement and felling of the same shall be punished with middlemost amercement. In case of plants which bear flowers, fruit or provide shade, half of the above fine shall be imposed. The same fine shall be levied in the case of trees that have grown in places of pilgrimage, forests of hermits, or cremation or burial grounds. For similar offences committed in connection with trees which mark boundaries, or which are worshipped or observed or trees which are grown in the king's forests, double the above fines shall be levied.[9]

***The Forests***

Kauṭilya also prescribes that (Book-II) that Brāhmaṇas shall be provided with forests for Soma plantation, for religions learnings and for the performance of penance; such forests being granted with safety for animate or inanimate objects and named after the tribal names (*gotra*) of the Brahmaṇas residing there.

A forest as extensive as the above, provided with only one entrance, rendered inaccessible by construction of ditches, all round, with plantation of delicious fruit trees, bushes, bowers, and thronless trees, with an expansive lake of water, full of harmless animals and with tigers (*Vyālas*) beasts of prey, male and female elephants, young elephants and bisons—all deprived of their claws and teeth—shall be formed for the king's sports. Then there is a mention of the commodities produced by the forests and their usages in commercial terms. The text also provides for the planting of elephant forests in the extreme limits of the country.[10]

**List of Trees and Plants detailed in Chapter-27 of *Arthaśāstra* by Kauṭilya**

**(a) *Trees of Strong Timber***

(1) Śāka—(teak)

(2) Tinisa—(*Dalbergia ujjeinensis*)

(3) Dhanvan (?)—(*Alhagi maurorums*)

(4) Arjuna (*Terminalia arjuna*)

(5) Madhūka (*Bassia latifolia*)

(6) Tilaka (*Clerodendrum phlomoides*)

(7) Tala (*Borassus flabelliformis*, Linn.) (Palmyra)

(8) Simsapa (*Dalbergia sisu*)

(9) Arimedha (*Vachellia farnesiana*, Linn.)

(10) Rajadana (*Buchanania latifolia*)

(11) Śirīṣa (*Albizzia lebbeck*, Benth.)

(12) Khadira (*Abutilon catechu*,Willd.)

(13) Sarala (*Pinus roxburgii*)

(14) Talasaraja (Sāla tree or *Shorea robusta*)

(15) Aśvakarṇa (*Vatica robusta*)

(16) Somevalka (a Kind of white *Khadira*)

(17) Kasmara

(18) Priyaka (Yellow Sāla tree)

(19) Dhava (*Anogeissus latefolia*, Wall.)

**(b) *Group of Bamboo Trees***

(20) Utaja, a kind of bamboo, very hollow, of soft horms and of very hard lower part.

(21) Chimiya, is not hollow, but is of soft bark and has no thorns.

(22) Chav

(23) Veṇu is very hard and has very small central aparture

(24) Vaṁśa has thorns and the space between the two knots is long.

(25) Satina is smaller than Vaṁśa.

(26) Kanaka is very big and bears seeds like wheat.

(27) Bhalluka is very big, long and has no thorns.

**(c) *Creepers***

(28) Vetra (Cane)

(29) Śākavallī.

(30) Vase—this bears flowers like those of Arjuna tree.

(31) Śyāmalatā (*Ichnocarpus*)

(32) Nāgalatā (Betel).

**(d) *Fibrous Plants***

(33) Mālatī (*Jasminum grandiflorum*)

(34) Dūrva (panic grass)

(35) Arka (*Calotropis gigantea*, R. Br.)

(36) Sona (Hamp)

(37) Gavadhuka (*Coix barbata*)

(38) Atasi (*Linum usitatissimum*, Linn.).

**(e) *Trees Yielding Leaves* (patram)**

(39) Tali (*Corypha Taliera*)

(40) Tāla (Palmyra or *Borassus flabelliformis*, Linn.)

(41) Bhurja (birch).

**(f) *Flower Yielding Trees***

(42) Kiṁśuka (*Butea frondosa*)

(43) Kusumbha (*Carthamus tinctorius*, L.)

(44) Kumkuma (*Crocus sativus*)

## (ii) 2nd Century B.C

### (a) *Manusmṛti*

Manu has classified the plants, trees and creepers etc. into different categories, viz— Vanaspati, Auṣadhi, Vṛkṣa, Gucha, Gulma, Tṛṇa, Pratana and Valli—

उद्भिज्जा स्थावराबीजकाण्डप्ररोहिण:।
ओषध्य: पालपाकान्ता बहुपुष्प फलोपगा।।
अपुष्पा फलवन्तो ये ते वन्स्पतय: स्मृता:।
पुष्पिण: फलिनश्चैव वृक्षा: स्तूभयत: स्मृत:।।
गुच्छ गुल्म च विविधं तु तथैव तृण जातय:।
बीज काण्डरुहाण्येव प्रताना वल्ल्य एव च।। —(1.46-48)

(All plants, propagated by seed or by slips, grow from shoots; annual plants (are those) which bearing many flowers and fruits, perish after the ripening of their fruits. (Those trees) which bear fruit without flowers are called *Vanaspati* (lords of the forests), but those which bear both flowers and fruit are called *"Vṛkṣas"*. But the various plants with many stalks growing from one or several roots, the different kinds of grasses, the climbing plants or creepers, spring all from seeds or from slips).

Manu conceives for the first time that the plants have not only life in them but also possess internal consciousness—

तमसा बहुरूपेण वेष्टिता: कर्म हेतुना।
अन्त: संज्ञा भवन्त्येते सुख दु:ख समन्विता:।। —(1.49)

(These plants which are surrounded by multi form darkness, the result of their acts (in former existences), possess internal consciousness and experience pleasure and pain).

### (b) *Svapnavāsavadattā* by Bhāṣa

In the first Act of the play, the chamberlain informs the king that Padmāvatī was there and she had been to visit the queen-mother, Mahādevī, who had made her home in a hermitage, where she was pleased to stay for a day. He further desired that some holy water, fuel, flowers and sacred grass be brought for the use of the queen-mother.

Subsequently, a student observes the presence of a penance grove, where "the deer are quietly grazing, free from fear, in a place where they feel safe. All the trees tended with loving care, have their branches loaded with fruit and blossom. There is a great wealth of tiny kins, but no fields are tilled on any side. Undoubtedly it is a penance grove, for this smoke rises from an altar". Indeed this passage projects the picture of a hermitage in the forest.

In fact the scene of Act-II of *Vāsavadattā* is supposed to have been enacted in the palace garden at Magadha. There the princess is reported to be playing with ball near the Jasmine bower. The scene of Act-III is also enacted in the palace garden, with a difference that Vāsavadattā is deeply brooding over the past events, when a maid servant enters to disturb her. Act-IV also has direct relation with the palace garden. The same author describes the beauty of Vāsavadattā's palace in Act-IV, thus—

"What a magnificent gateway embellishes the palace of Vāsavadattā. The pavements have been sprayed with water, swept clean, and decorated with green plants. The threshold strewn with fragrant flowers of every variety, cast down as offerings. The summit of the gateway soars aloft as though it were fain to look into the very depths of the sky. Down the walls; hang long garlands of Jasmine that sway like the restless trunk of Indra's elephants."

## (iii) 4th-5th Century A.D. Works of Kālidāsa

### (a) *Abhijñānaśākuntalaṁ*

First of all the king Duṣyanta goes to the forest and reaches the hermitage of the sage Kaṇva. Reaching there he overhears the conversation betweens Anusūyā and Śakuntalā in which the former praises the plants of the hermitage—

हला सउन्दले। तुव त्तोबि तादकस्सवस्स
अस्समरुक्खआ पिअदरेसि तक्केमि। जेण
णोमालि आकुसुमपेलवाबि तुपं एदाणं
आलवाल पूर्णे णिउत्ता। —(Act-I)

("Dear Śakuntalā, I think father Kaṇva must love these hermitage shrubs even more than yourself. How else could he enjoin you, yourself as delicate as the fresh blown Jasmine flower, to water the flower beds").

Now Śakuntalā has been compared with the delicate flowers of *Malikā* or Mallikā. This is a kind of double Jasmine with large flowers, sometimes called 'Arabian jasmine', its delicious fragrance or perfume and abundant nectar attract the bees. By Alvala the flower beds are meant or the channel for watering the plant, excavated round the roots of the tree.

The charming and delicate figure of Śakuntalā has been compared with tender plants and blossoming flowers thus—

अस्या खल अधर: किसलय राग: कोमल विटपानुकारिणौ बाहु।
कुसुममिव लोमवीयं यौवनमङ्गेषु सन्नद्धम्।। —(Act. I, 21)

"Her lip is purple like the bird, her arms appear like tender shoots, charming youth is like a bloom attached upon her graceful form."

The attachment of Śakuntalā with the plants and trees of the hermitage of saga Kaṇva is highlighted in the fourth act of the play, when Śakuntalā intends to depart from the hermitage. She, while departing, expresses to her father that she must bid farewell to Jasmine, whom she had named as "the hight of the Grove." Approaching the creeper, she observes that though it was wedded to a mango tree, the creeper had extended the arms towards her.

In the *Abhijñāna-śākuntalaṁ* the height of poetic imagery could be witnessed when the poet compares the tender limbs of Śakuntalā with the beautiful plants thus—

अधर: किसलय राग: कोमल विरपानुकारिणौ बाहु।
कुसुममिव लोभनियं यौवनमङ्गेषु सनद्धम्।।
—*Abhijñānaśākuntalaṁ*, I.21

"Her arms are tender shoots; her lips are blossoms red and warm, Bewitching youth begins to flower in beauty on her form".

In fact the limbs of a women are, as it were, the sprays and flowers of creepers. On the other hand, his plants, trees and animals too are capable of human feelings and response. The mango tree and the Navamallikā creeper are a happy couple wedded to each other in love as per the imagination of Kālidāsa.

The young fawn refuses water from the hand of a stranger in the hermitage of sage Kaṇva, but clings to the apron of Śakuntalā at the time of her departure. She had tended him like an adopted child and healed his bruised mouth with *Ingudi* oil. She also had bestowed her motherly care on the plants and trees of the hermitage and was both to slake her thirst before having watered them first under her fostering care. They grew and bloomed. They two responded with equal ardour by sending their gifts for her nuptial adornment—

क्षौमं केनाचदिन्दुपाण्डु तरुणा गाङ्गल्यगाविष्कृत
निष्ठयूतश्चरणोपभोग सुलभो लाक्षारस: केनचित्।
अन्यभ्यो वनदेवता करतलैरापर्व भागो त्थितैर्
दत्तान्या भारणनि न: किरालयोद्मदे प्रति द्वन्दिभि:।।
—*Abhijñānaśākuntalaṁ*, IV.5

Śakuntalā's love for plants of hermitage of Kaṇva is well projected in the dialogue between Anusūyā and the former which is as follows—

Anusūyā— Dear Śakuntalā, I think father Kaṇva must love these hermitage shrubs even more than yourself. How else could he enjoin on you, yourself as delicate as the fresh blown Jasmine flower, to fill their trenches with water.

Śakuntalā— Dear Anusūyā, it is indeed, not only by my father's command I do this task. I feel as sisterly love for these plants."

**(b) *Kumārasaṁbhavaṁ***

Kālidāsa adored nature as also the flowers, which are part of the same natural beauty. He had a great fascination for the flowers visualised on the moonlit nights. His love for lotus flower was immensely illustrated in his works. For example, the tender form of Pārvatī is conceived to have been derived from the delicacy of the lotus flower thus—

क्लमं ययौ कन्दुक लीलयापि या तया मनीनां चरितं व्यगाह्यत।
ध्रुवं वपुः काञ्चन पद्मनिर्मितं मृदु प्रकृत्या च स सारमेव।।

—*Kumārasaṁbhavaṁ*, 5.19

"She would not spare her form, so fair and frail, if sterner penance could perchance prevail. Oft had sweet pastime wearied her, and yet Pain would she match in toil the anchorite. Sure the soft lotus at her birth had lent Dear Umā's form its gentle element; But gold, commingled with her being, gave. That will so strong, so beautifully brave."

The same poet in the same text describes the pine (Devadāru) trees thus—

भागीरथीनिर्झरसीकराणां बोढा गुहुः कम्पित देवदारुः।
युद्धायुरन्विष्टमृगैः किरातैरामेण्यते भिन्नाशखण्डितर्हः।।

—*Kumārasaṁbhavaṁ*, I.15

"With pearly dewdrops Gaṅgā loads the pole that waves the dark pines towering over the vale, and breathes in welcome freshness over the face, of wearied hunters when they quit the chase."

Indeed in the *Kumārasaṁbhavaṁ*, Kālidāsa has made the minutest observation of the floral wealth in India to which reference has been made about sixty times. He has described all the known varieties of lotus—an achievement which has been highly eulogised by many great authorities on aesthetics and literary criticism.

**(c) *Mālavikāgnimitra***

In Act-III of the *Mālavikāgnimitra*, the king with the clown enters the *Pramādavana*, which is infested with the fragrance of numerous flowers blooming there. Kālidāsa's flaming love for the country bordered on obsession and found expression in the exquisite description of rivers, cities, majestic trees like *Devadāru*, *Sarala* and *Sāla*.

### (d) *Meghadūtaṁ*

In verse seven of Purva Māgha, Kālidāsa mentions the location of a Śiva temple in a garden situated outside the city of Alakā—

बाह्योद्यानस्थितंहरशिरश्चन्द्रिका द्यौतहर्म्या।

(A place with its edifices bathed in moon which shines as an ornament on the head of Śiva, who resides in a garden outside the city).

In this verse two aspects are evident. The temple of Śiva had a grove, and secondly, it was located outside the city. It therefore appears to be a practice to construct temples of Śiva inside the groves and gardens. In verse II, mention is made of *Kandalī* flowers, which are traditionally supposed to burst into bloom at the sound of lowering monsoon clouds. An abundance of these flowers is a happy augury for the crops. Further the flocks of Swans are imagined to fly with the lotus stalks in their beaks—

कर्तु यच्च प्रभवति महीमुच्छिलीन्ध्रामवन्ध्यां
तच्छुत्वा ते श्रवणसुभगं गर्जितं मानसोत्का:।
आकैलासाद्बिसकिसलयच्छेदपाथेयवन्त:
सपत्स्यन्ते नभसि भवतो राजहंसा सहाया:।।

(That rumbling which makes the earth filled with Silindhra or *Kandalī* blooms and is capable of rendering her fertile, having heard that delightful rumbling noise, the Swans impatient for approaching Mānasa lake with shreds of lotus stalks in their beak for their journey will be your companions as far as Kailāśa).

Then there is a mention of village women well versed in agricultural lore, who know that the cloud is the dispenser of their wealth.

त्वय्यायत्त कृषिफलमिति भ्रूविकारनभिज्ञ:। —(Verse-16)

Besides there is frequent mention of flora and fauna in this work of the great poet.

The cloud's rumbling is an indication of his consent to accept the offer of the Yakṣa. The cloud seems to have been intrigued enough. The *Kandalī* flowers are traditionally supposed to burst into bloom at the sound of the lowering monsoon clouds. An abundance of these flowers is a happy augury for bumper crop. The clouds need not worry for the lonely journey. There are flocks of swans, impatient to migrate to the snowy regions at this time of the year. With succulunt stalks of lotus in their beaks, the swans will keep company of the clouds all the long and weary way to Kailāśa. The *Kandalī* flower is not merely comely to the eyes but is a symbol of well being, as a precursor of agricultural prosparity. In verse 18 of the *Meghadūtaṁ*, Kālidāsa invokes the sensual image of the Earth, which is beautiful, infested with thick forests. The rivers like the Gaṅgā, Yamunā and Sarasvatī are her necklaces; the sea is her blue dress, the Himālaya with all its greenery and snowy mountains is her crown. Verse 21 conceives that at the approach of the clouds the herds of deer will indicate in their passage through the forest the track left by the cloud by the fine rain. The soil has a distinctive smell. The stags sniffing with twitching noses this strong aroma and not content with the mere gratification of their sense of smell, their unfailing instinct will bade them on to wet patches around lakes where *Kandalī* blooms will offer them a rare treat to their palates. The first showers have nearly opened the buds of *Kadamba*. The harmony of green and brown, which the flowers made up is delightful to the eyes. In verse 27 of the text, a height of imagination is reached when the poet seeks the indulgence of the cloud to provide some shade offering a timely relief to the girls picking flowers. He further believes that the cloud on it way would shower a spray of fine rain on the delicate buds of wild Jasmine. There are other flowers too in that forest. The young women, out there to pick flowers, sweat profusely in the heat. While wiping this sweat, their hands inadvertently will touch the lotuses they have suspended from their ears. Thus the poet has been able to create a situation in which the flowers in the region dominate with the natural fragrance.

### (e) *Raghuvaṁśa*

Kālidāsa had a flaming love for the country bordering on obsession and found expression in the exquisite description of rivers, cities and the majestic trees like the Devadāru, Sarala and Sāla. In one of his most beautiful descriptions in the *Raghuvaṁśa* the poet gives expression to his innermost love for the preservation of plant kingdom—

अमुं पुरः पश्यसि देव दारुं पुत्रीकृतोऽसौ वृषभध्वजेन।
यो हेम कुम्भस्तन निसृतानां स्कान्दस्य मातु पयसां रसज्ञः।।

—*Raghuvaṁśa*, 2.36

"Yonder Devadāru tree that you seest has been adopted as a son by the bull-bannered god. It has the good fortune of having been nursed by Pārvatī, the mother of Skanda, from out of her own golden watering pots representing her nursing breasts as it were."

The Birch trees situated at the highest altitude always refresh the visitors as well as the warriors having broad chests and bull like shoulders—

भूर्जेषु मर्यरीभूताः कीचक ध्वनि हेतवः।
गङ्गाशीकरिणो मार्गे मरुस्तस्तं सिषेविरे।।

—*Raghuvaṁśa*, 4.73

"The soft breeze of the river Gaṅgā rustled on among the dry leaves of the birch trees, making the wild bamboo resound and charged with the particles of its waters, refreshed Raghu on the way."

The poet also become conservant with the Sandal trees so abundanlly available in the forests of Malaya. This too has been immortalised in fresh, vital and shining phrases. The reddish brown sacrificial cow has been compared with Lodhara tree—

सपाटलायां गवि तस्यिवांसं धनुर्धरः केसरिणं ददर्श।
अधित्यकायामिव धातुमय्यां लोध्रद्रुमं सानुमतः प्रफुल्लम्।।

—*Raghuvaṁśa*, 2.29

"The poet bow-man behold the lion sitting upon the rosy cow like the full blossomed Lodhara tree on the table land of a mountain full of red metallic substance".

He is manifestly a lover of flowers, the labyrinth of leaves of trees and of blossoms of every hue exulting in the floods of sun-shine in all the seasons. Among the creepers Atimuktalatā was a symbol of charm for Kālidāsa, while the creepers in general carried the eternal music of nature to the poet—

श्रुति सुखभ्रमरगीतयः कुसुम कोमलदन्तरुचो बभुः।
उपवनान्त लता पवनाहतैः किसलैः सलयैरिव पाणिभिः।।

—*Raghuvaṁśa*, 9.35

"With songs consisting of humming of bees charming to the ear, and with pleasing splendour of teeth in the form of flowers, the creepers of the planted forest looked beautiful with their sprouts shaken by the breeze, as if they were hands gesticulating in consonance with the song."

The mango tree carries a great message for Kālidāsa as could be evidenced from a passage from *Raghuvaṁśa*. In its sixth Sarga, Indumatī is desirous of prince Aja, but when she finds him, does not want to meet other princes present in the court in the same way as the swarm of bees does not long for any other tree when it finds the full blossomed mango tree—

तं प्राप्य सर्वावयवानवद्य व्यावर्ततान्योपगमात् कुमारी।
नहि प्रफुल्लं सरकारमेत्य वृक्षान्तरं काङ्क्षति षट्पदाली।।

—*Raghuvaṁśa*, 6.99

"At last the princess approached Aja who was faultless in all his limbs and gave up the idea of going to any other prince, as a swarm of bees does not long for any other tree when it reaches a fully bloosomed mango tree."

Kālidāsa had a special attraction for flowers and he particularly enjoyed the company of such flowers in moonlight. His love for lotus has been no less important. He conceived a golden lotus for the enjoyment of his Muse (*Kumārasaṁbhavaṁ*-5.19).

Kālidāsa's liking for the Aśoka and Campaka flowers is proverbial and in his works he has quite often brought out situations in which one has to admire the flowering fragrance of these trees.

### (f) *Vikramorvaśīyaṁ*

Like other works, the natural beauty is appreciated by the poet in his works. His young maidens have moon-like faces, creeper like arms, lotus like hands and feet and flower-like exuberance of youth pervading every limb of their bodies. The various epithets, describing the limbs of the heroines, are both significant and striking. While describing the beauty of his heroine in the aforesaid work, the poet says—

अस्याः सर्गविधौ प्रलापटिरभूच्चन्द्रो लु कातिप्रदः।
शृङ्गारैकरसः स्वयंभु मदनो पासो न पुष्पाकरः।
वेदाभ्यासजडः कथं नु विषयव्यवृत्त कौतूहलो।
निर्मातुं प्रभवेन्मनोहर मिदं रूपं पुराणो मुनिः।।

—*Vikramorvaśīyam*, I.10

"In the act of her creation was the Moon of lovely splendour that became the progenitor? Was it cupid him self who concerns himself solely with the sentiment of love? or was it the month of spring which has a weath of flowers. For how could the old sage, grown dull by the study of the *Vedas* and who is devoid of all the desires of worldly pleasures, has produced this beautiful form from that captivates his mind."

## (iv) 6th Century A.D.

### *Kirātārjunīyam* by Bhāravi

Bhāravi while relating Arjuna's penance for the procurement of the *Pāśupata* weapon describes the forests in considerable detail. The Pāṇḍava brothers have already been exiled and are residing in the *Dvaitavana*, a thick forest. Then Yudhiṣṭhira gets the news of Duryodhana's acquiring some deadly weapons. Yudhiṣṭhira then desires Arjuna to visit Indrakila hills for performing penance to obtain *Pāśupata* weapon. In Chapter-V of the Mahākāvya, a detailed description of the Himālayas and the forests decorating the same has been provided. The beauty of the Himālayan botanical scenery is available in the following verse—

पृथु कदम्बकराजितं ग्रथित्मालतमाल वनाकुलम्।
लघुतुषार तुषार जलश्च्युंत धृत सदान सदान नदन्तिनम।।

—*Kirātārjunīyam*, 5.9

## (v) 7th Century A.D

### (a) *Mālatīmādhava* by Bhavabhūti

In the Act-IX of the famous play *Mālatīmādhava* of Bhavabhūti, the beauty of various types of trees and plants has been described in the following piece of prose—

एताश्चन्दनाएवकर्णसरलपाटला प्रागः तरुगरनाः परिणत
मालूर सुरमयोऽरण्यगिरि भूमनाः स्मारान्ति तरुणकदम्ब
जम्बू वनावबद्धान्ध कारगुरुगिरि निकुञ्जगुञ्जद्रम्भीरगप्रदो
द्रारधोरघोषण गोदावरी मुखरित विशाल मेखलाभुवो दक्षिणारण्यमूधरान्।

—*Mālatīmādhava*, Act-3

### (b) *Uttararāmacaritam* by Bhavabhūti

In the first Act of the play, Rāma points out to the Ingudi tree in the city of Śṛṅgavera, where his meeting with Niṣādas took place. Then Sītā points to a mountain to Lakṣmaṇa, full of Kadamba trees, and Lord Rāma, seated under one of these Trees. Then there is that well-known Malyavān mountain fragrant with the flowers of *Kakulha* trees on the top of which a fresh cloud is said to be resting.

In Act-II, Śaṁbūka points out to Rāma, the extensive Daṇḍaka forests, which are on the borders of Janasthana stretching in Southern direction. "The caves of the mountains are full of herds of wild and fierce animals, which cause the hair of all to stand erect. The borders of the wood are in some places noiseless and motionless, in other resounding with terrible roars of wild beasts, having fires kindled in them by the breath of serpents with bread hoods that have fallen asleep at their will, having a little clear water in hollows of their crevices, and in which the moisture of the sweat of the python is drunk by thirsty chameleons. While describing the beauty of the majestic middle forests, Śaṁbūka tells Rāma, that these are "studded with mountains having the soft beauty of the necks of peacocks tuneful from joy, adorned with clumps of young trees offering dark thick shades and densely planted, and where the herds of deer of different kinds wander fearlessly. Here flow torrents, the numerous streams which are noisy as they struggle through the bowers of Jambū trees, dark with their ripened wealth of fruit, having water that is pellucid, cool and fragrant with flowers of the Vanira creepers covered with impassioned birds perching on them. Here the growls of young bears, dwelling in the caves, deepened by the echo, swell and the perfume is spread abroad, cool, pungent and fragrant issuing from the joints of the Sallaki trees split and scattered about by the elephants."

Thereafter Rāma describes the beauty of the forest in a most lively manner by saying, "How do I behold the same wood once more today, wherein, we formerly lived for a long time, both as hermits and householders, devoted to our proper duties, and also tasting the flowers of worldly pleasure. These are those very mountains, with peacocks uttering their notes on them, these are the very forest lawns with intoxicated deer, these are those same river banks, covered with beautiful Vanjula creepers, and having Nichula trees that are deeply merged in water. This is the Praśravana mountain, where the river Godāvarī flows by, and which appears as if it were near and is like a garland of clouds. On the great summit of this very mountain was the home of the Vulture-King; beneath it we possessed pleasure in those leafy huts, where, there is the delightful border of the wood in which coo the noisy birds and where the dark beauty of the trees is reflected in the water of Godavarī. Here, therefore, must be that Pañcavaṭī where there are spots which on account of our long sojourn there witnessed an excess of all kinds of confidential intercourse, and where a dear friend of my beloved, a sylvan deity named Vasantī, dwelt... where formerly Hare was a current of river, there is now sand of the rivers; the density and thinness of trees have been exchanged, seen after a long time, I almost think that this wood to be a different one; but the situation of the mountain assures me that it is the same..."

In Act-III, Sītā again is exiled and goes to the forest where she had enjoyed her early years in the company of her husband and Lakṣmaṇa, delighting in nature and sharing company of the hermits. The atmosphere in the hermitage of Vālmīki is therefore not new to her. But in despair at her desertion and in her extreme pain, she throws herself in the

Gaṅgā. Shortly, thereafter, she emerges supported by the river goddess Gaṅgā, bearing one son on her arm and the goddess Pṛthivī, carrying the other. The two infants are entrusted in the care of the sage. In a violent attack of grief, Rāma visits the same holy wood. The third act, one of the most remarkable in Sanskrit drama, depicts his experience of the forest.

**(c) *Nāgānanda* by Śrī Harṣa**

Harṣa in *Nāgānanda*, beautifully describes the pleasure gardens, and the use of bark of the trees for clothing of the sages in the hermitages—

वासोऽर्थं दययेव नातिपृथव: कृत्तास्तरूणां त्वचा
भग्वाऽऽलक्ष्य जरत्कमण्डलु नभ: स्वच्छं पयोनैर्झरम्।
दृश्यन्ते त्रुटितोज्झिताश्च बटुभिमौञ्ज्य: क्वचिन्मेखला
नित्याकर्णनया शुकेन च पदं साम्नामिदं पठ्यते।। —(I.II)

(For clothing, the bark of the trees are cut thin, as if out of pity, for them, the water of brook is clear as the sky, in which broken old water pots are visible, and in some places girdles made of Muñja-grass, torn up and then thrown away by the boys are seen, and due to his constant hearing, a parrot recites this hymn of the *Sāmaveda*).

The hero with the jester (in Act-I), then enter the pleasure garden and the former describes the forest beauty of the *āśrama*, thus—

"So come, let us enter and sea. Both men enter looking in wonderment. O how screne is the charm of the penance grove, where many doubtful passages of the Vedas are being discussed by the self-contented sages, where very wet and fresh fuel-woods are being lopped off by boys reciting *mantras*, and where the hermit girls are filling up the basins of young trees with water."

"These trees give out, as it were a melodious welcome by the humming sound of the bees, seem to make a bow with their heads bent low on account of fruits and give to me a present by scattering showers of the flowers—how have they learnt the service to guests even. So this penance-grove is fit for residence."

Thereafter both the persons hide themselves behind a Tamāla tree, when the princess enters the grove with her maids.

In the interlude of the second Act of the play, a grove with the bower of sandal creepers, surrounded by tender Banana leaves is mentioned. The hero and the heroine visit this grove one after the other. When the hero leaves the grove, the heroine gets suspicious of the hero's love for some one else, she out of frustration entwines a creeper around her neck by the side of an Aśoka tree. The maid servant raises an alarm and hero comes to the rescue of the princess and removes the creeper noose from her hand and observes—

न खलु न खलु मुग्धे साहसं कार्यमीदृक्
व्यपनय करमेतं पल्लवाभं लताया:।

कुषुममपि विचेतुं यो न मन्ये समर्थः
कलयति स कथं ते पाश मुद्वन्धनाय।। —(2.19)

(O Simple-minded one, not indeed, not indeed such violent act should be done. Remove from the creeper thy hand as tender as shoot. How can thy hand, that is not worth collecting flowers, catch hold of the noose for hanging?)

In Act-III of the play, the beauty of the flower garden has been described where "the droppings of the sandal trees cool the creeper bower with its tassallated pavements and the peacocks dance to the still sound of the shower baths. The cascade, brown with the pollen of flowers, shaken from the trees by the impetuous foam, falls with a rush from the machine and fills the basins at the foot of the trees. Again, these bees, making the creeper bower resound with their attempts at song drink in abundant honey, the company with their wifes, covered with a perfumed dust by the pollen of flowers on every drinking festival."

**(d) *Ratnāvalī* by Śrī Harṣa**

In this play, Śrī Harṣa, projects the portrait of Pārvatī, who is attending to the adoration of Śiva, with a bunch of flowers in her hands which she intends to offer over the Lord's head. But gets nervous on reaching the Lord and in this nervousness, the bunch of flowers, which she is carrying, drops down on the ground—

पादाग्रस्थितया मुहुँः स्तनभरेणानीतया नम्रतां
शम्भोः सस्पृहलोचनत्रयपथं यान्त्या तदाराधने।
हीमत्या शिंरसीहितः सुपुलकस्वेदोद्‌गमोत्कम्पया
विश्लिष्यन्कुसुमाञ्जिलि गिरिजया क्षिप्तोऽन्तरे पातु वः।। —(I.I.)

The main point to be noticed here is the presenting of a bunch of flowers by Pārvatī to her Lord. This bunch of flowers is the novel imagination of the poet. Again in the first Act of the play the scenic beauty of *Makarandodyāna* (garden of Makaranda flowers) which is full of fragrance of full blossomed flowers has been described.

In the first Act of the play, the two maid-servants inform the clown that the queen was coming to garden of flower-honey ( मकरन्दोद्यानम् ) to worship the God of Love erected at the base of a red Aśoka tree. The hero on the other hand describes the said garden thus

उद्यद्विद्रुमकान्तिभिः किसलयैस्ताम्रां बिभ्रतो
भृङ्गालीविरुतैः कलैरविशदव्याहारलीलाभृतः।
घूर्णन्तोमलयानिलाहतिचलैः शाखा समूहैर्मुहु
र्भ्रान्ति प्राप्य मधु प्रसङ्गमधुना मत्ता इवामी द्रुमाः।। —(Act. I-17)

(On account of the dedicate foliage lovely as rising coral trees, bearing the reddish lustre, possessing the semblence of indistinct speech because of the melodious humming of rows of large bees, and shaking with their boughs moving at the strokes of the Malaya wind, these trees, giddy, seem to be intoxicated with the spring wine.).

A verse in the same Act (I.18), points out that a mouth full of wine when sprinkled by a damsel at the root is scented by Bākula by a shower of blossoms, as it were, the Campakas shine today after a long time as the moon face of a damsel is slightly red on account of the wine and hearing the excessive sound of the anklets at the time of kicking the Aśoka trees by the feet, the swarms of large black bees begin to sing in the imitation of the jingling sound—

मूले गण्डूषसेकासव इव बकुलैर्वास्यते पुष्पवृष्ट्या
मध्वाताम्रे तरुण्या मुख शशिनि चिरा चम्पकान्यद्य भान्ति।
आकर्ण्याशोकपादाहतिषु च रसितं निर्भरं नूपराणां
झङ्कारसस्यानुगीतैरनुकरणमिवारभ्यते भृङ्ग सार्थैः।। —(Act I-18)

## (vi) 8th Century A.D.

### (a) *Veṇīsaṁhāra* by Bhavabhūti

This play provides a vivid picture of the botanical trees and plants. In Act-II, Bhānumatī is seated in the *Krīḍāvātika* of the temple and the maid servant moves on to inform to the queen's presence in the *Vāṭikā* (grove) to the king. In the same Act, there is a mention of an *Abhinavavāṭikā* in the harem besides a grove of Bākula (मौलश्री) grove which is filled with fragrance.

### (b) *Mudrārākṣasa* by Viśākhadatta

Viśākhadatta in Act-I of the play describes that scattered seeds grow profusely in fertile soil, even when sown by an ignorant person. The clustered paddy does not expect any merit in the sower (Verse-3)—

चीयते बालिशस्यापि सत्क्षेत्र पतिता कृषिः।
न शालेः स्तम्बकरिता वप्तुर्गुणमपेक्षते।।

## (vii) 10th Century A.D.

### (a) *Naladamayantī* or *Nalacampū* by Trivikrama Bhaṭṭa

In this play (Act-I, Sec. III) the king has been equated with a gardner and all his wide realm is his garden. He has his plants to lay therein and his hedges to trim. He has to lop all wanton growth and to tend to same saplings from Cankar, so that the flower and fruit may be whole. He has to expend to feed the roots and quench their thirst, because on him depends all the garden's weal.

### (b) *Navasahasrāṅka Carita Mahākāvya* by Parimal Padmagupta

It treats the winning of the snake king's daughter Śaśiprabhā by the king Sindhurāja Nava Sahasrāṅka. Sindhurāja kills Vajrankuśa, the enemy of the Nāgas and brings the golden lotus from Vajra's pond and marries Śaśiprabhā, fulfilling the condition for the marriage.

In the second canto, one day the king desires to go to the forest for hunting. While wandering here and there, he finds a deer of rainbow colour with golden chain around his neck. The king shoots the deer with an arrow inscribed with his name. The deer though hurt takes to flight. The king and his retinue rest for the night in the same forest. The third canto highlights the search for the dear by the king and his soldiers. The kings spots a royal swan, white like the lotus stalk flying towards him with a garland of gems in its peak. When the king tries to capture the swan, it drops the garland on the grass, jumps into the pond and disappears. The king picks up the garland and finds the name of Śaśiprabhā inscribed on the beads (4th Act). The king notices a beautiful woman under a Tamāla tree and king desired to know his whereabouts. She is found to be the maid servant of Śaśiprabhā. Thereafter upto eighteenth canto; the description of forests, the plant kingdom and the forest beauty is beautifully described. The imposition of the eyes of the trees upon the flowers having bees in them in the following verse, is a novel imagination—

अमुनि पुष्पाणि महीरुहाणाम् आभान्ति लीन भ्रमरोदराणि।
विद्याधरी विभ्रमदर्शनार्थम् उत्तानितानीय विलोचनानि।। —(XIV.35)

## (viii) 11th Century A.D.

### *Vikramāṅkadevacaritaṁ* by Bilhaṇa

The text contains a number of fruits, plants and flowers, some of which are detailed hereunder—

| | | | | | |
|---|---|---|---|---|---|
| (1) | Granthiparṇa | (I.17) | (2) | Kesara | (I.21) |
| (3) | Gunjaphala | (I.25) | (4) | Pankeruha | (I.32) |
| (5) | Bimbaphala | (I.35) (IV.17) | (6) | Sarojinī | (I.36) (I.84) |
| (7) | Pārijāta | (I.,39, 86) | (8) | Nagara Khanda | (I.64) |
| (9) | Pugadruma | (I.64) | (10) | Saptacchaḍa | (I.84) |
| (11) | Durvāṅkura | (II.23) | (12) | Nīlotpala | (II.27) (VIII.72) |
| (13) | Kramuka | (V.20) | (14) | Ketaka | (V.21), (XII.51) |
| (15) | Nārikela Phala | (V.22) | (16) | Kadalī | (II.22) (VII.15) (IX.17) (X.10) |
| (17) | Guladruma | (VII.46) | (18) | Aśoka | (VII.47, 48) (VIII.54) (X.24) |
| (19) | Palāśa | (VII.47, 52) | (20) | Bākula | (VIII.48) (X.25) |
| (21) | Toda | (VIII.71) (XI.33) | (22) | Moca (Śalmali) | (VII.71) |
| (23) | Kunda | (VIII.68) | (24) | *Vaṁśavṛkṣa* | (VIII.79) |
| (25) | Karṇikara | (IX.93) | (26) | Sevala | (IX.112) |

## (ix) 12th Century A.D.

### (a) *Rājataraṅgiṇī* by Kalhaṇa

The writer of this *kāvya*, was well versed not only about the plant kingdom, but knew fully well the system of irrigation. In order to keep the produce in pace with the people's needs, proper irrigational measures were available. Floods and famines afflicted the people quite often. The first irrgational canal is said to have been built by the king Suvarṇa, known as Suvarṇamaṇi Kutya (I.97)—सुवर्णमणि कुल्याया: कराले प्रवर्तक:। The king Dāmodara is said to have erected the *Gadelasetu* to bring waters to the plateau of Dāmodara Suda (I.156). Lalitāditya is credited with the erection of water-wheels (*aragahaṭṭa*) near the village of Cakradhārā, for irrigation facilities (IV.191). Avantīvarman's minister, is said to have embellished all the regions with proper irrigation facilities which lowered down the price of rice (V.116). This and other references from *Rājataraṅgiṇī* show that proper irrigation was the great necessity of the valley of Kashmir, which was fulfilled by the king from time to time.

The following fruits and trees have been mentioned in this text—

| | | | | | |
|---|---|---|---|---|---|
| (1) | Kacchaguccha | (I. 21) | (2) | Umkha | (II. 60) |
| (3) | Śiśira | (II. 83) | (4) | Mallikā | (II. 167) |
| (5) | Tamāla | (III. 326) | (6) | Tāḍī | (III. 326) |
| (7) | Bimba | (III. 416) | (8) | Vaṭa | (III. 430) |
| (9) | Dūrva | (IV. 18) | (10) | Ketaka | (IV. 113) |
| (11) | Nāriyala | (IV. 155) | (12) | Svarṇaketakī | (IV. 151) |
| (13) | Candana | (VI. 156) | (14) | Drākṣa | (IV. 192) |
| (15) | Kesara | (IV. 196) | (16) | Kapittha | (IV. 219) |
| (17) | Bākula | (IV. 450) | (18) | Utpalāsaka | (V. 49, 50) |
| (19) | Sevara | (V. 94) | (20) | Āmalaka | (VII. 526) |
| (21) | Aśvattha | (VIII. 1071) | (22) | Sanla | (VIII. 143) |
| (23) | Kadalī | (VIII. 843) | | | |

**(b) *Pṛthvirāja Vijaya* by Jayanaka**

This *kāvya* is said to have been composed by Jayanaka, to commemorate the victory of Pṛthivirāja over Sahabuddin Gauri which took place in A.D. 1191. The work mentions the following trees, plants and flowers—

| | | | | | |
|---|---|---|---|---|---|
| (1) | Rasona | (I. 23) | (2) | Darbhāṅkura | (IV. 12) |
| (3) | *Mṛṇāla* | (IV. 12) | (4) | Kiravini | (V. 19) |
| (5) | Sandala | (V. 33) | (6) | Kunda | (IX. 21) |
| (7) | Gostani | (X. 44) | | | |

**(c) *Dvyasrayamahākāvya* by Hema Candra Sūri**

The work deals with the history and achievements of the Cālukyas of Gujarat and also serves the purpose of teaching Sanskrit grammar. He was also well versed with the plant kingdom, since a large number of them are found mentioned in his work—

| | | | |
|---|---|---|---|
| (1) Lavalī | (XIII. 51, 55) | (2) Tāla | (XV. 65) |
| (3) Śāmi | (XV. 66) | (4) Mallikā | (XV. 78) |
| (5) Śirīśa | (XV. 79, XVI. 82) | (6) Badarī | (XV. 80) |
| (7) Plākṣa | (XV. 81) | (8) Aśvattha | (XV. 81) |
| (9) Kapittha | (XV. 83) | (10) Nada | (XV. 91) |
| (11) Kumuda | (XV. 91) | (12) Vetasā | (XV. 91) |
| (13) Sadavala | (XV. 91) | (14) Khandu | (XV. 96) |
| (15) Cūta | (XVI. 73) | (16) Bākula | (XVI. 75) |
| (17) Mādhavī | (XVI. 77) | (18) Kadalī | (XVII. 79) |
| (19) Ketaka | (XVI. 83) | (20) Jati | (XVI. 87) |
| (21) Kasa | (XVIII. 61) | (22) Alābu | (*Kumārapāla Carita, I. 46*) |
| (23) Nīlotpala | (*Kumārapāla Carita*, I. 56) | (24) Kiṁśuka or Palāśa | (*Kumārapāla Carita*, I. 67) |
| (25) *Madhūka* | (*Ibid.*, I. 75) | (26) Kusmāṇḍa | (*Ibid.*, I. 76) |
| (27) Karaṇikara | (*Ibid.*, II. 13) | (28) Śāka | (*Ibid.*, II. 16) |
| (29) Ullukha | (II. 30) | (30) Khadira | (II. 30) |
| (31) Pannaga | (II. 33) | (32) Nipa | (II. 54) |
| (33) Campaka | (II. 56) | (34) Bisini | (II. 56) |
| (35) Maruvaka | (II. 78) | (36) Kainkeli | (III. 4, 17) |
| (37) Sindhuvara | (III. 16) | (38) Tilaka | (III. 27, 62) |
| (39) Patali | (III. 31) | (40) Kurbaka | (III. 32) |
| (41) Vicakila | (III. 45) | (42) Japa | (III. 63) |
| (43) Hintala | (III. 63) | (44) Kachanāra | (IV. 3) |
| (45) Bandhujiva | (IV. 20) | (46) Kharjūra | (IV. 23) |
| (47) Priyala | (IV. 23) | (48) Rajadana | (IV. 23) |
| (49) Panāsa | (IV. 23) | (50) Śriphala | (V. 7) |
| (51) Dradimi | (V. 8) | (52) Kutaja | (V. 9) |
| (53) Arjuna | (V. 9) | (54) Tapiccha | (V. 9) |
| (55) Langalin | (V. 10) | (56) Bimba | (V. 10) |
| (57) Pārijāta | (V. 13) | (58) Cincini | (V. 14) |
| (59) Sāla | (V. 16) | (60) Nicula | (V. 20) |
| (61) Sallaki | (V. 20) | (62) Kalhara | (V. 22) |
| (63) Tulasika | (V. 25) | (64) Malura | (V. 37) |

**(d) *Gīta Govinda* by Jayadeva**

Jayadeva, the author of immortal Sanskrit poem *Gīta Govinda*, was the court poet of Lakṣmaṇasena (A.D. 1170) the last Hindu king of Bengal, *Gīta Govinda* or the song of the cowherd is a type of oriental opera in which the love of Rādhā and Kṛṣṇa is described. In brief the story of the poem is as follows—

After paying homage to the ten incarnations of Viṣṇu, Jayadeva describes how the love of Rādhā and Kṛṣṇa started on the bank of the Yamunā. Then follows the delightful description of the spring and he describes the sports of Kṛṣṇa with the cowherdesses. Rādhā, who is heart broken, having been ignored by Kṛṣṇa, retires to a bower and her maid describes how Kṛṣṇa wanders with pretty cowherd girls. After some time Kṛṣṇa feels penitent, searches for Rādhā, and on not finding her, laments for her. The female messenger now goes to Rādhā, and tells her of the longing of Kṛṣṇa for her. Ultimately she pursuades Rādhā to meet Kṛṣṇa in a bower. The above situation is displayed in the following lines of Canto-I of *Gita Govinda*—

क्षितिरति विपुल तरे तिष्ठति तव पृष्ठे
धरणि धारण किणचक्र गरिष्ठे।
केशव धत कच्छपरूप
जय जगदीश हरे।।

"Keśava, I bow to thee in Thy Tortoise form holding the great earth on Thy back which has become horny in consequence".

The scene of Kṛṣṇa playing with the cowherdesses in the forest has been beautifully depicted in a Basohli painting in which Gopīs are shown adoring Kṛṣṇa; one of them is reverently touching his feet, another is holding a white *chowrie* and yet another is bowing to him with folded hands. Kṛṣṇa is embracing two of the Gopīs. In the background is the charming forest of Vṛndāvana with many trees of different species. All these are artistically delineated.

**(e) *Abhilāṣitārtha Cintāmaṇi (Mānasollāsa)* by Someśvaradeva**

The fruit and flower culture is treated in considerable detail in the fourth *Prakarṇa* of *Abhilāṣitārtha Cintāmaṇi*, which is quite interesting. Though agriculture and plantation of fruits and other trees besides the flowers are mentioned by Kauṭilya in his *Arthaśāstra*, but there is no reason to believe that it is entirely Aryan in its origin. Some of the features of the botany as discussed in the above work are brought out hereunder—

**(i) Treatment of Seeds**—Ripe fruits without any blemish are selected and dried in the Sun, having coated them with cowdung each day for five days, one has to perfume and fumigate them with smoke caused by burning of Vidaṅga (*Erycibe paniculata*) and clarified butter. The seeds of all sorts of trees are thus treated before they are sown.

Another way of treatment consists of soaking the seeds for ten nights in cow's milk; and when they are dried in shade and mixed with ashes of barley, corn and wheat, they are given a coating of cowdung. This is a special treatment of seeds of milky trees.

The form of soil fit for fruit and flower trees is that which contains sweet water and is free from pebbles. It must be greasy and unexposed to frost. A luxirious growth of Tila (*Sesamum indicum*, DC), Māsa (*Phaseolus radiatus*) and Mudga (*Phaseolus Mungo*, L.) is a sure indication that the soil is fit for fruit cultivation.

Trees or fruit trees are to be planted leaving a space of 16, 11, 8 or 14 *hastas* (feet) according to the size of these trees, which could be of huge, medium and small sizes. The pit of the trees should be four feet deep and four feet square.

The treatment of pits consists in filling them with plenty of bones and cowdung or any other dung and burning the same. Then the ashes have to be removed and the pit has to be filled up with a layer of sand at the bottom and over it a layer of flesh and serum of goats with water should be made. If seeds are now sown in such pits, they will produce very good trees. Instead of seeds, young trees with roots could be brought and planted.

The trees that are to be planted in parks are—Aśoka (*Saraca asoca*, Roxb.), Nimba (Nimba tree), Punnaga (*Colophyllum inophyllum*), Bākula (*Manilkara hescandra*, Roxb.), Nāgakesara (*Mesuaferrea Roxburgii*), Śirīśa (*Albizzia lebbeck*, Benth) and Tilaka. These trees are conducive to health and happiness besides being auspicious. Trees such as Palāśa, Kachanāra, Chincha, Slesamataka should never be planted (in public parks).

**(ii) Watering of Plants**—In the snowy and dewy seasons trees are watered on alternate days. In the spring and summer, they are always watered in the morning. In the rainy and autumn seasons, they are watered only when the pits are found dry. When the water poured into the pit is not absorbed quickly, it is indicative of indigestion and as such no water should be poured further. Under such circumstances the pit should be dug up and the stagnant water should be made to evaporate, when the earth appears dry, it should be watered again.

Care should be taken to protect the trees from snow, heat, smoke, fire, insects and birds, weeds and creepers growing around the pits should be regularly uprooted.

The soil of the pit of a tree struck with lightning may be brought and put into the pits of the trees in the parks with a view to avoid the trouble of the snowball, cooked white rice mixed with curd and rock salt is also thrown into the pits with a view to attract hailstorm.

Perfuming the trees with the smoke caused by burning Nisa (*Turmeric curcuma*), Vidāṅga (*Erycibe paniculata*), Siddhārtha (white mustard), the flowers of Arjuna plant, mixed with fish and flesh of *rohita* (a kind of deer) is not conducive to abundance of flowers and fruits but destroys all the worms and insects besides removing the ailments of plants.

The perfume caused by the burning of Vidāṅga, Hingu (*asafactida*), Sindma, Marīca (pepper), Ativiṣa (?), Vaca (a kind of aromatic root), Bhallataka(ki) (*Semecarpus anacardium*, Linn.), and buffalo's horn mixed in euqal quantities will destroy all kinds of insects and worms.

**(iii) Watering of Plants**—For a year after the seeds are sown, trees are to be watered with a *kumbha* of water daily. Thereafter the number of *kumbhas* (Vessels) should be as many as the number of years upto twenty years. Thus the trees watered after the requirements of seasons also will produce fruits and flowers always.

Perfumes caused by burning of fish, clarified butter, white mustard and plantain trees will be conducive to the abundance of fruits.

The trees when watered with the decoction of Ankola (*Alangium salvifalium,* Linn.) mixed with clarified butter and honey and with the fat of *Kira* (parrot) and Kuranga (a kind of deer) will produce abundance of fruits in all seasons.

Milky trees watered with milk mixed with Vidāṅga (*Erycibe paniculata*), honey and clarified butter and perfumed with the smoke caused by burning of Nata (*Tabernaemontana*) and Kuṣṭha (*Costus speciosus*) will always produce fruits.

Creeping plants (?) watered with decoction of Kuṣṭha (*Costus speciosus*) mixed with milk and fat of cows and Agaru (*Aporusa agallocha,* Roxb.) will always bear fruit.

Trees scratched with Vṛścika (a thorny shrub *Vangueria spinosa*) or by animals with horns, perfumed with clarified butter of cow and water with the grease of Kira (parrot) and Mūṣaka (mouse) will always bear fruits.

Vine plants manured with the dung of cocks and the serum of horse flesh will bear abundant sweet fruit. Pomegranates manured with the grease and flesh of a deer, elephant, cat, Krodha (wild boar?) and Chapa(?) will bear abundant sweet fruits. Oranges manured with any flesh (*palala*) mixed with cow's milk and jaggery will bear plenty of fruits.

Mango trees watered with the decoction of Ankola (*Alangium salvifalium,* Linn.) mixed with honey and fat of a wild boar will always bear fruits. Jaka trees be smeared with paste made of mud, mire, clay and lotus plants, will bear fruits abundantly. Kapitha and Śridruma trees will have abundant fruits if they are manured with honey, clarified butter, jaggery and milk mixed together.

The Āmalaka tree (*Emblica officinalis,* Gaertn.) will produce abundant sweet fruits, if before the appearance of flowers if it is scratched by Kumara(?) and the scratched surface is besmeared with the paste made of the powder of Japa (China rose), Tila (*Sesamum indicum*) and an equal quantity of honey and clarified butter. The mixture thus prepared will have the appearance of mire.

The plantain trees scratched with Vahni (the thorn of *Grisba tomentosa*), at the place from which its fruits are to come out will bear innumerable fruits if the scratched place is besmeared with the paste made of the dung of a boar or a horse.

The coconut tree will bear sweet big fruits, when it is besmeared with the paste made of Vidāṅga (*Erycibe paniculata*), Māsa (*Phaseolus radiatus*), Madhūka (Liquorice root), Surabhi and Jati (?) mixed with *Madirā* (liquor) and salt.

The Jujube tree will bear fruit at once when its roots are besmeared with the paste made of the powder in equal quantities of Sesamum and liquorice-steak mixed with honey and water.

The plantain tree will bear big fruits when its root is scratched with a gold needle heated in fire caused by burning of the powder(?). The plantain tree even in miry soil will bear fruits as big and long as the trunk of an elephant, when the paste made of the powder of an ivory piece and elephant's tooth is thrown at its roots and the whole is well covered with fresh soil.

The Bijapura tree will produce big fruits when it is manured with oil cake, liquor (?), fish (?) and water.

The Tamarịnd tree will produce sweet fruits when it is watered with the decoction made of Jambūdala (the leaves of rose-apple), Usira (*Andropogon muricatus*) and Muṣṭha (*Cyperus rotundus*) or with the decoction made of milk and flesh and serum of deer, parrot, antilope, jackal and water.

The Aśoka tree will bear tender leaves soon after it is kicked by a maiden with her leg wearing *nūpura*—a bracelet. The Tilaka tree will bear flowers when it is beaten with cane.

The Śyāma creeper will have flowers when it is married to a tree in the vicinity. The Mallikā creeper will bear flowers when it is burnt with fire caused by burning chaff (*palala*). The Japa plant will bear white flowers when its roots are watered with milk. The Karnikara, Japa, Aśoka, Bana and Amritaka plants will bear white flowers when they are similarly treated. Plants producing flowers having no smell will bear fragrant flowers where they are manured with the soil taken from the pits of plants bearing fragrant flowers and are watered with the water mixed with the powder of Gandhapatra (sandal tree leaves), Muṣṭha (*Cyperus rotundus*), Tagara (*Tubernaemontana cornata*, Willd.) and Usira (*Andropogon muricatus*). Plants which are besmeared with honey and clarified butter, when young and wound round with a thread, will grow singly and produce red bunches of flowers.

Creepers besmeared with honey will produce red flowers, and when besmeared with clarified butter will produce white flowers.

The Kumuda and Ambhuja varieties of the lotus plant will bear two flowers on one twig when they are besmeared with honey and clarified butter and manured with the powder of *Kalaya* and mire.

The Campaka tree will bear flowers when it is watered with milk. The Madhūka (*Bassia latifolia*) will bear flowers when it is watered with the water mixed with the powder of Yaṣṭi (Liquorice stick), Kulutha (horse gram) and honey.

Seeds kept in the oil of Nṛsamsa (?) and Ankola for seven days and sown in a soil wetted with the thunder stone will sprout at once. The pomegranate seeds wetted 21 times with the blood of Kukura (a kind of fish) will at once sprout and bear fruit.

Any seed wetted with milk mixed with flesh and serum of *matsya* (fish) and Kroda (boar) will quickly sprout and bear fruits.

A single tamarind seed kept for seven days in milk mixed with the serum of fish and boar and with sesamum oil, and besmeared with the ashes of *bṛhati* and *tilakand* will, when sown, sprout at once and bear fruits. The fruits will be quite big when the seed is soaked with water mixed with human flesh and serum, and the powder of ivory.

The plantain tree will bear a pomegranate fruit when it is watered with the flesh and serum of Kroda (boar) and decoction of Ankola.

A small ball made of the flower of Yaṣṭimadhūka (*Bassia latifolia*), the white variety of Kuṣṭha (*Costus speciosus*) and honey will, if placed into the root of a tree, cause it to bear fruit quickly.

When the root of a mango tree is manured with the decoction of Ankola, the flesh of a dog and the milk of a goat mixed with oil cake, the tree will take the form of vine and will always bear fruits.

When the branch of a fruit bearing tree is wound round with cloth in seven rounds and at the joint of the trunk and the branch, it is also covered with the skin of a deer and an elephant then the tree will have its fruit evergreen.

When at the joint of the trunk and the branch, a tree is well covered with the skin of a goat, it will have its fruits evergreen.

When the seed of Vartaka(ki) (Egg Plant, *Solanum melongana*) mixed with honey and clarified butter is kept inside a *Kusmāndā* (a pumpkin gourd) for some time, and then sown, it will have the leaves of the pumpkin gourd bearing however, Vartaka fruits.

## Notes and References

1. *Arthaśāstra* by Kauṭilya, 2.24.1.
2. *Ibid.*, 2.24.3.
3. *Ibid.*, 2.24.4.
4. *Ibid.*, 2.24.5.
5. *Ibid.*, 2.24.14.
6. *Ibid.*, 2.24.15.
7. *Ibid.*, 2.24.17.
8. *Ibid.*, 2.24.18.
9. *Ibid.*, 3.19
10. *Ibid.*, 2.2.

# 3
# Trees and Plants Kingdom in Buddhism

## 1. Literature

### (i) *Lalita-Vistāra*

Though there are innumerable references to the Trees and Plant kingdom in the Buddhist art and literatue, still to start with the subject one has to refer to the birth of the Śākya prince in the Lumbinī garden under a Sāla tree. There are principally three versions of this legend, the Pāli version describing the tree as Sāla and the *Lalita-vistāra*[1] and *Mahāvāstu*[2] versions as Plākṣa. According to the *Mahāvāstu* version, Queen Māyā stood supporting her arm on a branch of Plākṣa, while *Lalita-vistāra* version tells us that only a branch of the noble Sāla lowered its top to salute her. It is the Pāli version which says that only a branch of the noble Sāla-tree bent down so low as to be within her easy reach. But for the difference of tree which is taken to be Plākṣa, there appears hardly any difference between the two versions. According to the Pāli legend, when Queen Mahāmāyā's time of confinement drew high, she desired to see her people in Devadaha. On the way she wished to have a walk in the delightful forest-grove of Lumbinī, where all the trees bore flowers out of season. On reaching the foot of an imposing Sāla tree, she playfully tried to catch hold of one of its branches and it bent itself, to the amazement of all, so far down as to be well within the grasp of her hand. Hardly had she held it than her pain began.

The childhood of the Buddha is also related to the Śākyan plough festival. According to *Lalita-vistāra*[3], five *ṛṣis* when flying above the hallowed spot where the prince Siddhārtha then only a young boy was lost in ecstasy under a rose-apple tree during the Śākyan festival, annually celebrated in connection with a ploughing match. The other that may be suggested in *Paṇiya-Jātaka*[4] giving a description of the flight of the five Pratyeka Buddhas from their distant Himālayan abode towards the city of Vārāṇasī and return. The five *ṛṣis* according to the *Lalita-vistāra* story, are typically the ascetics. On enquiry they came to know that their power of miraculous force was counteracted by the overpowering majesty of the Bodhisattva meditating in the forest grove beneath. The Bodhisattva then appeared to the ascetics, a luminary shining light emitted from his body, and to Śuddhodana, his father, as a fire issuing from a mountain peak and glorious like a lamp.

After the great Renunciation the prince enters the forest and gets seated under the Bodhi-tree for meditation. He attains his supreme knowledge under the Bodhi tree.

### (ii) *Mahāvāstu*

The Buddha is also associated with the Jetavana which was purchased from the prince Jeta and later converted into a Buddhist monastic residence with all new additions of cottages made by Anathapindika and Prince Jeta. According to the Ceylonese version of the story, in agreeing to sell the garden, the prince made a bargain with the banker that he must pay him as many gold pieces as would be required to cover the whole site. Accepting this bargain, the banker ordered his men to cut down the trees and level the ground. Meanwhile the prince changed his mind and was not willing to sell his garden. As the banker would not give up his right, the matter was referred to the minister incharge of justice and the case was decided against the prince. The Ceylonese version further testifies that the prince and the banker went together to the garden, and saw that all the useful but the mango and sandal were cut down and the whole place was made perfectly level. The banker commanded his treasurer to bring out from his stores of wealth as many gold pieces as would be necessary. He emptyed seven stores in sending eighteen crores, which sufficed to cover the whole of the site minus the entrance. According to the same source, the banker built a seven storied building at each one of the four corners of the garden.

## (iii) *Saddharma Puṇḍarīka*

In Chapter-5 of this text[5], the botanical species have been mentioned as under—

"In a case, Kaśyapa, similar to a great cloud big with rain, coming up in this wide universe over all grasses, shrubs, herbs, trees of various species and kinds, families of plants of different names growing on earth, on hills, or in mountain caves, a cloud covering the wide universe to pour down its rain every where and at the same time. Then Kaśyapa, the grasses, the shrubs, herbs and wild trees in this universe, such as have young and tender stalks, twigs, leaves and foliage, and such as have middle sized stalks, twigs, leaves and foliage, and as such as have the same fully developed, all those grasses, shrubs[6], herbs and wild trees, similar and greater (other) trees, will each, according to its faculty and power, suck the humid element from the water emitted by that great cloud, and by that water, which all of one essence, has been aubndantly poured down by the cloud, they will each according to its germs, acquire a regular development, growth, shooting up and bigness; and so they will produce blossoms and fruits, and will receive, each severally, their names. Rooted in one and the same soil, all those families of plants and germs are drenched and vivified by water of one essence throughout. In the same way, Kaśyapa, does the Tathāgata, the Arhat, and others appear in the world."

## (iv) *Bodhicaryāvatāra*

This is a Buddhist[6] Sanskrit work by Śāntideva (7th Century A.D.) composed in the form of a *kāvya*. It deals with the ideal of attaining Buddha-hood as opposed to the narrow Hīnayāna ideal of Sainthood. It describes eloquently the botanical properties in some of its verses, a few of which have been brought out here—

यावन्ति पुष्पाणि फलानि चैव भैषज्यजातानि च यानि सन्ति।
रत्नानि यावन्ति च सन्ति लोके जलानि च स्वच्छमनोरमाणि।।

महीधरारत्नमयस्तथान्ये वन प्रदेशाश्च विवेकरम्याः।
लता सुपुष्पा भरणोज्ज्वलाश्च द्रुमाश्च ये सत्फल नम्र शाखा।।
देवादि लोकेषु च गन्धधूपाः कल्पद्रुमा रत्नमयाश्च वृक्षाः।
सरांसि चाम्भोरुहभूषणानि हंसस्वनात्यन्तमनोहराणि।।
अकृष्टजातानि च शस्य जातान्यन्यानि वा पूज्य विभूषणानि।
आकाशधातुप्रसरावधीनि सर्वान्य पीमान्य परिग्रहानि।।

—*Bodhicaryāvatāra*, II. 2.5

(I offer whatever kinds of flowers and fruit there are, all kinds of medicines or medicinal herbs, whatever types of jewels exist in this world, a clean limpid waters; so also mountains abounding in jewels, forest groves beautiful and quiet, creepers resplendent with ornaments of beautiful flowers, and trees bent with the weight of fruits-laden branches; fragrance and incense from the celestial realms of gods, wish fulfilling trees, ponds with lotus ornaments bedecked and the enchanting, sweet-warbling music of swans; harvest sprung without the ploughshare and those properly cultivated, all other sacred establishment, all things known and unknown and as vast as the limitless sky.)

## (v) *Milindapanho*

This is a Buddhist[7] non-canonical work; the original work, which was probably in Sanskrit or some north-Indian Prakrit and was composed about the beginning of the Christian era in north-west India. It is currently lost and the present one is only the Pāli translation of the original made at a very early date in Śrīlaṅkā. It contains in the style of Sutta a dialogue between the king Milinda or Menandar (1st century B.C.) and Nāgasena about the most important problems of Buddhism. In Book-5 of this work, while describing the Bazar of perfumes of the Blessed One, it is ordained that—

*"No flower's scent can go against the wind,*
*Nor sandal wood's, nor musk's nor jasmine flower's;*
*But the sweet perfume of the good doth go*
*Against the wind and good man pervades,*
*On every side, the sweetness of his life."*

*"Red sandal-wood, musk, and the lotus and jasmine—*
*The perfume of goodness surpasseth them all.*
*Abundant the sweet scent of musk and of sandal wood—*
*Still stronger the scent of the good mounts to heaven."*

Verse 8 of the same book, while discussing about the fruit Bāzār of the Buddha says—

"Certain fruits have been made known. And they are these—

The fruit of the first stage of the Excellent way (Conversion) and of the second stage and other third stage and of the fourth (Arhatship)—the fruit of attainment of emptiness—the fruit of attainment of the absence of three signs (of an unconverted life, lust, malice and dullness) and truth of attainment of that stage in which no low aspirations survive. And whosoever desires any one of these, he gives his *karma* as the price and buys the fruit he longs for."

"Just O king, as any man who has a mango tree bearing fruit, all the year round, he does not knock down the fruits until a buyer comes. But when a buyer has come, and the fruit grower has taken the price, then he says, 'Come my good man, this tree is always in bearing (it has therefore fruits in all stages of growth), take the kind of fruit you prefer, whether unripe or decayed, hairy, sour or ripe'. And the buyer for the price takes the kind he likes the best (II.9)."

## (vi) *Prākṛta-Dhammapada*

In Chapter-18 entitled[8] 'Puspa Vagga' the flowers and plants are used as simlies with human behaviour in many ways, *viz.*—

यथ बि रुयिद पुशु वणमद अगन अ।
एमु सुभषिद वय अफल (भोदि) अकुवदु।।
यथ विरुचिद पुशु वणमद सगन अ।
एमु सुधविद वय सफल भोदि कुवदु।। —(Verses—290-291)

(Like a beautiful flower full of colour but without fragrance, are the fair words which bear no fruit in action. Like a beautiful flower, full of colour and full of fragrance, are the fair words, full of action).

यथ वि भपरु पुष्प वर्णगण अहेडई।
परिदि रसम अदइ एमु गमि मुणि चर।।
यद वि पुष्प रशिस, कय-मल-गुण बहो।
एमु जदेण मचेण कतिव (कुशल बहो।।
पुषणि येव पयिणदु वसित मनस नर।
सुतु गमु मरोरो व (अदय मचु गछदि।। —(Verses—292-94)

(Even as a bee without injuring the colour or fragrance of the flower, takes the nectar and departs from the flower, so let a sage dwell in a village without doing any harm to the village. As many garlands can be made from a pile of flowers, so many good moral works should be achieved by a mortal once he takes birth. As a flood carries off a sleeping village, so death carries off a man, who is gathering flowers whose mind is distracted).

The text further goes on to say that the fragrance of flower never goes against the wind, be it the scent of Tagara or Mallikā or even of the sandal wood tree. But the fragrance of holy saints goes even against the wind. The holy men pervaded in all directions. The fragrance of virtue is unsurpassed, even among the perfumes of sandalwood, lily flower or jasmine.

### (vii) *FO-SHO-HING-TSAN-KING* (Life of the Buddha)

The text[9] was originally composed by Aśvaghoṣa, and translated from Sanskrit into Chinese by Dharmarakṣa in A.D. 420. In Chapter-I of the work, queen Māyā commanded her kinsfolk to repair to the garden shade, realising that the time of her child bearing had come.

Then there is a mention of the prince's proceeding to the gardens from the city, where the road was well prepared, smooth and wide, the trees were bright with flowers and fruit (I.5.329). The text further testifies to the agricultural activity thus—

(i) Now by the road side as he beheld the plough men, plodding along the furrows, and the writhing worms, his heart again was moved with pitious feelings (I.5.330).

(ii) To see those labourers at their toil struggling with painful work, their bodies bent, their hair dishevelled, the dripping sweat upon their faces, their persons fouled with mud and dust (I.5.331).

(iii) The plough oxen, too, bent by the yokes, their wotting tongues and gaping mouths (I.5.332).

In Book-4 (varga-30) the Buddha's visit to Jetavana-vihara, is described which had fountains, streams laid throughout the garden, which glittered with flowers and fruit trees and rare birds (IV.20.1613). Then there is the mention of scattering flowers (IV.20.1615).

## 2. Sculptural Art

### (i) Bharhut

In the Bharhut sculpture there is an astounding admixture of the plant kingdom with a variety of symbolism. As for example, we find the seated elephant from whose mouth proceeds the creeper or floral scroll with its serpentine course symbolising *Pranidhana* or the firmness with which an individual becoming a Bodhisattva forms the resolution, while the creeper itself represents the career of the individual as Bodhisattva. The *Jātaka* scenes and the fruit, flower and other ornamental compositions alternating with each other are not without significance of their own. The former represent the nature of the effort made, and the latter, the nature of the fruition obtained. In other words, the Elephant and the scroll with these *Jātaka* scenes and ornamental composition in the coily folds of the creeper symbolises the earlier process of the evolution of the Buddha-personality through the practice of three *Caryas* and the development of ten perfectionary virtues (*paramis, parmitas*), *dana* (liberty), *Sila* (moral restitude), and the rest. The Bo-tree Aśvattha and the Vajrāsana, introduced in some of the ornamental compositions, symbolise the attainment of Buddhahood, while the Tri-ratna symbol figures as an emblem of the Buddhist faith. The efforts made by the fruitions obtained may be classified according to their similarities. But even where they are similar, they are not identical.

## (ii) Lotus Symbolism in Bharhut Sculptures

The Lotus-symbolism[10] is an all pervasive element. It plays its role in the creeper-work on the coping; it has its impress also in the medallions and half medallions. Each medallion is theoretically a lotus medallion. If the details are absent for want of space, the outline remains to indicate what the medallions, full or half, are. So the symbolism to be discussed and explained here is only a special symbolism within the general symbolism of the Lotus. The medallions and half medallions in which the lotus is present and those in which the lotus is comparatively having lesser importance, have to be kept in mind while dealing with the subject—

(a) In one of the lotus medallions[11], the Vedic Agni is represented on a lotus blossom (Plate-1). There is no human figuration of the Āryan god as yet. The flames only are represented by the leaves and shoots. Even there is no chariot, though the idea of a chariot is suggested by the two winged steed attached to the shooting plant on its two sides. A duplicate of the same is also available in a half medallion.[12]

(b) In two other medallions[13], the tree or the tree of life is represented by an imaginary lotus-plant. The bulbous root figures as an ornamented pot out of which the plant grows up. The presence of water and its level are indicated by a pair of geese or swans that rest on a lotus in one instance and by two pairs of them in the other. In each design there are three lotuses of the same kind on each side, one of which blooms below the water-level, one rises to the surface and the third remains above water. The explanation of this particular symbolism lies in a famous utterance of the Buddha, which is met with in the Pāli text, *Majjhima-Nikaya* and *Vinaya-Mahāvagga*. In both, the Buddha is represented as saying, "Just as among the *uppalas* or the *podumas* or the *puṇḍarīkas*, some are born in the water, grow in the water and remain submerged in the water, being reared under the water; some, though born in the water, etc. remain on a par with the water; and some, though born in the water, etc., remain untouched by the water, rising up above the water, even so, O Bhikhhus, when I surveyed the world, with the all seeing eye of a Buddha, I saw among the beings, some with little defilement, and some with much, some possessed of astute faculty; and some of moderate, some well-formed and some deformed, some intelligent and some un-intelligent, and some of them acting with a clear perception of the fear of life hereafter and of sin.[14]"

In the first medallion, the geese remain face to face as if to signify the joy of a happy pair in nature, while in the other, the geese of each of the two pairs are busy eating the lotus stalk from opposite directions. Kālidāsa, in giving a practical description of the movement of the swan towards the Mt. Kailāśa in his *Meghadūta*, says that before they begin their flight through the sky, they gather lotus stalks as a provision of vegetable food during their long journey.

(c) In one medallion[15], the bees are seen collecting honey from lotus blossoms, one from each flower, while a devotee remains with folded hands. This may be taken to signify the ideal of a saint as inculcated in the following passage from the *Dhammapada* (No. 49)—

*Yathapi bhamaro puppham Vannagandham ahethayam paleti rasam*
*adaya evam gami muni Chare.*

(Just as the bee having collected honey, flies away, without causing any harm to the colour and fragrance of the flower, even so should an ascetic move about in a locality).

(d) In one medallion, the lions,[16] in one elephants, in a third tortoise perambulate in the outer zone of the lotus without touching upon one another. These may be taken to symbolise a pure state of joy, when the beings of one mind and form are busy performing an act of merit.

(e) In some of the medallions, the Sun-god[17] figures in human bust, the rays being represented by the stamens of lotus. In some a goddess, either holding a mirror or a lotus in the left hand. The descriptive name of the goddess as suggested in the *Mahāvaṁśa*, is the mirror holding deity. Having regard to the fact that the mirror symbolises "a thing that reflects other things", and there are stamens to represent rays, there can hardly be any doubt that the divinity is the goddess of the moon (Figs. 1-4).

(f) In one medallion, the Vedic goddess[18] Śrī remains seated cross-legged on a lotus with folded hands, while two elephants standing on two lotuses on two sides are pouring water on her head from opposite sides from a pot held up by the trunk of each (Plate-2). In the other medallion the goddess remains[19] in the standing posture placing her right hand on her breast. There is another representation of the goddess in a coping panel. In the Bodhagayā representation of the goddess[20], the elephants are conspicuous by their absence. The iconography of the Bharhut representation of Srī is fully explained by the description of the goddess in the Śrīsukta of the *Ṛgveda*.

(g) The Bo-tree (Aśvattha) on the architrave of the gateways at Bharhut, symbolises primarily an incident in the life of Buddha Śākyamuni, namely the attainment of Buddhahood. Thus the Bo-tree was subsequently taken not only for the Buddha, but also to represent the entire Buddhist faith and was adored by humans, animals and birds alike.

## (iii) The Bodhi Trees of the Seven Buddhas

The most important aspect of Buddhism has been that the life of the Buddha and even the earlier life of the Buddha, as portrayed in the *Jātaka* stories and other texts are closely connected with the plant kingdom. As for example, the birth of Buddha, Śākyamuni, took place under a Śāla tree in Lumbinī garden. According to *Mahāvāstu*, Queen Māyā stood supporting her arm on a branch of Plakṣa tree, while *Lalitavistāra* tells us that the Śāla tree lowered its top to salute her. In the Pāli version it is only a branch of the noble Śāla tree which bent down to be in her easy reach at the time of the birth of the Buddha. The Śākya prince renounced the world, retiring into the forest, followed by attainment of his enlightenment under the Bodhi tree. This tree according to the Buddhist tradition has been variously interpreted. The Buddhist texts describe seven Buddhas and of these, Vipascit Buddha's Bodhi-Tree is Patali (*Bignonia suavealens* or the famous Trumpet-flower). This is also called *Bhagavato-Vipasino Bodhi* or "The Bodhi-tree of the Divine Master of Vipascit."[21] This has been depicted in a medallion in the Stūpa at Bharhut. According to Cunningham[22], there is a full front view of the tree in flower preventing its identification from its representation. The throne of the Buddha at the foot and in front of the tree is an elevated platform of solid structure, apparently a cubical jewel seat, bearing some auspicious leaf and flower marks, the leaves and flowers appearing to be no other than those of a Bodhi-tree. The trunk and the upper part of the tree are seemingly of the same height. The sculptor has delightfully represented its wealth of flowers blossoming in bunches, distributed harmoniously over the balloon shaped foliage. The scene presents on two opposite sides of the tree and throne, two groups of worshippers, one group of five on the right side, and the other one of six on the left. All the worshippers appear to be male deities, one of whom, on each side, is bowing down, one is offering a garland, two on the right side are making demonstrations of the wealth of flowers of the tree, while the remaining worshippers are in a standing attitude of devotion. Possibly these worshippers are performing the act of circumambulation, making offerings, bowing down by turns, one after another, holding garlands of the bunches of flowers. In so far as the representation of Bodhi-tree is concerned, it has reference to the subject matter of such a discourse as the *Mahāpadana*, while the act of devotion cannot be explained by any other canonical discourse than the *Atanatiya.*

Compared with Buddhaghoṣa's[23] description, the Bharhut picture of the Bodhi-tree is rather a simple representation. According to his description, the trunk of Patali, the Bodhi-tree of Vipascit, rose aloft to a height of fifty *ratnas* (cubits), while its branches towered fifty *ratnas* high, giving the tree a total height of a hundred *ratnas.* it was uniformly decked with bunches of flowers from its bottom to its top, looking like a dome shaped ear-ring. All the Patali trees in the ten thousand World-systems were at the same time adorned with flowers, all the trees and plants shared this glorious fortune. The earth and oceans were covered with varieties of lotuses of all colours. Here and ther one might sea sights of celestial groves, the *Citralatā*, the *Nandana*[24] and the rest.

The Bodhi-trees of all the Buddhas were, according to Buddhaghoṣa's description, of equal height. The Bodhi-tree of Sikhi was Pundarvka (*Mangifera indica* or the White Mango). It was adorned with fragrant flowers. It was rich in its wealth of fruits. Its fruit in the same stage of development hung together from the same stalk. The ripe fruits were all juicy and delicious, as if celestial delicacy[25] was put into them.

The representation[26] of the scene of the Bodhi-tree of Buddha Viśvabhṛt is similar to that of the Bodhi-tree of Vipascit, with the difference that it represents more elaborate ornamentation and similar group of worshippers, just a pair of male and female deities.

The bunches of flowers and fruits of tree are abundant and many garlands are hanging from different branches and joints. The pair of worshippers is so represented that on each side, if the male worshipper is standing up, the female worshipper is bowing down, i.e., kneeling besides the Bodhimanda, and vice-versa. The offering made by the male worshipper is a basket of flowers and that by the female worshipper is a long piece of garland. Here we find a double artistic device for representing kinds of offerings made and act of adoration by turns. The act of circumambulation is an essential part of *Bodhi-vandanā*. The upper part of the trunk is adorned with ornamental *Dharmacakra* and *Triratna* symbols, both of which are crowned by umbrellas with hanging garlands. There is in front the throne of the Buddha, which is a solid structure, similar to that of Vipascit, and is encircled with impressions of the fruit and garland offerings placed on it. The frieze on the front side of the throne contains a double chain-work design with three parallel rows of lotuses, set up or suspended like bells.[27]

The Bodhi tree of Buddha Kakutsandha is, according to tradition, Śirīśa or Acacia. Like the Puṇḍarīka and the Śāla, the Śirīśa abounded, as Buddhaghoṣa[28] describes it, in fruits and flowers. The lower portion of the throne in front of the tree is lost, while the upper part presenting the full view of the tree and the scene of worship[29] is in good state of preservation. The small leaves and large bunches of flowers are characteristic of the tree *Acacia Sirissa*. But branches of fruits and hanging garlands are also among its notable features. The throne usually bears the leaf and flower marks. The trunk of the tree is not adorned with any auspicious symbol. Here too, one has just one pair of worshippers, the male kneeling besides the throne in an act of devotion, while the female standing besides the tree is making a demonstration of the length and size of the garlands, precisely as on the opposite side, the female is bowing down in kneeling posture, while the male is standing up demonstrating the abundance of flowers and fruits. Excepting the inverted order of the groupings the scene of worship and circumambulation is similar to that in the preceding representation.[30]

The Bodhi-tree of Buddha Konagamana[31] is Udumbara (*Ficus glomerata* or the fig). According to the Buddhaghoṣa's[32] description, it can not have flowers. This well accords with an old Indian belief as expressed in saying, "He who sees the flowers of Fig-tree, becomes a monarch." The Bharhut representation faithfully adheres to this Indian tradition. The leaves of the tree are well marked with garlands hanging from its branches. But its bunches of fruits are clearly visible. No one can mistake that the tree is the well known *Ficus*. It has, as usual, a Bodhi-maṇḍapa in front, in this instance. In this case, the throne is supported upon pillars with octagonal shafts. Though the worshippers, as in the preceding two scenes, are a pair of male and female devotees, it is only the female, who on both sides, is actually making the offering of figs and bowing down in kneeling attitude, while the male remains standing, either holding in one hand the fruits of the tree put into a receptacle for offering, or drawing attention to the garland. In spite of these alterations, the same device for representing the scene of worship and circumambulation is there.

Nyagrodha[33] (*Ficus bengalensis* or Banyan) is the Bodhi-tree of Buddha Kaśyapa, which does not bear flowers according to *Sumaṅgala Vilāsinī* (Siamese Ed., II, p. 13). Like the Fig-tree, it is adorned with hanging garlands. The *Bodhi-maṇḍapa* in front is a soild structure. There is a pair of worshippers, the male devotee remaining on both sides, in a standing position, either simply showing a piece of garland or holding it with joined hands. On the right side, the female worshipper seated on a *mohrā* is eagerly and amazedly looking up and watching how beautifully the garlands are hanging down as these are being set up by the male. She on the left side is kneeling[34] before the tree as an act of homage touching its trunk with her hands.

The Bodhi-tree of Buddha Śākyamuni[35], the last of the Mahapadana list of the seven Buddhas, *i.e.,* the Buddha who is truely a historical personage, has naturally received most elaborate treatment. Two umbrellas are placed, one above the other on its top. Some garlands are suspended from its branches. Its up-spreading branches, well-shaped leaves and small round fruits without any flowers all have distinct effects in the composition of its charming foliage clearly show that it is Aśvattha (*Ficus religiosa*, Linn.) or the holy Pīpala. The heights of its trunk and foliage are, as represented, disproportionate, in disregard of the received tradition of equality. But the observed disproportionateness is more an appearance than a reality. The apparent longer height is accounted for by the fact that here the sculptor is required to represent, within the limited space at his disposal, the tallness of trunk prominently behind the gate chamber of a circular edifice which serves as a wall surrounding the tree and provided with a barral-shaped vaulted roof. This roof, even in its highest elevation or altitude. *i.e.,* the portion of it covering the gate chamber, remains below the foilage. The circular edifice is a two storeyed building, the upper storey of which is a superstructure upon an open pillared circumambulation-hall presenting a colonnade of two rows of pillars with octagonal shafts and bell-capitals, the upper story is provided, on both sides, with corridors and large bands of Buddhist railings. There are many arched windows or niches, each containing an umbrella, with hanging garlands and set up, to all intents and purposes, over the Bodhi-maṇḍapa or *Vajrāsana* representing, technically known as *ratna-vedica* or elevated square platform, the jewel-seat. Several of these are to be seen on two sides, while the wing-niches have, on each side, the standing figure of a female worshipper with some kind of offerings in her hands. It is apparent that the corridors with protecting railings are intended to be used as passages for perambulation and worship of *Vedikā* at regular intervals. The gate chamber presents, in its lower storey, an opens pillared square hall.

## (iv) Gateways of Sanchi

The representation of natural trees and flowers were tending to become more realistic at Sanchi, where the decorative designs were becoming more stylised. The fresh spontaneous designs, such as those emerging in the lotus panel (Pillar 23 of Stūpa-2), gave place to a richer and more elaborate kind of decoration in which human beings, animals and birds are frequently mingled with the floral and foliate forms, and the whole modelled with a skill and feeling for plastic form which distinguish them at a glance from the ornamental carvings of an early age. The finest specimens of this decorative relief work on the gateways at Sanchi, are to be found on the outer faces of the gateway pillars.[36] The motif underlying all these decorations is the same, as is found recurring in varying forms among the reliefs on the balustrate of Stūpa-2, *viz.*, the Tree of life and Fortune. In its simplest form, this tree is figured on the balustrade of the second Stūpa as a highly stylised lotus sometimes with birds resting among its leaves[37], sometimes without[38] them (Fig.-5). In another form it appears as the Kalpalatā or Kalpadruma—a wish-fulfilling creeper or a wish-fulfilling tree; and its boughs are hung with garlands and jewels—symbolic of the riches and riches which it has to confer on the mankind[39] (Figs.-6, 6a, 7). In some of the leaves again, there are animals usually lions, or bulls, or deer emerging from its stem[40], or ensconced among its foliage like the bull in Pillar 5-6 and sometime in addition to, or in place of, the animals there are male or female figures in the middle or at the top of the tree. Generally the root or source of the tree does not appear, but in some cases it is shown issuing from the naval of the Yakṣa and in another it issues from the mouth of a *makara* and in some cases the bottom of the stem is grasped by an elephant, which, in its role of Nāga stands for water in which all life was believed to have its origin with single or double horns-descendants of the unicorn heads which are shown starting from the stem of a Pīpala tree on one of the pre-historic seals from Mohenjodaro, and of the pairs of animals shown against the stem of the Lotus-tree on the balustrade of Stūpa no. 2 at Sanchi. From the latter, they differ, in that they are attended by riders, either on their backs or standing at their side; these could mean a logical combination of the Yakṣa-Tree spirits and their *vāhanas*. What makes this composition so dignified and attractive, is not only the effective modelling of beasts and the riders and the admirable taste shown in the stylisation of the lotus flowers and the leaves, but the peculiarly happy blending and balancing of plant and animal forms in such a way that neither seems quite to dominate the other. On the corresponding face of the opposite north pillar, though slightly different, is evidently a copy of this one, but falls short of it in technical beauty of workmanship[41](Figs.- 8, 9, 10 & 11).

The same Lotus-tree[42] of life is drawn on the top and bottom architraves in the front of the Southern gateway—on the former in its simple form with birds resting among its foliage and Māyā in association with it, as she also appears twice on the pillars of Stūpa[43] no. 2. In one case, it is a piece of crude, stencil-like carving which has little to commend, while the other one is of high order. The serpentie stalk of the lotus is a defect that detracts from the strength of the composition, and clashes at the same time with the slight curvature of the architrave; but in spite of this, there is a vigour and freedom about the drawing and a vitality about the figures which strike a new note in this class of decoration and shows that the artist who conceived it was a man of extraordinary taste. For the position they occupy, the workmanship is so fine and beautifully finished that the eye is attracted to notice rather the individual patterns than the effect of the composition as a whole. In the carvings on the eastern gateway, the one on the north pillar corresponds with the formal, full blown lotus pattern on the eastern pillar of the older gateway, the one on the southren, with the serpentine lotus on the opposite side of the same gateway. In executing the second panel, however, the sculptors have evidently been fully aware of the insignificance of their model, and have been at pains to produce a bolder design more in keeping with the proportions of the gateway, and to this end they have had to eliminate the medley of jewelled ornaments, amulets etc. crowd the earlier composition, and to fill the embrasures between the windings of the lotus stem with leaves and flowers only and a few birds among them.[44] Although this composition suffers from the same conspicuous shortcomings as the serpentine lotus on the Southern gateway, seeming by its undulations to diminish the stability of the pillar, nevertheless in point of proportion and symmetry it is by far the most satisfactory of the Tree of Life designs that have been discussed earlier. The designers of the Western gateway produced not only the most brilliant of all the decorative designs at Sanchi, but also the most notable to be found in the whole field of Indian art. In this composition, the central stem of the Lotus-tree is straight and thick enough only to link together the component parts, without itself being too obtrusive. Along this stem, flowers and leaves are arranged in five successive groups, and, between them, pairs of heraldic-looking animals and riders. The more to enhance their reality, the lotus blooms are shown as if seen slightly from above, and their leaves are treated in the conventional manner. The animals comprise unicorns, griffins and lion-like beasts of the Buddha; at its summit is the *tri ratna* device symbolising the Buddha, the law and the Church. The tree here is conceived as having its root in the Buddha himself represented by his foot-prints at the foot of the tree—just as in some of the reliefs it has its roots in a Yakṣa; and it is reasonable to suppose, therefore, that the tree itself was taken to represent either the Faith which emanated from the Buddha of the mystic indissoluble Trinity of the Buddha, the Law and the Order, together with all the priceless gifts which he had to offer to the mankind. Indeed the Tree of life and Fortune went back to an age long anterior to the Buddha and that, like the Yakṣas or the Yakṣīs, Nāgas and a multitude of other elements taken from the popular religion of the people, it was only in the process of being absorbed into Buddhism.[45] It is quite likely therefore, that at this time the Tree bore virtually the same meaning in the eyes of the Buddhists as it had done for generations past in the eyes of their non-Buddhist ancestors, the only significant difference being that, like similar trees on the balustrade of Stūpa No. 2 of Sanchi, it was now stamped with the symbolic hallmarks of Buddhism and thereby recognised as belonging to the Faith. The corresponding tree on the right pillar of the Northern gateway bears no distinctive Buddhist emblems. Its main stem, in the middle band of the panel, issues from a "Jar of Fortune" (*bhadra-ghaṭa*) and among its foliage, and starting out, as it were, from the stem are pairs of Yakṣas and Yakṣiṇīs like those present in the Life Trees on the balustrade on the Stūpa No. 2; but, here, males and females are arranged in alternate pairs while the two sinuous creepers in the border spring from the mouths of the *makaras*. Both of these panels mark a decided improvement on the serpentine design of the southern gateway, not merely because of their carving; for though the details are disproportionably small. On the right pillar of the Southern gate at Sanchi, which is the oldest of the four, it takes the highly stylised form of a row of full-blown flowers, between the borders of sinuous lotus creepers; on the left pillar of the same gateway it appears as a creeper with undulating stalk, held at the base by a *Kumbhanda*; and from it hang fruits and flowers, garlands, necklaces, while amidst its foliage are seen pairs of male and female figures, animals, birds, jewelled ornaments and amulets. One of the female Yakṣīs is holding a necklace of amulets in her hands, while her companion Yakṣa is playing on a musical instrument. From the artistic point of view the design could hardly be more inappropriate or feeble. In such a position, a design of this kind with undulating stem of lotus standing out so prominently, is entirely out of place. By carrying the spectator's eye from side to side, it has the inevitable effect of weakening, or seeming to weaken, the gateway support, instead of adding to its strength. But this is not the only fault. The figure and the objects which fill the embrasures formed by the windings of the stem, are merely a confused medley thrown together by the sculptor, without regard to symmetry or composition, and are strikingly disproportionate to the size of the pillar, so that, even if their composition were more orderly and pleasing, they could not escape appearing insignificant.[46]

In the design of the Northern gateway, these defects are avoided and another type of decorative Life-tree is resorted to—*viz.*, a series of formalised lotus plants one above the other, with artificial brackets in the borders from which hang jewelled garlands and necklaces of lucky talismans betokening both worldly and spiritual riches. At the root of the tree are the foot prints of *Śrīpada* on the Bharhut balustrade, the Lotus-tree of Life is commonly depicted issuing from the mouth of an elephant. From the top of the tree emerge, as a rule, one or other of the chief emblems of the Buddhist faith, i.e., the figure of Māyā symbolising the Birth of the Blessed one, the holy pīpala tree symbolising Enlightenment, the *dharmacakra* supported on a pillar, symbolising his First Sermon, and the Stūpa symbolising the *Mahāparinirvāṇa*. Although not specifically Buddhist in origin, these four emblems were recognised as sufficiently representative of the Faith, to give the tree a predominantly Buddhist character. It must be kept in mind that the Tree of life went back to a time ages before the advent of the Buddha. Among the seals of Mohenjodaro dating from the third to fourth millenium[47] B.C. is one depicting a stylised pīpala tree with two heads of unicorns emerging from its stem and on others is shown the same tree with a dryad (*Vṛkṣa-devatā*) standing in its branches. In those remote days we may surmise that the pīpala tree, like the secred cedar of the Chaldeans, was looked upon as the Tree of knowledge as well as the Tree of Life. In the historic times, although commonly regarded as the most holy and protected among the trees, the pīpala has functioned in a special sense as the Tree *par-excellence,* of knowledge rather than as the Tree of Life; the tree beneath whose shadow Gautama himself attained Supreme Enlightenment.

With the Buddhists as well as with other sects, the Tree of Life—of birth and immortality—was the Lotus, which has scarcely figured at all in the Chacolithic age, and this plant became symbolic not only of the birth of the Blessed One but also of the religion which he promulgated.

What idea originally underlay the animals emerging from the stem of the tree, is uncertain, but it seems likely at least that they were associated with the tree in the same way as animals as *vāhanas* were associated with the tree spirits, which they support, for example, on the balustrade of the Bharhut Stūpa. Be this however, as it may, there can be little doubt that the heraldic looking lions or bulls or deer, which sprang from the sides of the lotus stains on the balustrade of Stūpa No. 2, are the lineal descendants of the unicorn heads emerging from the pīpala tree on the Mohenjodaro seals and the same idea is carried on by the unicorns and other hybrid animals in the reliefs on the Sanchi gateways. As to the male and female figures appearing among the foliage, the simplest explanation seems to be that they too are the lineal descendants of the tree spirits which we first meet on the prehistoric seals of the Indus, or which two or three thousand years later are still a prominent feature among the sculptures of Bharhut; nor does it require any stretch of imagination to see in the pairs of male and female figures which are so commonly found in association with these trees, an allusion to the fertility and reproductiveness of nature, of which they were the acknowledged guardians. That Yakṣas and Yakṣīs should sometimes be shown emerging from the foliage of the tree, at others standing beneath its boughs on their animal *vāhanas*, at others as the source from which the tree itself has its being, can hardly occasion surprise; for, as the tree spirits became matamorphosed into human shape, it could hardly have happened but that popular ideas regarding the relation in which they stood to the tree or to their own animal *vāhana* would take divergent forms, and that seeming anomolies and contradictions would creep into the representations of these figures in plastic and pictorial art.

## (v) The Tree Spirit

The Yakṣas displayed on the Sanchi gateways were looked upon as veritable spirits of magic Kalpalatā[48] or rather perhaps as the plants themselves transformed into human shape, is clear from the way in which they are depicted, sometimes as sources from which the plants take their birth, at others as themselves growing like buds from the parent stem and then in turn sprouting forth leaves and blossom from their mouths or holding jewelled treasures in their hands. The idea of a Tree of Life springing from a Yakṣa can be traced as far back as the Vedic[49] age, but it becomes specially familiar during the medieval times in the representation of the birth of Brahmā, who is shown seated on a lotus flower, the stem of which issues from the navel of Nārāyaṇa.

## (vi) The Lotus Medallion

As in the art of Bharhut, the lotus designs are quite popular at Sanchi as well (Plates 3 & 4), the salient features of which are listed here—

(a) Lotus designs[50];

(b) Lotus stylised with birds[51];

(c) Lotus serving as throne[52];

(d) Lotus leaves[53];

(e) Lotus supporting Buddha[54];

(f) Lotus medallion with human heads[55];

(g) Lotus as Tree of Life[56];

(h) Lotus supporting Brahmā[57];

(i) Lotus with Nandipada[58];

(j) Lotus as emblem of immaculate[59] birth;

(k) Lotus supporting Māyā[60];

(l) Lotus grasped by elephant[61];

(m) Lotus springing from *Bhadra-ghaṭa*[62];

(n) Lotus springing from Tortoise[63];

(o) Lotus supporting elephant[64];

(p) Lotus as attribute of Avalokiteśvara.[65]

## (vii) Art of Amarāvatī

The art of Amarāvatī, though basically projecting the Buddhist themes, portray plant kingdom in a variety of forms including the Bodhi-tree, the lotus medallions, in and the carving of the *Jātaka* stories. As for instance, the Tree-spirit is found depicted in a stone panel. This panel depicting the *Losaka Jātaka*, suggests a village by the thatched huts. A man and a woman are standing behind a fence and each with a baby. On the other side of the fence, the man stands with two boys, the babies grown up. A boy is eating food as a woman is cooking—the same woman as the one with child in previous scene (Plate-5). Cavaliers and foot soldiers approach the scene and a warrior seated with a sword in his hand is addressing the elephant, with the *mahaut* on his head, that is stuck in the mud or wading in the pool, besides a tree in the hallow[66] of which is to be seen the head of a tree spirit. Indeed the tree represents the village *Caitya-vṛkṣa* with deity represented in the trunk hallow close by out side the village, which corresponds to the description of *Caitya-vṛkṣa* given by Kālidāsa in the *Meghadūta*.

Besides the above, there are numerous lotus medallions and half-lotus-medallions, and the *pūrṇughuṭas* filled by lotus stalks, which provide ample testimony to the artist having been conversent with the Tree-spirits, plants and lotuses (Fig.-12).

## (viii) Bodhagayā Art

The lotus medallions and half-medallions are also found on the Bodhagayā[67] railing, which are minutely carved displaying high quality of workmanship. Some of them are quite unique in design and execution (Plates-6 & 7).

## (ix) Mathura Art

The railings from the Stūpa at Mathura[68], have railings identical with the art of Bharhut, Sanchi, Bodhagayā and Amarāvatī (Figs.-13 & 14, Plate-28d).

# Notes and References


1. *Lalita-vistāra*, Ch. VIII.
2. *Mahāvāstu*, II, p. 19.
3. *Lalita-vistāra*, Ch. IX.
4. *Paniya Jātaka*, F. 459.
5. *Saddharma Puṇḍarīka*, Atlantic Publishers and Distributors, New Delhi (1990), p. 119.
6. *Bodhicaryāvatāra*, II.2.5.
7. *Milindapanho*, Book-5, verse 8.
8. *Prākṛta Dhammapada*, Chapter-18, 290-291. (Prakrit Bharat Academy, Jaipur), 1990.
9. *FO-SHO-HING-TSAN-KING* (Life of Buddha).
10. Barna, B.M., *Bharhut*, Book-III, Section-II, pp. 45 ff.
11. *Ibid.*, Pl. XXXIII (25).
12. *Ibid.*, Pl. XXVI.
13. *Ibid.*, Pl. XXXII.
14. *Hajjhima Nikaya*, I. pp. 169, 277; p. 16, III.
15. Barna, B.M., *op. cit.*, Book-III, p. 46, Pl. X.
16. *Ibid.*, Section-III, p. 47, Pl. XXXII.
17. *Ibid.*, Pl. XXXI.
18. *Ibid.*, Pl. XXXI, Fig. 79.
19. *Ibid.*, Fig. 80.
20. *Ibid.*, Fig. 80a.
21. *Ibid.*, Fig. 80b.
22. *Ibid.*, II, p.1, Scene-26; Pl. XXIX-1.
23. *Ibid.*, II, p. 2.
24. *Ibid.*, II, p. 2, Scene 27.
25. *Sumaṅgala Vilāsini*, (Siamese text), II, p. 13.
26. Barna, B.M., *op. cit.*, pp. 2-3, (Section-28), Pl. XXIX-2.
27. *Ibid.*
28. *Ibid.*, Pl. XXIX, 3, Scene-29.
29. *Sumaṅgala Vilāsinī*, (Ceylonese Ed.), p. 13.
30. Barna, B.M., *op.cit.*, p. 3, Pl. XXIX-3, (Secne-29).
31. *Ibid.*, p. 4, Pl. XXIX, 1.
32. *Sumaṅgala Vilāsinī*, (Ceylonese Ed.), II, p. 13.

33. Barna, B.M., *op. cit.*, p. 4, pl. XXX, 1, Scene-31.
34. *Ibid.*, Pl. XXXI Section-31.
35. *Ibid.*, pp. 5-7, Pl. XXX, 4, (Secne-32).
36. Marshall John, *The Monuments of Sanchi*, Vol. I, p. 142, Pls. 19, a, c; 37-a, b; 50-b, c; 66-b, c.
37. *Ibid.*, Pillars, 23-a, 44-a, p. 142.
38. *Ibid.*, Pillar-23c.
39. *Ibid.*, Pillar 5-b.
40. *Ibid.*, Pillars, 44a, c; 49-a, b; 66-b.
41. *Ibid.*, p. 146.
42. *Ibid.*, Pls. 11, 1-3.
43. *Ibid.*, p. 146, Pls. 83, 87.
44. *Ibid.*, p. 146.
45. *Ibid.*, p. 144.
46. *Ibid.*, pp. 144-145.
47. Marshall John, *Mohenjodaro and Indus Civilization*, Vol. III, Pl. CXII, No. 387, Vol. I, p. 64, note-3.
48. Marshall John, *The Monuments of Sanchi*, Vol.-I, p. 147.
49. *Ṛgveda*, X.82.5; *Atharvaveda*, X.7.
50. Marshall John, *op. cit.*, (Sanchi) I, pp. 42-43.
51. *Ibid.*, I, pp. 42, 73.
52. *Ibid.*, I, pp. 58, 73, 252.
53. *Ibid.*, pp. 59, 73, 75, 77.
54. *Ibid.*, pp. 59, 74
55. *Ibid.*, pp. 106, 184.
56. *Ibid.*, pp. 144-146.
57. *Ibid.*, p. 147.
58. *Ibid.*, pp. 172, 184.
59. *Ibid.*, p. 177.
60. *Ibid.*, p. 184.
61. *Ibid.*, p. 184.
62. *Ibid.*, p. 184.
63. *Ibid.*, p. 184.
64. *Ibid.*, p. 185.
65. *Ibid.*, p. 185.
66. Sivaramamurti, C., *Bulletin of Madras Govt. Museum*, 1956, p. 219.
67. *A.S.R.*, III, Pls. XXVI, XXX.
68. Ghosh, A., *Jain Art and Architecture*, I, Pl. 8.

# 4
# Trees and Plants in Jainism

## 1. Literature

Jainism like Buddhism is a religion of considerable antiquity and Lord Mahāvīra, happens to be contemporary of the Buddha. The Jaina literature and art frequently mentions the flora and fauna. The texts while describing the distinguishing features of Jaina figures point out the presence of the Aśoka tree or the particular tree under which Tīrthaṅkaras attained Supreme knowledge. One of the eight *Pratihāryas* consist of (i) heavenly tree, (ii) a throne seat, (iii) Trilinear umbrella and a lion throne, (iv) *Aura* of beautiful radiance, (v) Drum, (vi) showers of celestial blossoms, etc. All these symbols are seen in a complete image of a Tīrthaṅkara.

Besides Aśvattha is the tree associated with the enlightenment of Anantanātha as would be evident from *Uttarapurāṇa*,[1] a Jaina text—

अश्वत्थ पादपोपान्त कैवल्यमुदपीपदत्।

The teeth of Suvidhinātha resemble flower buds, earning him the tile of Puṣpadanta. Śītalanātha has Aśvattha tree as his emblem according to the Digaṁbara sect, while Śvetāṁbara sect, Śrīvatsa (wishing tree) for the same. The floral designs appear on many of the Jaina images as well as the *Ayagapaṭṭas* and miniatures of Jaina *Kalpasūtras* and other manuscripts from Gujarat and other places.

In fact all the twenty-four Tīrthaṅkaras are closely associated with one tree or other, as would be evident from the following tabulation—

Trees connected with the Jaina Tīrthaṅkaras

| *S.No.* | *Tīrthaṅkaras* | *Tree* |
|---|---|---|
| 1. | Ādinātha or Ṛṣabhanātha | Nyagrodha (Vaṭavṛkṣa) |
| 2. | Ajītanātha Saptaparṇa | Nyagrodha (Saptaparṇā) |
| 3. | Saṁbhavanātha | Sāla tree (*Artocarpus locuca*) |
| 4. | Abhinandanātha | Sāla or Piyāla tree |
| 5. | Sumatinātha | Priyaṅgu |
| 6. | Padmaprabha | Chatrabha |
| 7. | Supārśvanātha | Śirīśa (*Uttarapurāṇa*)<br>सुपाश्र्वो मौनमास्थायच्छादमस्थे नववषक:।<br>सहेतु कवने मूले शिरीषस्य द्वयुपोषित:।। |

*(Contd...)*

| *S.No.* | *Tīrthaṅkaras* | *Tree* |
|---|---|---|
| 8. | Candraprabha | Nāgakesara |
| 9. | Suvidhinātha | Nāgakesara |
| 10. | Śītalanātha | Śvet.—Bilva *(Aegle marmelos,* C & Roxb.)<br>Dig.—Aśvattha (*Ficus religioso*, Linn.) |
| 11. | Sreyāṅṣanātha | Tumbara |
| 12. | Vasupūjya | *Abhidhānacintāmaṇi*—Patalika<br>*Uttarapurāṇa*—Kadamba |
| 13. | Vimalanātha | Jambū (Black berry)<br>निज दीक्षा वने जम्बूद्रुममूले द्रुयुपोषतः। |
| 14. | Anantanātha | Aśvattha (*Ficus religiosa,* Linn.) |
| 15. | Dharmanātha | Dadhiparṇa or Saptcchaḍa |
| 16. | Śāntinātha | Nandī Vṛkṣa |
| 17. | Kunṭhanātha | Tilaka tree |
| 18. | Aranātha | Mango tree |
| 19. | Mallinātha | Aśoka |
| 20. | Munisuvrata | Campaka |
| 21. | Naminātha | Bākula |
| 22. | Neminātha | Mahavanu or Vetasā |
| 23. | Pārśvanātha | Devadāru |
| 24. | Mahāvīra | Sāla (*Artocarpus locuca*) |

The Jaina texts also deal with the plant kingdom variously, a few examples of which are given here—

## (i) *Triṣaṣṭiślākāpuruṣacarita*

In the opening chapter of Book-II of this work[2], while describing the kingdom of Vatsa, it is stated that—

> "At evey step, there were large tanks with clear sweat water, just as if filled by canals coming out of the ocean of milk. Here and there were large, clear pools whose centres, like the minds of the noble could not be reached. At every step gardens with abundant green creepers gave the impression of variegated body decoration of the earth-goddess. In every village there were sugarcane plantations, which relieved the thirst of travellers, gleaning with large sugarcanes that resembled pitchers of water in the form of juice. On every road the fruit trees

with pairs of travellers seated by them looked like the wishing trees of the Kurus with twins."

The above description provides an evidence of the artificial water storage and water supply system besides the groves and parks in the cities and the roads having been provided with the shady trees. In the context of the visit to Sūri Arindama (67-142), the king is said to have reached a garden having the abode of much shade and resembling the Nandanavana. Then the sage (*muni*) is compared to the grove of the trees of *dharma* rejoicing in the Supreme spirit. In the verse 155, the palm tree is mentioned. In chapter-2, on the birth of Ajīta and Sāgara, the queen Vijayā Devī is described to have the appearance of being made of pieces of blooming lotuses from the looks of her hands, feet, eyes and face.

The use of Dūrva grass[3] by the astrologers is mentioned (in verses 86-108), whose bodies were shining with priceless ornaments like trees at the beginning of Caitra, with a few fragrant flowers. They on reaching the king's court threw Dūrva-grass, unhusked rice or paddy which conferred happiness on the king's head like the garden breeze throwing flowers.

In the third chapter, while describing the episode concerning Sāgara's going to Samayasarana (399-417) a garden of *sahasrāmravana*-garden[4] with a thousand mango trees is mentioned. While describing the Meru mountain[5] (554-565), it is stated that there are three terrace groves, called Nandana, Sumanasa and Sundara on it. Then on the peak of Meru, a garden called Pandaka, which is circular in shape, is mentioned. Besides the above, there are numerous other descriptions of the plant kingdom in this work. However, in Book-1, Chapter-IV[6], on the conquest of Magadhatīrtha, several trees have been mentioned—

> "The Deodār trees growing on the bank of Bhāgīrathī served as tying posts (ready) without effort for the king-elephants of the army. In a moment elephant keepers cut with axes, sprouts of the Pīpala tree, a Sallaki, Karṇikara and Udumbara for the elephants. The horses, tied in rows by the thousand, making arches, as it were, with their raised ear-shoots, had a beautiful appearance. Quickly the horse-keepers placed before the horses, like relatives, kidney beans, green gram, chick peas and barley.
>
> "Cross-roads, where three or four roads met, and two rows of shops sprang up at once in the camp, like those of *vinita*... The camels lopped thorny trees such as the sponge and jujube, as if showing the soldiers the removal of thorns to the state that had to be made... Some men brought fuel; some water from the river; some loads of Dūrva-grass, some vegetables, fruits, etc., some removed the husk from rice...".

In a sermon[7] (559), it is stated that the human birth of creatures bears fruit quickly by the attainment of *mokṣa*, like a tree by its desire to be touched by a woman at budding time.

### (ii) *Anuttaropattikādaśa Sūtram*

In the third varga of this Sūtra-grantha[8], names of several vegetables, trees and plants are mentioned, including Śālmali, Priyaṅgu, Badarī, Karkandu.

## (iii) *Ācarāṅga Sūtra*

The plants are conceived[9] as having life and any injury to them is prohibited—

"...All beings, those with two, three, four senses, plants, those with fine senses and the rest of the creation (experience) individually pleasure or displeasure, pain, great terror and unhappiness."

Here the plants have been equated with all the living beings who are said to experience all the emotions, pleasures and displeasures.

The fifth lesson[10] of the above text conceives that the plants are endowed with intellect—

"See there are men who control themselves, others pretend only to be houseless, for one destroys this (body of plant) by bad and injurious doings, and many other beings, besides which he hurts the means of plants, through his doing acts relating to plants... For this a man is longing when he destroys this (body of a plant) by bad and injurious doing acts relating to plants...".

In same[11] text, the tree is likened to a worldly man as under—

"As in a lake, a greedy leaf covered tortoise cannot rise up; as the trees do not leave their place (though shaken by storms, etc.), thus men born in various families, cry bitterly because they are attached to the objects of senses; on account of their sinfulness they do not reach liberation."

## (iv) *Uttaradhyāna Kalpasūtra*

Chapter-X of this text compares the leaves[12] of the tree with the human life in a variety of forms, a few of which are given hereunder—

दुमपत्तए पंडुयए जहा णिवडइ राइगणाण अच्चए।
एवं मणुयाण जीवियं समयं गोयम मा पमायए।।
कुसग्गे जह ओसबिन्दुए थोव चिट्ठइ लबंमाणए।
एवं मणुयाण जीवियं समयं गोमयम मा पमायए।।
इइ इत्तरियंमि आउए जीवियए बहुपच्च वायए।
विहुणाहि रयं पुरे कडं समयं गोयम मा पमायए।। —(X.1-3)

(As the days pass, the trees drop their fallow leaves, so it is the life of man, Gautama has no delusion even for a while; As dew drops hanging from *kuśa* tips have life but for a short while, so is the life of a man; As the life is so fleet and existance so precious, wipe out the sins acquired by thee, Gautama, have no delusions for even a while).

## (v) *Bhagavatī Sūtra*

In Book-2, Chapter-1, of this *Sūtra*, the suffering[13] of pain by the monk Skanda, who had been fasting, has been described as under—

"He became so weak that after speaking or in the course of speaking, or when he thought that he would have to speak, he had a weakness and the feeling of pain; like a cart laden with dry leaves, sesamum or any other dry object, or a cart laden with Eraṇḍa twigs, or a cart laden with charcoal, and when such a cart while moving makes a rattling sound, similarly when Skanda moved, his bones rattled."

Again the sun has been described as the one who makes the lotuses blossom—

"... When the sun, who helps the lotuses to blossom, who shines red like the red Aśoka flowers, red like Kiṁśuka flowers, or the red half of Guṅga fruits, who helps the cluster of lotuses to open, who holds a thousand rays..."

### (vi) *Jaina Jātakas*

Describing the events of the twelve previous births of Tīrthaṅkara[14] Ṛṣabhanātha, it is stated (Verse 512), that the lord—

"Sharing the same seat with (Svayamprabhā) he looked charming like a tree within a trench twisted round by a creeper."

Further her bright lips are compared with Bimba-fruit and her nose is equated with the stalk of lotus. Further it is stated that with a wealth of ornaments all over her person, she appeared like a moving Kalpa-tree.

## 2. Art and Iconography

### (i) Jaina Goddesses

Some of the Jaina goddesses are also closely connected with the trees. As for example Ambikā or Āmra happens to be the Yakṣiṇī of Neminātha. According to the Śvetāṁbara[15] description of the goddess, she rides on a lion and bears a bunch of mangoes, a noose, a child and a goad (Plate-39).

तत्तीर्थजन्मा स्वर्णकान्तिः सिंहवाहना आम्रलुम्बि
पाशसंयुक्तदक्षिण करद्वया पुत्राङ्कुश सहित वामकरद्वया
कुष्माण्डीतिद्वितीयनामधारिणी अम्बिका प्रभोः
शासन देवी समभवत्।

—*Guṇavijaya-gani's, Nemināthacarita*
(Nirṇayasāgara Press)

The Digaṁbara image of Yakṣiṇī is also described[16] as mounted on a lion, but bearing two hands with a bunch of mangoes and a child—

द्विभुजा सिंहमारूढ़ा आम्रा देवी हरित्प्रभा।।

—*Pratiṣṭhāsārasaṁgraha* (Mss. Arrah).

The goddess is sometimes depicted in Indian art seated under a mango tree laden with fruit.

## (ii) Yakṣas and Yakṣiṇīs and Other Minor Deities

The Jaina literature[17] prescribes various types of flowers and fruits as attributes to different forms of Yakṣas and Yakṣiṇīs, etc. which would be evident from the following tabulation—

| *(a)* | *Yakṣas* | *Śvetāṁbara* | *Digaṁbara* |
|---|---|---|---|
| (i) | Gomukha | *Mātulunga* | Fruit |
| (ii) | Mahāyakṣa | —do— | Lotus |
| (iii) | Trimukha | Fruit | |
| (iv) | Īśvara/Yakṣeśvara | Fruit | |
| (v) | Tumburu | | Fruit |
| (vi) | Kusuma | Fruit | |
| (vii) | Mātaṅga | Bilva fruit | |
| (viii) | Vijaya | | Fruit |
| (ix) | Ajīta | *Mātulunga* | Fruit |
| (x) | Brahmā | —do— | *Mātulunga* |
| (xi) | Īśvara | —do— | Fruit |
| (xii) | Kumāra | | Fruit |
| (xiii) | Saṇmukha | Fruit | |
| (xiv) | Pātāla | Lotus | Fruit |
| (xv) | Kinnara | Lotus | Fruit |
| (xvi) | Garuḍa | Lotus | Lotus and Fruit |
| (xvii) | Gandharva | *Matulunga* | Lotus or Fruit |
| (xviii) | Yakṣendra | —do— | —do— |
| (xix) | Kubera | | Lotus |
| (xx) | Varuṇa | *Mātulunga* | Fruit |
| (xxi) | Bhṛkuṭi | —do— | Lotus |
| (xxii) | Gomedha | —do— | Fruit |
| (xxiii) | Pārśva/Dharṇa | —do— | Fruit |
| (xxiv) | Mātaṅga | | *Mātulunga* |

*(Contd...)*

| *(b)* | *Yakṣīs or Yakṣiṇīs* | *Śvetāmbara* | *Digambara* |
|---|---|---|---|
| (i) | Cakreśvarī | *Mātulunga* | *Mātulunga* |
| (ii) | Ajītā or Ajītabālā | Fruit | |
| (iii) | Duritārī/Prajñapti | Fruit | Fruit |
| (iv) | Kālikā/Kālī or Vajraśṛṅkhalā | Lotus | Fruit |
| (v) | Mahākālī/Puruṣadattā | *Mātulunga* | Fruit |
| (vi) | Acyutā/Manovegā | —do— | *Mātulunga* |
| (vii) | Śāntā/Kālī | | Fruit |
| (viii) | Bhṛkuṭi/Jvālāmālinī | *Mātulunga* | |
| (ix) | Sutārā/Mahākālī | | Fruit |
| (x) | Aśoka/Mānavī | Fruit | Fruit |
| (xi) | Caṇḍā/Gāndhārī | Fruit | Fruit |
| (xii) | Aṅkuśā/Anantamatī | Fruit<br>Lotus | |
| (xiii) | Viditā/Vairoṭyā | Lotus | |
| (xiv) | Caṇḍā/Pracaṇḍā/Gāndhārī | Lotus | Lotus and Fruit |
| (xv) | Kandarpā/Mānasī | Lotus | Two Lotuses |
| (xvi) | Nirvāṇī/Mahāmānasī | Lotus | Fruit |
| (xvii) | Bālā/Jayā | Lotus | Fruit |
| (xviii) | Dharaṇī/Tārāvatī | *Mātulunga* | Fruit |
| (xix) | Vairoṭyā/Vijayā | —do— | —do— |
| (xx) | Naradattā/Bahurūpiṇī | | Fruit |
| (xxi) | Gāndhārī/Mālinī/ Cāmuṇḍa | | |
| (xxii) | Ambikā/Kuṣmāṇḍī/Āmrā | Bunch of Mangoes | Fruit |
| (xxiii) | Padmāvatī | Lotus and Fruit | Lotus and Fruit |
| (xxiv) | Siddhāyikā/Siddhāyinī | *Mātulunga* and Fruit. | |

*(Contd...)*

| (c) | *Mahāvidyās* | *Śvetāmbara* | *Digambara* |
|---|---|---|---|
| (i) | Rohiṇī | | Lotus and Fruit |
| (ii) | Prajñapti | *Mātulunga* and Fruit | |
| (iii) | Vajraṣṛṅkhalā | Lotus | Fruit |
| (iv) | Vajrāṅkuśā | Fruit | Fruit |
| (v) | Apraticakrā/Cakreśvarī/ Jambūnadā | | Lotus and Fruit |
| (vi) | Naradattā/Puruṣadattā | Fruit | Lotus and Fruit |
| (vii) | Kālī or Kālikā | Lotus | —do— |
| (viii) | Mahākālī | Fruit | Fruit |
| (ix) | Gaurī | Lotus | Lotus |
| (x) | Gāndhārī | Lotus | |
| (xi) | Mānavī | Lotus and Tree | |
| (xii) | Mahāmānasī | | Garland of flowers |

Evidently, besides other fruits, *Mātulunga* was quite popular with the Jaina deities, who held it as an attribute. The lotus on the other hand was used not only as an attribute but also as a seat for the deity.

## Notes and References

1. *Uttarapurāṇa* (Jaina), p. 177.
2. *Triṣaṣṭiślākāpuruṣacarita* (John, N.M.ed.), Book-II.
3. *Ibid.*, verses, 86-108
4. *Ibid.*, verses 399-417, p. 97.
5. *Ibid.*, verses 554-565, p. 100. The mention of terrace gardens and circular gardens here are quite conspicuous.
6. *Ibid.*, vol. I, p. 217.
7. *Ibid.*, I. p 199.
8. *Anuttaropattikadaśa Sutraṁ*, Lahore Ed., 1936, varga-3.
9. *Ācarāṅga Sūtra*, Book-I, Lecture-I, lesson-6, p. II (*Jaina Sūtras*, Oxford, 1884).
10. *Ibid.*, pp. 10-11.
11. *Ibid.*, Book-I, Lecture-6, Lesson 1, p. 53.
12. *Uttaradhyāna Sūtra*, K.C. Lalwani ed., Calcutta, 1977, pp. 96 ff.
13. *Bhagavatī Sūtra*, K.C. Lalwani ed., p. 174.

14. *Jaina Jātakas* (Lahore-1925), Punjab Sanskrit Series, vo. VIII, p. 55.
15. Bhattacharya, B.C., *Jaina Iconography*, p. 103, Vārāṇasī, 1974.
16. *Ibid.*, p. 104.
17. Tiwari, Maruti Nandana Prasad, *Jaina Pratimā Vijñāna*, (Vārāṇasī, 1981), pp. 256 ff.

# 5
# Plastic Art

Before the emergence of the Aryans on the Indian sub-continent, Punjab, Sindh, Rajasthan and Gujarat had an urban population flourishing on agriculture, powered by bullocks. While wheat, barley, peas and sesamum were their food crops, they also cultivated date-palm. The Harappan culture or the Indus valley culture as is also known, too, was quite familiar with the botanical properties as well as the horticulture inasmuch as a number of artifacts like seals, sealings, pottery and amulets, etc. discovered from the Harappan sites portray shrubs and trees, etc. The trees that could be easily identified from the available Harappan sites include, Pīpala[1], Arjuna[2], Banana[3], Banyan[4], Śīśam[5] (*Dilberga Sisu*), Palm[6], Nīma[7], and Acacia.[8]

## 1. Early Seals and Sealings

The most striking aspect of the Harappan artifacts has been that, they highlight the adoration of the trees at that early stage. As for example, a seal from Mohenjodaro (No. 430) shows a tree goddess or tree spirit, appearing in a Pīpala tree depicted with its leaves (Plate-8). A worshipper is kneeling before the deity, behind whom stands a human headed goat. In the register below there are seven ministrants or votaries, each dressed in a short kilt and wearing a long pig tail with the spray of leaves or feathers in hair. This tree spirit or goddess is apparently nude, but has a pair of horns, between which, there is a projection. The Nīma tree is also considered to be sacred in the Indian religions annals. Its leaves are supposed to repel snakes and they are consumed by humans for the cure various ailments. It is also believed to be the dwelling place for deities, especially the godlings of disease. Fragment of a stone disc from Rupar (B.C. 600) in which a female is shown standing besides a hut of leaves located under a Pīpala tree (Plate-9).

In seal no. 222,[9] a figure, probably that of a forest deity seated in what may be a Yogic posture with heals pressed together on a low dias, whose legs represent those of a bull, could also be interpreted as a forest deity. The arms are out stretched on either side, but it is possible that the hands are intended to rest on the knees and the seal cutter merely raised them slightly to obviate any confusion of the arms with the legs. On each arm there are a large number of bracelets. The head-dress is a twig with leaves like those of a Pīpala tree. The horns, if indeed they are horns, are definitely separate from the head. They appear to have been fastened to the base of a twig. The figure has three faces, one in front and two in profile. This figure has been variously interpreted to mean, as Śiva Paśupati, but the presence of the twigs decorating the headdress, makes one feel that it could as well be that of Vanaspati or the lord of forest. Another seal bearing S.No. 235, from the same site, depicts the same deity, but the stool is missing and the figure is apparently seated on the ground. The head-dress in this case consists of two horn-like objects, between which there appears to be a spike of flowers. A pigtail hangs down on one side of the head, which has a single face in profile, facing to the right. Another seal bearing No. 420, is pretty well preserved and has been interpreted by Marshall as Śiva.[10] Taking into consideration the adoration of the head-dress of the deities on seals 222 and 235, as detailed above, they could be termed as forest deities.

However, Marshall has traced the tree worship and Epiphany of the tree goddess in the antiquarian remains of Mohenjodaro, as follows—

> "That the animistic conceptions which have distinguished the worship of trees throughout the historic period were common also to the Chalcolithic Age is clear from several seals and sealings. One of the most interesting of this is the one in which Epiphany of the deity in the tree is portrayed in a half realistic, half conventional way. The tree itself at the right hand top corner of the seal is represented by two branches only, springing from a circle on the ground, and treated in a formal fashion. Between them appears a deity; a standing nude figure, with long hair, *triśūla* horns, and armlets, recalling those of the three faced images discussed above. In front of the tree is a half-kneeling figure of a suppliant, also with long hair, armlets and horns, but with the addition of half leaf-spray or plume between the horns. Behind this suppliant is a composite animal, part bull, part goat(?), with human face of a type commonly met with at Mohenjodaro and Harappa. In the field below is a line of seven small standing figures with dresses reaching to the knees, with a long plait of hair falling down the back, and a plume on the head. From its leaves the tree appears to be the Pīpala (*Ficus religiosa*, L.), a tree, which is still an object of universal worship throughout India, which no Hindu would willingly cut or injure, and beneath the shade of which he would be reluctant to tell an untruth. This is the tree of knowledge (*bodhi* or *bo* tree) under which the Buddha attained enlightenment. Whether it was regarded as the tree of knowledge at Mohenjodaro, we have no means of knowing. The conception of the tree of knowledge, like that of a 'tree of life' is traceable to a very early age in Mesopotamia and may have been equally ancient in India. The sacred Cedar of Chaldeans was not only the tree of life, but the revealer of the oracles of earth and heaven, and the name of Ea, the God of wisdom, was supposed to be written on its core. This point is not without significance, because the Pīpala tree, which in India is *par-excellence*, the "Tree of Wisdom" is conventionalised at Mohenjodaro and Harappa in a manner that can hardly fail to remind us of the schematized "Tree of Life" in Babylonia. An illustration of such conventional treatment is offered by seal no. 387... Here the tree springs from the conjugate heads of two 'unicorns', the same animals that are most commonly represented on seals and it is permissible to surmise that this particular animal may have been closely associated, perhaps as his *vāhanaṁ* or vehicle, with the Goddess of the Pīpala tree. The association of animal *vāhanas* with tree spirits (*Yakṣīs*) is a phenomenon that meets us at every turn in Hindu

and Buddhist iconography, and there can be no doubt that it was traditional from pre-Aryan times.[11]

"The nude deity appearing between the branches is very small and roughly portrayed, but the absence of any evidence of male sex, coupled with the fact that the tree deities in India are usually female and that the ministrant figures on this seal also appear to be women, all point to its being a goddess rather than a god. As to the human headed composite animal standing behind the suppliant, its meaning is not altogether clear. On some other seals whereon a similar scene is portrayed, its place is taken by goat or other horned animal, which sometimes stands behind the suppliant, sometimes between the suppliant and the deity, and which might therefore be taken as a sacrifice destined to that offered to the deity. That the animal is not a sacrificial one, is clearly proved by this seal, where it is of a mythical composite nature. My own view is that this animal stands in some such relation to the votary as does the intermediary deity on Mesopotanian seals, where he is frequently represented in anthropomorphic form taking the votary by the hand and pressing him or her to the higher deity; in other words I suggest that the composite human headed animal is a protecting local divinity of a minor type accompanying the suppliant into the presence of the Tree-goddess. ...The plume on their (ministrants) heads might be feathers; but it is more probable that they are small branches, such as in Kafiristan are still worn on the head by officiants at the worship of the *chili* or Himālayan Pencil Cedar (*Funiperus macropodes*), when branches are also burnt[12] in the honour of the tree spirit."

On certain sealings from Harappa, various sacred trees are represented, which the artists have been at some pains to differentiate one from the other, but the workmanship is too minute for several species to be identified with confidence. A point of some interest, however, is that two atleast of the trees from these sealings spring from what appears to be either an enclosing wall or railing such as commonly encircle the roots of the sacred trees on the later reliefs of the historic period and were regarded almost indispensible symbols of their sanctity[13] (Plates-10, 11 &12). Besides, there are several other seals and sealings from Indus sites which portray trees and plants.

Thus we have at both Mohenjodaro and Harappa two forms of tree-worship represented; one in which the tree itself is worshipped in its natural form, the other in which the tree spirit is personified and endowed with human shape and human attributes. This is precisely what is represented in the sculptures of Bharhut and Sanchi and others of the Early Indian school, but there is this difference that whereas in the later monuments the tree-spirit appears in a subordinate role as a dryad (*Yakṣī* or *Yoginī*), in the earlier she seems to have been already elevated to the position of an important goddess. This indeed is what might naturally be expected; for tree-worship was essentially a characteristic of the pre-Aryan, and not of the Aryan population and the tree spirit must have loomed far more important in pre-historic days among the people who originated this worship than it did later in an Aryanised India, where tree-worship invariably became subordinate to alien or semi-alien-cults.[14]

The personification of trees, or, indeed, of any natural object, has been quite common in India from the earliest times. In accordance with the ideas of the metampsychosis every tree or a plant has a personality or soul of its own, and is treated therefore, as a conscious human being. Thus for example, it is quite usual before cutting a tree to ask the pardon of the indwelling spirit; and the non-Aryan Gods will not shake a tree at night or pluck its fruit for fear of disturbing the sleeping spirit. This ideology has been well carried by the composers of Purāṇas, who seek blessings from the Tulasī plant while plucking its leaves for offering to Śāligrāma or Kṛṣṇa—

तुलस्यमृत नामासि सदा त्वं केशव प्रिया।
केशवार्थ विचिन्हामि वरदा भवतु शोभिने।।

Indeed the process of anthropomorphizing is carried even further than this among many non-Aryan tribes, brides being married to a tree, before being united with their husbands. According to Hewitt, Mariamma[15], the great Mother-goddess of South India, was a tree-goddess. The practice of marrying brides to a tree could be traced to matriarchal conditions, when true marriage was not recognized. The idea behind these marriages seems to be to bring the bride and the bridegroom into close association with the fertility of nature.

The trees may also be solemnly married one to another. As for example, the sacred Tulasī-plant is annually wedded, often with elaborate rituals and costly feasting to the Śāligrāma stone. In such cases as these, it is the tree itself that is personified as a deity and becomes an object of worship, when there can be no doubt that such personification is illustrated on the Indian seals. According to the present day notions a tree can also be tenanted by some alien deity, or, may be, by the spirit of a man of even of an animal, and not infrequently by a malignant spirit or demon, though whether these notions prevailed in prehistoric times, is difficult to say.

Apart from the other seals and sealings from Mohenjodaro, depicting the cult of Earth-Mother or Mother Goddess from Harappa[16] over which a nude female figure is depicted upside down with legs drawn apart and with a plant issuing from her womb. This figure is at the right extremity of the obverse face. At the left end of the same face and separated from it by an inscription in six letters are a pair of animal genii. This striking representation of the Earth-Goddess or the Goddess of vegetation, with a plant issuing from her womb is not unnatural, and is closely paralleled by a terracotta relief of the early Gupta age, from Bhita[17] in Uttar Pradesh, on which the Goddess is shown with her legs in much the same position, but with a lotus issuing from her neck instead of her womb (Plates-13 &14). This goddess of vegetation has also a reference in the *Devī Māhātmya* of the *Mārkaṇḍeya Purāṇa*, in which the Goddess in Chapter-XI declares thus—

ततोऽहमखिलं लोकमात्मदेह समुद्भवैः।
भरिष्यामि सुरा शाकैरावृष्टेः प्राणधारकैः।
शाकम्भरीति विख्यातिं तदा यास्यम्यहं मुवि।। —(XI.45)

(At that time, O gods, I shall support the whole world with the life sustaining vegetables, born out of my own body, until the rains set in again. Than I shall be famed on earth a Śākambharī).

Besides the seals and amulets, trees and plants are also found painted on the early pottery from Mohenjodaro and Harappa. Another Harappan site, *viz.*, Chanhudaro, also had artefacts depicting trees and plants (Plates 15, 16, 17, 18 & 19). The most common plant motif on the pottery of Chanhudaro, is the Pīpala tree, either represented in a natural manner or conventionalised. Sometimes the leaves of this plant are barely distinguishable from those of other plants, but when arranged in a stylized way, this Pīpala design forms a striking decoration. The jar-painter has painted on a jar, the tree displaying his skill, carefully avoiding all extraneous details. Many of the plants on the Chanhudaro pottery are represented as growing on the ground. Indeed in India, it is still a praisworthy act to plant and nourish one of the sacred trees and it may be that some of the painters of the pottery from Chanhudaro derived their inspiration from actual sacred trees in their own compounds.[18]

## 2. Early Pottery

Among the vegetable motifs depicted on the Harappan seals from Harappa and Mohenjodaro, the Pīpala tree occupies an important place. It is frequently painted on the earthen-wares and Pīpala leaf occurs as a symbol in the inscriptions on the seals. As a sacred plant, it is venerated by the Hindus even to the present day. The Aryans too venerated Aśvattha tree and used its dried twigs for making the sacred fire in the fire worship, indicating thereby that the veneration of the Pīpala tree was ot primarily non-Aryan.[19] The pottery at Lothal depicts the plant motifs quite often, in which the Pīpala tree has its own importance. As for example, an 'S' shaped[20] jar of medium fabric with a narrow flat base and flanged rim painted all over the exterior in Chocolate over a buffish slip. It is an excellent combination of Harappan and provincial style of painting. The surface is divided into four horizontal registers each separated by one or more horizontal bands. The lowest register is painted with Cross Latched leaf pattern derived from interesecting arcs while the one above it depicts a row of enclosed sun motif in a cross hatched panel, which is succeeded by a criss-cross pattern. The uppermost register is divided into four vertical panels, two of which are noted for semi-naturalistic forms such as the Pīpala tree, the palms and the rosette-like floral units. Two birds with partly open wings and holding fishes in their beaks are shown perched on a palm tree. Legs of two birds flying high up in the sky indicate other birds have flown away, after dropping the fishes. Three falling fishes are shown above one of the birds fluttering on the tree. Still more significant is the fox-like animal which is about to run away with the fish dropped on ground. Obviously the artist has tried to narrate the story of the "Cunning fox and the Crow", wherein the fox is trying to run away with the fish snatched from the birds after flattering them. In the lower half of the adjacent panel a caprid is shown gazing on a palm-like tree. The animals and plants are painted in free style which is more realistic than the normal Indus style. In the uppermost register of a pot shed from Lothal[21], Palm and Pīpala trees have been shown, with wave patterns, peacocks holding leaf in the beak, of Sun motif, filled triangles and hatched leaves. Indeed the jars from the Indus sites of Harappa and Mohenjodaro, as well as from Lothal portray plant or flower patterns, which goes on to testify that these motifs were appreciated and liked by the common man of that era, because the pottery vases are considered to belong to the lowest strata of the society. Besides the above, several other potsherds from the early sites have the plant or flower designs. Similar types of tree motifs are also available on pottery of other Harappan, post-Harappan and historical sites as well (Plates-15-19).

## 3. Ring-Stones

The ring stones discovered from the Harappa sites also portray the trees and plants. These ring stones appear to symbolise the principles of virility and fecundity, because some of them have the symbols of *liṅga* and *yoni* embossed on them, both of which are highly esteemed in all the ages. Marshall has discovered several stone discs, three of which were unearthed from the Bhir Mound at Taxila. One of them was discovered from inside the structures near the foot of Haithal (Taxila) and one at Kosam. The Haithal one is described by Marshall[22], thus—

> "It is a polished sand stone 3¼" in diameter, adorned on upper surface with concentric bands of cross and cable patterns, and with four nude females alternating with honey-suckle engraved in relief around the central hole."

It will be of interest now to compare with the above Taxila discs a partially broken reddish steatite circular disc about 2½" in diameter, found at Rajghat, which contains on the outer side of its top surface a well-carved decorative design. This decoration consists of a palm tree with a horse by its side, beyond which a female figure holding a bird in her outstretched right hand and a taurine symbol near her left shoulder; then follow in successive orders—a long-eared and short tailed animal, a crane, the goddess again with her hands this time stretched downwards, some object which is broken, a second palm-tree, a bird, a small circular disc, the goddess again with a circular disc near the left shoulder, then a winged mythical animal and lastly a crane with a crab-like object near its legs. The goddess is thrice repeated, with the various accessory figures.[23]

Bharat Kala Bhavan[24], has in its collection fragment of a red steatite disc unearthed in the course of excavations at Rajghat, which is more similar to the Taxila disc. This has a hole through the centre, around which, as in Taxila ones, are engraved two nude female figures with their hands stretched downwards with probably a honey suckle in between them; on the flat surface of the disc between cable designs are two monkey-like animals holding a creeper(?) with a lizard (or an alligator) in between them; there is a partially defaced inscription in early Brāhmī script on its rim which is illegible. Similar stone discs were also discovered from Patna[25] and other sites in Bihar.

Profuse lotus flowers, Palmyra and date-palm trees, etc. cluster round the goddess in several disc stones and emphasise the vegetation aspect of the deity so prominent in the ancient Sanskrit literature including the Purāṇas.

On the popularity of tree worship with the people of the Indus valley, Banerjea[26] has observed—

> "The worship of trees or tree-spirits is the characteristic manifestation of animistic belief. I have already referred to a seal which seems to unite in its device the worship of the female principle as well as that of the tree spirit, where the Epiphany of the female deity in the tree is portrayed in a half realistic, half conventional manner. Many seals in the Indus Vally sites show the presence of two different forms of tree worship among the people of the locality; one in which the tree itself is worshipped in its natural form, the other in which the tree spirit in personified and endowed with human shape and human attributes. On several sealings of the Indus region (*M.IC.*, Vol. I, Pl. XII, Figs. 16, 20, 21, 25, 26), various sacred trees are represented which the artist has attempted to differentiate one from another. A few of these trees appear to be like those enclosed by wells or railings commonly surrounding the base of the sacred tree (*Vṛkṣacaityas*) depicted in the later reliefs of the historic period... This enclosed tree on the Indus seals can very well be compared with the above, and can justifiably be taken as distant prototypes of the *Vṛkṣacaityas* or *Sthalavṛkṣas* represented by the latter. The terracotta seal (No. 2410) found at Harappa has as its obverse deivce a deity wearing a kilt or short tunic and a three pointed head-dress (or a *triśūla* horns?), standing under an ornamental arch, which appears to be made of the bent bough of a Pīpala tree. The lower ends of this bough are rounded up to form loops, each enclosing a star. The head of the deity is turned a little towards the right and on both arms he wears a number of armlets."[27]

## 4. Early Indian Numismatics

The tree and floral motifs also appear on the early Indian coins and one of the earliest devices found frequently on tribal coins is Gaja-Lakṣmī, *i.e.*, Lakṣmī standing, rarely seated over a lotus being bathed by two elephants. It appears on an uninscribed coin from Kausambi (3rd century B.C.), coins of Viśākhadeva, Śivadatta and probably also of Vayudeva of Ayodhyā (1st Century B.C.) and on uninscribed coins of Ujjaini (3rd-2nd Century B.C.). Such was the popularity of this device that some of the alien rulers of northern India like Azilises, Rajuvula and Solasa adopted it on their coins.[28] The reverse device of some copper coins of Maues and Azes[29] has been described as a female figure standing to front with trees. Coomaraswamy in his article on early Indian Iconography

(Eastern Art, Vol. I, p. 178), refers to three varieties of Lakṣmī, the third one described by him being Padmavāsinī or Kamalataya type in which she is surrounded by flowering stems and growing leaves and very often she holds one of the flowering stems in each hand. Relief carvings illustrating this motif of Lakṣmī or Gaja-Lakṣmī are found on the early monuments in Madhya Pradesh; here is a close approximation of the numismatic and sculptural representations. The motif as it typifies the Indian idea of prosparity, frequently appears on coins and sculptures of later date and still in use by the vast manority of Indian populace. Goddess Lakṣmī again, without the attendant elephants, either seated on a full-blossomed lotus or standing with a lotus flower in her hand, or standing on a lotus with the flower in her hand, very often appears on the coins of Ujjaini, on those of the Hindu kings like Brahmamitra, Dṛdhamitra, Sūryamitra, Viṣṇumitra Puruṣadatta, Uttamadatta[30], her hand very often appears on the coins of Ujjaini, on the coins of satraps of Mathura like Śivadatta, Hagamasa, Rajuvula and Sodasa, on the coins of Rajanya Janapada and on the coins of Bhadraghoṣa of Pāñcāla.[31]

In some indigenous and foreign coins, Allan has noticed a *liṅgam* on a square pedestal with a tree in the railing on the left of the same coin (Fig.-15), which could stand for *Sthala-Vṛkṣa* in association with the particular Śiva emblem and has been described as under[32]—

"Building(?) on l; tree in the centre; or r. female figure to l."

Now, the reverse of these coins has a hill with trees growing from its two sides and a honorific parasol like emblem on the top. Here again these symbols, taken together, seem unmistakably to point to their cult connection. A Śivaliṅga on a pedestal placed between two different trees inside railings is also represented on the obverse of some coins from Ujjaini.[33] Mention may also be made of a symbol appearing on some of the un-inscribed cast coins, which represents a trident with broad flattened prongs issuing out of railings enclosing Bodhi-tree and other Buddhist[34] symbols as represented in the Bodhagayā and Amarāvatī reliefs.

On the observe of the Andumbara chief Dharaghoṣa's silver coins, the figure of Vispamitra (Viśvāmitra) as described by the Kharosthi legend across the figure, but on the reverse there are two symbols *viz.*, combined trident-battle-axe, on a pedestal and a tree within a railing.[35]

Among the coins of early foreign rulers of India, Śiva appears on some coins of Gondapharnes. He stands facing with his left leg slightly advanced and head bent a little towards the left, clasping a long trident in his right hand and a palm branch in his left which rests in *kaṭihasta* posture.[36]

A few copper coins of uncertain origin bearing fragmentary legends have on the reverse a symbol described by Allan as a bushy tree in railing, but these have been interpreted to symbolize columns surmounted by fan-palm capitals.[37] A comparison with the representations of ordinary trees which appear on some coins of Ayodhyā, lend support to this view. It may be recalled here that fan-palm capitals were discovered from

Besnagar and Pawāyā and it is quite probable that these symbols are based on the votive columns connected with Bhagvatism.

The Tree-motif, depicting a tree enclosed with a railing also appears on a few coins of Siri Satkarni and other Satvāhana rulers. The reverse of these coins has been described thus (Sankalia, H.D., *From History to Pre-History*, pp. 169-172)—

**Fig.—74, No. 1**

"Five solid-branched tree in a railing, to right; to left symbol of Brāhmī "yya" letter; over head a taurine; to lower left of railing traces of some other symbols; below railing two fishes going left; river indicated by a thin slightly wavy line below fishes.

**Fig.—75, No. 1**

"Five solid-branched trees in railing to left; to its right triangle-headed standard, below it probably a *triratna* symbol, over tree a taurine; below railing fishes and river motif.

Besides the above, branches of a tree also appear on coin No. 79.2. On the coins of Kunindas, invariably a stag appears alongwith a goddess standing on lotus and holding a lotus flower in her right hand. This goddess has been interpreted to represent the Vedic goddess Śrī on the basis of a passage in the Śrī-Sūkta. The obverse of the coin type No. 30 of Azes[38] has been described as goddess Lakṣmī, standing to front with flower in raised right hand. Gardner writes about the device, "a female deity facing, clad in himation; holds in raised hand, flower; stands on lotus, besides her lion, (Lakṣmī?). The forepart of the lion is recognizable from his plate and this makes it highly probable that this represents Durgā-Siṁhavāhinī, the consort of Śiva, but in that case the attribute of flower in the raised hand and the lotus serving as a pedestal of the goddess still remain to be explained."[39]

On some of the earliest coins in the punch-marked series and on the Eran money, dated as early as the 3rd century B.C., the figure of a lotus is found; in the latter the eight petalled lotus is clearly recognizable.[40] This lotus was also intimately connected with the sun from the very ancient time and it played a significant role in the mythology of Brāhmaṇism, and its association with the Sun is fully borne-out by the testimony of the Purāṇas which enjoin the execution in sculpture of a twelve petalled lotus on different petals of which figures of the different aspects of the sun-god are to be placed with the god Bhāskara on the central pericarp. The lotus symbolising the sun and the creative force come to hold a unique position in Indian art of all ages and all religious creeds. The author of the *Viṣṇudharmottara*[41] realised the importance of this motif in iconographic art and gave full and detailed instructions for the mode of its representation. In the sculptures of divinities lotus is the commonest symbol found in their hands. Some of the lotuses, atleast those in early coins, may be taken to represent the sun.

## 5. Sculptural Art

The early specimens in the sculptural art in India take us to the Mauryan period. During this period Aśoka raised massive monolithic columns, on the abacus of some of these pillars, honeysuckle designs or the geese holding the lotus stalks are found carved, which could represent the earliest specimens of the plants in sculptures. Besides the adoration of the Bodhi-tree, the forests infested by beasts, etc. are abundantly found in the sculptural art of Bharhut, Sanchi and Amarāvatī, which aspect has been discussed in detail in the chapter on 'Buddhist Art and Literature' in this work.

On the adoration of the plant and tree motifs in the art of Bharhut, V.S. Agrawala has observed as under (*Indian Art*, pp. 138-39)—

"It is clear that the religious world sculptured at Bharhut, partakes of these cult deities in an atmosphere of joy and freedom. As part of Vṛkṣa-maha or tree worship, we have a series of ponds showing worship of the following Bodhi-tree of the six Buddhas—

(1) Aśvattha or Pīpala (*Ficus religiosa*) known as the Bodhi-tree of Gautama Buddha. The Bodhi-tree is worshipped by royal persons, common citizens, elephants, Nāgas, etc. It is placed on a railing and at times the Bharhut sculptor has conceived its shrine (*Bodhi-ghara*) as a building of great beauty with an open pavilion below and a domed top with railings in both the stories.

(2) The Banyan-tree (*Fiscus indica*), Bodhi-tree of Kaśyapa Buddha; in several scenes it is depicted in a charming manner specially where a group of elephants are paying homage to the tree, the scene being labelled as *Bahu Hatthiko Nigoda, i.e.*, the Nyagrodha tree of the herd of elephants.

(3) Udumbara (*Ficus glomerata*), Bodhi-tree of the Buddha Kanakamuni.

(4) Patali (*Bignonia suavealens*), Bodhi-tree of the Buddha Vipassin, whose name is inscribed above the medallion. This tree is depicted in full flowering state.

(5) Sāla-tree (*Artocarpus locuca*), Bodhi-tree of the Buddha Viśvaloka.

(6) Śirīśa (*Albizzia lebbeck*, Benth.), Bodhi-tree of the Buddha Krakacanda. All these trees are fortunately inscribed with the names of their respective Buddhas. These constituted an essential feature in the ancient cult of tree worship as adopted to the needs of Buddhism.

"We also find here a side feature in the form of garden sports known as *Pracam Krīḍa* or *Udyān Krīḍa* in which women are shown standing under the various flowring trees, especially Sāla and Aśoka and enjoying such sport as *Sālabhañjikā*, *Aśoka-puṣpa-pracayika*. These are depicted at Bharhut and continued in the railings of later times."

In the *Jātaka* stories depicted on the Bharhut and Sanchi gateways, the trees, plants and wood are depicted in abundance. As for example, in a scene from *Śyāma Jātaka*, depicted in a panel on the Sanchi gateway, the blind parents of Śyāma are shown seated in front of their hermitage in the forest. There is also a tree of *Śrīphala* on the bank of the river, and the boy carrying the earthern pot is approaching the river to fill the same with the river water. In the left hand corner, the king has shot an arrow hurting the boy fatally. Next the king gets remorseful of the act and finally the hermits gain their eyesight and the boy his life with the intervention of Indra. The entire scene is enacted in the forest.[42] The story of *Mahākapi Jātaka* too hovers around the forest, which is also found depicted on the Sanchi gateways. In this case too, the trees bearing flowers and fruit can be seen.[43] Another panel from Sanchi is divided in four compartments beautifully depicting the ancient type of gardens. In each one of these compartments, two persons, a male and a female are seated on couches. Two of these groups are close to the water of a lake in which lotuses are shown growing; two others are on the terrace, in the centre of which a flight of steps leads from the lake. On each one of the couches, a man and a woman are seated in close conversation, and drinking at the same time. Below or in the front of the couch of the couple in the left hand upper corner is the wine pot or a jar. The entire panel is filled with floral as well as fruit plants. There are several other panels at Sanchi which portray trees, plants, gardens and forests. Sometimes the trees are laden with fruits and at least in one case the trees are shown half submerged in flood waters (Plates-20-26).

There are several *Jātaka* scenes from the gateways of Bharhut, which are supposed to have been enacted in the forest. One of them, concerning *Aramdusaka Jātaka*, portrays the real plant life. As the story goes, a gardner, incharge of a royal port had to proceed on a holiday leaving the care of the garden to the monkeys, who dwelt there enjoying royal courtesy in the garden. The monkeys as desired by the gardner, started watering the plants and trees with water skins and wooden watering pots. Anxious to prevent the wastage of water, their chief commanded to pluck each one of the young tree to examine the size of the roots. The Bodhisattva, however, cautioned them against doing this. But it was too late. By that time all the plants had dried up. So the foolish monkeys inspite of their best intentions destroyed the plants. This story has been portrayed in a panel from Bharhut,[44] showing a man and two monkeys in the forest. One of the monkeys carries away a pair of water vase slung from the ends of a *behaṅgi* pole. The other monkey sits on the ground holding a plant(?) in both his hands. There are several other specimens showing the adoration of Bodhi-tree and Kalpalatā (Plates-29-31).

Another most interesting aspect of the plant life at Bharhut, Sanchi and Amaravatī, is the flower vases full of plants laden with birds and flowers with leaves. While those from Bharhut and Sanchi are simple flower vases, but the one from Amarāvatī represents *purṇaghaṭa* with lotus buds and flowers (Plates-32-36).

It is not during the Aśokan period alone that the abacus of the pillars was decorated with honey-suckle design, but during the Śuṅga period the same practice was followed. To such specimens one belonging to the Aśokan period and other to the Śuṅga period were discovered from Vidisha. The abacus of each one of these two capitals is adorned with honey-suckle design. The best specimen of the abacus having the flowery decoration resembling a Kalpalatā, is available on the Elephant capital from Saukisa, which belongs to the Aśokan period, while those belonging to the Śuṅga period project similar type of floral designs on the Heliodorus pillar, and the *makara*-capital from Vidisha (Figs.-16-19).

Besides the above, the Śuṅga period is also known for its fan-palm capitals. Two such capitals have been discovered, one from Pawāyā while the other one from Survaya. The presence of these fan-palm capitals belonging to the Śuṅga period goes on to testify the importance the palm tree enjoyed with the contemporary rulers. The fan-palm capital from Survaya (Gwalior Museum) is also known as the *Taladhvaja* and is believed to be closely related to the cult of Balarāma, the elder brother of Kṛṣṇa. It projects a bunch of palm-leaves somewhat in a realistic manner. But the figure of a donkey projected on the capital itself makes it quite interesting. This could possibly represent the ass-demon Dhenuka, who was killed by Balarāma by hurling it against a palm tree (Figs.-20 & 21). Another *Taladhvaja* capital is also reported from Lucknow Museum which has been mis-reported as coconut capital. The Fan-palm capital from Pawāyā, however, projects possibly the date-palm capital with the leaves handing down words and a heap of dates is seen over the abacus (Plates-37 & 38).

The most striking evidence of the conception of mythical tree known as *Kalpavṛkṣa* was conceived during the Śuṅga period and a unique and rather the earliest specimen of the same was discovered from Vidisha, which possibly served as the capital of a column, now extinct. This *Kalpavṛkṣa* has been prescribed in the *Matsya Purāṇa* to be planted as under—

ऋत्विग् मण्डप सम्भार भूषणाच्छादनादिकम्।
काञ्चनं कारयेत् वृक्षं नाना फल समन्वितम्।।
नाना विहग वस्त्राणि भूषणानि च कारयेत्।
शक्ति तस्त्रिपलादूर्ध्वमासहस्त्रं प्रकल्पयेत्।।
अर्ध कलृप्तसुवर्णस्य कारयेत कल्पपादपम्।
गुड प्रस्थोपरिष्टाच्च सितवस्त्र युगान्वितम्।।
ब्रह्मविष्णुशिवोपेतं पञ्चशाखं सभास्करम्।
कामदेवमधस्ताच्च सकलत्रं प्रकल्पयेत्।।
संतानं पूर्व तस्तद्धत्तुरीयांशेन कल्पयेत्।
मन्दारं दक्षिणे पार्श्वे श्रिया सार्धं घृतोपरि।।
पश्चिमे पारि जातन्तु सावित्र्या सह जीरके।
सुरभी संयुतं तद्धत्तिलेषु हरिचन्दनम्।।
तुरीयांशेन कुर्वीत सौम्येन फल संयुतम्।
कौशेयवस्त्र सम्बीता निक्षुमाल्य फलान्वितान्।।

—*Matsya Purāṇa*, 277.3-9

(The *Ṛtvikas, maṇḍapa,* ornaments, clothes, etc., should be arranged. A golden-tree yielding all desires (*Kalpa-Pādapa*) of the various kinds of fruits should be made; and also birds perched in the tree; the clothes and various ornaments should also be made. The tree should be four to one thousand *pals* according to the means of the devotee. The gold used should be mixed with alloy. It should be located on a mound of molasses of 32 seers, covered with a piece of white cloth. Five branches alongwith images of Brahmā, Viṣṇu, Śiva, Sūrya and cupid should be made, on the lower branch should be placed Kāmadeva with Rati. Satanaka tree ¼ of the golden Kalpavṛkṣa should be located to the east of the latter. Mandāra trees together with the image of Lakṣmı should be located on a pot of clarified butter in the south, Pārijāta tree alongwith the image of Sāvitri should be located to the west on a mound of Cumin seed (*Jīraka*). Similarly Haricandana tree should be located to the north alongwith Surabhi Cow on a mound of sesamum. The tree is to be adorned on one-fourth part of it with flowers. All the other trees should be made 1¼ of the golden *kalpa*-tree and covered with a yellow cloth and decorated with flowers, sugarcane, etc.)

The Vidisha *Kalpavṛkṣa* is represented as a Banyan tree (*Ficus indica*) with long pendent roots, from which untold wealth, in the form of square pieces of money is dropping in such quantities that all the vessels placed below are full and overflowing. The upper part of the tree which is spherical in shape is covered with large leaves and small berries and the stains of pendant roots are represented on a cylindrical neck below which they divide into eight compartments. In these are placed four vessels full of money and four skins full of wine(?). The skin bags could also be intended for large bags of money as each one is tied with a band round the neck (Plate-27).[45]

In the Kuṣāṇa art of Mathura, the woman and the tree motifs are found in abundance in the red sandstone sculptures. The Kuṣāṇa kings Kaniṣka, Haviṣka and Vasudeva had their capital at Parśupura (modern Pesawar) and also ruled over Mathura from A.D. 78-200. The leaves and flowers of Aśoka tree (*Saraca asoca*, Roxb.) were the most popular motifs with the Mathura sculptors. Numerous such sculptures were found in Lucknow and Mathura Museums, where Aśoka-tree is shown associated with female figures. These specimens do not represent dancing girls, but are tree-deities or symbols of fertility who were adored for gifts of offsprings to barren ladies (Plate-28 a-28 c).

The most important aspect of the Kuṣāṇa art of Mathura has been the correlation of the females with the tree, the tradition which has come down to us from the art-motifs of Bharhut and Sanchi. The fact with respect to the beauty of the female figures was noticed in ancient times showing that the art connoisseurs also were conscious of this supreme fact which has made Kuṣāṇa art of Mathura such a loving object in the eyes of the modern critics. "The *Salabhanjikā* woman carved on the Vedikā pillars were standing in various graceful poses, well supported (on crouchant figures), beautifully ornamented wearing garments painted in various colours and necklaces of various designs, slender waisted, having round prominent breasts, eyes with red corners and black curly hair, standing under Aśoka Trees and holding their distended boughs, stealing the hearts of the gods as it were with their rallying glances and teasing as it were with the play of their eyes" (*Rayapaseṇiyasutta*). These *Sālabhañjikās* are found depicted on both the Buddhist and Jaina railings to which the above description equally applies. The *Rayapaseṇiyasutta* further testifies that there were sixteen railing pillars (at Mathura) on each side of the *torṇa* gateways which gives a total of sixty-four pillars for the entire railing, a number which has actually been confirmed by the pillars of the Bharhut railing.

> "The term *Sālabhañjikā*, as already pointed out earlier, denoted a female sport implying the gathering of Sāla flowers by women standing under the blossoming Sāla trees. This was a kind of garden sport which was popular in the eastern India referred to as *bracam-krīḍa* by Paṇini (6.2.74). A graphic account of the *Sālabhañjikā* festival is found in the *Avadānaśataka* which was undoubtedly based on earlier tradition. "Once lord Buddha dwelt at Śrāvasti in Jetavana, the garden of Anantapiṇḍa. Now at that very time the festival called *Sālabhañjikā* was being celebrated at Śrāvasti. Several hundred thousands of beings assembled there and having gathered Sāla blossoms, they played, made merry and roamed about." And again quoting from Nidānakathā, the description of Sālabhañjikā festival celebrated in Lumbinī garden. "Now between the two towns (Kapilavastu and Devadahu), there is an auspicious grove of Sāla trees belonging to the people of both the cities, and called Lumbinī grove. At that time from the roots to the top most branches swarms of five-coloured bees, and flocks of birds of different kinds roamed, warbling sweetly. The whole of the Lumbinī grove was a wood of variegated creepers, or the well-decorated banqueting hall of some king. The queen beholding it was filled with the desire of disporting herself in the Sāla-grove (*Sālavanakilam kilitu-kamata*); and the attendant entered the wood with queen. When she came to the root of an auspicious Sāla tree, she wanted to take hold of a branch of it. The branch, bending down like a reed heated by steam, approached within reach of her hand. Stretching out her hand she took hold of the branch and then her pains came upon her."

On the question of the Mathura sculptor dealing with the theme of *Sālabhañjikā* in the sculptural art of Mathura of the Kuṣāṇa period, V.S. Agrawala[46] has observed—

> "The Mathura artists were called upon to re-organise the railing in which work they displayed much originality and broad outlook, mainly introducing a variety of motifs based on female postime and sports known as *Krīḍa-Vihāra*. It was a popular belief that Indra in heaven enjoyed with his nymphs and this form of pleasure became archetypal for kings on earth. These festive celebrations were ordered by royalty and were comprised of the following items—(i) tree and garden sports (*Udyāna-krīḍa*); (ii) Water sports (*Udaka-Krīḍa*); adorning the body with unguents, clothes and ornaments (*gatra-mandana*), including painting minute scroll designs on cheeks, forehead, chin, etc. called as *Viśeṣaka-racana*; (iv) drinking wine in company of females (*apana-goṣṭhi*)... The Mathura railing

pillars present a rich canvas of these scenes which are made permanent in stone as reflecting the joyous feeling and buoyancy of life in that age...".

Someof the motifs associating the females with trees, during the Kuṣāṇa period at Mathura and Gandhara, have been described as under—

(i) A scene[47] of *Aśoka-dohada* is carved on an ivory plaque found at Begram or ancient Kapisa. The head-dress has a spirally rolled scarf on the top of which hangs a braid of hair tied in a loop. It was this special hair style referred to by Aśvaghoṣa as *Suklamsuka attalaka keśa* (*Saundrānanda*, 7.7) or *Kokila Keśapāṣa* because of the upper black tuft of hair on the pile of white silk resembling the black bird seated on a tree of white flower.

(ii) Another motif[48] is that of weaving heavy flower garlands and making a *sekhara* of the same. These were the popular scenes of *Puṣpacchaāya*, *Puṣpagrathana* and *Puṣpabharaṇa*, *i.e.*, floral amusements of which the artist made the fullest use.

(iii) The next[49] motif is that of a female figure standing aganist the stem of a tree and holding a large lotus flower in her hand. The lotus had prominent association with these sports. This motif corresponds to the one described by Kālidāsa as *Lāla-kamala* held in the hands by the sporting girls. In later Orissan art, the same was known as *Padmagandhā*, *i.e.*, a young woman smelling a lotus or holding it, whereas in western India, she was known as *Padminī* or *Saugandhikā*.

Besides the above, in the Kuṣāṇa art of Mathura, various other motifs, like the Kalpalatā, vases filled with flower stems, lotus medallion are depicted.

The Gupta art was evolved at Mathura and the evolved classical formula travelled subsequently towards the east and found another congenial atmosphere for its further refinement at Sarnath. During the Gupta period flourished a number of artists, writers, philosophers and poets, whose creations earned global applause. These artists created the cave art at Ajanta, Ellora, Blumra and Bhitargaon, etc. Some of the most beautiful monuments representing the very acme of India's artistic achievement are a cultural heritage of Gupta period. Amongst them the immortal Ajanta frescoes and art specimens of several other sites represent richly drawn floral and plant motifs, which still remain masterpieces of Indian plastic art. The importance of plants and flora is evidenced from a reference to Baṇa's *Harṣa-carita* (Cowell-p. 124), which testifies that at the time of Rājā Śrī's marriage multitude of (*Lepyakaras*) modellers were engaged in moulding clay figurines of fishes, crocodiles, coconut trees, plantain and betel trees. Besides, female clay figurines holding auspicious fruits, technically named *Añjali-Karika* were fixed in the sides of the altar. A beautiful plaque from Rajaghat shows a lady gracefully seated on a swing suspended from an Aśoka tree. The breeze in the garden is wafting her flowing garment and is gently shaking the foliage of the tree (Fig.-22). It is a masterpiece[50] combining the plants, movement and poise.

Coming to the Gupta sculptures, it may be stated that they differed in style slightly from their counterparts in the Kuṣāṇa art. The halo in the Kuṣāṇa images of the Buddha was plain with only a short scalloped margin, but the halo[51] of the Gupta Buddha is elaborate and covered all over with concentric bands showing rope design, rosettes, foliated scroll with peacock design, a coiled garland of flowers, beaded border and scallops (*hastnikha*). Besides, several sculptures of Gupta era show elaborate floral carvings on them. The Gupta period is also known for its Kalpalatā motif. In this connection it may be recalled that *kalpalatā-dāna* is considered to be quite virtuous for a devotee and the methodology to be adopted for such a gift has been described in chapter-286 of the *Matsya Purāṇa* as under—

> "On an auspicious *tithi*, after *Svastivācana*, the devotee should arrange for *Ṛtvikas*, *paṇḍāla*, articles, ornaments, clothings as before and invoke Lokapālas as is done in the *Tulādāna*. The *Kalpalatās* (creepers yielding all desires) of gold should be made and they should be adorned with various types of fruits, flowers and several kinds of birds, couples of Vidyādharas, pairs of golden birds, Siddhas culling flowers and fruits. Birds should be made alongwith images of ascetics living only on fruits; as well as the images of the Devas, the attendants of Lokapālas should also be made. Beneath the two creeper twigs on the mound of salt should be placed an image of the goddess Brāhmī of infinite power and glory, holding lotus and conch, as well as the goddess of Ananta Śakti. On the mounds of molasses to the east should be located the image of the goddess Indrāṇī seated on an elephant holding a thunderbolt in her arm. The consort of Agni should be located in the south-east corner on a mound of turmeric holding a sacrificial ladle in her hand. In the south should be located the image of goddess Godini mounted on a buffalo; in the south-west corner should be placed the image of goddess Nairityo on a mound of clarified butter..."

The *Matsya Purāṇa* has been believed to belong to the late Gupta period and the practice of the gift of Kalpalatā has been recognised by its author, which was possibly based on the prevailing tradition. In view, however, of the importance shown to the gifting of the Kalpalatā and the merits derived out of such a gift, it is no wonder the Gupta artist continued the practice of depicting the Kalpalatā motif in the sculptural art. On a door jamb from Garhwa, the Kalpalatā motif is treated in an exquisite style[52] (Fig.-23). According to Cunningham, it is an undulating stem of a creeper with large curly and intertwining leaves, and small human figures, both male and female, climbing up the stem and sitting on the leaves in various attitudes. The whole scroll is deeply sunk and very clearly and carefully carved; and... is one of the most pleasing and graceful specimens of Indian Architectural ornament.[53] The foliated scroll was a special trait of Gupta art. As Smith has observed, "The intricate scroll work on the western face of the celebrated Dhamekh Stūpa at Sarnath, is one of the most successful examples of the decoration of a large wall surface to be found in India. The artist who traced wonderfully complex spirals must have undergone prolonged and rigorous training".[54] The Kalpalatā motif is also carved on a jamb of the doorway of the Deogarh temple (Fig.-24). It also illustrates the deep cut foliated scroll work called *patralatā*. The creepers issues from the navel of a squatting Yakṣa figure. The narrower band consists of a series of smaller rosettes called *Phullavali*.

The conception of Kalpalatā is also referred to in the *Jātakas* (Vol. IV, p. 352), *Rāmāyaṇa* (Kiṣkindhākāṇḍa, 43.48), *Mahābhārata* (Bhīṣmaparvan, VII.9) and the Purāṇas.[55] The Ayodhyākāṇḍa of the *Rāmāyaṇa,* speaks of the divine damsels as fruits from trees in the garden of Kubera (दिव्यनारीफलम्).

In literary description several names resembling the Kalpalatā can be traced; *viz.*—

*Patralatā, Patrāṅguli, Patraphaṅgarcna, Anekabhaṅgakuṭilapatrāvali.*

The fondness for curves and twists grew in the later times with an increasing tendency for deep cutting until at last it reached its logical sequence in almost perforated work of the screens and reliefs in the medieval art of A.D. 11th-12th century, as seen in Chandela and the Dilwara marble temples.

During the post-Gupta era, the earlier traditions of carving scroll work and Kalpalatā motifs in sculptural art were continued. During this period India was visited by the famous Buddhist traveller Hieun Tsiang. He, in his accounts, mentions about the botanical properties in the country, as witnessed by him—

1. *Āmala (Ngan-mil-lo)*
2. *Madhūka (mo-to-kia)*
3. *Bhadra (po-ta-lo)*
4. *Kapittha (kie-pi-ta)*
5. *Mocha (mau-che)*
6. *Nārikela (na-li-ki-lo)*
7. *Panāsa (panna-so)*
8. *Wild palm (nai)*
9. *Apricot (hang or mui)*

About the cultivators and the crops they raised, Hieun Tsiang states as follows—

"In cultivating the land, those whose duty is to sow and reap, plough and harrow (weed), and plant according to the season; and after their labour they rest awhile. Among the products of the ground, rice and corn are most plentiful. As regards the edible herbs and other plants, ginger and mustard melon and pumpkins, the *Heun-to* (*kendu*?) plant and others may be mentioned. Onions and garlics are little grown, and few persons eat them. It anyone uses them for food he is expelled beyond the walls of the town. The most usual food is milk, butter, cream, soft sugar, sugar candy, the oil of musterd seeds and all sorts of dishes made of corn are used for food."

The medieval period witnessed more of plant and flower patterns in India, as compared to the earlier times. Earlier Hindu art was centred upon temples and the multiplicity of deities involved. It was based on the concept of an organic growth, on the rhythms of nature and of humanity, always complex and individual. Sculptures were the natural expression of such beliefs, and the temples themselves related perhaps more nearly to sculptures than to any architectural approach to structures. Nature and the human figures are displayed lifelike groupings of great beauty and often of surprising eroticism. Flowers like lotus and plants were part of worship and were shown with intricate details in carvings. Some of the deities were portrayed carrying the flowers like lotus. Besides that, the deities were shown standing or seated under trees laden with fruit. The images from Somnathpur and Mysore could be quoted as examples. This trend is also available in the sculptures in Greater India. With the appearing of the Muslims on the Indian historical horizon, the cultural life of India witnessed far reaching changes. In the thirteenth century, they introduced handmade paper from central Asia thus providing an impetus to painting and writing which rose to perfection during the time of Akbar (Plates-40-43).

The Indian terracottas, like other specimens of art represent the rich cultural heritage of the country. Several excavated sites produced specimens of terracottas, which were exquisitely designed with numerous floral and plant motifs. Indeed, the terracotta art in India is of great antiquity and is mentioned in epics as well as subsequent literature. According to the *Mahābhārata*, Ekalavya, the pupil of Droṇa, made a replica of his teacher in clay for worship.[56] Aśvapati, king of Madra, as child made small clay figurines of horses as toys.[57] In the *Bhaddhasala Jātaka*, reference is found to princes receiving presents of elephants, horses and other toys from their mother's father.[58] The *Mṛcchakaṭikam* play derives its title from a clay toy cart with which Rohasena, the infant son of Carudatta used to play. In the *Mārkaṇḍeya Purāṇa*[59], a clay image of Durgā (*mahi-mayi mūrti*) is installed for worship. Kālidāsa refers to a painted clay peacock in the hands of child Bharata.[60]

The earliest terracotta art emerges from the Indus Valley and besides several hand modelled figurines, the flora and fauna are found represented on the terracotta seals and sealings, which have been discussed earlier. The clay figurines of the historical period although separated from those of Indus Valley by an interval of over twenty centuries, nevertheless show links of styles and subject matter. The most remarkable of these terracotta figurines belonging to the Mauryan period were unearthed from Pāṭaliputra consisting mainly of a type of female figure with elaborate floral head-dress and panniered Skirt somewhat oversized and spread out on two sides of legs like wings. A remarkable specimen comes from Patna Museum. On the technical side, the figure is entirely modelled by hand except the head pressed out of a mould, later on a good number of applique decorations were attached; for example a couple of falling locks on the protuberance of the head-dress, surmounted by a full-blown lotus, another disc or lotus in the centre of two projections.[61] There are several other such figures with floral designs, Basarh (ancient Vaiśālī) was another centre of clay figurines flourishing during Śuṅga period. A plaque shows a figure in the midst of lotuses with two long stems held in hands on waists. The abundance of lotuses points to a kindred Śrī Lakṣmī type.[62] Another specimen allied both in type and ornamentation was found at Kosam,[63] currently preserved in the Allahabad Museum. It is undoubtedly a male figure although wearing the same type of conspicuous ornaments. Its hands hang by the side between lotuses springing from below. A wing preserved at the back of the right shoulder connects it with the winged cupid type. In Indian mythology, Kāmadeva is believed to be the son of Lakṣmī and this may explain the intimate association of this figure with lotuses, its formal affinity with the Śrī-Lakṣmī type. Several other terracotta figurines decorated with floral patterns have been unearthed from several other sites including Sonkh from Mathura (Figs.-25-29).

A notable example in the group of Kosam terracotta is the female figurines now preserved in the Indian Institute Oxford. Its headgear is distinguished by two prominent side-rolls of a turban flanking a central boss stamped with floral designs and a band of sacred symbols, probably to mark the divine nature of the figure.[64] Similar figures also come from other sites like Mathura, and Ahicchatra.

The best specimens of the late Gupta terracotta art of Eastern India is preserved at Paharpuṛ in Bengal, which provides a rich display of flora, fauna and other wild life. The plants like Palm, Plantain, Campaka and Kadamba are the favourite themes of illustration.

The tradition of illustrating terracotta objects with plant and floral designs travelled from the post-Gupta and medieval ages to modern times, as could be witnessed from Mahasthana (Bogra district), Mathurapur (Faridpur district), Brick temple at Deuliya in Burdwan district and other sites in Bengal, like Birbhum.

## 6. Murals

While tracing the presence of early murals in the country, one is reminded of the world renowned frescoes of Ajanta, wher many of the ceilings and wall surfaces have been decorated with floral designs and plant motifs. As for example, such floral designs are executed in the ceiling of the verandah of Cave No. XVII. The entire ceiling of the verandah is divided into panels containing creepers and floral designs of great variety. The artist in his love for rich patterns has not only delineated flowers in different stages of bloom, but has also added ornamental features to the stems and leaves of the creepers bearing these flowers.[65] Similarly, the design of the middle two brackets is quite noteworthy in which two small boys are holding a pair of ducks.[66] The drawing of the lotus-vine on the architrave, in front of the door, is quite attractive.

The floral designs are also painted on the ceilings of the hall and the corridors of Cave No. XVII. The ceiling of this *vihāra,* like its walls and columns, is lavishly painted, the designs comprising of floral patterns, fancy creeper-motifs, besides the mythical animals. One of the designs represents a lotus creeper and the artist has painted large cap-shaped leaves between branches of flowers. Another band contains a simple fine petal floral design, so that the eye of the onlooker may not be tried by a succession of too many intricate patterns.[67] The ceilings of the corridors are also magnificently painted, and on the ceiling of the front corridor, there is a design in which three peafowl are represented around a lotus creeper with flowers and birds. The design is highly conventional, even the tail feathers of birds have been changed into ornamental scrolls, yet the position of their necks and other parts of the body are very characteristic of the habit and carriage of the peafowl.[68] Similarly on the ceiling of the left corridor is delineated a lion to whose tail the painter has given an artistic loop. the raised head and the shape of the muzzle are also unreal, but the entire motif, set in a circle of lotus creeper, is extremely attractive, and shows well the ingenuity and taste of the artist. During the fifth century A.D. the decorative aspect of art from the creative point of view reached its climax. afterwards, perhaps there was retrograde movement, whether for political or other reasons is not known. The ceiling of Cave No. XXI, at Ajanta has floral and geomatric designs. It also shows a charming lotus creeper in which freshness of blue colour is particularly striking. On the creeper, there are also white flowers and buds, the delicate petals of which have been painted with great care. In the choice of colour too, the artist's refined taste may be admired, the flowers being both white and blue, the stem green; while the creeper, with a view to throwing it into relief has been painted on a dark background.[69] The ceiling on the right side of aisle of Cave XIX has also some charming creeper designs among which dwarfish human figures are interspersed.[70] (Cl. Plates-1-3).

Besides the above, the following trees and plants are also found depicted in the Ajanta paintings—

| | |
|---|---|
| (1) Aquatic plants[71] | (2) Areca nut[72] |
| (3) Banana[73] | (4) Banyan[74] |
| (5) Cotton-wool tree[75] | (6) Creepers[76] |
| (7) Ferns[77] | (8) Gūlar (Wild Fig)[78] |
| (9) Lilly[79] | (10) Lotus[80] |
| (11) Mango[81] | (12) Osier[82] |
| (13) Palāśa[83] (Flame of forest) | (14) Pine-apple[84] |
| (15) Pumpkin[85] | (16) Reeds[86] |
| (17) Samalu[87] (*Vitex trifolia*) | (18) Teak[88] |
| (19) Sugarcane[89] | (20) Mūlīs (Radish)[90] |

The frescoes of Bagh Caves come next, which are taken to be contemporaneous with later Ajanta murals. In the craftsmanship they are similar. They have the same mood of reserve even in the midst of joy. The only difference between the two is that while the Ajanta frescoes are deeply religious, those of Bagh caves are more human, depicting the life of the time with its religious associations. The richness of floral designs can be witnessed in the frescoes of Bagh Caves as well. Possibly, the best of these floral designs are available from Bagh Cave No. 4. These floral designs decorate the pillars and walls of the Cave delicately displaying flowers, fruits, buds, and half opened buds (Figs.-30-32) (Col. Plate-4).

Besides the Ajanta and Bagh Caves, the fresco paintings of exquisite beauty showing floral patterns are also available in the Buddhist monasteries of Lahaul Spiti in Himachal

Pradesh and Ladakh in Kashmir. A number of temples in the South, the Mattanchery Palace, Cochin; Hazardwari palace in Murshidabad, Royal Palace, Patiala, Brijarajabehari temple, Nurpur, Narbadeśvara temple at Tira Sujanpur, are some of the numerous temples in the country, which preserve such motifs in abundance. A painting from the ceiling of the Sikh shrine (now demolished) at Amritsar, depicts the greatest of the Sikh *Gurus* seated under a tree, with other ten *Gurus* flanking him.[91]

The use of flower patterns was restricted to those drawn on the walls and ceilings of temples and palaces alone, but was adopted by the Christian artist in India. Many Christian churches in the country, particularly in Goa, adopted profusely decorated floral designs, which could be rated as some of the best not only in the country, but also in the global context. These motifs, created mostly under foreign influence, still portrayed the involvement of Indian artists and artisans, who while working on such edifices could not set aside the traditions of the ancient art, which was handed down to them by their ancestors. The pillars in the Dolly Jesus Church at Daman and other churches in Goa and other parts of the country stand testimony for the same (Cl. Plate-5).

The frescoes in the Jaina Caves of Sittanvasal in the South, dating back to the middle of 9th century A.D. depicts a lotus pool, beaufitully displaying the natural environments and the humans plucking the flowers (Skira Albart, *Paintings of India*, pp. 42-43).

## 7. Miniatures and Paintings

The artist of the miniature paintings, like other professional artisans in the field of Indian art, did not neglect the beauty of nature so abundantly available in the country in the form of the plant life or floral pattern. He never missed an opportunity to express his concern for the plant life whenever he had the occasion to do so, be it, fresco, manuscripts or miniature paintings. These representations in the art of paintings are so lively that one cannot help admiring the same even during the modern times.

### (i) Jaina Art

One of the earliest expression of *kalpa-latā* motif is available over a painted book-cover from Bhandara, Jaisalmer, which depicts the lotuses besides a human and a monkey. The miniature has been dated to the eleventh century A.D. though it could be of earlier date.[92] Another specimen of the similar type, though dating back to the twelfth century A.D. comes from Bhandara, Jaisalmer, and depicts in the Kalpalatā, the figures ducks, giraffe, a female, a demon and flowers[93] (Cl. Plate-6). An illustration from a palm leaf manuscript from Bhandara dating back to the 12th century A.D. depicts four armed goddess Sarasvatı, holding a lotus in each one of the upper two hands. Two lotuses also appear flanking at the top, one on each side.[94] Another book-cover from the same site and of the same age have Kalpalatā motifs with animals, birds and fabulous creatures appearing in the artistically decorated circular abodes.[95]

A miniature from the *patli* of Deva-Sūrī-Kumudacandra, displays a beautiful tree encased by a railing, from which a snake emerges, besides a stream, female musicians and a bearded male following a nude female.[96] A miniature from the *Kalpasūtra* (from National Museum), dating back to A.D. 1417, displays the scene of Mahāvīra's renunciation of the world seated under an Aśoka-tree. In this scene the tree above as well as the flowers on the ground have been realistically depicted[97] (Cl. Plate-7). In another miniature from *Yaśodharā-Carita* (A.D. 1494), there is a forest scene, represented by five trees and other shrubs (Cl. Plate-8). In the environments of this wild forest, a porcupine is showing killing a snake and in turn is being attacked by a wild animal. The background surface is painted blue with small white flowers spread over the entire surface.[98] In a scene from the same text (*Yaśodharā-Carita*), the activity for the preparation for sacrifice to the goddess by the king Maridatta is lively projected. Here some of the attendants are standing holding thick garland of flowers, or human skulls (third register) (Cl. Plate-9). In the second register, two priests each holding a garland in his hand, followed by the butcher with a sword held in right hand and two goats, have been depicted. In the topmost register,the king is seated over a throne, with four lotus flowers decorating it.[99] A folio from *Ādi Purāṇa* illustrates Vidhyādhara Amitābha, seated over a raised throne in a cabin decorated with flower patterns (Cl. Plate-10). The uppermost register portrays a number of trees including palm, banana and others. Both the palm trees have been painted with a somewhat realistic touch, while those of the banana, have subdued realism.[100] Another[101] folio from the same text displays the scene of Srenika extolling the beauty of *Samayasarana*, seated over raised pedestal (Cl. Plate-11). The environ of forest is suggested by half a dozen trees appearing on the right side of the deity and two on the left. The foreground too stands for the dense forest infested with wild life. But the depiction of the wild animals like elephant, lion, horse (?) and a cock are disproportionate to their sizes; for all the animals are shown of almost similar height, which could be far from reality. The painting belongs to the fifteenth century A.D.

### (ii) Mughal Miniatures (Col. Plates 14-20)

After the conquest of northern India by Muslim invading army, the Hindu culture received a big jolt. The Hindu mind was so tarnished that it became oblivious of the beauty around. Their poetry and fine arts became stereo-typed, loosing the glamour of Kālidāsa, Bhaṭṭi and Bhāravi. It denigrated into devotional songs and ultimately became colourless. What to say of fresh creations, it became an uphill task for them to preserve even the earliest texts, which quite often were targetted for wholesale destruction by the invading army.

During the period from the eleventh to fifteenth century, very little was done in the field of miniatures, though some good specimens of the same are available, but by the large miniatures of this period are quite rare.

However, during the Mughal period the art of making miniatures, got a boost. During this period, the artist created in his miniatures lively scenes of plant-life, floral patterns and even wild-life, displaying height of imagination.

In a miniature from *Bābur Nāmā*[102], Babur is shown standing holding a hawk, near a reservoir, on the top of the painting the artist has painted a peacock, dancing with its tail spread out in a gorgeous fan, admired by a pair of pea-hens. In the rocks are a pair of mountain goats. An other illustration from *Bābur Nāmā*[103] by artist Ram Das, Babur is shown seated under a pavilion in a garden, outside Agra, receiving Uzbeg envoys on 18 December, 1528. The inscription of the painting reads, "They brought plants wrapped in red, white and black piece of cloth. I ordered them to plant the plum which was wrapped in red cloth."

During the time of Akbar, a large majority artists working on paintings of *Akbar Nāmā* were Hindus, some of whom include Keśu (Keśava), Madhu, Mukund or Makand, Nand, Nārāyaṇa, Paras (Paraśurāma), Śankara, Sūradāsa and Rāmadāsa. Some of them are believed to be Ahīrs from Mathura regions. Some Muslim painters like Miskīna, Mansūr, Qutb Chela and Farrukh Beg are also said to be collaborating with them.

The depiction of trees in the *Akbar Nāmā* is of much richer variety and intensity than could be derived either from Persian or European miniatures. In his detailed analysis of nature in the Mughal paintings, Iron Stchoukine has suggested that these trees of local origin such as banana, the Banyan, the Pīpala, tha Aśoka and the mango tree each derive from a familiarity with the landscape[104] They are to be found scattered through a large number of the *Akbar Nāmā* paintings with banana groves appearing in an orchard across the pontoon bridge in Agra or with date-palm appearing *en route* the pilgrimage to Agra.

A painting from *Diwan-i-Anwari* depicts a tree house on which the poet Anwari is seated alongwith his cup-bearer (*Saqi*). Anwari is reading a poem from a piece of paper and the *Saqi*, a handsome youth, is offering him a cup of wine. A servant is barbecuing a *kabab*, while another one is taking a flask of wine to the poet. The Chinār with birds perching on its branches provides a touch of nature.[105]

In a miniature dating back to A.D. 1564 of the reign of Akbar,[106] the prophet Eleas is depicted as saving the prince from the river, an episode of *Amir Hamzah.* The swirling waters of the river, the large fish, the monsters emerging from the river and the fantastic dream-like landscape with huge trees and shrubs are the special features of this miniature. In another Mughal[107] miniature appearing in a folio of *Anwar Saheli*, dating back to A.D. 1570, the rocky landscape is Persian in style but the monkey and trees are Indian in their realism. The lush green lawns and the tree leaves are remarkable. Another page from the same[108] text has an illustration, which is dominated by a forest scene, with a number of shrubs and trees. The rocks are infested with bears and a monkey. The leaves of the trees with yellow edges are pointing to their withering effect. In the same[109] text a hunting scene is depicted in a forest infested with dead animals killed during a hunting operation, besides skins of a panther. The landscape is Persian in appearance while the animals both dead and alive are Indian.

A painting from *Razam Nāmā*[110] depicts a scene of battle between Kauravas and Pāṇḍavas. Since the battle scene is intended to be depicted with destruction, no tree or plant could escape such a tornado. Still a few trees appear on the rocks, away from the battlefield.

The size of the Ranthambor[111] Fort in the late sixteenth century is depicted in a miniature currently preserved in the Victoria and Albert Museum. Though this represents the scene of destruction, yet the artist could not resist the temptation of projecting several trees on the scene. Similarly[112] in a miniature depicting the seige of Chittorgarh by Akbar, the artist did portray some trees inspite of the otherwise gloomy atmosphere depicting the burning of ladies of the *haram*. In a portrait of Ibrahim-II, Adil Shah of Bijapur[113], the background is illustrated with lush green trees. The Sultan is shown in a shady garden while the great palace is in the background, which is the regal residence. The minute treatment of details in the vestments, ornaments and flowers is characteristic of the Bijapur school. In a miniature of Mughal[114] art of the late sixteenth or early seventeenth century an ascetic is shown in *yoga* posture with his head placed on the ground and rest of the inverted body raised up. This *yoga* text was translated into Persian under title of *Bahr-al-rayat*.

This illustration is far removed from the schematic representation of the same in Hindu art, for it is naturalistic not only in the body but also in the surrounding landscape showing cocks, a cottage door with a tree having green leaves.

During the time of Akbar, the great epic of *Rāmāyaṇa* by Vālmīki was also translated into Persian and the translated text contained a number of illustrations depicting scenes from the same epic. In a miniature[115] from the said text, preserved in the Freer Gallery of Art, Washington, Rāma is shown pursuing the golden deer in the forest, which has a thick growth of trees and shrubs. It has, however, not been possible to identify the biggest of the trees having widespread roots, with birds perched on it. In an other miniature[116], the scene of Rāma's meeting with Aṅgada, the son of Bāli in the forest is painted. The powerful and abstract treatment of rocks, the detailed rendering of the soldiers of fortune, the army of monkeys and the adviser Hanumat, and the effect of lush green lawns and trees is remarkable. Another miniature[117] from the same source, shows a scene of Rāma's battle with the demons in the forest against the background of high rocks (Cl. Plate-12). The union of Sītā with Rāma in the forests of Laṅkā is also depicted[118] in a miniature from the same source. Huge rocks and palaces are shown in the background. But the rest of the space is mainly covered with different varieties of trees and plants, besides a chariot and a palanquin possibly carrying Sītā.

An illustration from *Akbar Nāmā*[119] (A.D. 1590) displays the scene of thick forest at night. The emperor Akbar is shown seated on a boulder under a huge tree, and a stag is lying dead in front of him. He pronounces his courteers that the slaughter of animals at a *quamargah* near Bhera should cease. The animals and trees are amassed in the upper left corner. In another miniature[120] of the same period, Idris is shown seated on a deer-skin spread on the hillock under a huge tree laden with fruits. The courteers around him bring scroll of cloth. In the foreground various stages of weaving cloth, including the washing, drying of wool, spinning, skeining and weaving are projected. The top surface is covered with plants and bushes.

Jahangir (1605-1627), though was the greatest builder of gardens in India, but the representation of plant life during his reign is somewhat on the down side, as compared to his father, still some beautiful paitings highlight this. In a miniature of this period, Noor Jahan[121] is depicted seated over a carpet with other ladies in a garden. The carpet itself has a beautiful floral design and the tree at the back of the queen is blossoming with flowers. There are lotuses in the lake with a *sikhara* in the background, indicating the environment of Kashmir. Rows of trees are seen even beyond the lake. In a miniature[122] currently preserved in the British Museum, London, Jahangir is seen shooting in the Dal lake near Śrīnagar. In the middle of the lake is a small island with a grove of cedars, and in the foreground are the lotuses. Like most visitors with aesthetic sensibility, Jahangir was struck with the beauty of the lotuses of Kashmir lake. A folio from Anwar[123] Saheli belonging to A.D. 1606, beautifully depicts the landscape of the forest with huge rocks with shrubs and trees. Two stags are also shown in combat, while another is lying dead. In a folio of Jahangir's[124] reign (A.D. 1620), a yellow Narcissus and a butterfly is artistically shown by an artist, named Muhammad Nadir from Samarkand. Another miniature[125] of the same date shows a white breasted kingfisher with flower plants in the foreground. The motif of squirrels[126] in a *Chinār* tree, attributed to Abul Hasan Nadir-al-Zaman (A.D. 1610), is shown (Cl. Plate-13). In the upper branches of the *Chinār* tree a group of squirrels is gambolling, two of ten young ones looking out from a hole in the trunk to front one of two adults. In the foreground a barefooted man dressed in two shades of brown with a fur-lined cap, is striving to climb the base of the tree. The surrounding landscape with rocks is filled with animals and birds—four wild goats are in a glade on the right, and the outer branches of the tree shelter numerous species of birds. The five pointed leaves are in many cases turning yellow and orange. The gold sky has a narrow strip at the top. The picture is accepted as the one of the most delightful of all Mughal miniatures.

The art of painting miniatures was continued during the reign of Shah Jahan. A miniature[127] preserved in Victoria and Albert Museum, London, represents the emperor Shah Jahan riding with prince Dara in the forest. There are hills in the background with pine trees and green grassy soil. A miniature[128] dating back to A.D. 1630-40, depicts a party of asceties camping under a tree, receiving eatables from a disciple, while others are staying scattered busying themselves in one way or the other. Two birds are shown perched in the tree. A miniature in the collection of Cowasji Jahangir depicts a group of asceties seated by the fire under a tree in the forest environments. Another miniature[129] from the same source depicts a wandering mendicant, holding a staff in his right hand and a small bag in the left, walking in the same forest environment.

The Mughal paintings of the later half of the seventeenth century, also did not neglect the plant life or tree motifs, and they continued to portray these aspects in their creations. In a portrait of a nobleman of this period, he is shown seated in the background of a cyprus and mango trees with flowery plants.[130] Another miniature[131] depicting *Śrī-Rāga*, shows a prince seated and attended by a flywhisk bearer, facing possibly Nārada and a horse-headed Gandharva. There are plantain and other trees in the backgound. The carpet on the floor too has beautiful floral design. Two ascetics[132] are shown seated beneath a tree facing each other in another miniature. The tree has scanty leaves, denoting the spring time, when many of the branches shed their leaves. The forest scene is lively depicted in a Mughal[133] miniature belonging tothe same period. The theme of the painting relates to the story when Khusro espies Shirin in the forest pool, while bathing. There is a horse grazing on the river bank, while Shirin is rinsing her hair beneath a water fall. There are trees in the background. Khusro has already alighted from his horse is looking at the lady stealthily. An interesting miniature[134] of the same period comes from *Khabar Nāmā*, in which two soldiers are shown fighting with an elephant headed deity, Gaṇeśa, who is carrying a huge boulder for a counter attack, in the forest, which is evident from the rocks, trees and shrubs scattered all over the surface.

## (iii) Rajasthani Schools

### (a) Mewar School

The former state of Udaipur in Mewar is a land distinguished for its forests, hills and lakes and its old palaces nestling by water's edge. The miniatures of Mewar are highly conventionalised in the western schools, affording clear proof of the composite culture of Gujarat and Rajasthan. With spread of Vaiṣṇavism in the area, Kṛṣṇa became the supreme god and several illustrated copies of the *Bhāgavata Purāṇa* of the Mewar school were composed. The paintings in the *Bhāgavata Purāṇa* text are quite lively and depicted freely the forests, trees, plants and flowers, quite abundantly. In fact no painting was considered complete without the plant life and floral landscape. Some distinguished features in the paintings could be noticed, which portray the influence of Mughal artist.

A miniature from this school depicts *Mathyadhira*—a heroine seated on a cushioned bed, holding a lotus leaf welcoming the hero outside the pavilion against a deep black background,[135] is a characteristic of flowering plant. The existence of wood land is suggested by shrubs and trees with the presence of birds. In the right foreground the heroine is standing in a grove swarming with bees, casting off her ornaments. In a miniature[136] depicting the dance of Kṛṣṇa in a grove, which is suggested by the presence of seven trees, one of which is a palm tree, two are of banana; the remaining four cannot be distinguished. The miniature[137] displaying the scene of *Gajendra-mokṣa* closely follows the description of texts with the Trikuta mountains having iron, silver and golden peaks. The green verdure of the forest, the winged *apsarās* engaged in dancing, the music of the Gandharvas, the serpents and the wild animals, provides the natural effect of a forest, with lush green surface, trees and shrubs around, represent the natural surroundings. In another[138] miniature, the love lorn Rādhā is being addressed by a friend in the forest. Kṛṣṇa too is standing between two trees, holding a lute. Rādhā is shown seated over a lotus under a tree. The presence of cypress trees in the painting points to the influence of Mughal school. In a painting of Mewar[139] school, portraying a scene from *Gīta Govinda* (A.D. 1550) provides the environment of the forest, which is suggested by two big trees and creepers. Kṛṣṇa is shown seated with milkmaids under these trees. Another miniature of the same school, currently preserved in the British Museum,[140] London, which displays forest scene, Sītā is shown seated in the hut, while Rāma is performing *homa* and Lakṣmaṇa is busy in performing other rites. Rāma and Lakṣmaṇa are shown seated on the ground, having their meals on the plates of leaves. Sītā is serving them, sitting in front. The entire surface is covered with trees and plants (Cl. Plates-21-25).

### (b) Bundi School

The Bundi School of Rajasthani paintings flourished in and around Bundi and Kotah and is second to Mewar school in age as well as in excellence and quality of output. The Bundi School from its very inception seems to have been receptive of Deccani influences and this is not suprising considering the intimate contact between the Hara rulers both of Kotah and Bundi, with Deccan from the beginning of the seventeenth century. Like other schools, the Bundi miniatues are found infested with plant life, flowers, trees and shrubs. A miniature[141] from Bundi, currently preserved in the Prince of Wales Museum, Bombay, has two lovers standing over a stool in the centre of the picture. The trees and plants are blossoming with flowers of pure and brilliant colours. They are painted against a patch of green that contrasts with the luxuriant foliage. Another miniature[142] from the same place depicts a courtesan standing gracefully in front of a plantain tree, a leaf of which is held by her hand. The attendant who offers white flowers stands in front of her. The trees of the garden are painted against the monochrome green background. The fountain and flower beds in the foreground recall Deccani painting and flowering creepers are done with quiet formality. The use of mases plantain trees is noteworthy being a constant favourite of Bundi painter. A miniature dating back to the seventeenth[143] century A.D. currently preserved in the Prince of Wales Musuem, depicts a princess with her companions having retired to the garden, which is full of dense vegetation and surrounded by water flowing from fountains into channels. At the top, there is a row of trees; next the row of women, their arrangement in a straight line echoing the grouping of trees, followed by yet another row of horizontally placed trees and fountains in the foreground. In fact the Bundi painters display a particular fondness for horizontal disposition of trees which stretches from one end of the picture to another (Cl. Plates-26 & 27).

### (c) Kotah School

In the foregoing account it has been possible to point out the close contact between the Kotah and Bundi states and the consequent influence of the artist of one school over the other, but in Kotah school sometimes new dimensions were added in portraying the forest scenes as well as the plant-life. As for example in a miniature of Kotah school, dating back to A.D. 1830, depicting Raja Ram Singh-II, the autumn season is depicted. Here the subject is hilly jungle around Kotah, untamed and thick with bamboọs. The hunters are often reduced to a head peering intently through the leaves. The jungles of Kotah artist are dangerous places and the men of Kotah were shown quite familiar with them. In the painting under reference the hunters occupy more of the picture. The jungle and its animals are drawn with a soft and delicate naturalism, quite sage with the artist in high shooting box, beyond the reach of either. The other conspicuous aspect of the forest scene is that the arrival of autumn season, when most of the tree shed their leaves. There are two trees beyond the pavilion. The first one has most of the leaves green, but some of them have been painted white giving an impression that the leaves have just started falling due to onset of autumn. On the next tree; half of it is laden with green leaves, while the other half of it projects the withering effect on the leave which have become pale and are about to fall off. The entire set of trees in the background have no leaves at all and the artist has been successfully able to project the onset of autumn by not drawing the leaves and the trees are projected with trunks only. The other trees and grass also project withering effect in spite of maintaining the greenery in leaves, though partially.[144] (Cl. Plate-28).

## (iv) Pahari Schools

### (a) Garhwal School

Under the patronage of the Mughals, the age-old practice of painting flowers received a fresh impetus and Garhwal school developed as a separate branch of painting, alongwith Calligraphy. The Garhwal artists, with their Mughal background, took delight in painting flowers and flowering trees. Exquisite flowers were also painted on the walls of the palaces of the princes and the houses of the gentry.

In a Garhwal painting[145] of A.D. 1800, the theme of Kṛṣṇa washing the feet of Rādhā is portrayed. Both Rādhā and Kṛṣṇa are shown seated under the flowering Mandāra tree and the characteristic screens on the door are recurring motifs in all Mala Ram's works. In another[146] Garhwal painting from the same artist, the decorative trees *viz.*, mango and Mandāra form the grove in which the cowherd Kṛṣṇa is talking to Rādhā. In a miniature[147] Kṛṣṇa is shown watching Rādhā from beind the flowering trees and plantains and admiring her beauty. In a painting[148] projecting the scene of the temptation of Śiva, the lord is shown cross-legged in an attitude of deep contemplation, with his hair raised up, flame-like. Śiva is seated over a leopard skin under shade of three huge trees, two at his back and one to his right (Cl. Plate-29). A lake with water lillies appears on the foreground. A painting from Garhwal portrays the scene from the *Bhāgavata Purāṇa*,[149] in which Balarāma is seen drawing water for Kṛṣṇa. There are two mango trees and against the rock a flowering Mandāra tree. A beautiful[150] mango tree is seen besides a well, in a miniature from Garhwal, showing the theme of Akbar at a village well.

### (b) Basohli School

Basohli miniature could be divided into the following three classes—

(i) Paintings from the *Bhāgavata Purāṇa;*

(ii) Paintings from *Rasamañjarī*; and

(iii) Painting from *Gīta Govinda* text.

All these miniatures, though executed with height of delicacy quite common in the Pahari miniatures, still the treatment of plant-life and floral patterns are done in an exquisite manner as would be evident from the following—

(i) *Paintings from Bhāgavata Purāṇa and Epics*

In a miniature showing the gods approaching Viṣṇu, with a prayer to relieve the earth of the burden of the *Rākṣas,* Viṣṇu and Lakṣmī are shown seated on a lotus, with all the gods standing in prayer, with earth in the form of a cow. The willows with their graceful branches provide a delightful contrast with the stylised trees with circular crowns. The picture is perfectly composed and the centre of interest is Viṣṇu seated on a full blown lotus with Lakṣmī.[151] In a miniature showing the swallowing of the forest fire by Kṛṣṇa, the forest is shown by a circle of trees. The trees as well as the cattle are highly stylized.[152] In a miniature showing the theme of Kṛṣṇa's stealing the clothes of cowherdesses, a huge *Kadamba* tree is shown on the bank of the Yamunā.[153] The theme of the lifting of the mountain Govardhana by Kṛṣṇa is depicted in another miniature, but the trees are shown on the surface of the raised mountain. Interestingly, since the trees are expected to be seen from the surface, they are projected in tiny sizes in a naturalistic manner.[154] According to the Purāṇic story, the Yamunā,[155] once paid no regard to Balarāma, who wanted the river to flow close to him. Balarāma then angrily drew her towards himself with his plough, and bathed. The bend in the river at Gokula is believed to remain to this day. This theme is projected in a Basohli miniature. The Yamunā is shown as a small girl draped in red, holding a golden vase, following the plough of Balarāma, Kṛṣṇa is embracing a cowherd lad in the right corner. The cows are grazing peacefully in the pasture land painted green in the background. The presence of five huge and a few small trees, point to the fact that the environment of forest is required to be created. Then there is a lively depiction of the story of Kṛṣṇa's bringing the Pārijāta-tree from Indra's heaven. In the painting Kṛṣṇa[156] is shown riding on a Garuḍa, warding off Indra's attack. Behind Kṛṣṇa, Nārada is seated holding the Pārijāta-tree as well the *Vīṇā.* Indra is riding on the elephant. There is a lotus pond in the foreground with skylarks moving in a natural manner. In a miniature[157] showing the scene of the departure of Rāma and Lakṣmaṇa, to the forest with Viśvāmitra, the tree separating the chariot of Rāma and Lakṣmaṇa from the palace of Daśaratha is highly stylized. The green background is symbolic of the forest to which the princes are departing. Another miniature from Basohli shows Kṛṣṇa seated on a throne surrounded by *gopīs* in a beautiful environment of a grove (Cf. Plates-30-32).

(ii) *Rasamañjarī*

*Rasamañjarī* is an important record of the social life of the upper classes in India during the fifteenth century A.D. It reveals an ease-loving society in which plurality of wives was common. The theme of this social life in *Rasamañjarī* was also interpreted for spiritual aspiration of the Hindus in the age. The *Rasamañjarī's* theme of *Nāyikās* was drawn in the miniatures, decorating them with plant-life and floral designs. In a miniature depicting *Prauda-Adhira*[158], currently preserved in the Dogra Art Gallery, Jammu, Pārvatī, is shown reproaching Śiva in the mountain abode, with a big tree at his back. The hero, the heroine[159] (*Parkīya Prodha*) with her friend are shown in a grove. The trees are beautifully arranged around a pond with lotus flowers in another miniature portraying *Parakiya Gupta.*[160] Another miniature from the same source, shows the heroine walking in a grove of beautiful trees.[161] Another miniature from *Rasamañjarī* of poet Bhānudatt, currently preserved in the Victoria and Albert Museum, shows Kṛṣṇa seated in a pavilion over a carpet of floral designs and Rādhā stands outside instructing a cowherd standing under a tree (Cl. Plate-33).

(iii) *Gīta Govinda*

In a miniature depicting the theme from *Gīta Govinda*, has the Gopīs adoring Kṛṣṇa in the environment of the forest. One of them is reverently touching his feet, another holding a white *chauri*, and yet another is bowing to him with folded hands, Kṛṣṇa is embracing two of the Gopīs. In the background is the charming forest of Vṛndāvana, and in the foreground is the Yamunā.[162] In a miniature[163] from the same school, Kṛṣṇa and Rādhā are shown seated under a tree on the bank of the Yamunā gazing at each other with passion filled in the eyes of both of them. The pointed crown of the trees and the sun drenched landscape are symbolic of the passion of lovers. In a miniature from *Gīta Govinda*,[164] Kṛṣṇa is shown seated on leaves in a grove, waiting for Rādhā, who comes accompanied by her maid. In another miniature[165] from the same source, both Rādhā and Kṛṣṇa are shown seated flanked by trees on both sides (Cl. Plate-34).

## (v) Kangra School

Like the Basohli School, the Kangra school selected various topics to be portrayed in the miniatures in which the plant-life, forests, flora and the floral designs received utmost attention. Indeed barring a few, these floral designs form part of most of the Kangra miniatures. M.S. Randhawa has drawn an elaborate list of the "Trees, Shrubs and Climbers", so often projected in the Kangra paintings.[166] The subjects dealt with by the artist of the Kangra School are listed as under—

### (a) Miniature on Love

In a miniature[167] showing the Love-lorn lady, the heroine is standing besides a banana tree. The bunch of bananas with a bud, is found hanging. A flowery plant appears on either side of the railing in the picture. The Love-lorn[168] lady is shown on the bank of a water tank which is filled with lotus leaves, buds and lotus flowers. A trunk of a tree with little of leaves and branches lies in her front, while a number of trees encircling a palace appear in the above background. A miniature[169] depicts a lotus lady shown plucking flowers of plum blossom. She has the odour of lotus, thus attracting bees. In a miniature from Guler[170], preserved in the Municipal Museum, Allahabad, and dating back to A.D. 1820, two lovers—Rādhā and Kṛṣna—are meeting near a lotus pool. The reflection of the orb of the moon in the pool and trees covered by flowering creepers create an enchanting scene, which provide appropriate background to the romance of Rādhā and Kṛṣṇa. There are trees, giving an impression of a forest and the pool is studded with lotus leaves and blossoming lotus flowers. Another miniature[171] from Guler, of the same date and belonging to the J.K. Modi collection, Bombay, portrays *Suklābhisārikā Nāyikā* in a forest. She stands near a pool full of lotus leaves and flowers, the trees with flowering creepers hanging over them. There is a pair of birds, *Chakravāks,* who usually dance at the sight of the moon. though the moon appears in the sky and its reflection can be seen in the river flowing besides the pool but the birds get attracted to the moon-like face of the heroine who is none else then Rādhā herself. In a miniature from Kangra School,[172] now preserved in the National Museum, a lady is shown on a swing tied to a huge tree (Cl. Plates-35 & 36).

**(b) Miniature from *Bhāgavata Purāṇa***

A miniature[173] from Punjab Museum, Patiala, dating back to A.D. 1800, depicts the scene of Rādhā and Kṛṣṇa sheltering from rain. Kṛṣṇa was grazing the cows alongwith the cowherd boys and *gopīs,* when it started raining suddenly. They all took shelter under a pair of trees. While a cowherd boy protected himself from the rain with a lotus leaf another one hid himself among the trunk. Kṛṣṇa and Rādhā sheltered themselves beneath a common black shawl. The cows are providing a cover to the lovers. A *Liana* is embracing a *tamāla* tree in its coils and the trees themselves are locked in embrace. On the crown of a tree, a peacock, the lover of the cloud is shouting exultantly. Two *gopīs* carrying pitchers on their heads are leaving the place. In another miniature from Kangra[174], dating back to A.D. 1810, in private collection, shows Rādhā and Kṛṣṇa, standing before a marble pavilion, studded with precious stones. It is surrounded by a grove of trees on which a pair of love-birds are warbling music. In the foreground is a lotus tank with half open lotus flowers and birds (Cl. Plate-37).

The theme of Kṛṣṇa's childhood stands depicted in many of the Kangra miniatures. In one of them, the salvation of Nalakubera[175] and Maṇigrīva is lively depicted. After steeling the butter, Kṛṣṇa broke the vases of curd and butter-milk. Snapping the churning stick and taking a pot full of butter, he ran out to play with the cowherd boys. He found a mortar placed upside down, on which he mounted and sat, and he caused his companions to sit down around him. All of them started consuming the butter. The monkeys and cowherd boys had a heavy feast of butter and spilt butter-milk. On discovering what had happened, Yaśodā chased Kṛṣṇa and threatened to beat him. She then tied him with a rope to the mortar. However, the rope was not long enough and was found to be two inches short. She joined another rope to it, and then another, and every time it was found to be a little short. Taking pity on the exertion of his mother, who was perspiring, Kṛṣṇa allowed himself to be tied to the mortar.

In the courtyard of the house were two Arjuna trees. Kṛṣṇa managed to crawl between them alongwith the mortar to which he was tied, and gave a jerk uprooting the trees. Out of the trees emerged Nalakubera and Maṇigrīva sons of Kubera. Hearing the crashing of the trees, the cows broke their ropes in panic and Nanda, Upananda, Yaśodā and all the neighbours rushed to the place. The two *Siddhas* bowed and circumambulated to Kṛṣṇa tied to the mortar and left for their abode.

In a miniature from Kangra, the panic created by the crashing of trees is shown in a dramatic manner. Balarāma is pointing towards Kṛṣṇa tied to the mortar and their foster parents are expressing a sense of relief that he has not been injured. Tied to the mortar, Kṛṣṇa is showing his resentment to Nanda for the manner in which he had been treated (Cl. Plate-38).

In a miniature from the same school, the scene of the killing of Bakāsura[176] has also been shown to have been enacted in a forest. In this painting, the story is told in two parts. On the left, Bakāsura is shown holding Kṛṣṇa in his beak and on the right Kṛṣṇa is depicted tearing the beak of the giant bird, while the cowherd boys are rushing to his help armed with sticks and even the calves are charging the *asura*. In the background is placid Yamunā with tall trees grown on its banks. The scene[177] of forest with trees and flowing climbers in abundance, is shown in a miniature displaying the search of *gopīs* for Kṛṣṇa in the forest (Cl. Plate-39).

In a miniature[178] displaying Kṛṣṇa's revealing his divinity to Akrūr, the thick forest in the background provides a sombre setting, breathing an atmosphere of devotion (Cl. Plate-40). The entire story is told in separate groups which comprise the picture—Kṛṣṇa and Balarāma arriving in a chariot. Akrūr seeing their reflection in the water and later on seeing Kṛṣṇa in his Viṣṇu form. On the bank one elderly herdsman wrapped in a dark blanket is watching a *papīhā* bird sitting on the branch of a tree and in the background Nanda is shown riding a bullock-cart preceded by a group of cowherd boys proceeding towards Mathura. In this painting one can see what Coomaraswamy calls the "transfigurining idealism of mystic art where everything is perfected by living imagination". Against the green valley floor the patient dignity of giant trees standing on the bank of the Yamunā, makes a great impression on mind. A miniature from Chamba shows Śiva and Pārvatī watching the celestial dance in the forest (Cl. Plate-41).

### (c) Royal and Other Scenes

In a miniature from Kangra, a princess and prince[179] are shown standing facing each other in the environment of the forest. Most of the trees depicted have no leaves, though one at the back of the prince has lush green leaves while the one at the back of the princess is laden with yellow flowers. This could be interpreted to mean that the artist intended to create an impression of the onset of autumn. The rainy season has greenary alround and the sky is overcast with clouds. The rainy season is reflected in another miniature[180] from the same source. The surface is by and large green, the trees are also laden with green leaves which are shown swinging as a result of the breeze. A prince and a princess are standing in the foreground, enjoying the natural beauty. A miniature in the Punjab Museum[181] collection depicts the theme of welcoming of the cloud by a prince and a princess standing over the terrace of the palace. Since the upper register is supposed to have the clouds, nothing much is visible. But in the second register trees with tender leaves and fruits are shown. In order to provide a realistic touch to the viewers, the artist has tried to create an impression that some of the leaves of the trees having brownish appearance are the newly grown tender leaves distinct from the other green leaves. In the lowermost register, three plants with blossoming flowers are shown. A forest scene has also been depicted in a miniature from the Punjab Museum collection[182], wherein Śiva and Pārvatī, carrying Gaṇeśa and Kārttikeya as children in their laps, are seated over a tiger skin, covering their bodies with the elephant hide. The entire landscape projects the mountains with trees in the upper foreground. The twisting curves in the tree under which the family of Śiva is seated, the artist possibly has tried to create an impression that the tree is quite an old one. The mounts of the entire family *viz.*, bull of Śiva, Lion of Pārvatī, peacock of Skanda and the mouse of Gaṇeśa are seated on a grassy land.

## (vi) The Art of Bundelkhaṇḍa

Bundelkhaṇḍa is also known for its wall paintings, rendered on the palace of Orchha and also the miniature, depicting in a lively manner, the plant-life. In a painting from *Mahirāvaṇa-carita* from Bundelkhaṇḍa, the scene of the arrival of Mahirāvaṇa in Laṅkā is projected with the rows of trees in the background. In another miniature of the same school, Mahirāvaṇa is shown carrying the Ayodhyā princes in the air, with a cluster of trees in a corner in the foreground. Trees and plants are also projected in other miniatures portraying scenes from the same story, but in the end of the story, portraying the honouring of Hanumān by Rāma, the banana trees are prominently shown (*India Magazine*, 1992, pp. 56-58).

## (vii) Other Schools

Moor has published a sketch based on a miniature from South, which depicts child Kṛṣṇa lying over the leaf of *Vaṭa-Vṛkṣa*, (Fig.-33) holding in his mouth the thumb of the left foot. This scene has been described in the *Purāṇas*.

In a miniature from a Sanskrit manuscript currently preserved in the National Museum, the scene of Pāṇḍavas (accompanied by Aśvatthāmā and Kṛṣṇa) meeting with Draupadī in the forest has been depicted. The artist has painted two trees, one on each side and six small trees in the background, in three pairs. Two other miniatures from the same source depict trees in one form or the other (Plates-44-47).

A miniature from manuscript no. 76.729, from the National Museum, Kṛṣṇa displays his universal form (*Viśva-rūpa*) to Arjuna. Here Kṛṣṇa is shown seated over a lotus, having four hands and holding all the attributes of Viṣṇu *viz.*, *śaṁkha*, *cakra*, *gadā* and *padma*. The head, however, combines the heads of Brahmā, Viṣṇu and Śiva, besides those of an elephant, horse, Garuḍa and several other human and animal heads. The entire painting is beautifully decorated with floral borders, front walls of the pavilion and the surface. Even Arjuna has the floral dress. Another manuscript bearing No. 63.27, displays the scene of *Gajendra-mokṣa* or the liberation of elephant Gajendra from the clutches of Grāha (Plate-48). The elephant is shown sheltering under a palm tree, whose left foreleg is caught by the Grāha in its mouth. Viṣṇu holding the usual attributes including a lotus, appears on the scene with Lakṣmī mounted on Garuḍa. The entire lake has lillies and lotus buds scattered everywhere.

In a Persian[183] miniature from Tuhfat-al-Hind, portraying *Madhyamadi Ragini*. Śiva and Pārvatī are seated in embrace (Cl. Plate-46).

## Notes and References


1. Mackay, E.J.H., *Further Excavations at Mohenjodaro*, I, pp. 220, 337, 338, 341, 351, 352.
2. *Ibid.*, I., p. 355.
3. *Ibid.*, I, p. 217.
4. *Ibid.*, I, p. 351.
5. *Ibid.*, I, p. 464.
6. *Ibid.*, I, pp. 220, 328.
7. *Ibid.*, pp. 240, 341.
8. *Ibid.*, I, p. 341.
9. *Ibid.*, I, p. 335.
10. Marshall, John, *Mohenjodaro and Indus Civilization*, pp. 63-66, Pl. XII, 17.
11. Typical examples are found on the railings of Bharhut and Sanchi (Second-first century B.C.) where the *Yakṣis* are portrayed standing on their animal *vāhanas* and clasping the tree with arms, legs, their pose indicating that they are part and parcel of the tree.
12. Marshall, John, *op. cit.*, I, pp. 64-65.
13. *Ibid.*, I, p. 65.
14. *Ibid.*, I, p. 65.
15. *JRAS*, 1899, pp. 38-39.
16. Marshall, John, *op. cit.*, p. 52.
17. *ASIAR*, 1911-12, Pl. XXIII, 40.
18. Brown Norman, W., *Chanhu-daro Excavations*, New Heaven, 1947, 1947, p. 93.
19. Rao, S.R., Lothal, *A Harappan Township*, I, p. 215, N.D. 1979.
20. *Ibid.*, II, p. 354, Figs. 41 and 42.
21. *Ibid.*, p. 354, Pl. CLXV, C-D.
22. *ASIAR*, 1927-28, p. 66, Pl. XX, Fig. 7.
23. Banerjea, J.N., *Development of Hindu Iconography*, p. 170.
24. *Ibid.*, p. 171.
25. Nagar, S.L., *The Universal Mother*, p. 103, Fig. 156.
26. Banerjea, J.N., *op. cit.*, p. 174.
27. Vats, M.S., *Excavations at Harappa*, I, pp. 43-44.

28. Banerjea, J.N., *op. cit.*, p. 110, Pl. XI, Fig. 1.
29. Whitehead, R.B., *Punjab Museum Catalogue*, Vol. I, p. 135, Pl. XIII, Fig. 333.
30. *Ibid.*, I, pp. 100-101, 122.
31. Banerjea, J.N., *op. cit.*, pp. 110-111.
32. *Ibid.*, p. 113.
33. Allan, J., *Catalogue of Coins of Ancient India in the British Museum*, p. 85, No. 2, XI. 2; p. 233, Nos. 154, and 154a, Pl. XXXV, 5.
34. *Ibid.*, pp. 87-91, Pl. 1, Fig. 5.
35. Banerjea, J.N., *op. cit.*, p. 118.
36. *J.R.A.S.*, 1903, pp. 285-86.
37. Banerjea, J.N., *op. cit.*, p. 131, Pl. 11, Fig. 3.
38. Whitehead, R.B., *op. cit.*, I, p. 129.
39. Banerjea, J.N., *op. cit.*, p. 135.
40. Allan, J., *op. cit.*, p. 143.
41. *Viṣṇudharmottara*, Book-III, Ch. 45, 1-8.
42. Nagar, S.L., *Jātakas in Indian Art*, pp. 166-167, Pls. 61-62.
43. *Ibid.*, p. 57, Pl. 11.
44. *Ibid.*, pp. 101-102, Pl. 31, Fig. 68.
45. *Ibid.*, pp. 67-68, Pl. 10.
46. Agrawala, V.S., *Indian Art*, p. 225 ff.
47. *Ibid.*, p. 228.
48. *Ibid.*
49. *Ibid.*, p. 227.
50. Agrawala, V.S., *Gupta Art*, p. 14, Pl. 11.
51. Kālidāsa compares the halo of his ideal hero Raghu with a full blown lotus parasol (*padmatapatra*) and says that the former was the exact shadow of the latter (*Raghu*, IV.6); this description seems truly confirmed in the elaborate halo of Pārvatī as स्फुरत्प्रभामण्डल on account of its radiating pencils of light (I.24).
52. Agrawala, V.S., *Gupta Art*, p. 19, Fig. 23.
53. *ASR*, X, p. 12.
54. *H.F.A.*, p. 168.
55. *Vāyu Purāṇa*, 45.12-50.
56. *Mahābhārata*, Ādiparvan, 123.12.
57. *Ibid.*, Vanaparvan, 294.13.
58. *Jātaka* No. 465.
59. *Durgāsaptasatī*, XIII. 7.
60. *Śakuntalā*, Act. VII.
61. Agrawala, V.S., *Indian Art*, p. 310, Fig. 227.

62. *Ibid.*, p. 311, Fig. 229.
63. *Ibid.*, p. 311, Fig. 230.
64. *Ibid.*, p. 312, Fig. 232.
65. Yazdani, G., *Ajanta*, IV, p. 30, Pl. IX (a-b).
66. *Ibid.*, IV, p. 30, Pl. IX(b).
67. *Ibid.*, IV, p. 107, Pl. LXXI(a).
68. *Ibid.*, IV, p. 108, Pl. LXXI(b).
69. *Ibid.*, IV, p. 108, Pl. XXI.
70. *Ibid.*, IV, p. 108, Pl. XIX.
71. *Ibid.*, IV, pp. 74, 97.
72. *Ibid.*, IV, pp. 20, 37, 46, 56, 63.
73. *Ibid.*, IV, pp. 37, 57, 75.
74. *Ibid.*, IV, pp. 34, 65, 104.
75. *Ibid.*, IV, p. 74.
76. *Ibid.*, IV, pp. 32, 97.
77. *Ibid.*, IV, pp.31, 32, 97.
78. *Ibid.*, IV, p. 24.
79. *Ibid.*, IV, pp. 32, 58.
80. *Ibid.*, IV, pp. 31, 42, 49, 52, 55, 72, 78, 88, 97, 107, 108.
81. *Ibid.*, IV, pp. 33, 37, 65, 96.
82. *Ibid.*, IV, p. 48.
83. *Ibid.*, IV, pp. 32, 53, 106.
84. *Ibid.*, IV, p. 25.
85. *Ibid.*, IV, p. 94.
86. *Ibid.*, IV, p. 23.
87. *Ibid.*, IV, pp. 74, 106.
88. *Ibid.*, IV, pp. 20, 53, 65, 74, 104, 106.
89. *Ibid.*, IV, p. 77.
90. Nagar, S.L., *Protection, Conservation and Preservation of Indian Monuments.*
91. Ghosh, A., *Jaina Art and Architecture*, III, Pl. 266.
92. *Ibid.*, III, Pl. 267.
93. *Ibid.*, III, Pl. 270.
95. *Ibid.*, III, Col. Plates, 23-B, C and D.
96. *Ibid.*, III, Col. Pl. 24.
97. *Ibid.*, III, Col. Pl. 27.
98. *Ibid.*, Col. Pl. 30(A).
99. *Ibid.*, Col. Pl. 32(B).

100. *Ibid.*, Col. Pl. 36(A).
101. *Ibid.*, Col. Pl. 36(B).
102. Randhawa, M.S., *Gardens Through the Ages*, p. 109, Pl. 16.
103. *Ibid.*, p. 111, Pl. 17.
104. Ivan Stchoukine, *La Peinture Indienne a l'Epoque des Grand Mughals*, Paris, 1929, p. 79.
105. Randhawa, M.S., *op. cit.*, pp. 115-116, Pl. 20.
106. Bussoli, Mario, *Indian Miniatures*, London, 1969, Pl. 11.
107. *Ibid.*, p. 40, Pl. 14.
108. *Ibid.*, p. 40, Pl. 15.
109. *Ibid.*, p. 40, Pl. 16.
110. *Ibid.*, p. 40, Pl. 17.
111. *Ibid.*, p. 62, Pl. 21.
112. *Ibid.*, p. 63, Pl. 22.
113. *Ibid.*, p. 86, Pl. 37.
114. *Ibid.*, p. 86, Pl. 38.
115. *Ibid.*, p. 104, Pl. 45.
116. *Ibid.*, p. 104, Pl. 46.
117. *Ibid.*, p. 104, Pl. 47.
118. *Ibid.*, p. 104, Pl. 48.
119. Folk Toby and Archer Mildred, *Indian Miniatures in the India Office Library*, p. 47, Colour Pl. 3, London, 1981.
120. *Ibid.*, p. 47, Colour Pl. 4.
121. Crowe Sylvia and others, *The Gardens of Mughal India*, London, 1972, p. 33.
122. Randhawa, M.S., *op. cit.*, p. 121, Pl. 22.
123. Khandalawala, Karl, *Miniatures and Sculptures*, Bombay, 1965, Pl. C.
124. *Ibid.*, Pl. E.
125. *Ibid.*, Pl. F.
126. Archer, W.G., *Indian Paintings*, Switzerland, 1956, Pl. 5.
127. Khandalawala, Karl, *Miniatures and Sculpture*, Pl. G.
128. *Ibid.*, Pl. 24.
129. *Ibid.*, Pl. 25.
130. *Ibid.*, Pl. 36.
131. *Ibid.*, Pl. 37.
132. *Ibid.*, Pl. 38.
133. *Ibid.*, Pl. 39.
134. *Ibid.*, Pl. 46.
135. Motichandra, *Mewar Paintings*, Bombay, 1957, Pl. 1.

136. *Ibid.*, Pl. 3.
137. *Ibid.*, Pl. 4.
138. *Ibid.*, Pl. 6.
139. Barrett Douglas and Grey Basil, *Paintings of India*, p. 71, Cleveland, Ohio, 1963.
140. *Ibid.*, p. 139.
141. Chandra Promod, *Bundi Paintings*, N.D. 1949, Pl. 2.
142. *Ibid.*, Pl. 3.
143. *Ibid.*, Pl. 4.
144. Barrett Douglas and Gray Basil, *Paintings of India*, p. 158.
145. Lal Mukundi, *Garhwal Paintings*, Colour Pl. VI, N.D. 1968.
146. *Ibid.*, Colour Pl. IX.
147. *Ibid.*, Colour Pl. X.
148. *Ibid.*, Colour Pl. XVI.
149. *Ibid.*, Colour Pl. XVII.
150. *Ibid.*, Colour Pl. XIX.
151. Randhawa, M.S., *Basohli Paintings*, Plate-6, N.D. 1959.
152. *Ibid.*, Pl. 8.
153. *Ibid.*, Pl. 9.
154. *Ibid.*, Pl. 10.
155. *Ibid.*, Pl. 13.
156. *Ibid.*, Pl. 15.
157. *Ibid.*, Pl. 16.
158. Randhawa, M.S. and Bhambri, *Basohli Paintings of the Rasamañjari*, p. 23, Delhi, 1981.
159. *Ibid.*, p. 29.
160. *Ibid.*, p. 30.
161. *Ibid.*, p. 68.
162. Randhawa, M.S., *op. cit.*, (B.P.), Pl. 19.
163. *Ibid.*, Pl. 20.
164. Sinha, R.P.N., *Gīta Govinda in Basohli School of Indian Paintings*, Pl. III.
165. *Ibid.*, Pl. IV.
166. Randhawa, M.S., *Kangra Paintings on Love*, pp. 200-01, N.D., 1962.
167. *Ibid.*, Fig. 57.
168. *Ibid.*, Fig. 58.
169. *Ibid.*, Pl. I.
170. *Ibid.*, Pl. V.
171. *Ibid.*, Pl. VII.
172. *Ibid.*, Pl. XV.

173. *Ibid.*, Pl. XVII.
174. *Ibid.*, Pl. XX.
175. Randhawa, M.S., *Paintings from Bhāgavata Purāṇa*, Pl. IV, N.D. 1960.
176. *Ibid.*, Pl. VI.
177. *Ibid.*, Pl. XIII.
178. *Ibid.*, Pl. XVIII.
179. Randhawa, M.S., *Kangra Valley Paintings*, Pl. 4.
180. *Ibid.*, Pl. 5.
181. *Ibid.*, Pl. 29.
182. *Ibid.*, Pl. 40.
183. Folk Toby, Mildred Archer, *Indian Miniatures*, Pl. 349 (IV).

# 6
# Industrial Art

India, from time immemorial, has produced highly skilled artists and artisans, who have created edifices earning global applause. The earliest creations of these artists and artisans are available in the remains of the Indian archaeological sites of Harappa and Mohenjodaro, where numerous structures, besides other antiquarian remains like pottery, seals, sealings, jewellery and ornaments; and the metal objects, including the dancing figure of a female, have been discovered. All these specimens point to the height of proficiency, the artisan or the artist of the bygone days had achieved at that point of time. Though the specimens of early stone sculptors, works are lacking, but during the Mauryan times we suddenly have to confront the sculptor, who created stone edifices like the Indra Sabha at Patna, lion and other capitals during the Aśokan period and the stūpas of Bharhut, Sanchi, Amarāvatī, Bodhagayā and Mathura.

In all the above cases, the stūpas were adorned with impressive railings, which were designed on the pattern of early wood-work in the country. But this stone carvings, at least at Sanchi are attributed to the ivory carvers of Vidisha on the testimony of the inscriptions on the Sanchi railings. But in the subsequent period the sculptural art in India was developed considerably (the emergence of which had started during the Mauryan period) and certain master artists were able to create highly commendable specimens during the Gupta and post-Gupta period.

## 1. Stone Jālī and Inlay Work

During the late medieval period, in India, the stone carvers developed an extremely impressive and exquisite work in stone, a specimen of which is available in the marble ceiling of the Dilwara temple of the Jainas at Mount Abu which belongs to the thirteenth century A.D. (Plate-53). This carving has exquisite pattern of flowers so delicately carved with tender petals. The said art of stone carving reached its zenith during the Mughal rule when delicate *jālī* became an essential part of Muslim tombs and other monuments. These *jālī* works were mainly based on the floral pattern and a good number of such specimens are available in the monuments at Agra (Plate-54). Besides the *jālī* work, the walls of the edifices were also at times decorated with the floral inlay work. As for example, the interior of the 'Diwan-e-Khas'[1] or the Hall of Private audience in the Red Fort, Delhi, have the columns, which are perhaps the most graceful of all products during the reign of Shah Jahan. Of particular interest are the lace-like marble fret work and the floral inlay decoration. The house of Turkish Sultan at Fatehpur Sikri also preserves delicate floral *jālī* work and stone carvings.[2] Another specimen of the tree and floral motif is available in the famous perforated stone window of Siddi Sayyed's mosque at Ahmedabad.[3] (Plate-55). In this case a column is shown standing in the centre, representing a palm tree with the leaves at the top. The boughs issuing from the tree take the form of creepers laden with flowers and cover the rest of the surface.

Indeed, "the stone carved architecture of the Mughals starts with the simple, dignified, yet grand structures of the time of Akbar, many of them showing signs of Hindu style. In the case of Govinda Deo temple at Brindaban, built by Man Singh of Jaipur, one of Akbar's ministers, this electicism was a distinct artistic triumph. This is especially to be seen in the introduction of animal and human forms, prohibited by Islam, but tolerated by the broad-minded Akbar. Even at Fatehpur Sikri, Akbar's red-stone-city, this influence is noticeable, especially in the Turkish Sultana's house and Jodha Bai's palace."[4]

> "The same design[5], or nearly the same, appears on each one of the bases, but that upon one side of the base is unlike that on the other. The inlay is so intricate that it looks quite intricate, but in mother-of-pearl instead of gold and copper wire. Each tiny peace of mother of pearl is exactly shaped and fitted into the allotted position previously prepared for it and then secured... with minute pins and shellac. The minute stripes of ebony between each piece of mother-of-pearl are secured in a like manner, and are perfectly flush with the face of mother-of-pearl. It is impossible to describe the effect of the mother-of-pearl as seen in the dim light of mortuary chamber. It is one of those things which almost defy description and in order to realise its beauty, it must be seen. The inlay is most beautifully and minutely executed."[6]

Probably it was in the reign of Jahangir[7] that *manbhat-kari* or stone inlay work came into fashion. Red sand stone gave place to marble and mosaic inlay. As H.S. Crosthwaite says, "In Akbar's reign the artist was content with beautiful but comparatively simple tracery work in red sand stone, with large mosaic in coloured marbles. In Shah Jahan's reign the ivory white marble of Makrana has replaced the sandstone of Vindhyas; the big mosaics and geometrical tracery of Futehpur Sikri have been refined till they reach a climax in the jewelled inlay and flowerlike tracery of the Taj, and in the virgin beauty of the Moti Masjid. There is probably no building in the world to which the Horatian 'Simplex munditis' can more truly be applied than to Moti Masjid. It is sometimes said that there is no carving in this mosque; but in reality it is a triumph of stone cutter's art. On the beautifully modelled pillars and dazzling while domes, on the marble inlaid prayer 'carpets' and delicately chiselled *mihrabs* the highest art of stone mason has been lavished. The Taj, though, like the Moti Masjid, a marvel of stone carvings is also the handiwork of the jewellers. With the Taj may be grouped the Jasmine tower, and the Diwan-e-Khas (Hall of Private Audiance—R.J.M.) in Agra Fort. In all the three the carving and inlay work are exquisite, elaborate and chaste. The floral patterns *(jhuh)* of tiger lillies, roses and other fruits and flowers are deeply cut and then smoothed, so as to give no impression of angularity and hardness to the eyes."

Thus it is evident that the inlay work as well as the exquisite *jālī* work was patronised by the Mughals, and the best examples of the same are available in the Taj and other monuments at Agra, Fatehpur Sikri and Red Fort, Delhi, though the Jālī work could be found at several monuments in the country.

## 2. Wood Work

Wood carving is quite an ancient craft in India, for it is certain that it was practised in this country long before sculpture in stone, which is of considerable antiquity, since its genesis could be traced to pre-Mauryan times. The wood carving was and is widely used even today, but the wood carvers of the ancient times not only carved bedsteads, wooden seats and thrones, but also chariots for the purpose of war, ornately and beautifully carved wooden doors can still be seen in many places testifying to the heights to which the craft had reached in early days.

One of the earliest example of the wood carvings with floral patterns is available on the doors of the Marcula Devī temple, Udaipur (Himachal Pradesh). The space between the figures of deities has been beautified with floral designs instead of leaving it blank.[8] These floral designs are also displayed in the ceiling of same temple. Indeed the tradition of wood carvings in Himachal Pradesh is of considerable antiquity, but only a few carvings mostly preserved in the wooden temples have survived to this date. The early wooden temples in the state have similarity with the Gupta style. The structural features of many early buildings in stone are taken from the wooden models, where the floral designs were quite popular. The excellence of wood crafts can be witnessed in many of the later carvings, with floral patterns preserved in the temples of Chamba, Kulu, Mandi and in the upper Shimla hills, besides the Sutlej Valley.[9] The carved panels on doors, brackets, pillars and ceilings are all reminiscent of wood carvings which earned applause in the country and abroad. James Ballie Fraser who visited Sirmaur and Kinnaur temples in 1815, observed—

> "The whole of the interior is sculptured in wood, with infinite labour and probably forms a detail of the exploits of the deity... but she seems to have been frequently engaged with monsters very uninviting shapes. The portion of the carving, however, which neither represents the human nor animal figures is cut into flowers and ornaments entirely in Hindu taste, with sharpness and precision, yet an ease that does honour to the mountain artist, and considering his tools, it is truly wonderful."

From temples, the carvings in wood work reached the places of the dwellings and several exquisite floral patterns were carved over the *Havelīs* of the elite of the society. The places best known for the highly skilled workmanship with tree and flower patterns, beautifully executed are Kashmir, Gujarat, Rajasthan, Mysore and even Kerala. The beautiful geometrical and floral designs carved in wood are known the world over. Similar is the case with the sandal-wood carvings of Mysore and of Kerala. Not only they ornament or carve floral patterns on small articles but also on river house-boats and ceilings. Mention may also be made of the elaborately carved temple cars (*rathas*) of South India, the beautiful carvings in the old palaces of rulers and the *Havelīs* of the high-up in Rajasthan and the wood work in the temples of Gujarat and Kathiawar (Plates-56-58).

## 3. Textiles

Of all the art manufactures in India, the textiles are certainly the oldest. The discovery of fragments of finely woven maddu-dyed cotton fabrics and bobbins at Mohenjodaro traces back the antiquity of the Indian textiles to over 3,500 B.C. long before the Aryan's emerging on the Indian horizon. The Seleucid envoy Megasthanis (328-302 B.C.) at the court of Candragupta, has mentioned in his detailed account of India, that in those days the Indians wore flowered muslins of the finest weave and the ruling classes were clad in robes embroidered in pure gold. "In contrast to simplicity they observe in other matters they love finery and ornament. They wore dresses worked in gold adorned with precious stones and also flowered robes of fine muslin."

The products of ancient weavers of India are compared to "exquisite poetry in colourful fabrics", and there was hardly a trick of the weavers art that was not known to them". It is now generally admitted that India was perhaps the first to perfect the art of weaving and that her gossamer-fine muslins and sumptuous brocades of gold are probably older than the code of Manu.

The Vedic literature too refers to the textiles in one or the other forms. In a hymn to Agni in the *Ṛgveda*, reference to the craft of loom is descernible. "Oh, I know not either warp or woof, I know not the web of they weave." There is also a reference to Uṣā, the goddess of dawn, "clothed with radiance" and in an other hymn it is recorded that, "Day and night spread light and darkness over the extended earth like two famous female weavers weaving the garments."

By the time of Epics, the art of weaving had reached a stage of high perfection. Reference to shining golden cloak, the *hiraṇya drapi* and the *maṇicīra* are found. In these epics woven cotton, silk and woollen fabrics are frequently mentioned.

The code of Manu makes frequent mention of the art of weaving and dyeing. The ancient temple sculptures represent women dressed in clothes, embroidered or brocaded in floral patterns and in muslins so fine as to expose the contours of their bodies, only the lines of their folds or the edges being represented to show that the female forms are fully garbed in reality.

In the magnificient frescoes of Ajanta, Ellora, Bagh and Sittanvasal caves, women's garments are often painted in different designs. "The prominent designs found at Ajanta are bands alternately filled in with geometrical patterns, such as chevrons, stripes, circles and checks; formal floral motifs or scrolls entwined with processions of *haṁsas* or *siṁhas*, dot patterns and diagonal bands of geese. Poets have waxed eloquent over these bird designs; Kālidāsa describes Pārvatī's dress, thus, "How shall I praise thy robe with gay flamingoes gleaning".

Sometimes in the early fifth century A.D. a temple was built in Malwa by the guild of silk weavers and dedicated to Sūrya, the sun-god. It was repaired in A.D. 473-474, and a memorial inscription of the time records as follows—

"A woman, though endowed with golden necklaces and betel leaves and flowers, does not go to meet her lover in a secret place unless she has put on a pair of silken clothes, so the whole of the region of the earth is adorned by silk weavers as if with a garment of silk, agreeable to touch, variegated in colours, pleasing to the eye."

The silk fabrics had been in use in India from the earliest times, and they were conspicuous by the presence of floral decorations in most of the cases. As for example an eighteenth century Kashmiri Shawl, preserved in the State Historical Museum, USSR, displays beautiful flower pattern. The same type of floral pattern is available on a *Masulipatman Kalamkārī* work. Another specimen from Manipur, displays the floral design in *tārakāśi* work (Plates-59-61).

## Notes and References


1. Mehta, Rustam J., *The Handicrafts and Industrial Arts of India*, Bombay, 1960, Pl. L.
2. *Ibid.*, Pl. LI.
3. *Ibid.*, Pl. LIII.
4. *Ibid.*, p. 53.
5. *Ibid.*, p. 54.
6. *Journal of Indian Art and Industry*, Vol. VIII, 1900.
7. Mohta, Rustam J., *op. cit.*, p. 54.
8. Nagar, S.L., *Temples of Himachal Pradesh*, Pls. 25-28.
9. Singh, Govardhan Mian, *Art and Architecture of Himachal Pradesh*, p. 23.

# 7
# Forests and Gardens

The forests or *vanas* in the Indian context, have been referred to in the *Ṛgvedic* text and Vanaspati has been conceived to be the Lord of forests. Besides, the lord of forests which makes one believe that there were numerous forests in the country. This is further supported by the *Ṛgvedic* assertion that the *Vanaspatis* emerged on the earth, at least five *yugas* before the human beings. In all probability most of the surface of the earth in the Indian subcontinent was covered with forest and wild growth at the time of the advent of the Aryans on the Indian soil. But the following terminologies used in the Vedic and post-Vedic literature relate to the forests, besides Vanaspati—

| | | *Vedic Literature* | *Āraṇyakas/ Brāhmaṇas* | *Upaniṣads* |
|---|---|---|---|---|
| (i) | Araṇya | *Ṛgveda*, VI.24.10; X.97.1-23; X. 146.1-6. | *Śatapatha Brāhmaṇa* V.2.3.5. | *Bṛhadāraṇyaka* V.II. |
| | | *Atharvaveda*, II.4.5. | | |
| (ii) | Udumbara Forest. | *Pañcaviṁśa Brāhmaṇa* XXVI. 6.4. | | |
| (iii) | Khāṇḍava | *Taittirīya Āraṇyaka*, V.I.I. | | |

Most of the discussions, discources and religious delibrations, as propounded in the *Āraṇyakas*, the *Brāhmaṇas*, the *Upaniṣads* and the *Gṛhyasūtras* were made in the environments of forests, where the ascetics, sages and *Ṛsis* had their abodes or *āśramas* in the dense forests.

Coming to the period of the Epics *viz.*, the *Rāmāyaṇa* and the *Mahābhārata*, it may be pointed out that most of the episodes described therein were enacted in the forests. The very fact that Rama, Lakṣmaṇa and Sītā, on the one hand and the Pāṇḍavas on the other were exiled for twelve years and were forced to wander extensively in the forests which were ordinary inaccessible to the human beings, having been infested with *Rākṣas* besides wild life. Both the texts provide ample testimony to the widespread nature of the forests in the country. The Purāṇas also highlight the extensive nature of the forest and most of them were composed in *āśramas* of the sages living in the solitude of forest environments. While some of the forests described in these texts have been mentioned in the chapter on "literary Evidence" in this work, the discussion in this chapter is confined mainly to the different types of forests as we know from these texts from the earliest times.

Though the plant life started appearing in the sculptural art of Mauryas followed by Śuṅga and Kuṣāṇa, but the period from second century B.C. to the 2nd Century A.D. enjoys a unique position in the plant history of the country, because it was during this period that the forest life is found extensively in the sculptural art of Bharhut, Sanchi, Amarāvatī, Vidisha and Mathura. It is during this peroid that the one could witness the intimate association of man with nature at large, in the thick woods infested with wildlife. Not only that, the Stūpas of Bharhut, Sanchi and Amarāvatī, also display the close proximity of the beautiful gardens and even the terrace gardens with the people. These depictions are so lively and pleasent and translate the aesthetic advancement the society had achieved in the forests and gardens, at that remote stage of the Indian history.

If the degree of development of a culture is judged by the aesthetic level of the people, it can safely be claimed that the Indian culture reached its heyday during the Kuṣāṇa-Gupta period, which is termed as the Golden Age of India. The art of sculpture and Sanskrit literature reached a high level of development. During this period flourished a number of outstanding scholars who added lustre to the annals of India. The laws of Manu, the *Arthaśāstra* of Kauṭilya and the tales of *Pañcatantra* were composed during this period. Besides, the outstanding poets and authors like Aśvaghoṣa, Kālidāsa, deserve particular attention. Aśvaghoṣa, the spiritual preceptor of Kaniṣka, mentions a number of beautiful trees in his work entitled *Soundrānanda* in which he describes the story of Nanda trying to conciliate his mistress to a Nāga-tree (*Mesua-ferrea*) in a garden which had broken down by the wind from its excessive burden of flowers. Describing the apathetic mood of Nanda pining for his beloved, he writes, "the Nāga-trees there, though studded with flowers with yellow interior as if with gold fitted caskets of ivory, no more drew the eyes of Nanda in his sorrow." In his description of a forest in the Sub-Himālayas, he describes waving Kadamba trees, and Pārijāta tree shining with all the qualities of majesty, and playing the king over the Mandāra trees and other trees laden with bloom of the day, water lillies and red lotuses. Aśvaghoṣa compares Nanda's mistress with a lotus pond, with her laughter for swans, her eyes for bees, and her swelling breasts for the uprising lotus buds.

Fa Hian who visited India during A.D. 401-411, tells us that in Pāṭaliputra the images of the Buddha were carried round in car processions with flotillas of dances and musicians and offering incense and flowers. The kings, chiefs and householders raised *vihāras* for monks and provided for their support by endowing them with fields, houses and gardens.

Kālidāsa, who is believed to have flourished in the court of Chandragupta-II, describes the Aśoka tree in most of his plays and in *Ṛtusaṁhāra*, "The Pageant of the Seasons", he gives charming descriptions of the indigenous beautiful trees of India which flower from month to month. In his description of spring, he describes the mango tree bent with cluster of coppery red leaves, and their branches covered with light yellow fragrant blossoms shaken by March breezes, which kindle the flame of love in the hearts of women. He describes also the Aśoka tree and the forests of Dhāk (Kiṁśuka) resembling a blazing fire, making the earth appear like a newly wedded bride with red garments.

Vātsyāyana, who also flourished during the Gupta times, mentions in his *Kāmasūtra* four types of gardens; *viz.*, *Pramododyāna* or pleasure garden, for the enjoyment of royal princes or princesses; *Udyāna*, the garden, where the royal elite passed their time, playing chess with their courtiers, enjoying company of dancers and court jesters; *Vṛkṣavāṭikā*, where ministers and courtiers enjoyed the company of the courtesans; and *Nandanavana*, which is dedicated to Lord Indra. Of these gardens, the *Pramododyāna* or pleasure garden has been depicted in a miniature from Kangra.[1] In a corner, a mango tree appears laden with fruit. The canopy of the pavilion is covered with Jasmines. The leaves of Plantain sway suggesting the breeze. Dark clouds in the horizon are indicative of mansoon.

Vātsyāyana[2] also recommends the building of a house close to a pond with an *udyāna* (garden) outside it. The lillies and lotuses are required to be grown in the pond. Geese, ducks and swans are also required to be kept in the ponds. In the garden attached to the house, a swing is also required to be provided from the thick branches of shady Nīma or a Pīpala tree and in the stifling monsoon weather when there is oppressive heat before the coming of rains and the weather is quite close, these swings could be useful. Describing the duties of a loyal house-wife towards the domestic garden, Vātsyāyana[3] writes—

> "A virtuous woman having affection for her husband, should act in conformity with his wishes as if he were a divine being, and with his consent should take upon herself the entire care of the family. She should keep the whole house well cleaned, and arrange flowers of various kinds in different parts of it, and make the floor smooth and polished so as to give the whole a neat and becoming appearance. She should surround the house with a garden. In the garden, she should plant beds of green vegetables, bunches of sugarcane, clumps of fig tree, the mustard plant, the parsley plant and the funnel plant. Clusters of various flowers such as the *Trapu bispinosa*, the Jasmine, the yellow Amarantti, the Nadyaworta, the China rose and others should likewise be planted, together with fragrant grass *Andropogan schoenanthus* and the fragrant root of plant *Andropogan muricatus*. She should also have seats and arbours made in the garden, in the middle of which, a well, a tank or a pool be dug."

A glimpse of trees and gardens of the post-Gupta period is available in the account of the Chinese traveller Hieun Tsiang, which has been detailed elsewhere in this work, but his views[4] on the cultivators and crops of India, are detailed below—

> "In cultivating the land, those whose duty is to sow and reap, plough and harrow (weed), and plant according to the season; and after their labour they rest a while. Among the products of the ground, rice and corn are most plentiful, with respect to edible herbs and plants we may name ginger and mustard, melon and pumpkins, the Heun-to (*kandu*) plant and othes. Onions and garlics are little grown; and few persons eat them; if anyone uses them for food, they are expelled beyond the walls of the town. The most useful food is milk, butter, cream, soft sugar, sugar-candy, the oil of the mustard seeds and all sorts of cakes made of

corn are used as food. Fish, mutton, gazelle and deer, they eat generally fresh, with respect to different kinds of wines and liquors, there are various sorts. The juice of the grapes and sugarcane, these are used by the Kṣatriyas as drink; the Vaiśyas use strong fermented drinks; the Śramaṇas and Brāhmaṇas drink a sort of syrup made of grape or sugarcane, but not of the nature of fermented wine."

## The Mysterious Trees and Lakes

A somewhat surprising aspect of the Purāṇic and other Sanskrit texts has been the concept of mysterious trees and lakes. In the case of plants, it was imagined sometimes that even the humans could be cursed to become trees. An example of this aspect is available in the *Devī Bhāgavata Purāṇa*[5], wherein Tulasī, the seduced wife of Śaṅkhacūḍa became a sacred plant to be adored by one and all. Similar is the case with *Yamalārjuna, i.e.,* two *Arjuna* trees which were turned into that state by the curse according to the testimony of the *Bhāgavata Purāṇa*[6]. Both of them were liberated by the infant Kṛṣṇa, by uprooting both the trees—

नलकूबरमणिग्रीव वासतु यमलार्जुनौ।

—*Bhāgavata Purāṇa*, X.10.23

There is another story in the *Devī Bhāgavata Purāṇa*[7] in which Nārada had his sex changed as a damsel after taking bath in the forest. As the story goes, Nārada once went to Viṣṇu's abode and seeing him approaching the Lord, Lakṣmī left the chamber. The sudden departure of Lakṣmī enraged Nārada who complained to Viṣṇu about the unbecoming behaviour of Lakṣmī, who had left the chamber, disregarding the great sage, Viṣṇu, somhow, pacified Nārada and persuaded him to go to a forest to enable Nārada to have the glimpse of Māyā. Both of them left mounted on Garuḍa and reached a forest near Kānyakubja, where they saw a beautiful lake surrounded by charming trees and attractive lotus flowers. Viṣṇu then induced Nārada to have a dip in the sweet waters of the lake. Nārada handing over his *viṇā* and clothes to Viṣṇu entered the lake and on coming out after the bath, he found his sex changed. He became a beautiful damsel. Viṣṇu in the meantime left the place. Nārada then started roaming about in the forest clad in divine clothes and ornaments, till he met the king Tāladhvaja, who proposed to marry the damsel who was named as Saubhāgya Sundarī. She started enjoying the married life and had several sons and grand-sons.

Once as luck would have it a king attacked the kingdom of Tāladhvaja and many of his soldiers, sons and daughters were killed. The wailing Saubhāgya Sundarī (Nārada) then reached the same lake in utter grief. She entered the waters of the lake and as soon as she came out, she gained her true form.

## The Mysterious Gardens

The Sanskrit texts quite often speak of some mysterious types of gardens with some supernatural environmental effects which are at times beyond imagination. Take the case of the pleasure garden of Śiva, in which the Lord spent his time with his spouse Pārvatī. The entry to this pleasure garden for males was prohibited and whosoever visited this was instantaneously turned into a female. The *Matsya Purāṇa* has detailed this story in its Chapter-12, according to which sometimes after Ikṣvāku and the brother of Ilā went in search of him (Ilā) who had mysteriously disappeared in the woods. In their wanderings they chanced to pass by the Sarvana, where they came across a mare saddled with glittering jewels. On seeing the mare with the jewelled saddle, they all recognised it to be Chandraprabhā, the famous charger of the monarch Ilā. They were highly astonished to find the charger transformed into a mare, and questioned the priest Vaśiṣtha about it. The sage then began to explain to them the mystery, through his powers of *Yoga*. The sage said that Śiva and Pārvatī had ordained that "anyone entering the limits of their pleasure garden would instantly be transformed into a female figure"—

यः पुमान् प्रविशेदत्र स नारीत्वमवाप्यति।।

—*Matsya Purāṇa*, 12.6

Unfortunately, both the king and the horses happened to be males, and they were transformed into a female and a mare. The brothers of the transformed monarch then besought the great sage to devise some means by which Ilā could be restored to manhood. At the advice of the Sage Vaśiṣṭha, the brothers of Ilā adored Śiva who was pleased and alternately granted manhood and womanhood to Ilā for a month in turn. The same story with some modifications is aldo repeated in the *Devī Bhāgavata Purāṇa* (Skanda-I, Chapter-6).

Though it may not be possible to explain the turning of the male to a female, in terms of modern scientific standards, but the very conception of the idea of a male turning into a female upon the entry into a pleasure garden, is quite interesting. Even in modern times there are some gardens in the country which are meant for females only and some others are notified as children parks.

The medieval period witnessed more of forestry and garden activity in the country, as compared to the earlier times. The early Hindu garden was mostly centred around the temples and other abodes of deities. It was based on a concept of organic growth, on the rhythms of nature and of humanity. Complex and individual sculptures were the natural expression of such beliefs and the temples themselves related perhaps more nearly to sculpture than in any architectural approach to these structures. Ornaments largely consisting of statues, carvings and bas-reliefs, which portrayed subjects both religious and temporal, nature and the human figurines in increasingly life-like grouping of great beauty and of surprising eroticism, were created. Flowers like the lotus and plants formed part of worship and were shown with intricate details in the carvings. Some of the deities were portrayed carrying flowers like lotus etc.

With the appearing of the Muslims on the Indian historical horizon, the cultural life of India witnessed far reaching changes. In the thirteenth century handmade paper from central Asia was introduced. This provided an impetus to painting and writing which rose to perfection during the reign of early Mughals. Before the ascendancy of the Mughals, garden history of India does not have much to be mentioned, though during the time of Firoze Tughlak who reigned in Delhi between A.D. 1351 to 1388, the environments of Delhi were improved considerably by planting gardens. His biographer[8] testifies that the Sultan had a great taste for laying out gardens which he took pains to beautify.

However, Babur is considered to be the pioneer of Mughal gardens in India. Before his becoming the ruler of India, Babur found time in Kabul to indulge in his favourite hobby of gardening. The following ten gardens are stated to have been laid by him—

(i) Shahr-ara (town adorning) which contained very fine plane-trees.

(ii) The Charbagh.

(iii) The Bagh-i-Jalu-Khana.

(iv) The Aurta Bagh (middle garden).

(v) The Saurat Bagh.

(vi) The Bagh-i-mahtab (moonlight garden).

(vii) The Bagh-i-ahukhana (garden of the deer hours).

(viii) Three other smaller ones.

In A.D. 1508-09, he laid out a garden known as Bagh-i-wafa near the fort Adinapur which he described thus—"The garden lies high, has running water close at hand and a mild winter climate. In the middle of it, a one mill stream flows constantly past the little hill on which are four garden plots. In the south-west part of it there is a reservoir, ten by ten, around which are orange trees and a few pomegranates, the whole encircled by a trefoil-meadow. This is the best part of the garden, a most beautiful sight when the oranges take colour. Truly that garden is admirably situated."

On settling down at Agra, Babur's[9] first concern was to raise a garden where he could take refuge from heat and dust. He was disgusted with the disagreeable aspect of the country; the flat monotonous dusty plains of Agra. He not only laid a garden for himself but also encouraged his nobles to do so. In his memoirs, he records the laying of garden on the bank of the Yamunā presently called Ram Bagh—

> "It always appears to me that one of the chief defects of Hindustan is the want of artificial water courses. I had intended, wherever I might fix my residence, to construct water-wheels, to produce an artificial stream, and to lay out an elegant and regularly planned pleasure ground. Shortly after coming to Agra, I passed the Jamunā with the object in view and examined the country to pitch upon a fit spot for a garden. The whole was so ugly and detestable, that I repassed the river quite repulsed and disgusted. In consequence of the want of beauty, and disagreeable aspect of the country, I gave up my intention of making Charbagh; but no better situation presented itself near Agra, I was finally compelled to make the best of the same spot. I first of all began to sink the large well which

supplies the tank with water; I next fell to work on the piece of ground on which are the *ambli* (or Indian tamarind) trees, and the octangular tank; I then proceeded to form a large tank and its enclosure; and afterwards the tank and *talar* (or grand hall of audience) that are in front of the stone palace. I next finished the garden of the private appartments, and the apartments themselves after which I completed the baths. This way, going on without neatness and without order, in the Hindu fashion, I however, produced edifices and gardens which possessed considerable regularity. In every corner I planted roses, and narcissuses regularly, and in beds corresponding to each other."

Babur's son Humayun (A.D. 1530-1556) lived in one of the garden palaces of his father at Agra, but he was comparatively uninterested in creation of gardens. Though occasionally he visited his orange gardens, making long journeys to the mountain passes to do so, he was not so much interested in botany or gardening like his father. But after his death he was laid in Humayan's tomb at Delhi, which is credited with having the distinction of the first garden tombs in Delhi. This was built by his queen Haji Begum, which is a remarkable example of the contrast of red sand stone and white marble. The surrounding layout was perhaps started during the lifetime of the emperor and this is the earliest surviving plan of Mughal garden. It takes the Charbagh motif and enlarges and repeats it in an intricate pattern, almost wholly two dimensional and strongly Persian in character. The channels and tanks have the depth of mere engravings, but minute differences of level are created in order to exploit the ripple of water. This water supply was lifted from wells outside the garden.

Akbar was the first Mughal emperor to enter the Kashmir valley. At Srinagar, close to the Dal lake he planned a large garden, Nasim Bagh. The walls, fountains and irrigation channels of this garden have disappeared, but the *Chinars* still remain. It is generally believed that *Chinār* was first introduced into Kashmir by Akbar, but the fact that the locals call it *Bhuni*—a corruption of *Bhavānī*—blissful-mother, shows that is was already present in the valley. *Akbar Nāmā* mentions that Akbar took thirty-four persons inside the hollow trunk of an aged *Chinār*[10] tree. *Chinār* infact is the state tree of Kashmir and its leaf-design can be seen in embroidary, carpets, wood-carving and even in brass-work. No one can cut it without the permission of the Government. There is a *Chinār* in the old Mughal garden at Bijabehara fifty-six feet in circumference.

Mughal gardens are of two types—garden for the pleasure of kings and the garden intended to provide peace to the dead. The tomb gardens were laid by the Mughals for the same purpose. Akbar will be remembered for two tomb gardens. Humayun's tomb at Delhi with a garden in it was built during his reign in A.D. 1556, on the Charbagh plan. The most remarkable of the tomb gardens is at Akbar's tomb at Sikandara near Agra. It is laid out in Charbagh plan in a huge square enclosure surrounded by walls. In the centre is the tomb which was commenced by Akbar himself. The garden was planted with cypress, planes, and areca-nut palm. None of the original trees has survived.

After Akbar, Jahangir took up the reins. He too was fond of gardening. He records his impression of *Chinār* tree in Kashmir, in his memoirs[11] thus—

> "At two and half *kos* from the city (of Srinagar) in the direction of Hindustan, there is a *Chinār* tree in the village Rawalpur. (Its hollow) inside is burnt out. Twenty years earlier, I had myself entered the (hollow) mounted along with five caprisoned horses and two domestic servants. (But) every time this incident was mentioned the listeners showed signs of scepticism. (Therefore) I again ordered some of the men to enter that (hollow) and it turned out to be exactly as I had thought."

Jahangir (1605-1627) is credited with laying the gardens at various places in India or some gardens were laid during his reign—

(1) Shalimar Bagh, Kashmir.

(2) Achabal, Kashmir.

(3) Verinag, Kashmir.

(4) Nishat Bagh, Kashmir.

(5) Garden at the tomb of I'tmad-ud-Daula, Agra.

(6) Wah Bagh, Hasan Abdal, Rawalpindi.

(7) Khusru Bagh, Allahabad.

(8) Lake Palace, Udaipur.

(9) Shahdra, Lahore.

Jahangir laid out Shalimar garden at the site of the villa of Pravar-Sen-II, (A.D. 79-139) and founded the city of Srinagar in A.D. 1619. Connecting the garden with the waters of Dal lake is a canal about 36 feet wide. On each side are green pathways shaded by giant *Chinārs*. The garden is divided into three parts, the outer garden, the central or Emperor's garden and the inner ladies garden. The outer garden was open to the public and terminates at the large pavilion, the Diwan-i-am. The Emperor's garden consists of two shallow terraces with the hall of private audiance in the centre. The snow of Mahadeo provides a lively and grand backgound to the pavilion in the ladies garden. According to Villiers Stuart, "a subtle air of leisure and repose, a romantic indefinable spell, pervades the Shalimar with its dim Vistas, smooth sheets of falling water and canals with calm reflections." This garden bears an eloquent testimony to the Emperor's landscape and fine taste.

Nishat, the Garden of Delight was built by Asaf Khan, brother of Noor Jahan. From the shore of the lake, its twelve terraces rise dramatically higher and higher to the mountain side. Each represents a sign of Zodiac. At the head of every waterfall is a stone or marble seat. At the sides are cypresses and *Chinārs*. Iris, narcissus, daffodils and roses were grown in profusion.

Gushing out of a limestone mountain on which a forest of Deodars is growing is a spring of clear water which is the life and soul of Achhabal garden. The garden is laid out in terraces and is planted with apple, pear, plum, apricot, and cherry. There are also stately *Chinārs* for shade. A pavilion with rich cedar woodwork presides over the canal which issues from the mountain spring.

Wah garden at Hasan Abdal near Rawalpindi was built around the springs by Jahangir. This palace garden was used as an imperial camping ground on way to Kashmir. Verinag, deep spring, the source of the Jhelum river also attracted the attention of Jahangir. In 1609, he built there an octagonal tank round the spring flanked by twenty-four arched recesses. The emerald green spring swarms with carps which are considered as sacred. The water from the spring flows into a canal and at its sides garden was laid. The tomb of Itmad-ud-Daula at Agra, besides preserving beautiful garden, also has inlay work on its walls which project flower patterns and trees.

Jahangir dies in A.D. 1627, in the mountain pass above Verinag. His last wish was that he might he brought back and buried by the spring. As winter was approaching and the passes could be closed by snow fall, his body was carried to Lahore and laid to rest in a tomb at Shahdara, where a garden was also laid. Indeed Jahangir was the greatest garden builder in India. The romantic beauty of his Kashimir gardens has contributed enormously to the aesthetic reputation of the Mughals. Wherever he stayed for long, he left a garden for the posterity to admire.

Emperor Jahangir's relations with his sons, more so with the prince Khusro, were not quite congenial. Khusro was quite popular and commanded love of the public at large. Both Jahangir and his third son Khurram (Shah Jahan) saw in Khusro a threat to their own security and the prince's rebellion was put down with barbaric cruelty. His death was contrived by Shah Jahan and Asaf Khan, in whose custody he had been placed. After his death, the body of Khusro was brought to Allahabad and was burried at Khusro Bagh, a terrace garden tomb where he rests with his mother and sons.

The pleasure garden of Noor Jahan at Shahdara, Lahore, was known as *Dilkhus* garden. Jahangir visited this place during his life time, and he was buried here after his death. Noor Jahan designed his tomb on the pattern of I'tmad-ud-Daula, her parents' burial place at Agra. There is an outer courtyard which served as a *sarai* where the travellers could rest for the night. A tall gateway gives access to the inner garden which contains the tomb. The enclosure is on a huge scale with five interlocking patterns of raised causeways, canals and tanks in which bricks have been beautifuly used. The tomb of I'tmad-ud-Daula itself preserves besides the beautiful *jālī* work of flower pattern, also inlay work portraying cypress trees, plants and flowers etc.

Shah Jahan succeeded Jahangir and remained saddled as an emperor of India from A.D. 1627 to 1658. As a great builder, Shah Jahan raised a number of monumental buildings *viz.*, the Fort and the Taj at Agra and Red Fort at Delhi. The Red Fort, Delhi, contained magnificent garden which was irrigated by a canal, drawing water from the Yamunā, which at that point of time was close to the fort. Water flowed by a channel lined with marble. This channel passes through below a carved marble screen of great beauty bearing a representation of scales of justice. During his life time Shah Jahan laid the following gardens in the country—

(i) Dara Sikoh and Peri Mahal, Kashmir.

(ii) Chashma Shahi, Kashmir.

(iii) Char Chinar Island, Kashmir.

(iv) Shalimar Bagh, Delhi.

(v) Lahore Fort.

(vi) Shalimar Bagh, Lahore.

(vii) The Fort, Agra.

(viii) The Taj, Agra.

The most delightful creation of Shah Jahan was the Shalimar garden of Lahore, which was built in A.D. 1634 under the direction of Mardan Khan, the architect of Chasma Shahi garden at Kashmir. Lahore Shalimar garden was laid out on seven divisions, symbolising the seven stages of paradise, according to Islamic concept. In the second terrace towards the east are royal bathrooms which consist of arched chambers with beautiful reservoirs which were heated by fire wood from outside. Shalimar is a garden of great beauty characterised by balance and symmetry. The central terrace consists of a huge tank with marble pavilions on three sides. In the centre is a platform reached by the narrow stone causeways. A canal flows from the middle terrace into a large marble basin. From the canal and the basin rise four hundred and fifty jet fountains which fill the air with soothing spray. Pavilions and summer houses of marble and red stone are scattered about in the garden. Shalimar suffered considerable damage in the last quarter of the eighteenth century as a result of the confusion caused by the fall of the Mughal empire. Copper of its fountains found its way to the market, marble was dismantled and transported to Amritsar for the tank of the sacred tank of the golden temple. In A.D. 1806, Maharaja Ranjit Singh stayed here after a short illness. He got the canals and the channels reopened and cleansed. During the British rule the garden became a public place.

Shah Jahan created a great garden around the Taj at Agra, which enshrines the remains of Mumtaz Mahal, his queen. The garden is of Charbagh pattern, a four fold plat, with a tank of white marble in the centre. The irrigation canal which contains fountains, had cypresses whose columns harmonised with four minarets. The proportions of the garden have undergone a great change during the last century and subsequently. Besides the profuse display of the horticulture in the entire Taj complex, the place has the distinction of preserving flower patterns in the inlay works on the walls of the tomb and the graves. On a frieze of flowers on the base of the outside walls of the Taj, Tulips and irises are clearly seen besides imaginary plants reesembling Dillenia. Besides, the flower patterns were also introduced in the marble lattice work, which served also to percolate diffused light in the interior. Similar type of inaly work showing flowering patterns also appear on the wall of the Naubat Khana, Diwan-e-Khas and other places at Red Fort, Delhi.

Aurangzeb's contribution towards the erection of gardens was minimal. A Lahore, Aurangzeb's Badshahi mosque is unsual in having a garden while in the Red Fort, he built Moti Masjid with flower patterns on walls. His daughter Zebunnissa Begum built two gardens at Lahore, known as Chauburgi Bagh and Nawan Kot Bagh, but no traces of the same are forthcoming. Roshan Ara Begum, the daughter of Shah Jahan also laid out a garden at Delhi, which was treated to be her pleasure garden. Aurangzeb's foster brother laid a garden at Pinjore near Chandigarh.

In spite of his being an orthodox ruler, the floral beauty could not be neglected by him, because, the Moti Masjid in the Red Fort, Delhi, has enough of such motifs.

While describing the salient features of a Mughal garden, it may be stated that bricks and mortar alone do not make a garden. It is the trees, shrubs, plants, flowering herbs and the constant and regular water supply provide a soul to it. These requirements were well looked after in the Mughal gardens. Though most of the Mughal gardens in the country are a much neglected lot, with the tanks, canals and water channels remaining dry or choked, and most of the trees which were originally planted having died out or replaced.

The details of plants grown by the Mughals are available in the autobiographies of the Mughal emperors. They mostly planted mixed avenues of cypresses and flowering trees like white flowered variety of Kachanāra (*Bauhinia tomentosa*, Linn.) or the orange or lemon trees. In Kashmir, the plum or the apple trees were grown in place of oranges and Kachanāra. Bordering the square plots were pomegranates and oranges. Oranges, pomegranates, almonds, plum and white Kachanāras are symbol of youth and life whereas cypress symbolise death and eternity. The dark green cypresses provide a solemn background to the silvery sprays of plums and the pink blossoms of almonds, the emblem of youth, life and hope. The Mughals mostly planted the spring flowering trees, shrubs and herbs in their gardens. They grew white, purple and mauve iris near liliac bushes, daffodils and narcissi under the apple and quince trees, and tulips under pear and plum trees in their Kashmir gardens. In summer, they grew roses, carnations, jasmines, holly hocks, peonies, and dolphiniums. Sometimes, the flowers of one variety were massed in a garden, thus creating a beautiful colour effect.

After the Mughals, the British took over the reins of the country and the gardens of the British are beyond the purview of this study. However, Cunningham has listed the names of the ancient forests in the territory of the Surasenas, which were cleared, some of which survive, and has also drawn the plan of these forests—

(1) The Mahāvana or Mahābana, the great forest.

(2) Kadambavana or Kāmavana, the Kadamba forest.

(3) Pilu-vana, or Piluana forest.

(4) Madhuvana or Madhuban, the Mahwa forest.

(5) Khadiravana or Khairban or Khair forest.

(6) Tālavana or Tālaban, the Palm forest.

(7) Vṛndāvana or Brindāban, the Tulasī forest.

Out of the above list, the last mentioned forest *i.e.*, Vṛndāvana is closely connected with the exploits of Rādhā and Kṛṣṇa and was named after Tulasī, as testified by *Brahmavaivarta Purāṇa*—

राधाषोड़श नाम्नां च वृन्दा नाम सुतौ श्रुतम्।
तस्या क्रीड़ावनं रम्यं तेन वृन्दावनं स्मृतम्।।

—*Brahmavaivarta Purāṇa*,
Kṛṣṇa Khaṇḍa, 17.214.

(Out of the sixteen names of Rādhā, Vṛndā is also one of them. Since it was her play-forest, therefore it was called as Vṛndāvana).

वृन्दारूपाश्च वृक्षाश्च यत्रैकत्र भवन्ति च।
विदुबुधास्तेन वृन्दा कृष्ण प्रियां तां भजाम्यहम्।।

—*Brahmavaivarta Purāṇa*

(The place where the Vṛnda or Tulasī plants are found is known as Vṛndāvana).

## Notes and References


1. Randhawa, M.S., *op. cit.*, p. 67, Pl. 5.
2. Vātsyāyana, *Kāmasūtra*, translated by Richard Burton, and F.F. Arbuthnot, London, 1963, p. 197.
3. *Ibid.*, p. 197.
4. Beal Samual, *Chinese Account of India*, Vol. II, Calcutta, pp. 132, 134, 143.
5. *Devī Bhāgavata Purāṇa*, 9.21.
6. *Bhāgavata Purāṇa*, X. 10.23.
7. *Devī Bhāgavata Purāṇa*, VI. 13.
8. Elliot and Dawson, *History of India as told by its own Historians*, II, London, 1857-77.
9. *Memoirs of Zehir-ud-Din Muhammad Babur*, Trs. by John Leyden and William Erskina, London, 1926, Vol. II, pp. 257-58.
10. Abul Fazl, *Ain-e-Akbari*, Trs. by Blockman and Jerret, Calcutta, 1873-94.
11. *Memoirs of Jahangir*, Trs. by Beveridge, II, 2nd Ed., Delhi, 1969.

# 8
# Damsel and Tree Motif

The damsel and tree motif has been of great antiquity in Indian art and literature. It was also known as *dohada* in the ancient literature. *Dohada* means the fertilising of trees, plants and creepers by touch or physical contact of a damsel, direct or indirect, and has been an impressive subject not only with poets but also with sculptors in India.

The act of *dohada* is not supposed to be performed for all the trees and plants, but only for the selected ones and the nature of the rite, which is always enacted by a young beauty, differs from tree to tree. The term *dohada* has been defined in the *Sabdarnava* as follows—

तरुगुल्मलतादीनामकाले कुशलैः कृतम्।
पुष्पाद्युत्पादकां द्रव्यं दोहदं स्यात्तु तत्कृया।।

The *dohada* rite, however, performed for the sake of the tree, is not a fertility rite, but one of fertilization, so that the tree may have flowers and fruits in abundance. As to the actual performance of *dohada*, it may be pointed out that as per the provision of the texts, the creeper Priyaṅgu puts forth new blossoms by the touch of woman; Bākula *(Manilkara hescandra*, Roxb.), through the sprinkling of liquor gargled by them; Aśoka (*Saraca asoca,* Roxb.) through the kick of the foot; Tilaka (*Clerodendrum phlomoides*) and Kurvaka (red-Amaryuth) through their glance and embrace respectively; Mandāra (*Erythrina indica*), through their sweet words; Cūta (Mango) by blowing with their mouths; Nameru (*Elaeocarpus ganitrus*) through their music and Karnikara (*Pterospermum acerifolium,* Willd.) through their dancing in front of it.

The idea then is that some plants, trees and creepers would blossom in the off-season, when lovely damsels direct their activities towards them. These activities may be of two kinds; they may consist of the direct contact with the body of a woman, as when she kicks, touches or embraces or an indirect contact as when she spits liquor upon them or blows at them, laughs, glances at them or when she speaks sweetly or sings or dances in front of them. There are ten select members of the floral world assumed not merely to be animate—this is held of all trees—but to react to the sensous intentions of words of a woman. It has also been pointed out that these trees, plants and creepers are associated with *śṛṅgāra* or love, and it is not difficult to ascribe response to emotions to the members of the floral world, particularly because tree-worship was strong in India from very ancient times.

These acts of *dohada* have been described in a number of ancient texts. In Act-III of *Mālavikāgnimitra* of Kālidāsa, Mālavikā, the heroine of the play, performs *dohada* rite by striking the *Aśoka* tree with her foot. This act has been interpreted both by the king and the maid servant of Mālavikā, rather differently—

आदयः कर्ण किसलयमस्मादियमत्र चरणमपर्यति।
उभयो सदृशविनिमयादात्मानं वञ्चितं मन्ये।। —(3.16)

(Mālavikā has struck the *Aśoka* tree with her foot, both had equal contact and neither of them had been in loss).

But the maid of Mālavikā felt that the striking of the *Aśoka* tree by Mālavikā with her foot was a special favour for the tree, and in spite of that if it did not blossom, the fault rested with the tree and not the damsel who struk it.

हला, णात्थि दे दासो। णिग्गुणो अअं असोओ जइ
कुसुमोद्भेदमन्धरो भवे जो दे चलणसक्कारं लम्भिअ। —(Act-III)

Actually *padāghāta* type of *dohada* stands by itself for the reason that it is the only variety of which full description is available as it occurs in the *Mālavikāgnimitra* of Kālidāsa. This is the only type that is apparently not found described as such among the varied activities of men and women in their love mood. Thus for more reasons than one, this is a unique type of *dohada*. This aspect would become more clear, when considered in some detail. The lady who is described as performing *dohada* here is Mālavikā. She was a substitute for the queen who had sprained her foot and that Mālavikā was quite fit to perform that function. This difinitely suggests two things; in the first place *padāghāta* is not a mere contact but a vigorous kick, and this fact is made clear in the king's speech later on. Secondly it further testifies that all women were not fit to perform it. It is also quite interesting to note the preparations made by Mālavikā for the performing of the act. Mālavikā's maid Bākulavalikā painted that feet with *alaktaka* and adorned with *nupuras*, the jingling anklets. She was not fully clad and her ornaments were meagre. the reference to her dress and ornaments which the king makes, shows that she had the minimum of both and this made the king more passionate. Thus bedecked and only thus; Mālavikā kicked the *Aśoka*-tree with her left foot, as is made clear by the king's speech. It is believed that if "*dohada* were successful, the tree would blossom within five days." That was also the expectation of the queen. In view of this it is significant that Mālavikā should have been anxious, when the tree did not show immediate signs of blossoming and that the king should have remarked that *dohada* has been wasted on the tree. As a matter of fact the tree did blossom within the prescribed time to the great surprise of all courtiers.

As to the eligibility of a lady to perform *dohada*, it may be pointed out the description of the same provided by Kālidāsa in *Mālavikāgnimitra* shows that not all are fit to perform the act of *dohada;* this is clear from the statement of the queen as reported by Bākulavalikā, the maid. She says that Mālavikā was deputed to do so, because the queen thought her fit for the function. The question arises as to what exactly was the fitness that the queen found in Mālavikā? It cannot be said that the queen chose her because she was a princess, for she was quite ignorant of her anticedents until the last stage in the play. The fitness therefore must have consisted in something else—her physical charm and her general conduct and behaviour which probably served as index to her noble birth. In other words, the queen must have felt satisfied that Mālavikā answered all the conditions of *Uttamanañjika*.

In the actual function of the *padāghāta* no less than in the sex communication, which this symbolises, the legs play an important role. Naturally, therefore, great attention is bestowed in the matter of adorning the legs. The feet of Mālavikā are adorned tastefully by Bākulavalikā, and they elicit encomium both from the king, the Vidūṣaka and the maid herself. After the paint, *nupuras* were put on and Mālavikā strikes the tree with her left foot. It is significant that the commentator also gives the place of honour to the left foot in this particular variety of sex act. Consistently with this interpretation, there is a specific purpose served by the *nupuras* and these are referred to by the king as resounding in the act of striking. The resounding symbolizes the *Śrīkṛta* which Vātsyāyana associates with this type of sex act. There is, however, some contradiction between the statement of the queen who expected the tree to blossom within five days and the expectation of the king, the maid and Mālavikā that it should have blossomed at once. Side by side with the term *āghāta*, Kālidāsa also uses the terms as *samparka* and *nyasa* which mean mere contact. Possibly these suggest the conventionalisation of the idea in the days of Kālidāsa.

Besides *dohada* there are also a number of other conventions described by poets in their works, some of which are listed hereunder—

| | | |
|---|---|---|
| (i) | Kālidāsa | *Raghuvaṁśa, Kumārasaṁbhava, Meghadūta, Vikramorvaśīyaṁ.* |
| (ii) | Bāṇa | *Harṣacarita.* |
| (iii) | Abhinavagupta | *Locana* |
| (iv) | Appayyadīkṣita | *Kuvalayānanda.* |
| (v) | Venkaṭa Kṛṣṇa Dikṣitar | *Natesa Vijaya.* |
| | | (i) *Subhāṣitaratna Bhaṇḍāgāra* |
| | | (ii) *Bharatacampu.* |
| | | (iii) *Natesavijayamahākāvya.* |

The tree and woman motif which had been popular with the poets was carried to the sculptural art, as well. Though on the literary side, the act of *dohada* is initially mentioned by Kālidāsa, but the same was quite popular during the 2nd-1st century B.C. because it is frequently found in art of Bharhut and Sanchi.

In the art of Bharhut, Yakṣī Candra[1] stands with her right foot planted firm on a fish tailed horse, her left leg and arm entwine the stem of a tree, while with her right hand she lowers a bough (Plate-49). This is a clear instance of the *latāveṣṭika* type of *āliṅgana-dohana* and the tree must be *Kuravaka*. Equally typical of the same kind of *dohada* is the figure of Culakoka Devatā from the same site. Standing on an elephant,[2] she has entwined the stem of a tree with her left arm and leg as in the case of Yakṣī Candra while with her right hand, she clasped a bough. There is another woman under a tree from Bharhut but without a label. Coomaraswamy[3], however, calls her a Yakṣī or a Devatā. She stands on a caparisoned horse.

Bodhagayā offers a very interesting type of the woman and tree arrangement. In this a woman is being helped by a man to climb a tree. It is quite clear that the man is not performing the function of a pedestal but he is actually helping the woman to reach up the tree. This figure illustrates the *Vṛkṣārohaṇa* type of *āliṅgana-dohada*. Coomaraswamy[4] refers to another instance from the same area, which he describes, "the familiar woman and tree" motif, in this case a *Vṛkṣaka* embracing her tree like Devatā at Bharhut.

On the gateways of Sanchi the same motif of woman and tree appears quite frequently one of the most typical is found in the eastern gateway in which the right arm of the Yakṣī is passed round the stem, not so much for support as for entwining, while her hand has clasped a bough whomever this may represent, Yakṣī or a mortal woman, evidently she is performing *dohada* of the *padāghāta* type. On the same side as this figure, but just below the uppermost architrave, there is a woman leaning against a tree with one hand raised above the head to the branches. Six bracket figures in the northern gateway, below each one of the three architraves on either side are shown in varied postures and in association with trees (Plates-50-52).

On the figures under the uppermost architrave, the one to the left may be identified as *padāghāta* type of *dohada*, for the right leg is raised and placed on the stem of the tree, while the right hand is holding a bough. The right arm of the figure is lost, the left hand is placed along the *mekhalā*. The posture suggests that the missing right arm must have been amongst the branches. This figure could come under the category of *sparśadohada*. The figures under the second architrave are identically placed. Below the third architrave also, there are two figures, somewhat fragmentary, suggesting the *padāghāta* type of *dohada* (Cl. Plates-42-45).

At Mathura, the figures are nude or almost nude and these type of figures are not conducive to the *padāghāta* type of *dohada*. The *Parimitabharanatva* mentioned by Kālidāsa is to be particularly noticed in one case in which the lady has a single necklace adorning her neck[5] and in this respect the figure stands distinct from other figures in the same series from Mathura.[6] Naturally these are not associated with trees (Fig.-34a-d). Other examples of *padāghāta* type of *dohada* are from Mathura which have been identified as *padāghāta* type of *asoka-dohada*. The kicking with the left leg adorned with heavy *nupuras* and the semi-nudity are both consistent with this interpretation. The tree of one further figure Commaraswamy[7] calls *Aśoka* tree, but the fruit do not confirm this identification and the act associated with *Aśoka*-tree is quite absent. This figure could therefore be identified as performing *sparśa-dohada*. An interesting bracket figure belonging to the same school comes from Kankali Tīlā.[8] This figure represents *āliṅgana-dohada*. Vogel[9] found another figure from the same site. The sculptors of Mathura school seem to have been particularly fascinated by the variations that the *dohada* motif admitted. A figure showing *Aśoka-dohada* is found on a railing. The woman in this case is leaning on a stem of *Aśoka*-tree, her right leg kicking it. The tree has started blossoming. The fruit with a rather long stem which she holds in her hand is the fruit of the *Aśoka* tree. Still another figure from Mathura shows a half back view of a female sculptured on a pillar. In case the vase held by the lady in her hand is taken to be a drinking bowl, this could be an instance of *gandusaseka* type of *dohada*. It is to be noted that the figure is nude and she faces a tree, while she is plucking flower from it with her right hand. All these figures come from Buddhist art. But there is one available in the Jaina art from Mathura[10] as well. It is nude but for a waist-band, which could be identified as *Kinkini*. There are no ornaments except the *nupuras*. The right foot presses against a tree, while the right hand is plucking the flowers from the branches. The left hand carries an instrument which is identified by Smith as an old type of Indian broadsword. The sword is held in the left hand. It this is a sword, it does not fit in the above identification. Possibly this symbolises *āghāta*. It would, however, be evident from the above that in most cases of Mathura sculptures on the theme of 'damsel and the Tree', a pedestal in the shape of a human being, Yakṣa or an animal is found present in them. This serves the purpose of a *pādapīṭha* and is mainly found in those images in which the main figure is performing the act of *padāghāta* or *āliṅgana* type of *dohada* but not in the *naratana* or *gandusaseka* type of *dohada*.

A *naratanadohada* of Gandharva school from Yusufzāi[11] is shown on a panel of a pillar. Another figure from Yusufzai depicts a female holding a mirror, agrees with the former in all respects, and may also be characterised as *naratanadohada*. *Aśokadohada* is also found on a disc on the intermediate rails of the outer enclosure of Amarāvatī. Cave No. 3 at Badami (8th century A.D.) provides a very striking representation of the *naratana-dohada*[12] motif. The women in this case holds a mirror and gives the final touches to her toilet.

Of the later period, numerous examples of this motif are available. The *maṇḍapam* of the Rāmacandra temple at Rajini has its pillars beset with them. In *Mukha Liṅgam*, *dohada* panels flank the doors of the main temple. Indeed this motif travelled down to the eighteenth century A.D. during which period a representation of the *padāghāta* type of *naratana-dohada* is available on one of the pillars in front of the small shrine standing to the right of Nandī.

## Notes and References


1. Barua, B.M., *Bharhut*, Pl. IX.
2. *Ibid.*, Pl. LXIV (75).
3. Coomaraswamy, A.K., *Yakṣa*, Part-I, Pl. VI.
4. Coomaraswamy, A.K., *History of Indian and Indonesian Art*, p. 33.
5. *A.S.R.*, III, Pl. XII, b.
6. *Ibid.*, Pls. VI, VII, XI.
7. Coomaraswamy, A.K., *Yakṣa*, Part-I, Pl. VI (ii).
8. *Ibid.*, Pl. XXII (i).
9. Vogel, J. Ph., *Woman and Tree or Sālabhañjikā*, Pl. III (b).
10. Smith, V.A., *History of Fine Arts in India and Ceylon*, p. 144, Fig. 95.
11. *Ibid.*, p. 117, Figs. 67-68.
12. Banerjea, R.D., *History of Orissa*, 1930, Pl. XIX-C.

# 9
# Medicinal Plants

The body ailments and diseases are as old as the human beings themselves, and various devices were used by human beings for the cure of these ailments. The use of plants for the purpose has been known to the early Vedic-Aryans who quite frequently used the word *Ausạdhi* which in general denotes the medicinal plant. Evidently though the use of plants as medicines was quite well known to the *Ṛgvedic* people, but their application appeared to be of some limited purposes since besides the word *Ausạdhi*, mention of other medicinal plants was not quite widespread. Some of the medicinal plants, besides *Ausạdhis* as projected in the *Ṛgveda* are given below—

(i) Asvavatī (*Ṛgveda*, X.97.7)

(ii) Urjayantī (*Ṛgveda*, X.97.7)

(iii) Virudha (*Ṛgveda*, I. 67.9; I.141.4)

In fact Chapter-97 of the Tenth Maṇḍala of the *Ṛgveda* specifies the extensive usage of herbs, trees and plants for the cure of human ailments as would be evident from the following few of the many such examples—

(i) The herbs are three ages older than the human beings (X.97.1).

(ii) The herbs cure the patients from the disease (X.97.2).

(iii) The holy fig tree has medicinal properties (X.97.6).

(iv) Physicians store herbs for the cure of men (X.97.6).

(v) The healing virtues of herbs are highlighted (X.97.8).

(vi) One regains lost strength by the use of herbs (X.97.11).

(vii) Fruitful and fruitless plants relieve one of the pains (X.97.15).

(viii) The herbs serve as cure of the plague (X.97.16).

(ix) The herbs serve as safeguard for animal diseases (X.97.20).

The *Atharvaveda* on the other hand has been somewhat more vocal about the usage of plants and herbs for the human good. Evidently these plants and herbs, besides curing the ailments, were also conceived to be useful in a variety of charms, glimpses of which have been provided in the chapter on the Literary Evidence of this work.

However, in the *Atharvaveda* some of the trees, plants and herbs have been described to contain medicinal properties, some of which are detailed below—

| | | | |
|---|---|---|---|
| (i) | Apāmārga | *Achyranthes aspera* | *Atharvaveda*, IV. 7.16; IV. 18.7; IV. 19.4. |
| (ii) | Kapittha | *Feronia limonia* | *Atharvaveda*, IV.4.8. |

| | | |
|---|---|---|
| (iii) Arundhatī | A creeper | *Atharvaveda*, IV. 12.1; V.5.5,9; VI.59.1-2; V.5.7. |
| (iv) Kuṣṭha | *Costus speciosus* | *Atharvaveda*, V.4.1; VI.102.3; XIX.1.39; XXIX.39.1; V.4.10; V.2.8; XXIX.39.9. |
| (v) Citraparṇī | | *Atharvaveda*, II.25.3. |
| (vi) Jangida | *Terminalia arjuna* | *Atharvaveda*, XIX.34.10; XIX.34.9; II.4.5; XIX.34.35. |
| (vii) Dhava | *Anogeissus latifolia* | *Atharvaveda*, VI.5.5. |
| (viii) Lakṣmaṇa | A plant used against abortion. | *Atharvaveda*, II.25.3. |
| (ix) Patrakanda | *Uraria logopo diodes* | *Atharvaveda*, II.25.3. |
| (x) Haridra | *Curcuma domestica* | *Atharvaveda*, I.24.2. |

In the post-Vedic literature, the *Āraṇyakas*, *Brāhmaṇas*, *Upaniṣads* and the *Gṛhya Sūtras* also mention about the medicinal properties of the plant-kingdom, but in the Epics their application becomes too apparent. The herb *Sañjīvunī* is said to have been brought by Hanumān from the Himālayas for the cure of the wounded Lakṣmaṇa. In spite of all this the usage of the herbs was not so widespread, even upto the time of the Epics.

But with the advent of the Purāṇic literature on the Indian religious scene, there was a great boost to research in the use of drugs made from the plant kingdom, for curing of ailments but also for their life saving qualities.

Of the various *Purāṇas*, *Garuḍa Purāṇa* is the one which provides a list of the medicinal plants in chapters 146-236. The same pattern is followed in the *Harivaṁśa Purāṇa* (second *Viṣṇu Purāṇa*, Chapter-41. The list is available in the section on *Purāṇas* in the Chapter on literary evidence in this work). Another list of such medicinal plants has been provided by the *Agni Purāṇa* (Chapter-282), which also provides a practical approach to the problem, because instead of merely mentioning the names of plants having medicinal properties, it lays more stress on their application or usages, highlighting at the same time the combination of different herbs for the cure of a particular ailment. A somewhat identical approach is followed by the *Padma Puraṇa* (I.28.19.32).

According to the *Matsya Purāṇa* (8.2-3) the Moon-god was made the master of Medicinal herbs and plants, trees and creepers etc.—

यथाभिषिक्तः सकलाधिराज्ये पृथुधरित्र्यामधियो बभूव।
तदौषदीनामधिपं चकार यज्ञव्रतानां तपसाञ्च चन्द्रम्।।
नक्षत्र तारा-द्विज वृक्ष-गुल्म लता वितानस्य च रुक्म गर्भः।

(At the time when king Pṛthvī was anointed and made the lord of the world, Candramā—the Moon-god—was made the king of medicinal herbs, sacrifices, fasts, penances, stars, planets, Brāhmaṇas, trees, thickets, creepers...).

The *Matsya Purāṇa* (217.42-54), however, provides a novel idea of the storage of the medicinal herbs in forts—

उक्तानि चाप्यनुक्तानि राजद्रव्याण्य शेषतः।
सुगुसानि पुरे कुर्यात् जनानां हितकाम्यया।।
जीवकर्षभककोलमामलक्याटरूषकान्।
शालपर्णी पृष्ठिपर्णी मुद्‌गपर्णी तथैव च।।
माषपर्णी च मदद्वैसारिवेद्धेबलात्रयम्।
वारा श्वसन्ति वृष्या च वृहती कण्टकाटिका।।
शृंगी शृंगाटकी द्रोणी वर्षाभूर्दर्भरिणुका।
मधुपर्णी विदार्येद्वे महाक्षीरा महातपाः।।
धन्वनः सहदेवाहा कटुकैरण्डकं विपः।
पर्णी शताह्वा मृद्धीका फल्गु खरजरयष्टिकाः।।
शुक्राति शुक्रकाश्मर्यञ्छत्रातिच्छत्रवीरणाः।
इक्षुरिक्षु विकाराश्च फाणिताद्याश्च सत्तम।।
सिंही च सहदेवी च विश्वेदेवाश्वरोधकम्।
मधुकं पुष्पहंसाख्या शत पुष्पा मधूलिका।।
शतावरी मधूकेच पिप्पलन्तालमेव च।
आत्मगुप्ता कट्‌फलाख्यादार्विका राजशीर्षकी।।
राजसर्पपधान्याक मृष्यप्रोक्ता तथोत्कटा।
काल शाकं पद्मबीजं गोवल्ली मधुवल्लिका।।
शीतपाकी कुवेराक्षी काकजिह्वोरू पुष्पिका।
पर्वत त्रपुसौ चोभौ गुञ्जातकपुनर्नवे।।
केसरु कारु काश्मीरी बल्या शालूक केसरम्।
तुषधान्यानि सर्वाणि शमीधान्यानि चैव हि।।
क्षीरं क्षौद्रन्तथा तक्रंतैलं मज्जवसा घृतं।
नीपश्चारिष्टकाक्षोड़वातामसोमवाणकम्।।
एवमादीनि चान्यानि विज्ञेयो मधुरोगणः।
राजा सञ्चिनुयात्सर्वं पुरे निरवशेषतः।।

(The king desirous of the well being of all should also keep in stock well preserved, all the various articles of the state. The following medicinal plants and articles of purest quality should also be stored, *viz., Jivaka, Ṛṣbhaka, Kakoli, Āmalaki, Vāsaka, Vara, Śālaparṇī, Pṛṣṇiparṇī, Mudgaparṇī, Māsaparṇī, Sarivadvya, Vatatraya, Vārā, Śasanti, Vṛṣyā, Vṛhatī, Kaṇṭakārī, Śṛṅgī, Śṛṅgatakī, Droṇī, Varṣā, Darbhā, Reṇukā, Madhuparṇī, Vīdaridvaya, Mahākṣīra, Mahātapā, Dhanvan, Sahadeva, Kaṭuka, Eraṇḍa, Viśā, Parṇī, Śatāhā, Mṛdvīkā, Phalgu, Kharjura, Yaṣṭimadhu, Śukra, Atiśaka, Kāśmorya, Chhatra, Atichhatra, Viraṇa, Ikṣu, Ikṣuvikāra, Panitadi, Siṁhi,*

*Sahadevī, Madhūka, Puṣpahaṁsa, Śatapuṣpā, Madhulikā, Śatāvarī, Madhūka, Aśvattha, Tāla, Ātmagupta, Kaṭaphala, Darvikā, Rajaśīrṣakī, Rajasarṣapa, Dhanyāka, Ṛṣyaprokta, Utkaṭā, Kālaśāka, Padmabīja, Govallī, Madhuvallī, Śītapākī, Kāliṅgākṣī, Kākajihvā, Villa, Śāluka, Nāgakesara,* all sorts of husks, grains, *Samidhanya,* milk, honey, curd, oils, fat and marrow, *vasā, ghee, nīpa, ariṣṭhaka, akṣota, vātāmra, soma* and *Vāṇaka,* etc.).

In addition to these a number of other products, having medicinal properties, have been recommended for storage in forts by the kings (217.55-87).

The *Brahmavaivarta Purāṇa* (Brahma Khaṇḍa, 16.28-30), refers to the various types of ailments and the herbs used for their cure [Appendix-3(b)].

Besides the *Purāṇas,* there is a vast literature on Indian medicines, *i.e., Āyurveda,* the treatment under which is based on Indian herbs, trees, plants and creepers. Indeed the health in India and the therapeutics of *Āyurveda* are bound by a chain which the rival professions with all their pretensions of being based on sound and scientific basis are feeble to disunite. While much impressed by the eloquent accusations like the scientific and empirical against the *Āyurveda,* the observer who turns to compare the clinical results of the two systems, is astonished to see that a large number of the Indian population still receives treatment from the indigenous system of medicines, which are based on indigenous herbs, plants and creepers, etc.

## 1. Āyurvedic Literature

According to the Āyurvedic texts, the plant kingdom has been classified, the glimpses of which have been provided by Caraka, as follows (*Caraka-Jivatiya,* Ch.-1, Sūtrasthāna, verses 72-73)—

(i) *Vanaspati*—(The trees which bear fruits without blossoming, like the one of *Vaṭavṛkṣa).*

(ii) *Vanaspatya*—(Those which bear fruits after blossoming like the mango trees and others).

(iii) *Auṣadhi*—(Which are grown as annual crops, like wheat, etc.).

(iv) *Virudha*—(Those which spread their shoots like the creepers).

Śuśruta, while maintaining by and large the above classification, substitutes *Vṛkṣa* in place of *Vanaspatya* (Śuśruta Sūtrasthāna, I.37). Besides Caraka and Śuśruta, these classifications of the plant kingdom have been made by other texts, but those provided by Caraka and Śuśruta are more appealing.

*Āyurveda* is not only a science today, but also happens to be the source, that has given birth to many of the medical sciences of the globe, ancient as well as modern. It descends directly from those mighty *Ṛṣis* of yore, that sprung up from the soil of India and some of them happened to be the pioneers on the subject. These *Ṛṣis* who had made the forests as their abode were fully conversant with the medicinal properties of a large number of plants and herbs. They carried with them the long experiene of their minute observations of the herbs, their good or ill-effects and developed methodologies in turning these herbs into life saving drugs or to rid the human body of the varied types of ailments. Caraka, Śuśruta and Vāgbhaṭṭa, were the three great teachers of the past in the Indian medicines and their works are still relevant even in the modern times. However, some extracts from Vāgbhaṭṭa's *Aṣṭāṅgahṛdaya*, referring to the use of various types of herbs and plants are quoted here—

(i) सारिवोशीरकाश्मर्यमधूकाशिशिरद्वयम्।
यष्टी परूषकं हन्ति दाह पित्ताऽस्रतृड्ज्वरान्।। —(12.11)

[*Sāriva*, (creeping plant—*Ichnocarpus frutescens*), Khesa, Kāśmarī, honey coloured—(*Mellifluous melodious* or *Bassia latifolia*), white sandal, red sandal, Yaṣṭhi, Paruṣa (*Xylocarpus granatum*) are useful in high blood pressure, burning sensation and various types of fever.

(ii) परूषकं वरा द्राक्षा कट्फलं कतकाल्फलम्।
राजाह्वं दाड़िम शाकं तृण्मूत्रामयवातजित्।। —(15.13).

(Paruṣa—*Xylocarpus granatum*, Triphala, grapes, Kāyaphala, Katakaphala, Khirni, Pomegranate, fruit of Śāka-tree, prevent wind formation and are also useful for urinary diseases).

(iii) गडूचीपद्मकारिष्टधानका रक्त चन्दनम्।
पित्तश्लेष्मज्वरच्छर्दिदाह तृष्णाघ्नमग्निकृत।। —(15.16)

(Giloya, Padmakāṣṭha, Nimba, Dhania, and Red Sandal are useful in *pitta*, cough, fever, burning sensation. Their use increases appetite).

(iv) Chapter-VI of the *Aṣṭāṅgahṛdaya*, deals extensively with the usages of different types of cereals like wheat, rice and barley; pulses like black gram, gram, *Rājma*, *Kuttia;* and oilseeds; various types of fruits.

Ācārya Priya Vrata Sharma, however has provided five-fold classification of the herbal and other plants, *viz.*, Herbs, flowers, fruits, vegetables and cereals and a large number of species have been listed under each class (*Āyurveda Kā Vaijñānika Itihāsa*, Delhi-1981, pp. 350 ff)—

## (i) Herbal Drugs

### (1) *Akarkara* (*Anacyclus pyrethrum*, D.C.)

This plant was initially grown in North Africa, from where it travelled to Southern Europe and then to Algeria, where its roots were collected and used for commercial[1] purposes. It reached India during the Muslim rule. In *Bhāvaprakāśa*, it is mentioned as *Akarakaram*.

### (2) *Ayapana* (*Eupatorium pyapana*, Vent)

This plant initially came from South America and reached India through Brazil and Japan. It was freely used in Bengal.[2]

**(3) *Ahiphen (Papaver somniforum,* Linn)**

*Ahiphen* is the Sanskrit adaption of *Arabic*[3] word *Afun*. It is mentioned for the first time in *Dhamantri-Nighantu*, as *Aphuk*. Possibly it also reached India with the Muslims, and was patronised by the medical men in India. This plant was well known in many countries of Europe.

**(4) *Isabgol (Plantago ovata,* Forsk)**

This was initially a product of Persia and was commonly used by the Muslim medical people. This could be the reason why this is not found mentioned in early *Āyurvedic* Nighanṭu literature. However, in the later medical texts, it was found with slight variations like *aspgol*, etc.

**(5) *Kankota (Piper cubeba,* Linn.)**

This plant was initially grown in Java and reached India at quite an early stage, since it is mentioned in the *Caraka-Saṁhitā*.

**(6) *Pitakarvira (Thevetia nerii-folia,* Juss.)**

Two types of white and red *Karviras* are found mentioned in the *Nighanṭus* of Madanpal and Dhanwantri. It was initially grown in the western islands and America and reached India in the sixteenth century A.D.

**(7) *Karpūra***

This happens to be an ancient recipe which is mentioned in the *Bṛhattrayi*. The Karpūra plants are not usually found in India but in Borneo, Sumatra, Java and Mataya they are found in abundance. Karpūra is extracted from *Dryobalanops Camphora Calbr.* There is another tree known as *Cinnamomum Camphoru Nessaberm*, which is grown in China and Japan[4] and Camphu is extracted by boiling the branches of this tree.

**(8) *Kalamba***

"कलम्बक इति लोके" is mentioned in the *Bhāvaprakāśa Nighanṭu*. This is possibly extracted from *Jatcorhiza Palmata Micrs* or *Coscinium fenestratum*.

**(9) *Kupilu (Strychnos naxvomica,* Linn)**

Though it was known in India from quite an early stage, but the Muslims did not know of its usage. It went to Europe[5] in the sixteenth century. *Nux-Vomica* does not appear to have been used in early Sanskrit medicines, although it is quite possible that some part of the tree may have been used by the aboriginal tribes of India, from a very early date, since the wood is currently used as a common tonic over very extensive tracks of the country. In the more recent Sanskrit compilations it is found mentioned under its vernacular name *kuchila*. According to the Hindu practicioners of the present day the seeds in various combinations are used as a medicine for dyspepsia and the diseases for nervous system. On the Malabar coast the root is given in snake-bite and in Maharashtra, the wood is a popular remedy in the dyspepsia of vegetations. There are several other usages of this tree.[6]

**(10) *Kumari* (*Aloe sp.*)**

In Barbados, where the plant is systematically cultivated, the plants are set six inches apart in rows of one or 1¼ feet wide, the ground having been carefully prepared and manured. The plants are kept dwarf and free from weeds. The leaves are 1-2 feet in length and are cut annually. The cut leaves are rapidly put into V-shaped trough with the cut end downwards and so arranged that the juice may drain from all leaves into a jar below. These troughs are placed all over the plantation so as to be easily accessible to cutters. By the time five troughs have been filled the first is exhausted. The leaves are neither boiled nor pressed and no use is made of them after the juice is drained off.[7]

**(11) *Kulinjana* (*Alpinia galanga*, Willd)**

The rhizomes of this species are aromatic, pungent and bitter and are used in the form of an infusion in fever, rheumatism and catarrhal affections. As a drug they are supposed to improve the voice. The aromatic tubers are sometimes used as carminative or fragrant adjunct in complex prescriptions, but they have nothing particular in their properties or action. The root of this plant is hot and stimulating.[8]

**(12) *Kṛṣṇabija* (*Ipomoea hederacea*, Linn.)**

This is a common plant, widely cultivated in India, still this drug does not appear to have been known to the early Sanskrit physicians and is not described by Ainslie. The author of the *Makhsan-ul-Adwiya*[9] says that it is a drastic purgative, useful in treatment of bilious and phlegmatic humours and that it acts also as an anthelminitie. Roxburgh was possibly the first to make these seeds known to European physicias, and it may be said that they now hold an important position as a useful and cheap substitute for jalap.[10]

**(13) *Kesara* (*Rottleria tinctoria*)**

This was also known as *Balhika* because it was available in abundance. Balkh in Afghanistan, which was one of the most prospering town known for international trade in the ancient past. It is grown in Indian mountainous regions in abundance. As a medicine it is used in fevers, melancholia and enlargement of the liver. It has also stimulant and stomachic properties, is highly thought of as a remedy for catarrhal afflictions of children and is used in certain Indian dishes as a colouring agent. In overdozes it is generally reported to act as a narcotic[11] poison.

**(14) *Kṣira-Campaka* (*Plumeria acutifolia*, Poirel)**

A small tree, which is a native of Tropical America but is cultivated in many parts of India. By the Persians the bark known as *aachin* is recommended as a cure for *gonorrhoea.* The leaves are abundantly grown in Malabar. The leaves made into a poultice are used to dispel swellings; the milky juice is employed as a rubefacient in rheumatism, and the blunt ended branches are introduced into the uterus to procure abortion. The bark is grown in the Konkan area with coconut, ghee and rice as a remedy for diarrhoea; the flower-buds are eaten with betel leaves in ague and the juice with sandal-wood oil and camphor is employed as a cure for itch.[12]

**(15) *Candrasurā* (*Lepidium sativum*, Linn)**

The seeds are described by later writers on Sanskrit medicines as a tonic, alterative, useful in hiccup, diarrhoea and skin diseases arising from disordered blood.[13]

**(16) *Cavya* (*Scindapsus officinalis*, Schott)**

The fruit is considered stimulant,[14] anti-catarrhal and carminative, and much employed as an adjunct to medicine for coughs, cold and hoarseness.

**(17) *Chopachini* (*Smilax china*, Linn)**

The tuberous roots at one time were official in all the European Pharmacopoeias, and were considered as a most valuable remedy in venereal and rheumatic disorders. The drug has now fallen in disuse in Europe, although in Chinese and Indian medicines it still enjoys a very high reputation and is considered a valuable remedy not only in syphilis and rheumatism but also an excellent demulcent and an aphrodisiac.[15]

**(18) *Chauhar* (*Artemisia maritima*, Linn)**

The flower-heads of this plant are largely used for their authelminitic, deobstruent, and stomachic tonic properties. In the form of a poultice it is used to relieve the pain caused through the stings of insects and poisonous bites, Santonine is chiefly employed in the treatment of round and thread worms. It has the peculiar property of sometimes causing object to appear yellow to patients under its action.[16]

**(19) *Jayapak* (*Croton tinglium*, Linn)**

The seeds are used as a powerful drastic purgative and the oild is regarded as useful medicine. In overdoses they act as an acro-narcotic poison. When externally applied the oil is a stimulant rubefacient and counter irritant. Croton oil is supposed to possess powerful hydragogue cathartic properties. It is also useful in dropsy, obstinate constipation and apoplexy.[17]

**(20) *Jatiphala* (*Myrica fragrans*, Honth.)**

This nutmeg was known atleast to the later Sanskrit writers by the name of *jatiphala*, but the time of its introduction in India is uncertain. In the sixth century A.D., records exist of the nutmeg having formed an article of Arabian import from the East and export to Europe, and the medicinal properties of the nut were probably then known to the Arab and Persian physicians. Ainsile informs, that "It is considered by the natives of India as one of their most valuable medicines in dyspeptic complaints, and in all cases requiring cardiacs and corroborants, they likewise prescribe it to such puny children as appear to suffer much in weaning.[18]

**(21) Tobacco (*Nicotiana tabacum*, Linn)**

The tobacco plant was introduced into India by the Portuguese about A.D. 1605 during the later part of the reign of Akbar, though the practice of tobacco smoking was made known to the Europeans about A.D. 1492, having been first observed by the followers of Columbus when visiting the West Indian Islands. It has been asserted by some authors that the habit of smoking must have been practiced in Asia long before the discovery of the New World, but definite evidence in this regard is lacking. The entire absence of any allusion to the plant in the works of early travellers and in the latest Sanskrit writings, and the universal adoption of a foreign name for the plant are strong arguments in favour of the conclusions that the use of tobacco was unknown to the oriental nations before the beginning of the seventeenth century.[19]

**(22) *Tāmbūla* (*Piper betle*, Linn)**

Commonly known as *Pāna* in India, Tāmbūla is used by masses for chewing and other purposes. But the leaves of this creeper, as is well known, masticated by people who make their packets of betel with the addition of lime, catechu, and betel nuts. The elite of the society, however, add cardamoms, nutmegs, cloves, camphor and other aromatics. Betel leaf thus chewed acts as a gentle stimulant and exhilarant. Those accustomed to its use feel a sense of languor when deprived of it. According to Śuśruta, it is aromatic, carminative, stimulant and astringent. it has several other medicinal usages.[20]

It appears that betel leaves were quite common during the period of Kālidāsa, who has referred to the same in his works. For example in the *Raghuvaṁśa* (6.64), he has mentioned about the Betel creepers, climbing the betel-nut trees—

ताम्बूलवल्ली परिणद्धपूगा, स्वेलालतांलिगित चन्दनासु।
तमालपत्रास्तरणासु रत्तुं, प्रसीद शखन्मलयन्थलीपु।।

After conquering the Kaliṅga territory, the army of Raghu drank the juice of *Nārikela* (coconut) making vases of betel leaves—

ताम्बूलीनां दलैस्तत्र रचितापान भूमयः।
नारिकेलासवं योधाः शात्रवं च पपुयशः।। —(*Raghuvaṁśa*, 4.4.2)

In the play entitled *Ṛtusaṁhāra*, it is mentioned that like the autumn season, the damsels in winter season also, enter their bed rooms with betel in their hands—

गृहीत ताम्बूल विलेपनस्रजः, सुखासवामोदित वकत्र पङ्कजः।
प्रकामकालागुरु श्रूपवासितं, विशन्ति शय्यागृहमुत्सुकाः स्त्रियः।।
—(*Ṛtusaṁhāra*)

In the *Hanumannāṭaka*, the betel leaves have been described to be responsible for increasing conjugal love between the hero and the heroine—

वक्रते ततः फणिलतादलवीटिकां स्वे, विन्यस्य चन्दनघनावृत पूगगभाम्।
रामेऽबवीदय गृहाण मुखे न वाले, तच्छद्मनां तदक्षरं प्रमातुम्।।
—(*Hanumannāṭaka*, 2.13)

In the *Brahmavaivarta Purāṇa*, the use of betel has been prohibited for a widow, ascetics, *Brahmacārīs*, *tapasvīs*—

ताम्बूलं दिधवास्त्रीणां यतीनां ब्रह्मचारिणाम्।
तपस्विनां च विप्रेन्द्र गोमांसं सदृशं ध्रुवम्।।

**(23) *Tāla* (*Borassus flabelliformis*, Linn)**

The juice of this plant is used as a stimulant and anti-phlegmatic when freshly drawn, it is exceedingly sweet, and, if taken regularly for several mornings in succession, acts as laxative. It is also useful in inflamatory affections and dropsy. The fermented juice called *tari* or toddy is intoxicating. The ash of the dry spandix is an antacid in heart burns and is useful in spleen. The terminal seed of the *tāla* tree is regarded as nutritive, diuretic and tonic. The root is regarded as cooling and restorative.[21]

**(24) *Turuska* (*Styrax officinale*, Linn)**

A native of Levant, Asia Minor and Syria, this is a solid resin somewhat resembling benzoin, which was held in great estimation from the time of Dioscorides and Pliny down to the close of the last century. In the east Styrax was given internally as a stimulant. The tree is mentioned in *Dhanvantri-Nighaṇṭu* and the *Bhāvaprakāśa*.[22]

**(25) *Tuvrak* (*Hydnocarpus wightiana*, Bluma)**

This is a common tree found in the western peninsula from Konkan along the coastal ranges. The seed has long been employed as a domestic remedy by people of western coast in cases of skin disease and apthalmia and as a dressing for wounds and ulcer. At the present day the oil is used as an external application to scabby eruptions, after being mixed with an equal portion of *Jatophara Curcas* oil, sulphur, camphor and lime-juice. For scalel-bead equal parts of the oil and lime-water are used as a liniment. It has other medical usages as well.[23]

**(26) *Dhattura* (*Datura tatula*, Linn)**

This is a much more temperate species than others, but its shape of flowers and character of fruits can with difficulty be distinguished. The corolla possesses, however, ten instead of five teeth or petals; the leaves are pubescent and show a pronounced tendency to be cordate at the base. The stems are almost sub-villose, a character by which the plant may be recognised from all other Indian Dhatturas. It is a small species than any of others, its pubscence and ten-petalled corolla being its characteristic features.[24]

**(27) *Dhumapatra* (*Aristolochia bracteata*, Retz)**

Every part of this plant is nauseously bitter and is much used by *Āyurvedic* physicians on account of its anthelminitic and purgative properties. Two fresh leaves rubbed up in little fresh water and given to an adult for a dose, once in 24 hours, are considered a cure for purging with gripes. The leaves are applied to navel to remove the bowels of children and are also given internally along with castor oil as a remedy for colic. Juice of the leaves when squeezed on a wound, kills the worms.[25]

**(28) *Parṇabīja* (*Bryophyllum calycinum*, Salisb)**

The leaves slightly roasted are used for healing of wounds, bruises, boils and bites of venomous insects. It has other medicinal usages as well.[26]

**(29) *Parnayabani* (*Coleus aromaticus*, Benth)**

The plant is used in asthama, chronic coughs, epilepsy and other convulsive affections. It is a powerful aromatic carminative, given in cases of colic to children, in the treatment of which the expressed juice is prescribed mixed with sugar or other suitable vehicle.[27]

**(30) *Parsikayavani* (*Hyoscyamus niger*, Linn)**

In the oriental literature the earliest record of this plant is to be found in the works of Arabian and Persian writers and the evidently foreign origin of the Sanskrit names indicate that it was known to the ancient Indian medical man. The Muslim writers of Arabia and Persia regarded the plant as intoxicating, narcotic and anodyne, and employed it extensively as an external application in case of inflammatory swellings, rheumatism and gout. A mixture of the powdered seeds and pitch was employed to stop painful hollow teeth, and as a pessary in painful affections of uterus; that the juice or a strong infusion of the seeds was dropped into the eyes to relieve pain; and that the seeds made into a paste with mare's milk and tied up in a piece of wild bull's skin was supposed to prevent conception when worn by a woman.[28]

**(31) *Pudīnā* (*Mentha piperita*, Linn)**

Peppermint is most agreeable and powerful of all the mints, possessing aromatic, carminative, stimulant, antispasmodic and stomachic properties. These qualities are especially possessed by the oil which, as the most convenient and elegant preparation, is generally prescribed. The oil with its preparations, a water and a spirit are officinal in Indian pharmacopoeia, and the herb and oil of this and other species of mint are largely employed as carminative and flavouring agents by the vegetarian people.[29]

**(32) *Puga* or *Supārī* (*Araca catechu*, Linn)**

This is commonly used in betels by people. The tender nut is said to possess astringent properties and is prescribed in bowel complaints and bad ulcer. It contains a large proportion of tannic and gallic acids and hence its astringent property. The burnt nuts when powdered form an excellent dentifrice. It has also been found useful in urinary disorderes and is reported to possess aphrodisiac properties.[30]

**(33) *Gadhapriyaṅgu* (*Prunus mahaleb*, Linn)**

This is known as *Gahula* in Marathi language and the scented kernels are used as necklaces by ladies.[31]

**(34) *Babūla* (*Acacia arabica*, Willd)**

The tree gum Acacia is used in preparation of mucilage. Its bark is a powerful astringent, recommended to be made officinal. It occurs in coarse, fibrous pieces of a deep reddish colour. It could be used for external applications as a substitute for oak-galls. It has been found a valuable remedy in prolapeus and as an external applicant in leucorrhoea, and is recommended as a poultice for ulcers attended with sanious discharge.[32]

**(35) *Bol* (*Balsamodendron myrrha*, Nees)**

Myrrha is beuifecial in dyspepsia, amenorrhoea, and chlorosis, and is a useful stimulent and astringent to all ulcerations or congestions of the mucus membrane. It is a useful application to old, foul and indolent ulcer, and a valued Nash for the mouth and gums, and a gargle in ulcerated sore-throat. It is a stimulant, expectorant, much admired as a remedy for pulmonary affections, especially the asthama of the aged. It is hot and dry, and that the best quality when broken shows white marks like those at the root of the finger nails. Internally it is regarded as a tonic and antispasmodic, emmenagogue, astringent and expectorant.[33]

**(36) *Madhuyaṣṭhi* (*Glycyrrhiza glabra*, Linn) (*Mulahthi*)**

Liquorice root has been used in India from a remote past. It is mentioned by Śuśruta and described as sweet, demulcent, cooling and useful in inflammatory infections, cough, hoarseness, thirst, etc. It is much employed for flavouring medical decoctions, oils and *ghṛtas* and enters into a composition of several external cooling applications, along with sandal wood madder.[34]

**(37) *Mastagi* (*Pistacia lentiscus*, Linn)**

This is a dioecious evergreen shrub of Mediterranean region, which yields the mastic of Chios imported into India. It has been known and much valued in medicines and the arts from the earliest period and until comparatively late years has been obtained from Island of Scio.[35]

**(38) *Mahanimba* (*Ailantus excelsa*, Roxb.)**

Its bark is aromatic and used for dyspeptic complaints; it is also regarded as a tonic and febrifuge in cases of debility. It is an expectorant and antispasmodic, given in chronic bronchitis and asthama. The leaves and bark are used as medicine.[36]

**(39) *Majuphala* (*Quercus infectoria*, Oliver)**

This is middle sized tree grown mostly in Greece, Asia Minor, Bosnea and Syria. It produces cylindrical acrous and has leaves gray underneath. It yields the galls which are used in dyeing and medicines. These galls are most powerful of vegetable astringents and are recommended as an antidote in poisoning with vegetable alkaloids.[37]

**(40) *Mehndi* (*Lawsonia alba*, Lam)**

This is known as *Henna* in Persian and Arabic world and has been employed and esteemed as a medicine from the remotest times. It was used as astringent by the Egyptians. According to Pliny the most esteemed cypress of egypt comes from Canope, on the banks of the Nile; the second from Ascalon in Judaea and the third from the islands of Cyprus.

In India too *Henna* is esteemed by Hindus and Muslims both. Its leaves are applied also to the hands and soles of the bride at the time of marriage; and also in small-pox, and are supposed to prevent the eyes being affected by the disease. Applied to hair and nails, they have reputation of promoting healthy growth. An ointment made from the leaves is said to have valuable healing properties and the decoction is used as an astringent gargle. The bark is given in jaundice and enlargement of spleen, also in calculous infections, and as an alternative in leprosy and obstinate skin diseases; in decoction it is applied to burns. The seeds with honey and tragcanth are described as cephalic. An in fusion of flowers is said to cure headache and to be a good application to bruises; a pillow stuffed with them has the reputation of acting as soporific.[38]

Śuśruta (25.42) includes it in cosmetics and Dalhana has described it as *Mehndi* (मदयन्तिका मेंहदी इति लोके यस्या पिष्टै: पत्रैर्नखानां रागं स्त्रिय उत्पादयन्ति।) The composers of Nighaṇṭus have termed it as *Madyantikā* or *Mallikā*. In the *Rajamartanda*, there is a provision of washing the uterus with the compounds of *mo-carasa* and *Madyantikā*.

**(41) *Yavasarkara* (*Manna*)**

This is a saccharine exudation obtained from several plants naturally and from others on the bark or epidermis being inside. There is another sort of *Manna* called Tiriamiabun or Trūmgibun which they gather from other leaves and herbs; they cut them in small pieces as big as Hempe seed and somewhat bigger, which are red in colour. Some think this *manna* grows on bodies of trees as Gumme doth; it is much in Ormus and Persia for Purgations, but not in India so much as the first sort.[39]

**(42) Eucalyptus (यूकेलिप्टस) (*Eucalyptus globulus*, Labill)**

This is a lofty tree, gregarious in Victoria and south of Tasmania. Its introduction to India has met with complete success on the Nilgiris, where the plantations, which were started in 1863 are well established. It has also been successfully cultivated at other places in the country. The bark of this tree exudes an astringent gum resembling, both in appearance and properties, that which, under the name of *kino*, is yielded by *Pterocarpus Marsupium*. The leaves yield an essential oil used in medicine, and sometimes as a substitute for Cajeput oil.[40]

**(43) *Rakaspattah* or *Rāmabansa* (*Agave americana*, Linn)**

This tree is originally the native of America and grown in many parts of India. Besides its various commercial usages, the leaves, as also the root when cut yield a saccharine juice which by evaporation may be converted into a syrup having a peculiar nauseous odour and acid taste; it reddens litmus paper. It has various other medicinal properties as well.[41]

**(44) *Revanda-Cini* (*Rheum embodi*, Wall)**

The roots of this species constitute according to the *Pharmacopoeia of India*, the 'large' variety of Indian or Himālayan Rhubarb. It occurs in twisted or cylindrical pieces of various sizes and shapes, furrowed; cut obliquely at the extremities, about four inches long and an inch and a half in diameter; of a dark brown colour, feeble rhubard odour, and bitter astringent taste; texture radiated rather spongy. Himālayan rhubarb is used as purgative and astringent tonic in place of imported article.[42]

**(45) *Lavaṅga* (*Spzygium aromaticum*, Linn)**

Also known as clove, it has been in use in India from the earliest times since Carak, Śuśruta and Vāgbhaṭṭa have discussed about its usages. Though it was imported from outside still it was the species of common usage. Initially it was used alongwith betel for chewing purposes but subsequently, its medicinal properties were fully made use of in the manufacture of medicines.[43]

**(46) *Vatsanabha* (*Calamus rotang*)**

The mass of the root sold by Indian druggists is derived from this species, but several others are no doubt used as substitutes or adultrants.[44] It is mentioned in *Kalpasthān* dealing with poison, in Śuśruta. In the *Aṣṭāṅgahṛdaya* it is also included in the Chapter on poison.

**(47) *Bhāṅg* (*Gāṅjā*) (*Cannabis sativa*, Linn)**

The earliest synonym appears to be *bhāṅga* which occurs in the *Ṛgveda* (IX. 61.13) and *Atharvaveda* (XI.8.15), besides the *Kauśika-Sūtra* (XIV-28, XVI.16 etc.). The word is derived from a root which means to break and is supposed to imply the process of debarkation by which fibres of the plant were separated from the stem. This would indicate that even at the remote period when the Vedas were composed, the use of hemp as fibre yielding plant was well known and the knowledge fully utilised. The Vedas, however, recon it, alongwith *Soma* as one of the five plants which were liberators of sin, and this would imply that its narcotic property was also well known. In the *Kauśitakī Brāhmaṇa*, which is next in antiquity to the *Atharvaveda*, the attribute form of the word *bhāṅga*, occurs both singly and in combination. In the institutes of Manu, the plant is noticed for its fibre.[45]

An alcoholic extract is prepared from it, from which tincture is also made. The use of Indian hemp in European drugs is quite common. It is variously used in medicines.

**(48) *Sanaya* (*Cassia angustifolia*, Vahl)**

Senna was first made known by the Arabs in the ninth century. It is extensively employed as a simple and active purgative. The Alexsandrian is generally regarded as more powerful than Tinnevelly and the Arabian or Moka much inferior to either of these.[46]

**(49) *Sarpgandhā* (Ichneumon plant)**

This is found mentioned in the *Śuśruta Saṁhitā* (V.84) and has also been mentioned as *Susarpagandhā*. Its use was in vogue in the medicines from quite an early date, but its importance as a medicinal plant dawned on western medical men in the seventeenth century. The French botanist Lumier recognised its importance. It was suitably adopted for use in medicines.[47]

**(50) *Sinduri* (*Bixa orellana*, Linn)**

It is a graceful shrub with handsome white and pinkish flowers and echinate red capsules, which is originally a native of America and largely cultivated in India for the red and orange dye obtained from the pulp which surrounds the seed. The pulp gives a beautiful flesh colour, largely used in dyeing silks. It is altered by certain combinations into orange, deep orange, or red, the brighter orange or red colours.[48] It is mentioned in the *Madanpāl-Nighaṇṭu* as well as *Bhāvaprakāśa-Nighaṇṭu*.

**(51) *Sinkona* (*Cinchona Succirubra*, Pavon)**

This species thrives at a lower elevation than the others, but is comparatively poorer in quinine, though rich in Cinchonme and Cinchonidine. It yields its best bark when eight years old. Its bark is most useful for medicinal use. The bark of the root contains a larger proportion of alkaloids than that of the stem, and that the proportion of alkaloid diminishes upwards to the branches.[49]

## (ii) The Flowers

**(1) Rose (*Rosa centifolia*, Linn)**

It is known in Sanskrit as *Śatapatrī*, *Śataparṇī* or *Taruṇī*. The petals of this species are officinal in the pharmacopoeia of India. They possess a sweetish bitter, faintly astringent taste and a roseate odour. They are mildly laxative and are used in combination with *Senna*, *Cassia* pulp and *Chebulic myrobalan*[50]. *Śatapatrī* has also been mentioned in the *Matsya Purāṇa*[51] and has also been recommended for storage in the forts.[52]

**(2) *Gulabbasa* (*Mirabilis jalopa*, Linn)**

The plant is said to be the native of South America, where it is called as the 'Marvel of Peru'. It is cultivated in greater part of India and is plentiful in the hotter valleys of North-Western Himālayas. The flowers are yellow, purple or magenta and in cultivation variegated or double. The stems of the yellow flowered forms are yellowish, of the others red. The plant is often so prevalent near village sites as to exclude all other vegetation. The tuberous roots of this plant was once supposed to be the *jalap* of the shops. It has feeble purgative power. The bruised leaves form a favourite application for abscesses and boils.[53]

**(3) *Gainda* (*Tagetes erecta*, Linn)**

The flower yields a yellow alye, which is used by dyers for colouring the clothes. The flowers are used in diseases of eyes, and when given internally, are supposed to purify the blood. The juice of the petals, heated with an equal quantity of *ghee* is useful for the removal of piles. The flowers are much admired all over India, but perhaps especially so with the hill people of the Himālayas. They are largely used for decorative purposes, and are strung into garlands and are hung round the images or in temples or shrines.[54]

### (4) *Campaka* (*Michelia champaca*, Linn)

The flowers are used for infusion or tincture. The tincture of the flowers is much more efficient than their decoction and to ensure the best effects of the farmer in severe or special cases, it should be used repeatedly in the same way as other stimulents are used under similar circumstances. The bark of the plant is antiperiodic and is used in decoction, prepared by boiling[55] it in two pints of water, till the liquid is reduced to half of it.

### (5) *Japa* (*Hibiscus rosa-sinensis*, Linn)

This is a favourite ornamental bush, a native of China, but is found in most of the flower gardens in the country.[56] There are several varieties of the same, which include single, and double, red, yellow or white. The flowers are considered emollient, and an infusioney, the petals is given as demulcent. An oil made by the mixing of the juice of its fresh petals and olive oil in equal proportions, and boiling till all the water has evaporated, as a stimulating application for the hair.[57]

### (6) *Sūryamukhī* (*Helianthus annuus*, Linn) or Sun Flower

This is an annual herb bearing large, falat, circular flowers-heads and is said to be the native of Mexico and Peru, and was introduced into Europe in the sixteenth century. The *Āftābi* or the Sun-flower has been mentioned in *Ain-e-Akbari*, as a flower cultivated for ornamental purposes.

The oil obtained by expression from the seeds is the most important product of the sun-flower and is useful for many purposes. The oil is equal to olive or almond oil for culinary and table purposes. The leaves and stalk make a valuable manure, either directly ploughed in or a box manure after being used as cattle litter.[58]

### (7) *Sthula-Kamala* (*Habiscus mutabilis*, Linn)

This is a small tree without pricklets, which has flowers that change in colour, almost white in the morning and red at night. It is a native of China but is found in the gardens all over India. As with most of the members of the genus, the bark yields a strong fibre, of which, that from inner layer is soft and silky, that from the outer layer hard and of a lead colour.[59]

## (iii) The Fruits

### (1) *Añjīra* (*Ficus carica*, Linn)

This fig tree is grown in many parts of India and does not require very rich soil. The dried fruit is demulcent, emollient, nutritive, and laxalive. It is, however, only rarely employed medicinally. Persons suffering from habitual constipation find it useful as an item of diet. The fruit is also used in the form of a poultice to effect suppuration. The pulp of the fig, mixed with vinegar and sugar, is very useful in bronchitis affections, principally in children. The dry fig contains about sixty to seventy percent of grape sugar and the unripe fruit starch.[60]

This is found mentioned in ancient medical treatises in different forms.

**(2) *Anānās* (*Ananassa sativa*, Linn)**

This is native of America and is commonly found in India. Possibly it reached India through Persia. The leaves which are required to be steeped in water for eighteen days yield a beautiful fibre, which, but for the difficulty of extraction could be largely used. This fibre is in demand in India, for threading necklaces, as it does not rot and is very strong. Both the wild and cultivated pine-apple yield fibres, which when spun, surpass in strength, fineness and lustre those obtained from flex. It can be used as a substitute of silk and as a material for mixing with wood or cotton. The fresh juice of leaves is regarded as a powerful anthelmintic, and that of the fruit as an antescorbutic.[61]

**(3) *Amarūda* (*Psidium guajava*, Linn)**

The fruit is astringent, and is used as a cure for diorrhoea. The bark, especially of root is also astringent. The leaves possess similar qualities, though perhaps to a less marked extent, and also make excellent poultice. In cases of scurvy, a decoction is used as a mouth wash for swollan mums, and has been used in cholera with some success, for arresting the vomitting and diarrhoeic symptoms. The fruit is eaten by all classes and is a special[62] favourite for its strong aromatic flavour.

**(4) *Aruk* (*Prunus persica*, Benth) *Āḍū***

The flowers are purgative, the leaves and kernels yield prussic acid on distillation, the fruit is demulcent, antiscorbutic, stomachic, and also kills worms of intestines. An oil is obtained from the kernels resembling that of bitter almonds for which it could be substituted. Its oil is also used for illuminating purpose and a dressing for the hair.[63] Dhanvantri has called the fruit as *Caturavidha*. According to Jahangir, the fruit trees were first imported by Akbar from Kabul and planted in Kashmir, together with other trees imported from Afghanistan.

**(5) *Karmaranga* (*Averrhoa carambola*, Linn)**

This is a small tree with sensitive leaflets extensively cultivated in India for its apples. The unripe apples are astringent and are used as an acid in dyeing. The leaves, the root and the fruit are used as cooling medicine. The acid dried fruit is given in fever. The ripe fruit has a pleasant acid and a pleasent taste and is used for culinary purposes. It blossoms in rainy season, and the fruit which ripes in December and January, is eaten raw to some extent. The flesh is soft, juicy and refreshing, and is then very pleasant; it is also made into a tasteful jelly.[64]

**(6) Cashewnut (*Anacardium occidentale*, Linn) or Kājū**

This is a native of America and reached India through Europe. The plant produces gum from the bark. The juice issuing from incisions in the bark is used as an indelible marking ink. The bark may be used for tanning. The medicinal usages of this plant are many. The acrid oil is used as an anaesthetic in leprosy and as a blister in warts, corn and ulcer. The spirit distilled from the expressed juice of the fruit could be used as a stimulant.[65]

**(7) Kharbūja (*Cucumis mello,* Linn)**

This is extensively cultivated in the country. The flattened and elliptic seeds yield a sweet edible oil. In fact the seeds of most of the members of the melon, pumpkin, cucumber and gourd family, contain oil, but the only kinds which are utilised to any considerable extent are those of sweet-melon (*Cucumis Meko*) and the water melon (*Citrullus vulgaris*). The seeds are used as cooling medicine. They are edible, nutritive and diuretic and are given in painful discharge and suppression of urine.[66]

**(8) Kharjūra (*Pheonix dactylifera,* Linn)**

This is a tall tree attaining a height of over hundred feet. The tree yields a gum called *hukm chil* which is used in medicines. Mats, fans, baskets and ropes are made of its leaves. The dates are considered demulcent, expectorant, laxative, nutrient and aphrodisiac. They are prescribed in cases of cough, asthma and other chest ailments.[67]

**(9) Tarbūja (*Citrullus vulgaris,* Schard)**

This is cultivated for its cool refreshing fruit. The seeds yield a clear, bland, pale coloured limpid oil, once used for burning in lamps. The seeds are also used as cooling medicines. The fruits are considered to be cooling, diuretic and strengthening. The fruit is large, ovoid green and smooth, the flesh is whitish yellow or red.[68]

**(10) Nārikela (*Cocos nucifera,* Linn) or Coconut**

This is a pinati-leaved palm, with a straight and often gracefully curved stem marked with annular scars. It is planted in most of the coastal regions in the country. The stem of this tree yields gum having red-brown, translucent and transparent shades. The thick pericarp or outer wall of the fruit yields valuable coir fibre which is used for commercial purposes. The sliced kernel, dried at ordinary temperature, either in sun or artificially, contains considerable quantity of oil. The green fruit is given as a refrigerant; the flowers are used as astringent and the oil is used as a substitute for cod-liver oil. The milk of the nut, the juice from the flowering spike and the tomentum from the leaves are all used medicinally.[69] Besides there are several other usages of the fruit and its products.

**(11) Nāśapāti (*Pyrus communis,* Linn) or *Amṛtaphala***

This is a small thorny tree, cultivated in the northern Himālayas. The fruit of this tree is used for making spirit in Kashmir. In Afghanistan this fruit is dried and its flour is mixed with wheat flour to increase its bulk.[70]

**(12) Papaya (*Carica papaya,* Linn)**

The juice of the fruit is regarded as medicinal. The milky juice of the unripe fruit possesses powerful anthelminitic properties but the ripe fruit is excellent and wholesome. It is an invaluable remedy in dyspepsia of those recovering from attacks of Cholera. It is useful in bleeding piles and ulcer of urinary passage, ringworms besides several other medicinal usages.[71]

**(13) Pistā (*Pistacia vera*, Linn)**

A small tree, forming forest at altitudes of three thousand feet and above, usually on sandstone formations, in Syria, Afghanistan and Khorasan. Its leaves were used for dyeing silk. The nut is worm, digestive, restorative, tonic and aphrodisiac and is prescribed as a sedative in nausea and vomitting. The galls are used as astringent, and the resin for the same purposes as mastic. Its oil is employed medicinally as a demulcent and restorative. The bark is believed to be tonic useful in indigestion.[72]

**(14) Bādām (*Prunus amygdalus*, Baill)**

This is commonly known in English as almond and an interesting history of almond may be found in the *Pharmacographia*, from which it appears that it has attracted much attention and has been commonly known in Europe and many other countries. In India, on the other hand, it appears to have been but little esteemed. It is but rarely mentioned in Sanskrit works and is not described as of medicinal value by older writers on Hindu medicines. The Arabic and Persian writers, however, discuss its properties at considerable length. Two types of sweet almonds are described, a thick shelled and a thin shelled and the usages of these have been essentially the same as in Europe. They are also used in perfumes. The oil of almond when applied to head kills the lice, as a suppository they relieve pain in difficult menstruation.[73] Besides, there are several other medicinal usages of the almond.

**(15) Bihi (*Cydonia vulgaris*, Pers)**

It is cultivated in North-West Himālaya and Afghanistan and grows wild in the woods of North Persia. The seeds act as demulcent and are useful in diarrhoea, dysentery, sore throat and fever. The dried fruit is used as refrigerant. The sweet and sub-acid quinces are commonly eaten as a fruit by Arabs and Persians and are considered tonic, cephalic and cardiacal; they are also eaten baked. The leaves, buds and bark of the tree were domestic remedies among the Arabs on account of their astringent properties. In India they are considered cold, moist and slightly astringent, and are one of the most popular remedies for coughs, bowel complaints.[74]

**(16) Bijapura (*Citrus medica*, Linn)**

This citron is supposed to have been introduced into Greece and Italy from Persia and the warmer regions of Asia. It yields scantily an unimportant gum. The flowers yield on distillation a very fragrant oil used in perfumes. Citron rind is hot and dry tonic; pulp cold and dry; seeds, leaves and flowers are also considered to be hot and dry. The juice is refrigerant and astringent. The fruit is an expellent of poisons. To one who has taken a poison injurious to life, it may be given, producing a strong effect on bowels and the poison is drawn out. The distilled water of the fruit is used as sedative.[75]

**(17) Madhukarkati (*Citrus decumana*, Linn) (Chakotra)**

The fruit is nutritive and refrigerant. It contains sugar and citric acid with much essential oil in the peel. The leaves are useful in epilepsy, chorea and convulsive cough.[76]

**(18) Rasabharī (*Physalis peruviana,* Linn)**

The fruit which closely resembles that of winter cherry is enclosed in a large inflated Calyx. It is of the size and shape of a cherry and has a bright ember colour. This plant is native of America and is quite popular in India.[77]

**(19) Lavali Harforauri (*Phyllanthus distichus,* Muell?)**

The fruit is acid and astringent. The root is an active purgative and the seed is Cathortic. It produces numerous small reddish flowers at the beginning of hot season which are succeeded by fleshy fruits resembling the goose-berry. These are used as an article of food, either raw or dressed in various fashions.[78]

The plant *Lavali* is mentioned copiously in Sanskrit literature. Bāṇabhaṭṭa has mentioned it along with *lavaṅga, ela,* etc. as mouth refreshing or adjunct to betel leaves.[79] The fruits were juicy and were commonly collected and stored in forest hermitages, evidently to be taken as a drug or a mouth refreshing agent. Rājaśekhara in his *Kāvyamīmānsā*[80] has described it to be a plant growing in the eastern region which blossoms in winter.[81] But Bāṇa has not included it in the list of substances by the king of Kāmarūpa to Harṣa. Moreover, Vārāhamihir describes it to be growing in coastal regions along with *ela* and *lavaṅga.*[82] In the *Caraka-Saṁhitā* (Sūtra-27) *Lavali* is seen only once in the context of *phalavarga* along with astringent fruits. Its fruit has been said to be palatable, adjuvient to meals and *vata* aggravating because of its *Kṣaya* and *Visada* properties and aromatic nature.[83] Similarly in the *Śuśruta-Saṁhitā*, it is mentioned only once in the same context but alongwith fruits growing on high altitude in Himālayan region, like almond, walnut, etc. The fruit of *Lavali* has been described as *Visada, Kasaya, Kapapittaghna,* slightly bitter, appetiser, palatable and of aromatic nature.[84] Vāgabhaṭṭa in his *Aṣṭāṅga Saṁgraha*[85] repeated the same aspects in abbreviated form but not in *Astāṅgahrdaya.* Cakrapāṇi, the commentator of the *Caraka Saṁhitā* and *Śuśruta Saṁhitā*, is silent about its identity except that *lavali* is commonly known as *lavali prasiddha* while commenting on the passage of the later text, but in *Cakradatta* (50-54), he used *Lavalikaphala* with *ela,* etc. Dallana in his commentary on the *Śuśruta Saṁhitā* says that the fruit of *lavali* is of aromatic nature and commonly known.

There is a controversy over the identification of the plant *lavali* with *Cicca acida* and the opinion of P.V. Sharma[86] in this connection is brought out; the characteristic features of the *lavali* plant on the basis of the reference to texts is as follows—

(a) It grows in eastern as well as in South-Western region along with *lavaṅga* and *ela*. It has got particular affinity with *lavaṅga.*

(b) The plant has fragrant white flowers which blossom in winter and fruit is aromatic, astringent and slightly bitter in taste.

(c) The fruits were used along with *lavaṅga* and *ela, kakkola,* etc. with betel leaves as mouth refreshing agent.

The plant *Cicca acida*, it has been argued, could hardly be taken as *lavali* because it does not have the above characteristics. It is grown almost everywhere. Its fruits are not astringent and bitter but sour and have no aroma at all. The fruits are used as pickles and not a mouth refreshing agent with betel leaves. *Lavali* plant has, however, been supposed to be identified with *Lavunga scandans Hanv* on the gound that it is a native of Assam hills and is staut climber having white fragrant flowers which blossom in April-May. Another plant which could be identified with *Lavali* is believed to be *L. eleutherandra* found in Western India from Konkan to Travancore, on the grounds that (i) it is a climber, and *lavali* has also been stated to be a *latā*, (ii) its habitat agrees with that described by Bāṇa and Rājaśekhara; (iii) its association with *lavaṅga;* and (iv) it blossoms white flowers in winter.

But there is another possibility too. The *lavali*[87] plant has also been identified with Harfarauri, and its botanical name has been pronounced as *Phyllanthus distichus.* Its fruit has also been described as astringent and is said to be a native of South. There is therefore, every possibility that this identification could also be equally justified.

**(20) Līchī (*Nephelium litchi*, Camb.)**

This hails from South China and found in the hilly regions of India. In China the leaves are stated to be officinal as a remedy for the bites of animals. The fruit is nearly round and about an inch and half in diameter. The edible portion is the sweet semi-transparent jelly-like pulp or aril, which covers the seed and the whole is enclosed in reddish or brownish brittle shell, which is rough with warty protuberances.[88]

**(21) *Lokata* (*Eriobotrya japonica*, Lindl)**

This is a native of Japan and was imported to India, where it is widely grown.

**(22) Siterphala (*Anona squamosa*, Linn)**

The ripe fruit is considered medicinally a maturant and when bruised and mixed with salt is applied to malignant tumours to hasten suppuration. The seeds, leaves and immature fruits contain an acid princile fatal to insects and the dried unripe fruit, powdered and mixed with gram flour is used to destroy vermin. An infusion of the leaves is considered efficacious in prolapsus ani of children. The root is regarded as a drastic purgative.[89]

**(23) Seb (Apple) (*Pyrus malus*, Linn)**

This fruit is primarily a native of Europe and Anatolia and is widely grown in hilly regions of India.

## (iv) Vegetables

**(1) Alābu (*Lagenaria vurgaris*, Mol.)**

This is quite an ancient plant of India since its name is found mentioned in several *Saṁhitās*.[90]

**(2) *Aśvabala (Medicago Sativa,* Linn)**

It was called *fisfisat* in Persian and Alfafa in Arabic. This plant is believed to be the native of Southern Anatolia, Persia, Afghanistan and Baluchistan. It was quite common in Asia Minor and is grown in North-west of the Indian sub-continent. It is mentioned but only once in *Caraka Saṁhitā* (1.4.7), which makes one believe that by the time of Caraka, this plant though known to the medical men, was not quite easily available. Śuśruta has mentioned it thrice; twice in the medicinal herbs and once in the vegetables.[91]

**(3) Ālū (Potatoes) (*Solanum tuberosum,* Linn)**

This is a native of America and was brought into India by the Portuguese. It is, however, found mentioned in the *Śāligramanighaṇṭu* and *Siddhabheṣajamaṇimālā* as *Alluka*.[92]

**(4) Kūṣmāṇḍa, Kāśīphala (*Cucurbita maxima,* Duchesne or *Benincasa cerifera*) or Pitakūṣmāṇḍa**

This resembles pumpkin. The fruit possesses alterative stylipic properties and is popularly known as a valuable anti-mercurial. It is also said to be cooling and is considered to be a tonic, nutritive and diuratic and a specific for haemoptysis and other haemorrhages from internal organs. For this purpose the fresh juice from the fruit is administered, while a slice of the fruit is at the same time applied to the temples. According to Sanskrit authors, it is useful in insanity, epilepsy and other nervous diseases; the fresh juice is given either with sugar or as an adjunct to other medicines for these diseases.[93]

Possibly the earlier Sanskrit writers were not acquainted with its peculiar action on the cultivatory system by which it rapidly puts a check to haemorrhage from lungs. The *Raja Nighaṇṭu*, the oldest work on therapeutics, gives a long account of its virtues but does not allude to its uses in phthisis or haemoptysis. Neither does Śuśruta mention it in his chapter on the treatment of haemorrhage and phthisis, though the plant is alluded to by him elsewhere. Vāgbhaṭṭa, however, in *Aṣṭāṅgahṛdaya* describes *Kūṣmāṇḍa* and other similar species, thus—

कूष्माण्डतुम्बकालिंगकर्कविर्वारुतिंडिशम्।।
तथा त्रपुसच्चनाकच्चिर्भटं कफवातकृत।
भेदि विष्टभ्यभिष्यंदि स्वादु पाक रसं गुरु।।
वल्लीफलानां प्रवरं कूष्माण्डं वातपित्तजित्।
बस्तिशुद्धिकरंवृष्यम् त्रपुसं त्वति मूत्रलम्।। —(6.86-87)

Besides, Kūṣmāṇḍa has also been mentioned in the *Amarkośa* (120.156), *Kāmasūtra* of Vātsyāyana (IV.1.29), *Upavana-Vinoda* (8) and *Matsya Purāṇa* (96.5; 118.27).

**(5) Garjaru *(Doucers carota Var. Sativa,* D.C.) Śikhā-mūlaṁ or Gājar**

The carrot is not officinal in English nor Indian pharmacopoeias, but the seeds are considered as nervine tonic. Boiled with honey and fermented, they produce a spirituous liquor. A decoction of the leaves and seeds is said to be used as stimulant to the uterus during parturition. The roots are made into marmalade which is considered refrigerant. Besides food there are several other usages of carrot.[94]

**(6) Tomato (*Lycoperstium esculentum*, Mill)**

This is a trailing plant and hails from South America. It is currently quite popular in India.[95]

**(7) Gobhī (*Brasica oleracia*, Linn)**

This is much valued cold season vegetable commonly cultivated in many parts of the country. The seeds are diuratic, laxative, stomachic and anthelmintic. The leaves form a good application in gout and rheumatism.[96]

**(8) Pālak (*Spinacea oleracea*, Linn)**

This is an annual plant cultivated throughout India. The native country of 'garden spinach' is not known, but is cultivated throughout India. The seeds are reported to yield a fatty oil. The seed is held by the practitioners of the Unani systems of medicines as having cooling and laxative properties and efficacious in breathing troubles and biliary derangements (*Talif Sherif*). The green plant is believed to act as solvent for urinary Calculi. The leaves are used as favourite vegetable.[97]

According to Candolle[98], the species has no Sanskrit name, but by Caraka and Śuśruta it is mentioned as *Palangi*, and was possibly in use in India from ancient times.

**(9) Baingan (Brinjal) (*Solanum melongena*, Linn)**

This species is believed to be the native of Arabian countries but is currently grown in the entire country. The leaves are said to have narcotic (*Baden Powell*) and the seeds stimulant properties.[99]

**(10) Bhiṇḍī (Lady's finger) (*Abelmoschus esculentus*, Linn)**

This is believed to be the native of South Africa and is commonly grown in almost all parts of the country. It is mentioned as *Bhiṇḍa* in *Śāligrāma Nighaṇṭu*.[100]

**(11) Groundnut (*Afachis hypogses, Linn*)**

This is believed to be the native of South America and is commonly grown in many parts of India.

## (v) Cereals

**(1) Kuṭṭu (*Fagopyrum esculentum*, Moench)**

This is essentially cultivated in the Himālayan regions. The leaves and tender shoots are boiled as a spinech and the nuts are husked and ground into flour, which is eaten as bread. The unhusked nuts are regarded as superior food for poultry. As an article of human food it does not hold a high place.[101]

**(2) Canaka (*Cicer arietinum,* Linn)**

This is a common type of gram commonly used in the country for a variety of purposes.[102]

**(3) Cinaka (*Panicum miliaceum,* Linn)**

The grain is considered to be digestible and nutritious, and in many places is eaten and cooked like rice. The small millet is easily recognised by its bright green foliage and spikelets of flowers arranged in lax drooping panicles.[103]

**(4) Jwār (*Hordeum sorghum,* Linn)**

It is supposed to be the native of Egypt, where the testimony of its usages as early as 2200 B.C. It is mentioned in the *Saṁhitās* as *juru* and in medieval literature as *vavanela.*[104] This goes on to testify about its usages in India from quite an early period.

**(5) Tavakṣira or arrowroot (*Curcuma angustifolia,* Roxb.)**

A good quality of arrowroot is prepared from tubers. A sort of starch is prepared from them. The flour when boiled in milk, forms an excellent diet for patients and children.[105]

**(6) Triputa (*Lathyrus sativus,* Linn) or Kesari-pulse**

This is mentioned in Śuśruta (46.27) as Triputa and in Caraka as *Khandika* (27.27). It is known commonly as *Kesari-dal* and when used continuously as food, develop paralytic symptoms in consumers.[106]

**(7) Makkā (*Zea mays,* Linn) or Corn**

This is the most useful grass and is cultivated not only throughout India, but also in the foreign lands. Besides being staple food, it yields a fibrous material which is said to be capable of being spun, but the chief use of the fibre has been the paper material. It is considered by physicians to have properties similar to those of *S. Vulgari, viz.,* resolvent, astringent, and very nourishing. It is considered to be suitable diet in consumption and relaxed condition of bowels.[107]

**(8) Atiyava (*Avena orientalis,* Sehreber)**

This is the native of Europe and central Asia and was cultivated in Greece and Italy.[108] Though it does not find place in *Caraka*, it is mentioned in *Śuśruta Saṁhitā.*

**(9) Rājamāsa (*Vigna sinensis,* Savi)**

This has been mentioned by Caraka and Vāgbhaṭṭa as Rajmas, and in Śuśruta it is known as *Alasemdra.* In *Bhāvaprakāśa,* its all three varieties are mentioned.

**(10) Sābudānā (*Manihot utilissima,* Pohl)**

This is a native of America and it is found from Brazil to western Islands.[109] Starch is made out of its roots and is taken as food as well.

**(11) Soyābīn (*Dolichos soja*, Linn)**

It grows in the coastal regions from China to Japan and Java. It was grown in China and Japan. It is quite nutritious as food and is grown in India as well.

**(12) Coffee (*Coffea arabarica*, Linn)**

This is the native of Sudan, Gurnia, Mazambic and the contiguous regions and was in use in Egypt from quite early times, where it was called as Cafe and was subsequently known a coffee. It is brought to India in the eighteenth century in Mysore and thereafter its use became quite popular together with its cultivation in the South.[110]

**(13) Tea (*Thea sinesis*, Linn)**

This plant is the native of China,[111] from where it entered into Assam and Manipur. In China the use of tea is found mentioned in the manuscripts dating back to 2700 B.C. as well as 300 B.C.

(From the foregoing account, it would be evident that P.V. Sharma has included in his classified list of *dravyas* certain species which are modern in nature and are not included in the ancient treatises).

However, Vāgbhaṭṭa, in Chapter-16 of *Aṣṭāṅgahṛdaya* (Sūtrasthāna) has described the use of *vanaspatis* and plants in making of different types of wines—

**(i) *Vāruṇi***

It is a particular kind of spirit prepared from the hogweed mixed with the juice of date or palm and distilled. It is said to be useful in stomachache, cough, vomitting, breathing trouble and other ailments—

तद्गुणा वारुणी हृद्या लघुतीक्ष्णा निहन्ति च।
शूलकासवमश्वासविवधाध्मान पीनसान्।।

—*Aṣṭāṅgahṛdaya* (Sūtra 6.69)

**(ii) *Vibhitaka* (*Terminalia bellerica*) (*Bahecta*)**

The wine extracted from this plant is slightly intoxicating and is useful in jaundice, leprosy and other diseases—

नाति तीव्रमदालध्वी पथ्या वैभीतकी सुरा।
व्रणे पाण्ड्वामये कुष्टे न चात्यर्थं विरुध्यते।।

—*Aṣṭāṅgahṛdaya* (Sūtra 6.70)

**(iii) *Yava(ka)* (*Hordeum vulgare*) (*Barley*)**

The wine made of barley is quite rough, hard and intoxicating and has various side effects—

विष्टम्भिनी यवसुरा गुर्वीरूक्षा त्रिदोषला।।

—*Ibid.*, 6.71

**(iv) *Mardhika* (Wine distilled from grapes)**

The wine distilled from grapes is useful for heart, is slightly warm, sweet, laxative and is useful in ailments like jaundice etc.—

मार्द्वीकं लेखनं हृद्यं वात्युष्णं मधुरं सरम्।
अल्पपित्तानिलं पाण्डुमेहार्शः कृमि नाशनम्।।

—*Ibid.*, 6.73

**(v) *Date-palm***

The wine extracted from Date-palm has almost the same ingredients as one has from the grape-wine. It is, however, heavy and creates wind ailments—

अस्मादल्पान्तरगुणखार्जूरं वातलं गुरू।।

—*Ibid.*, 6.74

**(vi) *Sugar***

The wine distilled from Sugar is fragrant, tasty, useful for heart and is light—

शार्करः सुरभिः स्वादुर्हृद्यो नातिमदो लघुः।

—*Ibid.*, 6.75

**(vii) *Guḍa***

The wine prepared from *Guḍa* is laxative, curative and clears wind—

सृष्टमूत्रशकृद्वातो गौडस्तर्पणदीपनः।

—*Ibid.*, 6.76

**(viii) *Sugarcane***

The wine distilled from the Sugarcane juice is beneficial in various ailments—

वातपित्तकरः सीधुः स्नेहश्लेष्मविकारहा।
मेदः शोफोदरार्शोघ्नस्तत्र पक्व रसोवरः।।

—*Ibid.*, 6.77

Vāgbhaṭṭa has also highlighted the qualities of the various types of cereals in chapter 16 of *Aṣṭāṅgahṛdaya Sūtra*, a glimpse of which is provided here—

(i) Paddy (धान्य)—This has been differentiated into several categories and the qualities of each one of them have been defined—

**(a) *Śāli-dhānya***

रक्तो महान सकलमस्तूणकः सकुनाहृतः।
सारा मुखोदीर्घशूको रोध्रशूकः सुगन्धकः।।
पतंग तपनीयश्च चे चान्ये शालयः शुभाः।
स्वादु पाकरसाः स्निग्धा वृष्या बद्धाल्पं वर्चसः।।

—*Ibid.*, 16.1-2

### (b) *Ṣaṣṭi-dhānya*

This species is considered to be one of the best of all types of rice, and is quite fragrant, tasteful—

स्निग्धो ग्राही गुरुः स्वादु स्त्रिदोषघ्नः स्थिरो हिमः।
षष्टिको ब्रीहिषु श्रेष्ठो गौरश्चासित गौरतः।। —*Ibid.*, 6.6

### (c) *Mahābrihī* and Other Types

Under this species, various qualities of paddy are mentioned which is good for health and stomach—

ततः क्रमान्महाब्रीहिकृष्णब्रीहि जतूमुखाः।
कुक्कुटाण्डकपालाख्यपारावराकशूकराः।।
वरकोद्दालकोब्ज्वाल चीनशारद दुर्दटा।
गन्धना कुरु विन्दाश्च गुणैरल्पान्तराः स्मृताः।।

—*Ibid.*, 16.7-8

### (d) *Kaṅguṇī-Dhānya*

While describing the good quality of paddy or rice, Vāgbhaṭṭa does not neglect the inferior type of paddy or rice, which was possibly used by the lower strata of the society and defines it as under—

कंगुकोद्रवनीवारश्यामकादि हिमं लघु।
तृणधान्यं पवनकृल्लेखनं कफ पित्त कृत।।
भग्नसंधानकत्तत्र प्रियंगुर्बृहणी गुरुः।
कोरदूषः परं ग्राही स्पर्श शीतो विषापहः।।

—*Ibid.*, 16.10-11

Besides the rice, chapter-16 of *Aṣṭāṅgahṛdaya* deals with the various types of medicinal usages of the different types of cereals, including barley, wheat, pulses etc.—

(i) *Kulthi* (*Dolichos bifloras*, Linn.)[112]

(ii) *Mūṅga* (*P. Mungo*, Linn)[113]

(iii) *Arhar* (*Cojanus indicus*)[114]

(iv) *Gram* (*Cicer arietinum*)[115]

(v) *Peas* (*Pisum arvense*)[116]

(vi) *Masur* (*Lens esculenta*)[117]

(vii) *Rājamāsa* (*Vigna sinerisis*, Savi)[118]

(viii) *Māsa* (*Phaseolus radiatus*, Linn)[119]

It would thus be evident that the Indian practitioners of medicines not only were aware of the different qualities of plants and trees and their medicinal values, but were quite aware of the medicinal effects of different species of cereals, and could prescribe which particular cereal had to be avoided in any particular ailment and the one which could be taken without causing any harm to the patient. A chart showing the various medicinal plants referred to in medical treatises of the country has been provided as an appendix, which is illustrative and not exhaustive.

Besides the medicinal plants, there are certain species, which are considered to be poisonous. A chart showing such plants also appears as an appendix to this work.

## Notes and References


1. George Watts, *A Dictionary of Economic Products of India*, Vol 1.I, p. 237.
2. *Ibid.*, Vol. III, p. 293.
3. *Ibid.*, Vol. VI, Part-I, pp. 23-36.
4. *Ibid.*, Vol. II, pp. 84-87.
5. *Ibid.*, Vol. VI, Part-III, p. 380.
6. *Ibid.*
7. *Ibid.*, Vol. I, pp. 179-80.
8. *Ibid.*, Vol. I, pp. 192-196.
9. *Ibid.*, Vol. IV, pp. 486-87.
10. *Ibid.*, Vol. IV, pp. 486-87.
11. *Ibid.*, Vol. II, pp. 592-93.
12. *Ibid.*, Vol. VI, Part-I, p. 297.
13. *Ibid.*, Vol. IV, p. 629.
14. *Ibid.*, Vol. VI, Part-I, p. 256.
15. *Ibid.*, Vol. VI, Part-III, pp. 253-254.
16. *Ibid.*, Vol. I, p. 325.
17. *Ibid.*, Vol. II, p. 518.
18. *Ibid.*, Vol. V, p. 313.
19. *Ibid.*, Vol. V, p. 361.
20. *Ibid.*, Vol. VI, Part-I, pp. 254-55.
21. *Ibid.*, Vol. I, p. 496.
22. *Ibid.*, Vol. VI, Part III, pp. 384-385.
23. *Ibid.*, Vol. IV, pp. 309-310.
24. *Ibid.*, Vol. III, p. 40.
25. *Ibid.*, Vol. I, p. 314.
26. *Ibid.*, Vol. I, p. 543.
27. *Ibid.*, Vol. II, p. 504.
28. *Ibid.*, Vol. IV, pp. 320-321.
29. *Ibid.*, Vol. V, p. 230.

30. *Ibid.*, Vol. I, p. 292.
31. *Ibid.*, Vol. VI, Part-I, p. 348.
32. *Ibid.*, Vol. I, p. 24.
33. *Ibid.*, Vol. I, p. 368.
34. *Ibid.*, Vol. III, p. 513.
35. *Ibid.*, Vol. VI, Part-I, p. 270.
36. *Ibid.*, Vol. I, p. 148.
37. *Ibid.*, Vol. I, p. 383.
38. *Ibid.*, Vol. IV, p. 600.
39. *Ibid.*, Vol. V, pp. 165-166.
40. *Ibid.*, Vol. III, pp. 280-281.
41. *Ibid.*, Vol. I, pp. 142-143.
42. *Ibid.*, Vol. VI, I, p. 486.
43. Sharma, Acarya Priyavrata, *Āyurveda Kā Vaijñānika Itihāsa*, p. 373.
44. George Watt., *op. cit.*, Vol. I, p. 87.
45. *Ibid.*, Vol. II, pp. 103-108.
46. *Ibid.*, Vol. II, p. 214.
47. Sharma, Acarya Priyavrata, *op. cit.*, p. 374.
48. George Watts, *op. cit.*, Vol. I, p. 435.
49. *Ibid.*, Vol. II, p. 301.
50. *Ibid.*, Vol. VI, Part-I, p. 560.
51. *Matsya Purāṇa*, 95.25.
52. *Ibid.*, 217.46.
53. George Watts, *op. cit.*, Vol. V., p. 254.
54. *Ibid.*, Vol. VI, Part-III, p. 403.
55. *Ibid.*, Vol. V, p. 243.
56. *Ibid.*, Vol. IV, p. 242.
57. *Ibid.*, Vol. IV, pp. 242-243.
58. *Ibid.*, Vol. IV, pp. 210-211.
59. *Ibid.*, Vol. IV, p. 242.
60. *Ibid.*, Vol. III, p. 349.
61. *Ibid.*, Vol. I, pp. 237-238.
62. *Ibid.*, Vol. VI, Part-I, p. 353.
63. *Ibid.*, Vol. VI, Part-I, p. 349.
64. *Ibid.*, Vol. I, p. 360.
65. *Ibid.*, Vol. I, pp. 232-233.
66. *Ibid.*, Vol. II, pp. 227-228.

67. *Ibid.*, Vol.VI, Part-I, p. 204 ff.
68. *Ibid.*, Vol. II, p. 332.
69. *Ibid.*, Vol. II, p. 415 ff.
70. *Ibid.*, Vol. VI, Part-I, p. 374.
71. *Ibid.*, Vol. II, p. 157 ff.
72. *Ibid.*, Vol. VI, Part-I, pp. 274-275.
73. *Ibid.*, Vol. VI, Part-I, pp. 343-344.
74. *Ibid.*, Vol. II, pp. 666-667.
75. *Ibid.*, Vol. II, p. 351.
76. *Ibid.*, Vol. II, p. 349.
77. *Ibid.*, Vol. VI, Part-I, p. 225.
78. *Ibid.*, Vol. VI, Part-I, p. 217.
79. *Kādaṁbarī*, p. 279.
80. *Kāvyamīmānsā*, Ch. 17, p. 278.
81. *Ibid.*, Ch. 18, p. 308.
82. *Bṛhat Saṁhitā*, 27.4.
83. *Caraka Saṁhitā*, 46.189.
84. *Śuśruta*, 46.189.
85. *Aṣṭāṅga Saṁgraha*, 7.191.
86. Sharma, P.V., *Fruits and Vegetables in Ancient India*, p. 45 ff., Delhi-1979.
87. Sharma, Acarya Priyavrata, *op. cit.*, p. 381.
88. George Watts, *op. cit.*, Vol. V, pp. 346-347.
89. *Ibid.*, VII. I, p. 260.
90. *Ibid.*, Vol. VI, Part-I, p. 376.
91. Sharma, Acarya Priyavrata, *op. cit.*, pp. 382-83.
92. *Ibid.*, p. 383.
93. George Watts, *op. cit.*, Vol. I, p. 440 and Vol. II, p. 638.
94. *Ibid.*, Vol. III, pp. 50-51.
95. *Ibid.*, Vol. II, p. 100.
96. *Ibid.*, Vol. I, p. 533.
97. *Ibid.*, Vol. VI, Part-III, p. 331.
98. Candolle, A.D., *Origin of Cultivated Plants*, p. 373.
99. George Watts, *op. cit.*, Vol. VI, Part-III, p. 163.
100. *Ibid.*, Vol. I, p. 2; Vol. IV, p. 241.
101. *Ibid.*, Vol. III, p. 311.
102. *Ibid.*, Vol. II, pp. 282 ff.
103. *Ibid.*, Vol. VI, Part-I, p. 12 ff.

104. Gode, P.K., *Studies in Indian Cultural History,* Vol. I, pp. 277-282.
105. George Watts, *op. cit.*, Vol. II, p. 654.
106. *Ibid.*, Vol. V, p. 521.
107. *Ibid.*, Vol. VI, Part-IV, p. 330 ff.
108. Candolle, *op. cit.*, pp. 373-376.
109. *Ibid.*, p. 95.
110. George Watts, *op. cit.*,Vol. II, pp. 460 ff.
111. *Ibid.*, Vol. II, pp. 75-77.
112. *Aṣṭāṅgahṛdaya*, (Sū), 16.18.
113. *Ibid.*, 16.15-16.
114. *Ibid.*
115. *Ibid.*
116. *Ibid.*
117. *Ibid.*
118. *Ibid.*, 16.17.
119. *Ibid.*, 16.20.

# 10
# Divine Trees and Plants

The Purāṇic and other texts often speak of an event commonly known as *Samudra-Manthana*, a project co-sponsored by the Devas and the Asuras alike. In the Vedic and post-Vedic literature, churning or *manthana* is a symbolical term signifying agitation (*kṣobha*) of the first cause which has many equivalents *viz., Manthana, Samindhana, Jagarna, Pranana,* etc. churning is the rhythemic movement and in terms of creative process it is best expressed as *Pranana-Apanana* or *Samchana-Prasarana (Pranovai Samamcanamprasaranam, Śatapatha Brāhmaṇa,* 8.1.4.10); also *Yajurveda* (97.45) or *Sankocavikasa* in the Purāṇas. This rhythemic swing of the pendulam of time is the mighty process produced by cosmic churning.

The next result of the universal churning which is taking place in the most distant stars and the minute cells and atoms is the manifestation of energy or life-principle or consciousness which is immortal fire. It is called *Dakṣiṇāgni* for the reason that Dakṣa means the right sided one, namely the material world as against *Vāma* which is manifested upper-half of cosmological unit. thus Puruṣa is the left-half and Prakṛti is the right-half of the cosmic Egg. It is also said Bhṛgu produced Atharva and Atharva produced Aṅgirasa. In fact Bhṛgu and Aṅgirasa are symbols of water and fire, the principles of heat and cold. They both exist in primeval water, or first cause as stated in the *Gopatha*—

आपो भुग्वंगिरोरूपमापो भृग्वंगिरोमयम्।

Atharva is the mid-most point between Bhṛgu and Aṅgirasa and as a link agitates both of them.[1]

The Purāṇas, however, interpret the churning in a somewhat different way, concerning at the same time the story of the churning of the ocean or *Samudramanthana*. According to the *Matsya Purāṇa*, this exercise of the churning of the ocean (as well other texts) was accompalished jointly by the Devas and the Asuras. As a result of the churning of the ocean, Kalpavṛkṣa was one of the precious *vibhūtis* which emerged out of the sea-bed. This aspect has been interpreted by the various texts with slight variations here and there. Some texts believe that instead of one, five trees emerged out of the churning of the ocean and the names of these trees have been described[2] as under—

*Pañcavṛkṣas*—Mandāra, Pārijāta, Samantana, Kalpavṛkṣa and Haricandana.

Of the above five trees, both Mandāra and Pārijāta have been described by Monier Williams to be *Erythrina indica* or the Coral tree.[3] Evidently a single tree has been mentioned by two different names, and if this be so, the number of trees that emerged out of the sea is reduced to four, which would not justify the term *pañca-vṛkṣas*. In order, however, to have a clear idea of the issue involved, it would be quite pertinent to refer to the species mentioned by Monier Williams in his aforesaid work at different contexts—

(i) *Kalpavṛkṣa* or *Kalpataru* (p. 262)—One of the five trees of Indra's heaven.

(ii) *Pārijāta* (p. 620)—A coral tree, *Erythrina indica,* one of the five tree of paradise emerging out of the churning of the ocean.

(iii) *Mandāra* (p. 788)—A coral tree, *Erythrina indica,* one of the five trees of paradise.

(iv) *Samantana* (p. 1142)—*Kalpavṛkṣa,* one of the five trees of Indra's heaven.

(v) *Haricandana* (P. 1290)—A sort of Sandal tree, and one of the five trees of paradise, other four being Pārijāta, Mandāra, Samantana and Kalpa.

(vi) *Tulasi* (Holy basil or *Ocimum sanctum*)—This plant too has been claimed to have emerged out of the churning of the ocean (p. 451).

In addition to the above, the Aśvattha (*Ficus religiosa*) and Soma (*Asclepias acida,* Roxb.) are also considered to be trees of divine origin and have been included in the discussion that follows—

## 1. *Kalpavṛkṣa*

Kalpavṛkṣa has been quite popular both as regards its representation in early Indian sculptural art as well as in the ancient literature, though its true identity is still a mystery. It is variously known as Kalpataru, Kalpadruma, Devataru, etc. and from the point of view of art, the Kalpavallī and Kalpalatā are but extensions of the same motif treated with luxuriance. The central idea of the Kalpavṛkṣa is that it is wish-fulfilling tree and grants all one desires. As per the Purāṇic evidence, it was produced as one of the fourteen gems at the time of gigantic churning of the ocean by the gods and the demons. The auspicious conch and the auspicious cow, Pāñcajanya and Kāmadhenu were also produced with the wishing tree and they too went to the gods. They are kindred symbols with the Kalpataru as bestowing fruition of human desires, and have been employed in Indian art and poetry with remarkable success. The phrase *Kamadugha* is quite frequent in the Vedas and is applied variously to speech (*Vāk*), Sarasvatī, the goddess of learning, Aditi, the mother of the gods and Pṛthvī, the Great Mother personified. The Kalpataru is also called Kamadugha as it grants all desires and fulfills all wishes. So long as man is under its shade, whatever he conceives, he realises, wealth, woman and all kinds of enjoyment issue forth from its fructifying bows.

The Mind is the Kalpavṛkṣa, which gives us everything for the mere thought of it. In a very real sense, the Mind is the most powerful creator, the perennial source of all our enjoyment and the inexhaustible fountain from which pleasure (*rasa*) constantly oozes out. Thought (*Sankalpa*) is the nature and power of the Mind which makes life teem with innumerable blessings.

The idea that the tree deity gave the merchants water to drink, food to eat, beds to lie on, maidens to attend them and all treasures, has an irresistible appeal of which both the poets and the artists took full advantage to embellish their works. In the epic literature and in the *Purāṇas* the conception of the Wishing Tree producing all desires is extended to the idyllic land of Uttarakuru where Elysian conditions prevail, leading to an exceedingly delightful life of the inmates.

In the *Rāmāyaṇa*, Sugrīva directs the monkey chiefs to go in quest of Sītā in the northern direction. There, at the end of the earth they would find the land of the Uttarakurus. "Flowers of gold as resplendent as fire are seen there in eternal bloom imbued with divine fragrance. The beautiful trees produce garments of various kinds and costly gems which are pleasant for men and women to use in all seasons; beds with beautiful coverlets and pleasing garlands; costly drinks and food of many descriptions; and to crown this all, maidens endowed with beauty, virtue and youth".

जातरूपमयैआपि हुताशनसमगमैः।
नित्यपुष्पफलास्तव नगाः पवर याकुजाः।।४३।।
दिव्यगन्धरसस्पर्शः सर्वान् कालान् स्रवन्ति च।
नानाकाराणि वासांसि फलन्तग्रन्धे नगोत्तमाः।।४४।।
मूक्तावैदूर्य चित्राणि भूषणानि तथैव च।
सवोषां यान्धनुरूपाणि पुरुषाणां तथैव च।।४५।।
सर्वर्तु सुखसेव्यानि फलबधन्धे नगोत्तसाः।
महार्हमणिचित्राणि फलन्त्यन्धे नगोत्तमाः।।४६।।
शयनानि प्रसूयन्ते चित्राक्षरधवन्ति च।
मनःकान्तोनि आख्यानि फलत्यवापरे द्रुमाः।।४७।।
मानानि च महार्हाणि भश्यानि विविधानि च।
स्त्रियय गुणसम्पन्ना रुपयौवनलक्षिताः।।४८।।

—*Kiṣkindhākāṇḍa*, Ch. 43, verses 43-48

In the *Mahābhārata*, the Elysian land of Uttarakura with all kinds of plenty in fruits and flowers and with trees producing all objects of desire (*sarvakamaphala vṛkṣah*) is conceived in the northern direction adjacent to Meru. There, as fruits from trees are produced, garments, ornaments and youthful pairs of men and women (*mithumani*) who draw sustenance from the nectar-like milk of the milky trees and are perfectly matched to each other in beauty, dress and appearance. The human beings in that region are happy and contented like gods being free from all sorrows and ailments, and they do not suffer the pangs of separation.

उत्तराः कुरयो राजन् पुण्याः सिद्धनिषेविताः।।२।
तव वृक्षा मधुफलां भित्यपुष्पफलौपमाः।
पुष्पाणि च सुगंधीनि रसवन्ति फलानि च।।३।।
सर्वकामफलास्त्रव केचिद्वृक्षा अनाधिप।
अपरे क्षैरिणो वाम वृक्षास्तव नराधिप।।४।।
ये क्षरन्ति सदा क्षैरं षड्वंसं क्षामृतोपमम्।
यस्त्राणि च प्रमूयन्ते फलेष्वामरश्चानि च।।५।।

मिथुनानि च जायन्ते स्त्रियषामृसरसोपमा:।
तेषां तु क्षौरिणां क्षीरं पिवन्त्यसूवसन्निमम्।।८।।
मिथुनं जायते काले समं सन्ध प्रवर्धते।
तुल्यरूपगुणोपेतं समवेषं तथेव च।।९।।

—Bhiṣmaparvan, 7, 2-11

It appears that this description of Uttarakuru was an inherent part of the *Bhuvanakoṣa* as we find it repeated in a similar context in the *Vāyu Purāṇa* (Chap. 45, verses II-50). The description in the *Purāṇa*, besides recording what the Epic contains is much more elaborate, and we find there reference to streams of *'madhu'* and *'maireya'*, of butter and curds, to mountains of delicious food, to groves furnished with beds, seats, cosmetics, garlands, etc. and to a great many other items of pleasure. In that region there is sweet music of every description rising from lute, flute and tabors, and hundreds and thousands of Kalpavṛkṣas produce fine and beautiful garments agreeable to wear.

A representation of the Uttarakuru land is probably what we find at Sanchi carved on the western face of the western pillar of the south gateway of the Great Stūpa (Marshall, *Monuments of Sanchi*, Vol. I, p. 144; Vol. II, Pl. XIXa). In several undulating bands of a lotus creeper are youthful couples engaged in music and pleasure surrounded by birds and animals and seated under the shades of boughs overladen with costly ornaments of many kinds.

This representation is so close to the literary tradition given in the Epics that it appears to suggest a conscious treatment by the artist of a theme which literary tradition had made popular. The sculptors at Sanchi and Bharhut (p. 5, Figs. 2-5) have carved time and again the lotus creepers as producing ornaments (*muktāvaidūryacitrāṇi bhūṣaṇaṇi*). Kālidāsa in the *Kumārasambhava*, VIII. 68, refers to necklaces (*harayasti*) hanging from the top of Kalpavṛkṣa trees. In the *Meghadūta* he mentions the Kalpavṛkṣa as the one complete source of all objects of adornment and toilet used by the women in Alakā for example; beautiful fabrics, wine causing love glances in the eyes, flowers and tender leaves, a variety of ornaments, and lac dye for painting the feet are the Kalpavṛkṣa products used for female beauty make-up (*abalā maṇḍana*).

वासयिवं मधुनयनयोर्धिधमादेशदक्षं,
पुष्पोभेदं सह किसलयैर्भूषणानां विकल्पान्।
लाक्षारागं चरणकमलग्धासयोग्यं च यस्यामेक:
सूते सक्षलमयलामं फनं कल्पवृक्ष:।।

—*Meghadūta*, II.11.

Again the poet refers to the Kalpavṛkṣa *'madhu'* named *'anaṅgadīpika'* (also called *'ratiphala'* in *Meghadūta* II, 3) which was served in cups of scarlet coloured gems (*lohitārka-maṇi-bhājana'*, *Kumar.*, 8.75). The conception of fine garments produced from the Kalpataru is found not only in Kālidāsa (*kalpataru-lambi aṁśuka, Kumar*, 8.71), but also in Bāṇabhaṭṭa who refers several times to the *'Kalpadruma dukūla'* or the *'Kalpalatā dukūla'* which was diaphanous and transparent like the slough of a snake and so light as to be tossed about even by the air of human breath (धौते निश्वासहार्ये निर्मोकशुचिनी कल्पलतादुकूले, *Kādambarī*, text, ed. Vaidya, pp. 201-206).

The 277th Chapter of the *Matsya Purāṇa* describes about the Kalpavṛkṣa, which is conceived to be the wish fulfilling tree in several texts. This *Purāṇa*, however, spells out as to how the replica of "a Kalpavṛkṣa should be donated, which is believed to be Mahādāna, and the dispeller of all the sins." According to this *Purāṇa*, a golden tree yielding all the desires of various kinds of fruits should be made; and also the birds sitting on that tree; the cloths and various kinds of ornaments should be made. The tree should be of four to one thousand *pals* according to the means of the devotee. The gold to be used should be mixed with alloy.[4]

> "Five branches alongwith the images of Brahmā, Viṣṇu and Śiva, Sūrya and cupid should be made, on the lower branch should be placed cupid and Rati. Santanaka tree, ¼ of the golden Kalpa tree should be located to the east of the latter. Mandāra tree with an image of Lakṣmī should be located on a pot of clarified butter in the south, Pārijāta tree with the image of Sāvitrī should be located to the west on the mound of cuminseed (*jīraka*). Similarly, Haricandana tree should be located to the north alongwith Surabhi cow on a mound of sesamum. The tree is to be adorned on one fourth part of its flowers. All other trees should be made 1¼ of the golden Kalpavṛkṣa and covered with a yellow cloth and decorated with flowers and sugarcane etc. Then eight jars filled with water and provided with fruits should be placed under the tree with a pair of sandals, cooking utensils, lamps, shoes, umbrella, *chamara*[5] and scents." The same text also describes the merits accruing from the gifting a replica of Kalpalatā separately.

The conception of an idyllic land or clime in which all kinds of human enjoyments are available to man as in a garden of paradise, is of old standing in Indian literature. We read in the story of the *Mahāvaṇija Jātaka* (No. 493) that a group of merchants who had set out in search of a treasure came to a mighty banyan tree with cool and pleasant shade. From its eastward branch pure and clear water trickled for them; they washed and drank their fill. One of the branches on the south gave them all things to eat—

> *"Both rice and meat out in a stream it brings,*
> *Thick porridge ginger lentle soup and many other things."*

From the western branch—

> *"Out came a bevy of fair girls all pranked in brave array,*
> *And O the robes of many hues, jewels, and rings in plenty!*
> *Each merchant had a pretty maid, each of the five and twenty."*

From the northern branch likewise—

*"...Out came a stream of gold,*
*Silver in handfuls, precious rings and jewels manifold.*
*And robes of fine Benares cloth and blankets thick and thin,*
*The merchants then to roll them up in bundles did begin."*

This tree which produced water streams, food, pretty maids and good things is the ideal Kalpavṛkṣa so often extolled in literature. The following Pāli *gāthā* sums up the virtues of this auspicious tree—

वारिदा पुरिमा साखा अन्नपानश्च दक्खिणा।
नारिदा पश्चिमा साखा सब्यकामे च उत्तरा।।

—*Fausboell. Jāt.* IV. 352

The artist in India made the full and best use of the conception of Kalpavṛkṣa as one of the symbols which adorned art, both in respect of beauty of form and depth of meaning.

At Bharhut, bounteous curves of lotus creepers running in continuous bands, project each bud or bough carrying beautiful ornaments, ear-rings, *kuṇḍalas*, for the ears of the *Prakara-vapra*[6] design, necklaces set with pearls and beads, armlets and anklets and spiral rings of various designs, delicate scarfs with flower patterns woven in their texture, and Sārīs and *patli* ends of '*gomutrika* designs'.[7] The Kalpalatā decoration producing such ornaments is a conspicuous feature of early Indian art and is often repeated leaving a pleasant effect on mind.

The best example of Kalpavṛkṣa belonging to the Śuṅga period is available from Vidisha in the form of the capital of a column and has been discussed elsewhere in this work.

The same motif is developed and continued later in the art of Gupta period and even afterwards. The deep-cut foliage decorations and creepers with intertwining tendrils are familiar features of Gupta art, and in some of them objects of enjoyment are carved as the products of the Kalpavṛkṣa. In a beautiful pillar from Garhwa a creeper with youthful maidens springing from its branches has entwining leaves and off-shoots.[8]

From the literary tradition, it is apparent that the Kalpavṛkṣa was considered as the source of the following objects of decoration and enjoyment, which was indeed translated into plastic art—

(i) Ornaments—*Hāryaṣṭi, ābharaṇa, bhūṣaṇavikalpa.*

(ii) Garments—*Kalpatarulambhi aṁśuka* (*Kumārasaṁbhava*, 8.71); *Citra-vasa* (*Meghadūta*, 11.11); *Māṅgalya Kusuma* (*Abhijñānaśakuntalaṁ*, IV); *Kāsika* cloth and blankets of Uddiyana or Gandhara country, known as Pāṇḍukabala in the *Jātakas* and in Pāṇini.

(iii) Cosmetics for the body and foot—*Lakṣarāga, vicchiti* etc.

(iv) Wine and drinks—*madhu* (*Meghadūta*, 11.11); *anaṅgadīpikā*, (*Kumārasaṁbhava*, VIII. 75); and *ratiphala*, (*Meghadūta*, II.3).

(v) Beds—*Citruslaranavants Sāyanāni*, (*Rāmāyaṇa*, Kiṣkindhā, 43.47).

(vi) Garlands—*manaḥkanta mālya* (*Rāmāyaṇa*, Kiṣkindhā, 43.47).

(vii) Foods—*Vivida-bhakṣya* (*Ibid.*, 43.48).

(viii) Maidens and couples—*gamasampanna rupayauvanalakṣita striyah* (*Ibid.*, 43.48).

(ix) Water in regaling streams—(*Mahāvanaja Jātaka*, IV. 351).

## 2. The Pārijāta Tree (*Erythrina indica*)

This is one of the five trees of paradise, emerged out of the churning of the ocean and was taken possession by Indra, from whom it was later taken by Kṛṣṇa. It is frequently mentioned in the *Purāṇas* and other Sanskrit texts. The *Bhāgavata Purāṇa* goes on to testify that this tree emerged from the churning of the "Ocean of Milk" and was planted in the garden of Indra—

ततोऽभवत् पारिजात: सुरलोक विभूषणम्।
पूरयत्यर्थिनो योऽथै: शश्वद् भुवि यथा भवान्।।

—*Bhāgavata Purāṇa*, 8.8.6

(Then emerged the Pārijāta tree, which illumines the *Svargaloka* and fulfills all the desires of the people).

The *Matsya Purāṇa* (251.5) too corroborates the statement of the *Bhāgavata Purāṇa*, that the Pārijāta tree which emerged out of the churning of the ocean was taken by Indra—

पारिजाततरुं वायुजग्नाह मुदितस्तथा।

But the *Harivaṁśa Purāṇa* states that the Pārijāta tree was created by the Sage Kaśyapa in order to please his wife Aditi and its flowers would never lose fragrance, which would remain eternal—

पारिजातं ततोऽस्राक्षीददिव्या: प्रियकाम्यया।
सर्व काम प्रदै: पुष्पैरावृतं नित्य गन्धदै:।।

—*Viṣṇu Parva*, 67.63

It is further stated that the Sage Kaśyapa created the Pārijāta tree by removing the essence from the Mandāra tree, and because of this the tree is considered to be the most auspicious of all the trees

मन्दारादपि वृक्षाच्च सारमुद्धृत्य कश्यप:।
तस्मादेष तरु श्रेष्ठ सर्वेषां श्रेष्ठतां गत:।।

—*Ibid.*, 67.65

This possibly refers to the budding of the tree by using branches of two separate species of botany.

The *Purāṇas* describe the story of uprooting of this tree from the heaven of Indra and its transplantation in the palaces of Dvārakā by Lord Kṛṣṇa. According to the *Bhāgavata*

*Purāṇa* its bark was of gold and it was embellished with young sprouting leaves of copper colour and fruit stalks bearing numerous clusters of fragrant fruits. Nārada brought a flower of this tree to Dvārakā as a present to Kṛṣṇa who in turn gave it to Rukmiṇī, whereupon, wearing a sorrowful look on his face, he went straight to Satyabhāmā, Kṛṣṇa's other consort. On an enquiry by Satyabhāmā, as to why was he so sad, the sage replied that he had presented Kṛṣṇa with a flower of Pārijāta thinking, that he would in turn present it to her favourite wife, but he was grieved to find that the flower has gone to Rukmiṇī. Satyabhāmā's natural jealousy was then aroused and she asked Nārada, as to what could be done to spite Rukmiṇī. The sage advised her to ask Kṛṣṇa to bring the Pārijāta tree itself to Dvārakā and plant it in her palace.

Satyabhāmā, then complained to Kṛṣṇa about his neglecting her by presenting the Pārijāta flower to Rukmiṇī and further emphasised that she would not be satisfied by anything less than the Pārijāta tree itself. Kṛṣṇa then proceeded to Amarāvatī, mounted on Garuḍa, uprooted the tree and escaped with it to Dvārakā. Indra who tried to stop Kṛṣṇa in his adventure, was defeated.

Lord Kṛṣṇa was then faced with the problem of transplanting the tree in Dvārakā in a way which could please Satyabhāmā and Rukmiṇī both. This was solved by transplanting the tree in such a way that while its trunk and roots lay within Satyabhāmā's abode, its branches extended over the palace of Rukmiṇī.

The *Harivaṁśa Purāṇa* (Viṣṇu Parva, 65.20-23) spells out the qualities of the Pārijāta tree thus—

कल्याण गुण सम्पन्ने सततं भर्तृवत्सले।
अम्लानमेतत् सततं पुष्पं भवति कामिनी।।
सम्वतसरपरं कालं कालज्ञे गुण सम्मते।
ईप्सितानपि गन्धांश्च ददाति वदतां वरे।।
शीतोष्ण चेच्छिते देवि पुष्पमेतत् प्रमच्छति।
स्त्रवत्यपि रसान् देवि मनसा काङ्क्षितान् वरान्।।
सेव्यमानं च सौभाग्यं ददाति वर वर्णिनि।
स्त्रवत्यपि तथा गन्धानीप्सितान् प्रीतिभर्द्धनान्।।

(This flower remains fresh for one year and does not fade away, and also gives the desired fragrance throughout the year. The quality of coolness and warmth required in all seasons is provided by these flowers, besides the *rasas* desired by humans. The use of this flower regularly, increases fortune and emits fragrance which is pleasing to the mind).

The scene of the carriage of the Pārijāta tree by Lord Kṛṣṇa has been a popular theme with the artists of the country, who created lively pictures of the same. In an eighteenth century painting from Basohli, the scene of Kṛṣṇa's stealing the Pārijāta tree from the heaven of Indra has been beautifully depicted. Kṛṣṇa here has been projected as having four arms and is mounted with his consort Satyabhāmā over Garuḍa who has the head and beak of a parrot having a green head and red beak, besides green wings. Garuḍa has two hands and the human body except the head and the feet. The sage Nārada is shown seated at the rear holding a *vīṇā* in one hand and the Pārijāta tree in the other. Indra mounted over the elephant Airāvata having a number of eyes on his body, is engaged in combat with Kṛṣṇa.

## 3. Tulasī Plant

There are several trees in India, which are considered to be sacred and Tulasī-plant is one of them or rather the foremost of all, because its presence could be found in almost all the arthodox Hindu families even today. This is also one of the five divine trees, which emerged out of the churning of the ocean, according to *Bhāgavata, Vāyu* and *Padma Purāṇa.* The benifits of planting a Tulasī-plant in a household are spelt out in various religious texts including the *Purāṇas.* According to these texts, wherever the Tulasī-plant is there, the place is considered to be auspicious and can not be haunted by evil spirits. The whole of the Tulasī plant from top to bottom is auspicious and even the gods aspire to be relieved from the bondages by the use of the Tulasī plant—

तुलसीतरुमूले च पुण्यदेशे सुपुण्येव।
अधिष्ठानं तु तीर्थानां सर्वेषां च भविष्यति।।
तत्रैव सर्व देवानां समधिष्ठानमेव च।
तुलसीपत्र पतन प्राप्तो यश्च वरानने।।

—*Brahmavaivarta Purāṇa,* Prakṛti, 21.37-38

The importance of the Tulasī leaves is further testified by saying that bathing with water in which a Tulasī leaf is placed is highly meritorious, and the adoration of gods, particularly Viṣṇu with a Tulasī leaf is of great benefit—

सस्नातः सर्वतीर्थेषु सर्वयज्ञेषु दीक्षितः।
तुलसी पत्र तोयेन योऽभिषेकं समाचरेत्।।
सुधा घर सहस्रेण सा तुष्टि न भवेद्धरेः।
या च तुष्टि मवेन्नृणां तुलसी पत्र दानतः।।

—*Brahmavaivarta Purāṇa,* Prakṛti, 21.39-40

The *Padma Purāṇa* goes on to declare, that at the time of cremation if a piece of Tulasī branch is added alongwith other burning wood, the dead person is relieved of the million of sins—

यद्येकं तुलसीकाष्ठं मध्ये काकुशतस्य हि।
दाह काले भवेन्मुक्तिः कोटिपाप युतस्य च।।

—*Padma Purāṇa,* Uttara, 24.6

The importance of the Tulasī plant has been highlighted in the *Skanda Purāṇa* as well, by stating that the adoration of the Devas has to be done by fresh flowers, water but the presenting of Tulasī leaf even if may not be fresh, is quite permissible—

वर्ज्य पयुर्षितं पुष्पं वर्ज्य पयुषितं जलम्।
न वर्ज्य तुलसीपत्रं न वर्ज्य जाह्नवी जलम्।।

—*Skanda Purāṇa*, U.M.M., 8.9

The *Vāyu Purāṇa* declares the purpose for which the leaves of the Tulasī plant could be plucked and also forewarns that one who removes the leaves of Tulasī without taking a bath, he commits a sin and his worship becomes infructuous—

अस्नात्वा तुलसीं चित्वा यः पूजां कुरुते नरः।
सोऽपराधी भवेत् सत्यं तत्सर्व निष्फलं भवेत्।।

—*Vāyu Purāṇa*

In the *Devī Bhāgavata Purāṇa* (9.25.32), the Tulasī plant is eulogised by the names of Vṛnda, Vṛndāvanī, Viśvapūjitā, Viśvapāvanī, Puṣpasāra, Nandinī and Tulasī-Kṛṣṇajīvinī—

वृन्दा वृन्दावनी विश्व पूजितां विश्वपावनी।
पुष्पसारा नन्दिनी च तुलसी कृष्णजीवनी।।

## 4. Soma (*Asclepias acida*) Plant

The Soma plant is said to be a climbing plant, the stalks of which were pressed between stones, by the priests, then sprinkled with water and purified in a staiver (*pavitra*); whence acid juice trickled into jars (*kalaśa*) or large vessels (*droṇa*); after which it was mixed with clarified butter, flour, etc. made to ferment and then offered in libation to the gods. It was collected by moon-light on certain mountains. (*Ṛgveda*, X. 34.1), the mountain of Mujavat is mentioned. It is sometimes described as having been brought from the sky by a falcon (*Syena*) and guarded by Gandharvas. It is personified as the most important of the Vedic gods to whom praise all the 114 hymns of the 9th book of the *Ṛgveda* besides in the *Sāmaveda* and the *Atharvaveda*.

Soma is defined in the Vedic literature as the king of plants, but it enjoyed a coveted position among the gods as would be evident from the following—

### (i) Soma as the King of Ṛṣis

It is often stated that Soma is accepted by the Ṛṣis as their king (सोमो अस्माकं ब्राह्मणानां राजा। *V.S.*, 1.40; तत्र ब्रह्मविभिः प्रोक्तमस्मत् स्वामी भवत्वयम्। *Matsya Purāṇa*, 23.10). Soma symbolises the universal mind of the creator (cf. चन्द्रमा मनसो जातः। *Ṛgveda*, X. 90.13). The *Ṛṣis* represent the spiritual tradition, and Kṣatriya kings the temporal power. At the time of coronation when everyone takes an oath of allegiance to the anointed ruler, the Brāhmaṇa priests and the *Ṛṣis* who represent the spiritual authority in the state take their oath of allegiance to king Soma and not to the Kṣatriya ruler. Here Soma is clearly the divine mind which is universal and worshipped by the Brāhmaṇas and the *Ṛṣis* in terms of *Dharma*.

## (ii) Soma as King of *Pitṛs*

In a compact statement it is said that Soma was also accepted as their lord by *Pitṛs*, Devas, Gandharvas and *Auṣadhis*. The *Pitṛs* or manes are associated with Soma because they are attached to the region of matter as against that of Puruṣa which is Sūrya (पितृ देवत्य सोमः, *Śatapatha Brāhmaṇa,* 3.2.3.17). Darkness, Afternoon, Night, second half of the gear were made the symbols of Prakṛti and Soma or Candra in one of them. It receives its light as the shadow of the Sun. The *Pitṛs* who are inclined to life in matter therefore are associated with Soma.

## (iii) Soma as King of *Auṣadhis*

The *Auṣadhis* or plants represent the principle of life-sap as being produced from a seed and growing as a plant, to reproduce the fruit and the seed. This makes a complete cycle of the life-process being manifested through the plant world. From this point of view each human being or animal form is not different from a plant (सस्यमिव मर्त्यः पच्यते सस्यमिवाजायते पुनः। *Kathā*, 1.6.6). As the *Gītā* says, the Sap or Soma is the nourisher of plants and therefore called *Auṣadhiśa* (पुष्णाभि चौषधीः सर्वा सोमोभूत्वा रसात्मकः, *Gītā*, 15.13).

# 5. The Pīpala Tree (*Ficus religiosa,* L.)

The Pīpala tree is the foremost of the sacred trees in India, its plantation is considered to be quite meritorious.

According to texts, once Sūrya was conceived of as *Aśva,* the imagery of the Cosmic Tree as *Aśvattha* and of its dynamic energy as *Aśva,* Mare was easy. The cosmos is called as *Aśvattha* because a horse is at its centre; or because it stands like a horse on three legs and is free on the fourth; or because it itself depends on the power or a movement of the architypal Horse, *i.e.*, Sūrya. Thus the conception of Sūrya as Sūrya or *Aśva*, and *Aśva* as the parents of Yama and Yamī on the one hand and of the twin Aśvins on the other is consistent to Vedic thought. Yama also is Sūrya, the regulating principle and Yami his sister is the female power that is in the Sun. The two Aśvins also represent the principles of *Prāṇa* and *Apana* or two fold energy as it is becoming manifest on the plane of Matter in the conscious World. In the *Bhagavadgītā* (10.26), Lord Kṛṣṇa declares, that he has the form of the Pīpala tree in the botanical species (अश्वत्थः सर्व वृक्षाणाम्). It is believed that one who plants the Pīpala tree, achieves the same comfort after his death, as one gets in the thick shade of the Pīpala tree in hot summer. According to the *Skanda Purāṇa*, Viṣṇu resides at the roots of Pīpala tree, Keśava in the trunk, Nārāyaṇa in the branches, Hari in the leaves and all other gods reside in its fruits—

मूले विष्णुः स्थितो नित्यं स्कन्धे केशव एव च।
नारायणस्तु शाखासु पत्रेषु भगवान् हरिः।।
फलेऽच्युतो न संदेहः सर्वदेवै समन्वितः।

एवं विष्णुर्द्रुम एव मूर्तो महात्मभि सेवितपुण्यमूलः।
यस्याश्रयः पाप सहस्रहन्ता भवेन्नृणां काम दुधोगुणाढ्यः।।
—*Skanda Purāṇa*, Nagar Khaṇḍa, 247.41, 42, 43

The *Bhāgavata Purāṇa* further testifies that Lord Kṛṣṇa, before leaving the world, rested under the shade of this tree—

तं तत्र तिग्मद्युभिरायुधैर्वृतं
ह्यश्वत्थमूले कृतकेतनं पतिम्।
स्नेह प्लुतात्मा निपपात पादयो
स्थादवप्लुत्य सवाष्प लोचनः।।
—*Śrīmadbhāgavata Purāṇa*, 11th Skanda, 30.41

When the charioteer of Śrī Kṛṣṇa found the Lord lying under the Pīpala tree, he was filled with emotions and bowed at the feet of the Lord.

The useless cutting of a Pīpala tree has been denounced and the worship of the Pīpala tree is said to be greatly rewarding.

## Notes and References

1. Agrawala, V.S., *Matsya Purāṇa—A Study*, p. 160.
2. Williams, Monier, *Sanskrit English Dictionary*, p. 577.
3. *Ibid.*, p. 620 and p. 788.
4. *Matsya Purāṇa*, 277.1-5.
5. *Ibid.*, 277.6-10.
6. For the meaning of Prakara—Vapra design of ear-rings, see *ABOBI*, XXXIII, pp. 19-22.
7. Pāṇini in the *Guṇasūtra 'gomutra sacchadani'* on Sūtra V, 4.3 refers to a cloth called *gomutrika* which means to be a *Sārī* worn by men or women as depicted in the reliefs as Bharhut and Sanchi; it seems to have derived its name from a particular *gomūtra* form of its first pendant portion.
8. *ASR*, Vol.X, Pl. III.

# 11
# The Lotus

The louts has been extolled as the prime symbol of creation, the seat and substratum of universal creative force which springs from the navel or centre of primeval creator. Simplified to concise Sūtra of art, it is expressed as Brahmā, the genius of creation, seated on a full blown lotus with long stalk springing from the navel of Viṣṇu. The elaborate Purāṇic cosmogony could be treated with such mastery of form gives a vantage ground to the artist. As the lotus floats above the water, so the created cosmos emerged to surface of the primeval deep or choas which originally was the confused mass out of which the order of the universe was evolved. The Greek word choas is the same as Vedic *'Kha'*, *'Kham'* represented the great chasm or vacuum out of which was born the *kham*, or the Plenum. *Kha* is the navel from which the cosmic lotus takes its birth. Again *'Kha'* represents the centre of the wheel of which the spokes and the peripheral frame mark the universe. Lotus is primarily the Vedic symbol.

The birth of lotus has been described in the *Matsya Purāṇa*, according to which Hari the creator of all the worlds, plays for some time and brings forth out of his navel a wonderful lotus of a thousand petals shining like a sun. That beautiful lotus, looking like the hair of that high soul was brilliant like fire and bright like the autumnal sun. That lotus of extravagant beauty began to shine—

स्वयं क्रीड़श्च विधिवन्मोदते सर्वलोककृत्।
पद्मं नाभ्युद्भवं चैकं समुत्पादितवांस्तदा।।
सहस्रपर्णं विरजं भास्कराभं हिरण्मयम्।
हुताशन ज्वलितशिखोज्ज्वलत्प्रभमुपस्थितं शरदमलार्कतेजसम्।
विराजतेकमलमुदार वर्चसं महात्मनस्तनुरुहचारुदर्शनम्।।

—*Matsya Purāṇa*, 168.15-16

Thereafter Viṣṇu out of the golden lotus begot Brahma, the creator of the universe. This lotus was many *yojanas* wide endowed with the qualities of the earth, full of *guṇas* and all *tejas* and of golden colour. The learned describe the same lotus as *terra-firma* and ascetics call it the lotus born out of Nārāyaṇa. *Rasā* also known as Padmā Devī is the earth. The weighty portions of the lotus are mountains—

तञ्चपद्मंपुराणज्ञः पृथिवीरूपमुत्तमम्।
नारायण समुद्भूतं प्रवदन्ति महर्षयः।।
या पद्मा सा रसा देवी पृथिवी परिचक्षते।
ये पद्मसारगुरवस्तान् दिव्यान् पर्वतान् विदुः।।

—*Matsya Purāṇa*, 169.3-4

The above Purāṇa compares earth with lotus and says that the lotus shape earth has four oceans on the four sides. By the mere contemplation of Nārāyaṇa, this lotus shaped earth appears in existence. So this springing up of the earth has been termed as Puṣkara; therefore the lotus is also called Puṣkara and for the same reason the high seers have enjoined the use of drawing the lotus before the performance of any sacrifice (169.14-16).

The lotus after having emerged out of the navel of Viṣṇu, became the seat of Brahmā, who started practicing austerities. While he was doing so, seated on Lotus, the great *asuras* Madhu and Kaiṭabha appeared to cause obstacles; the terrified Brahmā prayed to Viṣṇu, for help, who ultimately killed the *asuras*.

The lotus has been held in high esteem in the Indian religious thought and iconography. Several of the deities have held lotus as an attribute while this sacred flower serves as a seat for many of them. The following account provides a glimpse of the same—

## 1. Male Deities

### (i) Brahmā

It has already been brought out that Brahmā comtemplated over the lotus emerging out of the navel of Viṣṇu. The texts also prescribe a lotus seat for Brahmā. According to the *Śilparatna*, in case Brahmā was to be shown standing, he should be on a *padma-pīṭha*. In the *Śilpa*-texts, it is prescribed that in the temples exclusively dedicated to Brahmā, besides other *Parivāradevatās*, images of eight *Dvārapālas* should be installed. Some of these *Dvārapālas* are required to hold lotus in their hands as detailed below—

| | | Attributes | |
|---|---|---|---|
| | *Name* | *Right Hands* | *Left Hands* |
| 1. | Satya | *Padma, Sruk* | *Pustaka, Daṇḍa* |
| 2. | Sadharma | *Pustaka, Daṇḍa* | *Padma, Sruka* |
| 3. | Priyodbhava | *Akṣmālā, Padma* | *Agama, Daṇḍa* |
| 4. | Yajña | *Daṇḍa, Agama* | *Sruka, Kavaca* |
| 5. | Vijaya | *Akṣmālā, gadā* | *Kheṭaka, Daṇḍa* |
| 6. | Yajñabhadra | —do— | —do— |
| 7. | Sarvakamika | —do— *pāśa* | *Aṅkuśa* —do— |
| 8. | Vibhana | *Daṇḍa, Aṅkuśa* | *Pāśa* *Padma* |

Incidentally it may be stated that though *Agni Purāṇa, Viṣṇu Purāṇa, Amsubhedagama, Śilparatna* and the *Rūpamaṇḍana* prescribe the lotus seat for Brahmā, which was duly projected in the plastic art as well, it was only *Rūpamaṇḍana,* which prescribes lotus as Brahmā's hand attribute, the other three being *Akṣmālā, Akṣasūtra* and a book.

The ancient doctrine of *Puṣkara* which rose from the depth of the primeval waters and became the seat of Brahmā, is celebrated in Chapters-164-178, of the *Matsya Purāṇa* with all the force of richness of the Purāṇic imagination. It is called *Padmavidhi*, *viz.*, the cosmos as lotus-born from the naval of Nārāyaṇa lying recumbent in the ocean—

कथंपाद्ये महाकल्पे पद्ममयं जगत्।
जलार्णवगतस्येह नाभौ जातं जनार्दन।।
प्रभावात्पद्मनाभस्य स्वपतः सागराम्भसि।
पुष्करे च कथं भूत्वा देवाः सर्षिगणाः पुराः।।

—*Matsya Purāṇa*, 164.4-5

This is the dialogue between Manu and Matsya and it appears that the subject of cosmogony is reintroduced in a very serious manner but taking into its orbit new motif elaborated specially in the Purāṇas by the Bhāgavata teachers worshipping *'Śeṣa-śāyī-*Nārāyaṇa.

In the *Matsya Purāṇa* (Chapters 164-171), Earth is conceived as the Golden Lotus. It is stated therein in clear terms that the Golden Lotus which became the seat of Brahmā and also his birth place, is the same as the Earth which Nārāyaṇa created from the depth of the ocean—

स्रष्टारं सर्वलोकानां ब्रह्माणं सर्वतोमुखम्।
यस्मिन् हिरण्यमये पद्मे बहुयोजन विस्तृते।।
तञ्च पद्मं पुराणज्ञाः पृथिवीरूपमुत्तमम्।
नारायणसमुद्भूतं प्रवदन्ति महर्षयः।।
या पद्मा सा रसा देवी पृथिवी परिचक्षते।
ये पद्मसारगुरवस्तान् दिव्यान् पर्वतान् विदुः।।

—*Matsya Puraṇa*

In the new setting of the World Lotus from the navel of Nārāyaṇa which was known as *Puṣkara Kalpa, Jambū-dvīpa* and the relative geography of the mountains etc. which had been earlier disposed of in a systematic manner as part of the Bhuvanakoṣa is now repeated in a very half hearted way—

एवं नारायणस्यार्थे महापुष्कर सम्भवा।
प्रादुर्भावोऽप्ययं तस्मान्नाम्ना पुष्कर संज्ञितः।।

—*Matsya Purāṇa*, 169.15.

Brahmā is also associated with the *Puṣkara-tīrtha* and Puṣkara means a blue lotus. The Puṣkara Māhātmya of the *Padma Purāṇa* highlights the importance of the place which was predominantly sacred to Brahmā. According to this account, once a king on the eve of an Aśvamedha sacrifice, went with his charioteer for a joy ride. There, on his way, he came across a beautiful and heavenly lake. In the middle of this lake, there was a nice large lotus. He being pleased with the beauty of the lotus, asked his charioteer to pluck it. The charioteer bowing to the command of the king jumped into the lake and as soon as he reached the flower, the very touch of his hand produced a loud roar (*Huṁkāra*) and the charioteer fell dead instantly. The king also lost his lustre and strength and became a leper. The king then realised his having committed a great sin. In order, however, to rid himself of this pitiable condition, he approached Vaśiṣtha, who informed him that "the place, the king tried to desecrate was a place of worship of Brahmā. The lotus, the king tried to pluck had emerged from Brahmā himself. Its very sight makes one attain the result of seeing other gods. There was an image of Brahmā in the midst of the water there. An image of the Sun-god was also found there." Vaśiṣṭha then advised the king to approach lord Brahmā at Puṣkara and propitiate him. He did accordingly and was cured of his ailment.

## (ii) Viṣṇu

Viṣṇu is one of the most powerful gods of the Hindu Trinity and is conceived to be responsible for the universal protection. Viṣṇu is an old Vedic god who is mentioned in all the four Vedas. In the iconography of Viṣṇu, which developed during the Epics and the Purāṇas, he was conceived to be four armed, hold a lotus as one of the four attributes, the other three being *śaṁkha, cakra* and *gadā*. He was also conceived to be Padmanābha because of the issuing of a lotus stem out of his navel and Padmalocana because of the resemblence of his eyes with the lotus flower. There is an interesting episode in the *Mahābhārata*, according to which Viṣṇu once offered one of his eyes in place of a missing lotus, to Śiva, whom he was adoring for the grant of *cakra*. According to this episode, Viṣṇu was once unable to overpower a mighty *asura* and prayed to Śiva for the grant of *cakra* which was in latter's possession. To his prayers he added a *pūjā* in which he employed a thousand lotus flowers daily. One day, while performing the *pūjā*, with the required number of flowers, but at the end he missed one flower, which, to test the strength of his devotion Śiva had concealed. Viṣṇu at once plucked one of his eyes (which are always compared with lotuses, earning him the title of *kamalalocana*) and threw the same in offering to Śiva who was immensely pleased with Viṣṇu and granted him the *cakra*, invincible.

The *Uttarakamskagama* and the *Śrītattvanidhi* give the description of the Cakradānamūrti of Śiva, in which an eye is offered by Viṣṇu in place of the lotus.

## (iii) Śiva

Though Śiva is not directly connected with lotus but the *Śilpa* texts while describing the iconography of his various forms have connected lotus with him as his seat or an attribute, as would be evident from the following tabulation—

| | *Form of Śiva* | *Text* | *Seat* | *Attribute* |
|---|---|---|---|---|
| (i) | Alamghanamūrti | *Śrītattvanidhi* | Padmapīṭha | Devī holding Nīlotpala flower in her right hand. |
| (ii) | Umāmaheśvara mūrti | *Viṣṇudharmottara* | | Śiva to hold Nīlotpala flower in his right hand. |

*(Contd...)*

| | *Form of Śiva* | *Text* | *Seat* | *Attribute* |
|---|---|---|---|---|
| (iii) | Kalarimūrti | Agamas | Padmapīṭha | |
| (iv) | Sarbhesamūrti | *Śrītattvanidhi* | | The god has to hold a lotus flower in one of the sixteen left hands. |
| (v) | Candesānugraha-mūrti | *Uttarakomikagama* | | Śiva is to hold a garland of flowers. |
| (vi) | Cakradānamūrti | *Śrītattvanidhi* | | Viṣṇu is required to offer to Śiva a thousand lotus flowers daily. |

## (iv) Sūrya

Sūrya, besides other Hindu deities, also has lotus as his attribute and has been duly defined by the *Śilpa*-texts. Besides Ādityas are twelve in number and each one of them is said to preside over a month of the year. Their names are variously described in various work, but the most popular among their names have been noted here, each one of whom holds the lotus as his attributes—

| | *Name of Āditya* | | *Hand Attributes* | | |
|---|---|---|---|---|---|
| | | *Back right* | *Back Left* | *Front Left* | *Front Right* |
| 1. | Dhatṛ | Lotus garland | *Kamaṇḍalu* | Lotus | Lotus |
| 2. | Mitra | Soma | *Śūla* | —do— | —do— |
| 3. | Aryaman | *Cakra* | *Kaumodaki* | —do— | —do— |
| 4. | Rudra | *Akṣmālā* | *Cakra* | —do— | —do— |
| 5. | Varuṇa | *Cakra* | *Pāśa* | —do— | —do— |
| 6. | Sūrya | *Kamaṇḍalu* | *Akṣamālā* | —do— | —do— |
| 7. | Bhaga | *Śūla* | *Cakra* | —do— | —do— |
| 8. | Vivasvan | *Śūla* | Garland | —do— | —do— |
| 9. | Pusan | Lotus | Lotus | —do— | —do— |
| 10. | Savitṛ | *Gadā* | *Cakra* | —do— | —do— |
| 11. | Tvastri | *Sruka* | *Homaja Kalika* | —do— | —do— |
| 12. | Viṣṇu | *Cakra* | Lotus | —do— | —do— |

*(Contd...)*

## (v) Gaṇeśa

The text which prescribe lotus as one of the attributes for different forms of Gaṇapati are noted below—

| | *Form of Gaṇapati* | *Text* | *Attribute* |
|---|---|---|---|
| (1) | Bālagaṇapati | *Śilparatna* | Mango, Plantain, Jack fruit and sugarcane. |
| (2) | Bhakti Vighnesa | *Kryakramadhyoti* | Coconut, mango, lump of sugar and *pāyasaṁ*. |
| (3) | Ucchista-Gaṇapati | —do— | Lotus flower, pomegranate, paddy and *akṣmālā*. |
| (4) | Gaṇadhisa | *Śilparatna* and other texts | Lotus, *paraśu*, tusk and *modakas*. |
| (5) | Urdhva-Gaṇapati | *Kryakramadhyoti* | *Kalhara* flowers, ears of paddy, sugarcane and *bana*. |
| (6) | Piṅgala Gaṇapati | —do— | Mango, bunch of flowers, sugarcane, seasamum, *paraśu*, *modaka*. |
| (7) | Bhuvaneśa Gaṇapati | —do— | —do— |
| (8) | Lakṣmī Gaṇeśa | —do— | Parrot, pomegranate, lotus, water-vase, *pāśa* and *kalpalatā* bud. |
| (9) | Moha-Gaṇapati | —do— | Lotus, pomegranate, vase, *gadā*, sugarcane, ears of paddy. |
| (10) | Mahotkala-Vināyaka | —do— | —do— |

## (vi) Composite Forms

Besides the above, there are some composite deities, which are also associated with lotus in one or the other way as would be evident from the following tabulation—

| | *Deity* | *Reference to texts etc.* | *Position of lotus* |
|---|---|---|---|
| (a) | Brahmā-Viṣṇu-Śiva | *Śārdātilakaṁ* | Lotuses are held as attributes in two hands. |
| (b) | Sūrya-Brahmā | Image preserved in V.R.S. Museum | Holds lotus in each one of the two hands. |
| (c) | Sūrya-Nārāyaṇa | | Holds lotus as one of the attributes. |

*(Contd...)*

| | *Deity* | *Reference to texts etc.* | *Position of lotus* |
|---|---|---|---|
| (d) | Sūrya-Buddha | National Museum, Bangkok | The image holds the attributes of Sūrya. |
| (e) | Sūrya-Lokeśvara | Āśutosh Museum | The right hand holds a full blown lotus. |
| (f) | Śivalokeśvara | —do— | Stands in *Sampadasthānaka* pose over a double petalled lotus. |
| (g) | Viṣṇulokeśvara | Discovered from Dinajpur district of West Bengal | *Gadā* and *Cakra* are placed over full blown lotuses which are held by the stalks in both the hands. |

## 2. Female Deities

| | *Deity* | *Reference to texts etc.* | *Position of lotus* |
|---|---|---|---|
| (1) | Lakṣmī | *Śilparatna* | Should be seated on lotus, holding lotus in her left hand and a *bilva* fruit in the right. (Cpl Nos. 47-48) |
| (2) | Khemaṅkarī | *Agamas* | Should hold a trident, a lotus, a vase and *varada mudrā* in four hands. |
| (3) | Vindhyā Vāsinī Durgā | —do— | She should be seated on a golden lotus. |
| (4) | Mahiṣāsuramardinī | *Śilparatna* | Her eyes should be resemble the *nīlotpala* flowers. |
| (5) | Bhadrakālī | *Agamas* | She has to be eighteen armed and is to hold a lotus in one of her hands. |
| (6) | *Ambā* | —do— | She has to hold a flower in one of the four hands. |
| (7) | Sarvamaṅgalā | —do- | —do |
| (8) | Gaurī | —do— | —do— |
| (9) | Umā | —do— | —do— |
| (10) | Kālī | —do- | She carries a lotus in one of her two hands. |
| (11) | Siddhi | *Agamas* | She is seated upon a seat decorated with lotuses. |
| (12) | Jayā and Vijayā | —do— | Each one is to hold a lotus in one of her four hands. |

## 3. The Buddhist Deities

The association of lotus with some of the Buddhist deities is tabulated below—

| | *Deity* | *Text* | *Position of Lotus* |
|---|---|---|---|
| (1) | Amitābha | *Ad-vaya-vajrasaṁgraha* | He has the lotus as a symbol. |
| (2) | Pandara | —do— | —do— |
| (3) | Padmapāṇi | | —do— |
| (4) | Tārā | —do— | —do— |
| (5) | Gaganagaṅgā | *Nispannyagavali* | Kalpa tree. |
| (6) | Avalokiteśvara | —do— | Lotus as a symbol. |
| (7) | Mahasthama prapta | —do— | Six lotuses as symbol. |
| (8) | Candraprabhā | —do— | Moon on lotus as a symbol. |
| (9) | Maitreya | —do— | Nāgakesara as a symbol. |
| (10) | Manjughoṣa | *Sādhanamālā* | Lotus in the left. |
| (11) | Siddhaikavira | —do— | Lotus as a symbol. |
| (12) | Lokanātha | —do— | —do— |
| (13) | Khasarapana | —do— | —do— |
| (14) | Simnada | —do— | Sword on lotus. |
| (15) | Sadakṣari Lokeśvara | —do— | Rosary and lotus as symbols. |
| (16) | Padmanartteśvara | —do— | Double lotus in all the hands. |
| (17) | Vajradharma | —do— | Lotus as symbol. |

The above tabulations are only illustrative and not exhaustive.

In the Vedic and post-Vedic literature, however, the following epithets have also been used for the lotus—

(i) *Puṇḍarika* denotes the blossom of the lotus in the *Ṛgveda* (10.142.8). The *Pañcaviṁśa Brāhmaṇa* states that the lotus flower is born out of the light of the *Nakṣatras* and the *Atharvaveda* (10.8.43) and *Chāndogya Upaniṣad* (8.1.1) compare the human heart with the lotus.

(ii) *Puṣkara* is the name of the blue lotus flower in *Ṛgveda* (6.16.13; 7.33.11) and *Atharvaveda* (9.3.8.; 12.1.24), *Taittiriya Saṁhitā* (5.1.4.1), *Vājasneyī Saṁhitā* (9.29), *Śatapatha Brāhmaṇa* (4.5.1.16), *Taittirīya Brāhmaṇa* (1.2.1.4), *Maitrāyaṇī Saṁhitā* (3.1.5). The *Atharvaveda* mentions its sweet perfumes (12.1.24). The lotus grew in lakes which were then called *puṣkariṇī* "lotus bearing." That flower was early used for personal adornment is evidenced from an epithet of the Asvins known as *Puṣkara-Sraja* (lotus crowned).

Possibly because of its likeness in shape to the flower of the lotus, the bowl of the ladle is called *puṣkara*.

# 12
# Agriculture, Pastures, Mountains and Irrigation in Vedic Age

The deity of the field *i.e.,* Kṣetrapati, "Lord of Field" in the *Ṛgveda* (III.57) is invoked to grant cattle, horses, and with sweetness to fill heaven, earth, the plants and the waters. In another hymn (*Ṛgveda*, VII.35.10), he is asked along with Savitr, the Dawn, the Parjanya to bestow prosperity, and elsewhere, it is said that worshippers are fain to have him as a neighbour. The full nature of the deity is made clear in the *Āpastambha Gṛhya Sūtra* (II.10.4) and other *Gṛhya Sūtras,* which prescribe offerings to him, when the field is ploughed, and he is, it is clear, no more than a deity believed to be in the ploughed land. The figure of this goddess naturally has more life in the *Paraskara Gṛhya Sūtra* (II.17.9) which deals with the operation of agriculture. She appears as the wife of Indra, which may be due to the fact that in the *Ṛgveda* (VIII.21.3) the god is once called as Urvarapati, "Lord of plough field", and her existence is manifested in the four Sondergotter, the deities Sītā, Āśā, Arada and Anagha, who are to be worshipped at the furrow sacrifice, the threshing floor sacrifice, the sowing of the crop and the reaping and putting of the crop into the born. (*Gobhila Gṛhya Sūtra*, IV.4.27). In the *Paraskara Gṛhya Sūtra,* we have besides the offering to Sītā and to three other deities, Yaja, Sama and Bhuti, who are less concrete than is she, offerings to the guardians of the furrow on the four sides; these guardians are called by various names which are hardly more than epithets, 'having good bows and quivers, on the east, not winking the eyes, and wearing armour', on the South, 'prosperity', 'earth', Parsni, Sunamkari, 'on the west' and those 'that are terrible' and 'like Vāyu in swiftness', 'on the north.' These figures are interesting, they cannot properly in any case be classified with abstract deities, even if that class of deities can properly be set up, they are not personification of human activities or feelings, but they are the deities involved in the operation of nature, Sondergotter who live in growth of the crop and may best perhaps be called animalistic spirit of vegetation, whose precise differentiation of character may plausibly be ascribed as in the case of Roman Indigitamenta to the work of the priests. On the other hand, Sītā has always a more elevated character than the rest of the spirits; she is called Sāvitrī in the *Taittirīya Brāhmaṇa* (II.3.10.1) and she becomes the heroine of the *Rāmāyaṇa* and the model of wifely truth and chastity, preserving nonetheless here and there traces of her links with the furrow and the fruitful earth.[1] Urvara, the plough field, is also a goddess in the Sūtra literature and is described as having a garland of threshing floors.[2] It would of course be absurd to suppose that these scanty remnants of worship represent the whole of the agricultural side of the Vedic religion, and as a matter of fact, in the ritual there are abundant traces of other ceremonies having to do with agriculture, but the deities of the field were evidently as such not of great importance in the Vedic religion, the care of the prosperity of men in these respects, having been taken over by the great gods. When the agriculture and vegetation spirits as such have great honour among the importance in the ideas of a people, then their figures are elevated to the rank of great deities, as in the case with Attis, Adonis, or perhaps even Osiris in the lands of Asia Minor and of Egypt and the fact that other deities are the great gods of Vedic religion cannot be explained on any other hypothesis that the specific religious instinct of the Vedic people chose those forms as the favourite object of its devotions. Curiously enough, the later religion of India gives a picture of differences of religious predilection. The young Kṛṣṇa is represented as opposed to the worship of Indra by the pastoral world according to the *Harivaṁśa* (II.16.2ff) and *Viṣṇu Purāṇa* (V.10.29ff). He prefers instead that they should worship the furrow. In cases like these, it is impossible not to observe a definite cleavage of religious sentiments, which it is perfectly legitimate to apply as a principle to Vedic religion. The theory which sometimes seems to be held that the Vedic literature does not really give us the popular religion of the time cannot be supported when it is applied to the literature as a whole, including the Sūtras of the domestic rituals, which are packed with the popular ideas and magic practices of every kind.[3]

The mountains with their greenery are, according to the young Kṛṣṇa, essentially the deities of pastoral folks. The *Ṛgveda*, which shows no trace of the old belief of mountains as the home of the gods, generally, also invokes here and there the mountains, nearly twenty times in the plural, and four times in the singular. They never, however, occur alone, but only with other deities such as Indra or Savitr or with water, trees, plants, heaven and earth. They are described as rejoicing in the sacrificial offering and abiding securely. *Parvats*, the mountains, is once found invoked as a companion of Indra in the dual compound Indraparvata, and asked to come to the offering. In the rituals, mountains make but little appearance, but we hear in the Brāhmaṇas, a good deal of a salve from a mountain Trikakubh which had wonderful powers. In the late *Hiraṇyakosi Gṛhya Sūtra* (I.1.5), there occurs an obscure phrase which suggests that it was thought in applying the salve to the eyes that the spirit of the mountain entered into the performer of the rites. The idea is ofcourse perfectly in harmony with primitive ideas.[4]

## 1. The Drought

For the purpose of raising crops or agricultural products, it would be essential to ensure regular water supply. While the supply of water was no problem in the rainy season, it would indeed be different in summer season and in drought. It may not be out of place to mention here that the Indian continent often witnessed severe drought conditions, a mention of which is made in the *Vālmīki Rāmāyaṇa* in the episode of Ṛṣyaśṛṅga, who was seduced by the courtesans of the king and brought to Aṅgadeśa, which was facing terrific drought. The only remedy to overcome the drought was to bring sage Ṛṣyaśṛṅga to the kingdom and the very arrival of the sage on the soil of Aṅgadeśa, was to cause heavy rains. This happened as predicted and Romapada, the king of Aṅgadeśa, married his daughter to Ṛṣyaśṛṅga—

अङ्गेषु प्रथितो राजा भविष्यति महाबलः।
तस्य व्यक्तिक्रमाद्राज्ञो भविष्यति सुदारुणा।।
अनावृष्टि सुघोरा वै सर्वभूत भयावहा।
अनावृष्ट्यां तु वृत्तायां राजा दुःख समन्वित।।

—*Vālmīki Rāmāyaṇa* (Bālakāṇḍa, 6.8-9)

In the *Devī Māhātmya* (II-43) of the *Mārkaṇḍeya Purāṇa* also, there is a mention of a drought lasting for a hundred years.

भूयश्च शतवार्षिक्यामनावृष्ट्यामनम्भसि।
मुनिभिसंस्तुता भूमौ सम्भविष्याम्योनिजा।।

—*Devī Māhātmya* (II.43)

(And again when rain water shall fail for a hundred years, propitiated by the ascetics, I shall be born on earth but not from a womb).

## 2. Water Supply and Irrigation System

In these drought conditions and in summer months the regular water supply for irrigation purpose is a problem even in the present times. But in the *Ṛgveda* there are certain passages which stand for the use of artificial means for the supply of water.

In this context one may refer to the use of the word *Avata* several times in the *Ṛgveda*, which denotes a well artificially[5] made in contrast to a natural spring (*utsa*), though the latter expression is also applied to a well, artifically[6] made—

अप्रक्षितं वसु विभर्षि हस्तयो षाढं हसतन्वि श्रुतोदधे।
आवृतासोऽवतासो न कर्तृभिस्तनूषु ते क्रतव इन्द्रभूरयः।।

—*Ṛgveda*, I.55.8

Such wells were covered by the makers[7] and are described as unfailing and full of water.[8] The water was raised by a wheel (*cakra*) of stone, to which was fastened a strap (*varatra*) with a pail (*kosa*)[9] attached to it. The *Ṛgveda*[10] further goes on to testify that this *cakra* when raised, poured (*Sinc*) water into the buckets (*ahava*) of wood—

इष्कृताहावमवतं सुवरत्रं सुषेचनम्।
उद्रिर्ण सिञ्चे अक्षितम्।।
प्रीणीताश्वान्हितं जयाथ स्वस्ति वाहं रथमिल्कृ ध्वम्।
द्रोणो हावमवतमश्मेचक्रमं सत्र कोषं सिञ्चता नृपाणाम्।।

(X.101.6,7)

Sometimes those wells have been used for irrigation purposes, the water being led off into broad channels (*Surmi Suṣira*)—

सुदेवो आसि वरुण यस्य ते सप्त सिन्धवः।
अनुरक्षन्ति काकुदं सूर्म्यं सुषिरामिव।।

—*Ṛgveda*, VIII.69.12

In the above passage, the word *Sūrmi* ( सूर्मि ) is quite important because it ostensibly stands for a system which arranges distribution or supply of water. In this connection, a reference may be made to Monier William's Dictionary (p. 1244) according to which the construction of the word *Surmi* is attributed to *Su+Urmi*, which denotes a pipe for conveying water. In the *Vājasneyī Saṁhitā* too the word *Surmya* has been interpreted to mean a tube, a pipe or a channel.

## 3. Use of Channels or Tubes for Water Supply

Again the word *Suṣira* has also been interpreted to mean "having a good tube or a channel" by Monier Williams (*Ibid.*, p. 1238).

## 4. The Farming and Use of Plough

Adverting to the question of field cultivation, it may be stated that the tilling of the soil was known to the early Indians as is indicated by the expression of *yavani kṛṣṇa* in the *Ṛgveda*. It is also not without significance that the expressions of ploughing occur mainly in the first and tenth *maṇḍalas* of the *Ṛgveda*. In the first *maṇḍala* forms of the root *kṛṣṇa* (to plough) are conspicuously present—

उतो स मह्यमिन्दुभिः षड्युक्ताँ अनुसेषिदात्।
गोभियवं न चकृषत्।।

—*Ṛgveda*, I.23.15

In the *Atharvaveda*, *Pṛthi Vainya* is credited with the origination of ploughing and even in the *Ṛgveda*, the Asvins are spoken of as concerned with the sowing of grains by means of a plough—

यवं वृकेणाश्विना वपन्तेषं दुहन्ता मनुषाय दस्रा।
अभि दस्युं बकुरेणा धमन्तोरु ज्योतिश्चक्रयुरार्याय।।

—*Ṛgveda*, I.117.21

सोदक्रामत् सा मनुष्या इ गच्छत् तां मनुष्या उपाह्वयन्तेरावत्येही।
तस्या मनुर्वैवस्वतो वत्स पृथिवी पात्रम्।।
तां पृथी वैन्यऽधोक् तां षिं च सस्यं चाधोक्।।
ते कृषिं च सस्यं च मनुष्या इ उपे जीवनन्ति।
कृष्टराधि रूपजीवनीयो भवति य एवं वेद।।

—*Atharvaveda*, VIII.10.24

(She ascended; she came to men—*manuṣya*; men called to her; O rich in cheer, Come ! of her Manu, son of Vivasvant was young, earth was vessel; her Pṛthi son of Vena milked; from her he milked both cultivation—*Kṛṣi* and grain; upon these two, both cultivation and grain, men subsist; successful by what is cultivated—*kṛṣta*, one to be subsisted upon, becometh he who knoweth thus.)

In the later *Saṁhitās* and the *Brāhmaṇas*, ploughing is repeatedly referred[11] to. Even in the *Ṛgveda*, there is a clear proof of the importance attached to agriculture.[12] In the *Pañcaviṁśa Brāhmaṇa*,[13] the *Vratyas* are described as not cultivating soil.

The plough land was called *urvara* or *kṣetra*. Manure (*Sakin* or *Karisa*) was used and irrigation was practiced (*Khanitra*). The plough (*langala* or *Sira*) was drawn by bullocks, teams of six, eight or even twelve were employed—

इमं यवमष्टायोगैः षड्योगेभिरचर्कृषु।
तेना ते तन्वो ई रपोऽपाचीनमप व्यथे।।

—*Atharvaveda*, VI.91.1

*Urvara* also known as *kṣetra*, as already stated, gives the regular expression from the *Ṛgveda* (I.127.6; 4.41.6; 5.33.4; 6.25.4) and *Atharvaveda* (10.6.33; 10.10.8; 14.2.14) denoting the piece of plough land, fertile fields (*Ṛgveda*, I.127.6) as well as waste fields. Intensive cultivation by means of irrigation is clearly referred to in the *Ṛgveda* (7.49.2) and the *Atharvaveda* (1.6.4; 19.2.2) while allusion is also made to the use of manure *Atharvaveda* (3.14.3-4) (19.31.3). The fields (*kṣetra*) are carefully measured according to the *Ṛgveda* (I.110.5). This fact points clearly to the individual ownership of land for the plough, a conclusion supported by reference of Apala in a hymn in the *Ṛgveda* (8.91.5) to her father's field (*urvara*) which is put on the same level as his head of hair as a personal possession. Consistent with this are the epithets "Winning Fields" (*Ṛgveda*, 6.38.1; 6.20.1; 2.21.1) while "lord of fields" used of a god (*Ṛgveda*, 8.21.3) is presumably a transfer of a human epithet (*urvara-pati*). More fields are spoken of in the same connection as children (*Ṛgveda*, 6.41.6) and the conquest of fields (*kṣetrani-sam-ji*) is often referred to in the *Saṁhitās* (*Taittirīya Saṁhitā*, 3.2.8.5; *Kathaka Saṁhitā*, 5.2; *Maitrāyaṇī*, 4.12.3). Most probably, as suggested by Pischel (*Vediche Studiem*, Vol.II, pp. 204-207), that the plough land was bounded by grass land (perhaps denoted by *khila* or *khitya*) which in all likelihood would be joint property on the analogy of property elsewhere. There is no trace in Vedic literature of communal property in the sense of ownership by a community of any sort, nor is there a mention of communal cultivation. Individual property in land seems also presumed later on. In the *Chāndogya Upaniṣad* (7.24.2) the things given as examples of wealth include fields and houses. The Greek evidence also points to individual ownership (*Diodorus*, II, 40; Arrian, *Indica* II; Strabo p. 73; Hopkins, *J.A.D.S.*, 13.87). The precise nature of ownership is of course not determined by the expression of "Individual ownership." The legal relationship of the head of a family and its member is nowhere explained and can only be conjectured. Very often a family may have lived together with undivided share in land. The rules for the inheritance of landed property do not occur before Sūtras (*Gautama Dharma Sūtra*, 18.5, *Bandhyana Dhama Sūtra*, 2.2.3; *Āpastambha Dharma Sūtra*, 2.6.14).

In *Śatapatha Brāhmaṇa* (8.6.2.18; 7.1.13.15) the giving of land as fee to the priests is mentioned, but with reproof; land was no doubt even then a very special kind of property, not likely to be given away or parted with.

## 5. Harvesting, Storage of Grains

The operations of agriculture are neatly summed up in the *Śatapatha Brāhmaṇa*, as ploughing, sowing, reaping and threshing—

ते हैतामिधतुमेधां चक्रिरे। यमिषामेतामनुश्रृण्वन्ति
कृषन्तो ह स्मैव पूर्वो वपन्तोयन्ति लनन्तोऽपरे
मृणन्तः शश्वद्धेभ्योऽकृष्टपच्या एवौषधयः पेचिरे।।

—*Śatapatha Brāhmaṇa*, I.6.1.3

"Those *asuras* then thrive in such a manner that they (the gods) heard of it; for even while the foremost of the *asuras* were still ploughing and sowing, those behind them were already engaged in reaping and threshing; indeed even without tilling, the plants ripened forthwith for them."

Then the grain was cut with a sickle (*datra* or *Sṛni*) bound into bundles[14] (*parsa*) and beaten out on the floor of the granary (*khala*)[15]. The grain was then separated from the straw by the use of winnowing basket or *Sūrpa.*[16] The winnower was called *dhanyakṛt*[17] and the grain was measured in a vase called *udara.*[18]

## 6. The Crops

As to the cereals known to the Ṛgvedic people, it may be stated that *yava* (barley) and *dhana* (corn) were possibly known to them. But in the later *Saṁhitās*[19], the position is somewhat different and rice (*vṛhi*), beans (*mudga-māsa*), sesamum (*tila*) and other species of grains like *aṇu, khalva, godhūma* (wheat), *nivara priyaṅgu, masūra* and *śyāmaka* are mentioned, while cucumbers (*uravaru*) of different kinds were known. It cannot be claimed with authenticity whether the fruit trees were grown by human efforts and cultivated or they merely grew of their own as wild growth, but jujube is frequently mentioned (*Karkandu, Kuvala* and *Badara*).[20]

The seasons for agriculture are mentioned in a passage in the *Taittirīya Saṁhitā*,[21] barley ripened in summer, being no doubt sown as in modern India, in winter; rice ripened in summer, being no doubt sown as in modern India, in winter; rice ripened in autumn, being sown in the start of the rainy season; beans and sesamum planted at the time of the summer rains, ripened in the winter and cool season. There were two harvests (*Sasya*) a year, according to the *Taittirīya Saṁhitā*[22], the winter crop was ripe by the month of Caitra (March-April) according to the *Kauśītaki Brāhmaṇa.*[23]

## 7. Factors Leading to the Damage of Crops and Insects

The farmer had plenty of trouble of his own; moles destroyed the seeds, birds and various kinds of reptiles (*upakvasa, jabhya tarda* and *pataṅga*) injured the young shoots; excessive rains or drought might damage the crops.[24] The other texts too speak of worms and insects which are not only injurious to crops and vegetation but are also injurious to the human and animal life. Some of the prominent species of insects or worms are noted hereunder—

### *The Insects*

(i) *Upa-kvasa*, according to *Atharvaveda* (VI.50.2) is the name of a noxious insect injurious to seeds.

(ii) *Kapana* is referred to once in the *Ṛgveda* (V.54.6) and represents a 'worm' that destroys the leaves of the tree and is so interpreted in *Nerukta* (VI.4).

(iii) *Kaskasa* designates a kind of worm in the *Atharvaveda* (V.23.7).

(iv) *Kīṭa* is the name of a species of worm mentioned in the *Atharvaveda* (9.4.16) and *Bṛhadāraṇyaka Upaniṣad* (6.1.19), *Chāndogya Upaniṣad* (6.9.3), *Kauśītaki Upaniṣad* (1.2).

(v) *Kṛmi*—Worm. In *Ṛgveda* (I.191), *Taittirīya Upaniṣad* (5.5.11.1), *Maitrāyaṇī Upaniṣad* (3.14.11), *Vājasneyī Saṁhitā* (24.30), *Mantra Bṛhadāraṇyaka* (2.7), *Taittirīya Āraṇyaka* (4-36), *Śatapatha Brāhmaṇa* (5.4.1.2), *Atharvaveda* (2.31-32). In the *Atharvaveda* worms play a considerable part (V.23). They are regarded as poisonous and are spoken of as found in the mountains, in forest, in water, in plants and in the human body. In accordance with the widespread primitive ideas, they are considered to be causes of disease in men and animals. *Atharvaveda* contains three hymns as charms directed against them. The first of these hymns is of a general character, the second is meant to ḍestroy worms in cattle and the third is intended to cure children of worms. When found in men, worms have their places in the head or ribs (*Atharvaveda*, 2.31.4) and to creep into the eyes, nose and teeth (*Atharvaveda*, 5.23.3). They are described as dark brown, but white in the foreparts of the body, with black ears and having three heads (*Atharvaveda*, 5.23.4). They are given many names.

(vi) *Plus* is the name of some noxious insect in the *Ṛgveda* (I.194.1); it is also included in the list of victims at the Aśvamedha sacrifice in the *Yajurveda* (*Maitrāyaṇī Saṁhitā*, III.14.8 and *Vājasneyī Saṁhitā*, 24.20) and is also mentioned in the *Bṛhadāraṇyaka Upaniṣad* (I.3.24). This possibly related to a species of ant.

(vii) *Yevasa*—is the name of a destructive insect in the *Atharvaveda* (V.23-7-8) and also in the *Kathaka Saṁhitā* (I), *Kapisthala Saṁhitā* (46.4).

(viii) *Vṛkṣa-Sarpi*—"tree creeper" is the name of the species of worm, or female serpent in the *Atharvaveda* (IX.2.22).

(ix) *Vyodhvara*—"Perforating' designates a worm in a passage of the *Atharvaveda* (II.31.4).

(x) *Śalabha*—Locust appears in the Paippatada recension of the *Atharvaveda* (IX.5.9).

(xi) *Saluna*—is found in the *Taittirīya Saṁhitā* (V.2.9.3) denoting a worm.

(xii) *Sipavitnuka* in the *Atharvaveda* denotes a species of worm (V.20.7).

(xiii) *Tṛṇa Jatayaka*—Caterpillar is mentioned in the *Bṛhadāraṇyaka Upaniṣad* (IV.2.4).

These worms and insects are stated to harm not only the human beings but also the crops, animals and forests—

ये क्रिमयः पर्वतेषु वनेष्वोषधीषु पशुष्वप्स्व ऽन्तः।
ये अस्माकं ठन्व ऽमाविविशुः सर्वं तद्धन्मि जीवमक्रिमीणाम्।।

—*Atharvaveda*, II.31.5

(The worms that are in the mountains, in the woods, in the herbs, in the cattle, within the waters, that have entered ourselves (*tanu*)—that whole generation (*janiman*) of worms, smite).

Some of the insects mentioned in the *Atharvaveda* are noted below—

*Karuru*,[25] *Algandus*,[26] *Salunas*,[27] *Pis*,[28] *Avaskara*,[29] *Vyadhvara*,[30] *Sṛ*,[31] *Kṣullaka*,[32] etc.

The *Atharvaveda* further goes on to testify that rats and locusts are also responsible for the destruction of crops and grains, and should be crushed—

हत तर्दं समङ्कमाखुमश्विना छिन्तं शिरो अपि पृष्टीः शृणीतम्।
यवान्नेददानपि नह्यतं मुखमथाभयं कृणुतं धान्याऽप।।

—*Atharvaveda*, VI.50.1

(Smite, O Asvins, the borer, the *Samaṅka*, the ant; split their heads; crush in their ribs, lest they eat the barley, shut up their mouth; then make fearlessness for grains).

तर्द है पतङ्ग है जभ्य हा उपक्वस।
ब्रह्मेवासस्थितं हविरनदन्त इमान् यवानहिंसन्तो अपेदित।।

—*Ibid.*, VI.50.2

(Hey borer, hey locust, hey grinder, *upakvasa* as a priest an unfinished oblation, not eating this barley, go up away, doing no harm).

## 8. Insecticide

While dealing with the insects one would naturally think of the insecticides or the methodology used by the Vedic seers for the destruction of insects. At the outset it may be mentioned that Vedic people were quite aware of the fact that the sun rays could serve quite well to destroy the worms and insects—

उद्यन्नादित्यः क्रिमीन् हन्तु निम्रोचन् हन्तु रश्मिभिः।
ये अन्तः क्रिमयो गवि।।

—*Atharvaveda*, II.32.1

(Let the Sun—Āditya, rising, smite the worms, sitting, let him smite them with his rays—the worms that are within the cow).

The same text, however, prescribes several spells for the destruction of insects and worms.

## 9. Manure

As to the use of manure for the fields, it may be pointed out that the *Atharvaveda* makes an allusion to the use of manure. The nature of manure has been defined to be dry cowdung (*Karisa*) in the *Atharvaveda* (III.14.3-4) highlighting the value of the natural manure of animals for the purpose of farming. The same aspect has been emphasised in the *Śatapatha Brāhmaṇa* (II.1.1.7).

Another word used for the cowdung in the *Ṛgveda* (I.161.10) is *Sukṛt* and has been repeated in the *Atharvaveda* (XII.4.9) and *Taittirīya Saṁhitā* (VII.1.19.3). The *Taittirīya Saṁhitā* (V.7.23.1) and *Vājasneyī Saṁhitā* (37.9) have, however, used he word *Sakan* to denote the cowdung manure. It would thus be evident that the cowdung manure was quite popular for farming purposes in the Vedic and post-Vedic age.

## Notes and References

1. Keith, A.B., *Religion and Philosophy of the Vedas and Upaniṣads*, Vol. I, pp. 186-187.
2. *Paraskara Gṛhya Sūtra*, II, 17.9.
3. Keith, A.B., *op. cit.*, p. 187.
4. *Ibid.*, pp. 187-188.
5. Keith, A.B., *Vedic Index of Names and Subjects* I, pp. 39-40, Vārāṇasī, 1958.
6. *Ṛgveda*, I, 85.10, 11; I.1169, 22; I.130.2; IV. 17.16; IV. 50.3; VIII. 49.6; VIII.62.6; VIII.72.10; X.25.4; X.101.130.
7. *Ibid.*, I.55.8.
8. *Ibid.*, 101.6.
9. *Anisatra*, (*Ṛgveda*, X.101.7) is best thus rendered. For the stone wheel (*asamacakra*) which was above *uccacakra*. It was also mentioned perhaps as a *ku-cakra* in *Ṛgveda*, X.102.11, which is another name of the wheel.
10. *Ṛgveda*, X.101.6.
11. *Kṛṣi* is found in *Atharvaveda*, II.4.5; VIII.2.19; 10.24; X.6.12; XII.2.27; etc. *Taittirīya Saṁhitā*, VII.1.11.1; *Maitrāyaṇī Saṁhitā*, I.2.2; III.6.8; *Vājasneyī Saṁhitā*, IV.10; IX.22; XIV.19, 21 etc. *Śatapatha Brāhmaṇa*, VII.2.2.7; VIII.6.2 ?; *Taittirīya Brāhmaṇa*, III.1.2.5.
12. *Ṛgveda*, X.34.13; X.117.7.
13. *Pañcaviṁśa Brāhmaṇa*, XVII.1.
14. *Śatapatha Brāhmaṇa*, VIII.78.10, X.101.3.
15. *Ṛgveda*, X.48.7.
16. *Ibid.*, X.71.2; *Atharvaveda*, XII.3.19.
17. *Ibid.*, X.94.13.
18. *Ibid.*, II.14.11.
19. *Vājasneyī Saṁhitā*, XVII.12, contains a list of such grains.
20. The pulling of ripe fruit is referred to in the *Ṛgveda*, III. 45.4.

21. *Taittiriya Saṁhitā*, VII.2.10.2.
22. *Ibid.*, V.1.7.3.
23. *Kauśitaki Brāhmaṇa*, XIX.3.
24. *Atharvaveda*, VI.50.142.
25. *Ibid.*, II.31.2.
26. *Ibid.*
27. *Ibid.*
28. *Ibid.*, II.31.1.
29. *Ibid.*, II.31.4.
30. *Ibid.*, II.31.4.
31. *Ibid.*, II.32.2.
32. *Ibid.*, II.32.5.

## 13

# Implements of Vedic Age and Their Usages

For the purpose of work in the fields and plants, it is but natural to have some implements, which could be used for tilling and other purposes. However, while dealing with the subject of the use of garden and other implements during the Vedic and post-Vedic time it may be stated that there is no essential distinction between the worship of natural objects conceived as living beings and worship of objects made by human hands, but it is obvious that the worship of such objects tends to be restricted in effect and importance. This is borne out by the evidence of Vedic religion where such religion is of very minor importance despite the fact that it receives further attention in the *Ṛgveda* than might *a priori* have been expected. Thus we find there that two implements of ploughmen, *Suna* and *Sira* may be the plough share and the plough, are invoked and that in the *Śatapatha Brāhmaṇa* a cake is assigned to them in the sacrifice. The warrior invokes his arrow as divine (*Ṛgveda*, VI.75.11, 15, 16) and begs it to protect him and assail his foes. In the *Atharvaveda* (V.20) a whole hymn is devoted to his praise. But the priest is naturally prominent in the *Ṛgveda*, where not only the sacrificial post is deified, perhaps as the remnant of tree worship, is called divine and is asked to allow the sacrifice to go to the gods, but also the sacrificial grass (*Ṛgveda*, II.3.4) and the divine doors (I.142.6) are celebrated. The pressing stones have three hymns (*Ṛgveda*, X.76.94, 175) assigned to them. They are unaging, immortal and more powerful than the heaven itself. They are compared to steed or bulls, their sound in pressing reaches the sky. They bestow wealth and offsprings, driving away the demons. The mortar and pestles were deified also and the *Atharvaveda* adds to the category various ladles and invents a great deity in the remnant of the sacrifice, the Ucchista.

> "It is possible to draw a distinction between the cases of the deities of plougher and the warrior and those of the priest. In his case, the presence of the deity at the sacrifice may be held to be the cause why the instruments of the sacrifice are treated with so much reverence, while in the former cases, the reverence is not due to anything save the essentially valuable character of the objects, the mystic powers which are deemed by man to lie within them. It is particularly easy to understand, how the arrow or even the plough can be regarded as animate; the worship of the priestly implements would thus be rather more fetishistic in origin than real and direct worship of the implements for themselves."

However, from Vedic literature one can not have enough of information about the implements used in the fields or the gardens, though some stray references here and there could be traced. As for example, there is a mention of *Abhri* (Spade) in the *Atharvaveda*—

परिग्राम मिवाचितं वचसा स्थापयामसि।
तिष्ठा वृक्ष इव स्थाम्न्यभ्रिरवाते न रूरुपः।।
पवस्तैस्त्वा पर्यक्रीणन्दूशोभिरजिवैरुत।
प्रक्रीहसि त्वमोषधेऽभ्रिखाते न रूरुपः।।

—*Atharvaveda*, IV.7.5-6

(What a spell we cause to stand about, thee, as it were a collected troop; stand than, like a tree in its station; spade-dug one, thou rackest not;

For covers they brought thee, also for garments, for goat-skins; purchasable art thou, O herb, spade-dug one, thou racket not.)

In the above passages, *Abhri* has been used to represent the spade, which is so commonly used even during the present times for field works as well as for gardening. It also occurs in the *Vājasneyi Saṁhitā* (V.22; IX-10; XXXVII.I) and *Pañcaviṁśa Brāhmaṇa* (II.3.5.15). It could be made of a bamboo or the wood of Vikankara or Udumbara. It may be a span or a cubit in size. It is hollow and each one of its edges are sharp. From this it would appear that the handle was made of wood and the head of some metal.

For any field work the use of plough is essential and this plough has been mentioned in the *Ṛgveda* as *lāṅgala. Astra* stands for the plougher's goad or the badge of agriculturist meant for driving cattle.

शुनं वाहा शुनं नरः शुनं कृषतु लाङ्गलम्।
शुनं वरत्रा बध्यन्तां शुनमष्ट्रामुदिङ्गय।।

—*Ṛgveda*, IV.57.4

From the above verse, it could possibly be surmised that the plough was used for the purpose of agriculture, (which was driven by bullocks) and *astra* was used for driving the bullocks. the use of plough has also been testified in the *Rāmāyaṇa* of Vālmīki, wherein Janaka tells Viśvāmitra, that he had found Sītā, when he was ploughing the field—

न्यासभूतं तदा न्यस्तमस्माकं पूर्वके बिभो।
अथ मे कृषतः क्षेत्रं लाङ्गलादुत्थिता मम।।
क्षेत्र शोधयता लब्धा नाम्ना सीतेति विश्रुता।
भूतलादुत्थिता सा तु व्यवर्धत ममात्मजा।।
वीर्य शुल्केति मे कन्या स्थापिते यमयोनिजा।
भूतलादुत्यितां ता तु वर्धमानां ममात्मजाम्।।

—*Bālakāṇḍam*, 66.13-15

The Vedic texts also mention abut the agricultural labourer working in the field, who was known as Kinasa. This term was used for a plough man or a cultivator.

The same term has been used in *Taittiriya Brāhmaṇa* (II.4.8.7), *Vājasneyī Saṁhitā* (XXX.11), and *Atharvaveda* (IV.11.10 and VI.30.1). The ripe grain was cut with a sickle (*datra* or *Sṛṇi*). The grain was separated from the straw by means of sieve (*titau*) or a winnowing fan (*sūrpa*).

# 14
# Deities Connected with Forest and Vegetation

In the *Ṛgveda* the forest deities are conceived both as male and female, *i.e.*, Vanaspati, the Lord of the forest and Araṇyāṇī, the goddess of the forest. Araṇyāṇī is invoked as the forest goddess or 'Mother of Beasts', abounding in food without tillage. In the northern regions of the country, she is also known as 'Banaspati Ma' or the 'Mistress of Woods', who could be kept under control by placing a branch or a stone on her as a tribute to her dreaded presence. Sometimes, however, a cock or a goat or a young pig may be offered to her by a village herdsman, who graze their herds in the forest. Forest Mothers tend to acquire more and more characteristics of the Village Mothers. Thus among the Panyan forest tribe a goddess called *Kad Bhagavatī* (who also occurs as a sexless deity called Kulli) is a malignant and terrible being dwelling in a tree, resembling in its sinister features the terrific types of Dravaḍian godlings, because the Mother-Goddess was adored throughout the ancient world.

## 1. Vanaspati

The Indian religious texts have for long been emphasising for the adoration of trees and plants, but have also conceived of certain male and female deities having association with forest and wood. This is found translated in the *Ṛgveda* and the subsequent Vedic and post-Vedic literature. There is a frequent mention of Vanaspati or the Lord of Forest.[1] But the true identity of the deity is not mentioned. In this connection it may be stated that from the excavations at Mohenjodaro, some seals were enearthed which displayed a male nude figure, with a horned head-dress (Seal No. 222) seated cross-legged in the enviornments of a forest, with heels pressed together on a low dias, like a *yogī*, whose legs represent those of a bull. The arms are outstretched on either side, but it is possible that the hands are intended to rest on the knees and that the seal-cutter merely raised them slightly to obviate any confusion of the arms with legs.[2] On each arm there is a large number of bracelets. The horn, if they are intended to be so, are definitely separate from the head; they are moreover, represented as fastened to the base of the twig. The figure has three faces, one in the front and two in profile. Another seal bearing No. 235 has a somewhat similar figure. The stool is omitted and the figure is seated on the ground. The head dress consists of two horn like objects, between which, there appears to be a spike of flowers. A pig tail hangs down on one side of the head, which has one face only, in profile, facing to the right. There is another seal bearing the same features but is numbered as 420.[3]

A comparison of the three seals shows many points in common. Apart from the ornaments they wear, the figures are nude, save for a cincture worn round the waist and possibly also passed between the legs, a garment very similar to the *laṅgot* worn in the present times. On seal No. 222, the figure has only one head but on Nos. 235 and 420, it has three. On two of seals (Nos. 222 and 420), the god is seated on a low stool, whose supports in one case are shaped like the legs of a bull and in the other somewhat like an hour-glass. On Nos. 222 and 235, the headdresses are very similar but surmounted by a plant motif with three branches in one case and only a single branch in the other. The larger figure on seal No. 420 lacks this spray of foliage, but has instead a fan shaped ornament so commonly associated with the pottery female figurines. Though all the three figures appear to be male deities, but the pigtail of No. 235 is also worn by tree spirits and the worshippers and the seven ministrants on seal No. 430. Possibly both male and female deities were represented with pig-tails. The presence of wild beasts, has made Mackay to identify the figures as representing *Paśupati*—lord of animals, but the foliage decorating the head-dress and the environment of the forest, could tempt one to consider them to represent Vanaspati or the 'Lord of the Forest'.

## 2. Śākambharī

There is no derth of the female forest deities as well. In this connection, a passage from the *Devīmāhātmya*, brings out the Śākambharī aspect of the goddess—

ततोऽहमखिलं लोकमात्मदेह समुद्भवैः।
भरिष्यामि सुरा शाकैरावृष्टेः प्राण धारकैः।
शाकम्भरीति विख्यातिं तदा यास्याम्यहं भुवि।।

—(II.45)

(At that time, O gods, I shall support the whole world with life sustaining vegetables, born out of my own body, until the rains set in again. Then I shall be famed on earth as Śākambharī).

In this context, it may be pointed out that the deities of vegetation as depicted in seals discovered from Harappa, Mohenjodaro and Bhita highlighting this aspect, have already been discussed elsewhere in this work.

The name Śākambharī has been interpreted by Dr. V.S. Agrawala to be of historical significance. According to him in Verses 43-45 of the *Devīmāhātmya*[4] no human enemy is involved. "But people suffered from a natural calamity by the failure of rains for a period of one hundred years and the goddess on that occasion assumed two forms, namely, Śatākṣī and Śākambharī. It seems it refers to the hundred years, Śaka-Kuṣāṇa rule under the emperors Kaniṣka, Huviṣka and Vasudeva, whose last year of reign is recorded in an inscription as 98 Śaka Era. Several facts are quite significant here, firstly, the goddess is called Śākambharī here, literally the supporter of the Śakas; secondly no *asura* or enemy of the goddess is mentioned here and there is the complete absence of any battle motif in this case. It seems that the memory of the Śakas was quite fresh amongst the people and they were not yet regarded as hostile intruders and that their goddess was also accepted in the cult worship by the local people. This is made definite by the name Śatākṣī and the very strange description and etymology given of her form and name. She is called Ayonija, which has a reference to her heavenly birth; secondly, she is said to have been invoked by *Munis*, whom she looked at with her hundred eyes because of which she was named as *Śatākṣī*—

मुनिभिः संस्तुता भूमौ संभविष्याम्योनिजा।
ततः शतेन नेत्राणां निरीक्षिष्यामि यन्मुनीन्।
कीर्तयिष्यन्ति मनुजा शताक्षीमिति मां ततः।।

—*Devīmāhātmya*, II.43-44

"The *munis* here imply a technical reference to whom in the Nārāyaṇa Parvan of the Śānti Parvan are known as Citra Sikhaṇḍinah. This new epithet was the counterpart of Indian Saptarṣis. *Viṣṇudharmottara*, 3.8.25); but in fact it was coined to designate seven *Amesapendas* of the Pahlavi religion and they are specially mentioned in that Chapter of the *Nārāyaṇīya Dharma* which is loaded with many other Sasanian symbols. Above all the Iranian Goddess named Asa, the Pahlavi form of Sanskrit Rita, is also mentioned in the *Nārāyaṇīya Dharma*. Her Sanskritized name seems to have been which on the coins of Kaniṣka and Huviṣka appears as the goddess Ordokso (Whitehead, *Catalogue of Scythian Coins*, pp. 194, 207). She is named here Śatākṣī.

One fact agrees very well with the genesis of the whole text for the glorification of the goddess Mahiṣāsuramardinī or Kātyāyanī. It is for the first time that the images of the goddess Mahiṣāsuramardinī appear in the Kuṣāṇa art and there are many such sculptures in the Mathura Museum. In fact in no other period of Indian art history, Mahiṣāsuramardinī images are so numerous and popular as in the Kuṣāṇa art. The religious synthesis of this period is fully reflected in the names Asa, Ṛta, Ṛtākṣī, Ordokso and Śatākṣī. She was truly regarded as not of human birth, but of heavenly origin as the divine light of Ahura Mazda, who himself was known as Hari-Medhadeva in this period as is mentioned several times in the *Nārāyaṇīya Parvan*."

## 3. Tree Goddess

One of the most interesting seal from Mohenjodaro is that of a tree goddess, which has been discussed in detail elsewhere in this work. The tree *Aśvattha* is recognizable from its leaves shown on the right top corner and it is represented by only two branches springing from a circle on the ground; between the two branches, the nude deity, having long hair, a pair of horns, behind whom stands a human headed goat; seven ministrants stand in a row below.

## 4. Araṇyānī

The concept of the forest as the mother is evident from the use of the epithet Araṇyānī, the Jungle Goddess—She has been eulogized in the *Ṛgveda* as under—

अरण्यान्यरण्यान्यसौ या प्रेव नश्यसि।
कथा ग्रामं न पृच्छसि न त्वा भीखि विन्दतीँ३।।

वृषाखाय वदते यदुपावति चिच्चिकः।
आघाटिभिखि धावयन्नरण्यानिर्महीते।।
उव गाव इवादन्त्युत वेश्मेव दृश्यते।
उतो अरण्यानिः सायं शकटीरिव सर्जति।।
गामङ्गैष आ ह्वयति दार्वङ्गैषो अपावधीत्।
वसन्नरण्यान्यां सायमक्रुक्षदिति मन्यते।।
न वा अरण्यानिर्हन्त्यन्यश्चेन्नाभिगच्छति।
स्वादो फलस्य जग्ध्वाय यथा कामं पद्यते।।
आञ्जनगन्धि सुरभि बह्वन्नामकृषीवलाम्।
प्राहं मृगाणां मातरमरण्यानिमशंसिष्रम्।।

—*Rgveda*, X.146.1-6

(Goddess of wild and forest who see most to vanish from the sight. How is it that thou seekest not the village? Art thou not afraid? What time the grasshopper replies and swells the shrill Cicala's voice, seeming to sound with tinkling bells, the lady of the wood exults. And, yonder, cattle seem to graze, what seems a dwelling place, or else at eve the lady of the forest seems to free the wains. Here one is calling to his cows, another here hath felled a tree; at eve the dweller in the wood fancies that some body hath screamed. The goddess never slays, unless some murderous enemy approach. Man eats of savoury fruits and then takes even as he wills, his rest. Now I have praised the forest queen, sweet-scented, redolent of balm, the mother of all sylvan things, who till not but hath stores of food.)

## 5. Vanadurgā

This is also a forest deity included in the list of Navadurgā goddesses. The texts provide eight hands for her, in the seven of which should be held the *śaṁkha, cakra, khaḍga, kheṭaka, bāṇa, dhanuṣa* and *śūla*, the eighth hand being in *tarjanī mudrā*. The complexion of this kind of the goddess is said to be grass green—

अरिशङ्खकृपाणखेटवाणान्सधनुश्शूलकतर्जनींदधाना
मम सा महिषोत्तमाङ्गसंस्था नवदुर्गा सदृशी श्रियेऽस्तु दुर्गा।।

—*Amnaya*

## 6. Vindhya Vāsinī Durgā

The goddess resides in the Vindhyas and as such could be claimed to be the forest deity. Her complexion is conceived to be bright as the lightning, should be seated on a golden lotus; should have three eyes and four arms. Two of her hands should carry a *śaṁkha* and a *cakra* and the remaining two are to be held in *abhaya* and *varada* poses respectively. She should be adorned with crescent moon, the *hāra, aṅgadas, kuṇḍalas* and other ornaments.

Surrounding her should be Indra and other gods praising her. The lion, her vehicle, should be made to stand near her.

सौवर्णाम्बुजमध्यगां त्रिनयनां सौदामिनी सन्निभां।
शङ्खं चक्रवराभयानि दधतीमिन्दोः कलां विभ्रतीम।
ग्रैवेयाङ्गदहारकुण्डलधरामाखण्डलाद्यैस्तुतां
ध्यायेद्विन्ध्यनिवासिनीं शशिमुखीं पार्श्वस्थपञ्चाननाम्।।

—*Śārdatilakaṁ*

## 7. Kātyāyanī

The goddess derives her name from the fact that she emerged on earth in the *āśrama* of sage Kātyāyana in the forest and was responsible for slaying Mahiṣāsura and other demons in the forest. As per the account provided by the *Vāmana Purāṇa*, "the gods having been defeated by Mahiṣāsura, foresook their abodes and hastened, with Brahmā as their head, to implore the protection of Viṣṇu. Śaṅkara was also present there, and when the gods finished relating their grievances, Viṣṇu, and at his command, Brahmā, Śaṅkara and all the gods so emitted flames of anger from their eyes and countenances that a mountain of effulgence was formed. From this mountain came forth Kātyāyanī, resplendent as a thousand suns, with three eyes, with hair black as night and with eighteen arms. To her Śiva gave a trident, Viṣṇu a disc, Varuṇa a conch, Agni a dart, Yama an iron rod, Vāyu a bow, Sūrya a quiver and arrows, Indra a thunderbolt, Kubera a mace, Brahmā a rosary and a water pot, Kāla a sword and a shield, Viśvakarmā a battle-axe and other weapons, Himvān a lion, and other gods various arms and ornaments. Being thus named and adorned by all gods, Kātyāyanī proceeded to the Vindhya mountains. There two *asuras* Caṇḍa and Muṇḍa, saw her and immediately hastened to their king and informed him that a most lovely goddess had singly fixed her abode on the Vindhya mountain. On hearing the high encomiums which Caṇḍa and Muṇḍa bestowed on the beauty of this goddess, Mahiṣāsura determined to obtain possession of her and ordered his army to by collected and marched to the vicinity of the mountain where the goddess was residing. He himself entered into a lofty pavilion and then sent the son of Maya to summon the goddess to her presence. Dundubhi, the son of Maya, then proceeded to her, and standing respectfully at a short distance informed her that he was the messenger sent by Mahiṣāsura with a proposal of a union with her. This was rejected by the goddess, who ultimately killed the king of demons and other *asuras*, for the sake of gods.

In the *Harivaṁśa Purāṇa*, it is said that the mother goddess living in the Vindhyas has the following of fowls, goats, rams, lions and tigers.

## 8. Vanaśaṅkarī

An image of Vanaśaṅkarī goddess who is believed to be a forest deity, belonging to the medieval times was discovered from Annagiri in Karnataka and is a unique specimen of its kind. Literary reference in this respect coult not be traced (Pl. 62).

# Notes and References


1. *Ṛgveda*, I.166.5; III.34.10; V.7.4; V.41.8.
2. Mackay, E.J.H., *Further Excavations at Mohenjodaro*, I, p. 335; seal 222; Pl. LXXIII-4.
3. *Ibid.*, Pl. LXXV-1, 6, 21; LXXVI-9, 21.
4. Agrawala, V.S., *Devimāhātmya*, pp. 218-19.

# 15
# The Fruit Trees

Ever since the advent of humans on earth, they have been closely associated with fruits. At the early stages of the Indian culture, when the people led the life of aboriginals, they had to depend on fruits and roots of the trees which were so abundantly available in the forests, as a part of natural growth. At that stage there was no sense of cultivation available with them and they had per force to depend on the forest fruits, besides hunting for food. However, with the development of human mind, a crude form of cultivation was started during the Neolithic period. But even after that the importance of fruits as means of staple diet did not decrease. Besides the farming, planting of the trees like *dāḍima* (*Punica grantum* Linn), Jambū (*Xylia xylocarpa*, Roxb.), Nārikela (Coconut) and *Kadalī* (Plantain) were also resorted to for diet. However, the early Dravadian population was quite conversant with the fruit trees like *Panāsa* (*Artocarpus heterophyillus*, Lam), *Cinca* and *puga.*

As to the literary evidence concerning the use of fruit, one has to refer to the Vedic literature. In the *Ṛgveda*[1], there is a mention of the fruit trees like Aśvattha (*Ficus religiosa*). In the *Yajurveda*, the fruits like Badara (*Zizania jujuba*, Mill), Bilva (*Aegle marmelos*, C.& Roxb.), Kharjūra (Date-palm), Karīra (*Capparis aphylla*), *Udumbara* (*Ficus glomerata*), Jambira (*Citrus limonum*, Linn), Vikankata (*Flacourtia ramontchi*). The *Atharvaveda* has added Kapithaka and Purusahva or Purusaka (*Gardenia Latifolia*, Act) to the last.

The list of fruit trees is further improved by addition of more of the species in Brāhmaṇa literature. In the *Śatapatha Brahmaṇa*,[2] however, it is possibly for the first time that the mango tree (āmra) is mentioned, besides Aśvattha and Udumbara. The *Jaiminīya Brāhmaṇa*[3] mentions of āmala or āmalaka and is repeated in the *Chāndogya Upaniṣad.*[4]

During the Epics the use of fruits to be served as food items had been quite popular and it is recorded that the sages or ascetics who had made the forests as their abodes, were not only living on fruits and roots of the trees, but also served them to their guests or visitors. As for example it is stated in the Bālakāṇḍaṁ of the *Vālmīki Rāmāyaṇa*, that when once Viśvamitra visited the *āśrama* of the sage Vaśiṣṭha, the latter welcomed him in his hermitage offering him fruits and roots to eat—

उपविष्ठाय च तदा विश्वामित्र धीमते।
यथान्यायं मुनिश्वर: फलमूलमुपाहइत।।

—Bālakāṇḍaṁ, 52.3

In the Ayodhyākāṇḍaṁ of the same text, it is mentioned that when Rāma together with Sītā and Lakṣmaṇa reached the hermitage of Bhāradvāja, they were greated with offerings of various types of forest fruits and roots—

नाना विधानन्नरसान वन्यमूलफलाश्रयान्।
तेभ्यो ददौ तप्ततपा चैवाभ्यकल्पयत्।।

—*Ayodhyākāṇḍam*, 54.18

Besides the epic while describing Citrakūṭa, a number of fruit trees have been mentioned, which include *āmra, jambū, priyāla, panāsa, aṅkola, bhavya, bilva, tinduka, kāśmarya, badari, āmalaka, nipa* and *dhanvan*—

आम्रजम्ब्वसनैर्लोध्रैः प्रियालैः पनसैधतैः।
अंकोलैगर्ण्यति निशैर्बिल्व तिन्दुकवेणुभिः।।
काश्मर्यरिष्टवरणैर्मधूकैस्तिलकैस्तथा।
बदर्यामलकैर्नी पैर्वेत्रधन्वनबीजकैः।।
पुष्पवद्भिः फलोपेतैरछायावद्भिमनोरमैः।
एवमादिभिराकीर्णः श्रियं पुष्य त्ययं गिरिः।।

—*Rāmāyaṇa*, II.94.8-10

The *Mahābhārata*, on the other hand, speaks of fruits and fruit trees quite frequently, one of the books *viz.*, the Araṇyaparva, deals exclusively with the Pāṇḍavas, stay in exile in the forests. Thus while describing Dvaitavana, Vaiśampāyana speaks of the fruit bearing plants like *tāla, madhūka, nīpa* and *kadamba* etc. elsewhere, the *kadalī* forest (yielding banana fruits) has also been mentioned—

तच्छालतालाम्रमधूकनीपं, कदम्बसर्जार्जुन कर्णिकाहैः।
तपात्यये पुष्पधरैरुपेतं, महावनं राष्ट्रपति ददर्श।।
महाद्रुमाणां शिखरेषु तस्यु, मनोर्मां वाचमुदीरयन्तः।
मयूर दात्यूह चकोर संधा, स्तस्मिन्वने कानन कोकिलाश्च।।

—*Mahābhārata*, III.25.17-18

(The king saw the huge forest covered at the end of the summer, with *sālas*, palms, mangoes, *madhūkas, nīpas, kadambas, sarajas, arjunas* and *karṇikāras* clothed with flowers; and peacocks, *dātyuhas, cakravāka, varhins* and *kokilās*, sat on the top of the highest trees and emitted their sweet notes.)

During the Purāṇic period the use of fruits became quite common and there are numerous references pointing to the use of fruits for a variety of purposes. They not only formed part of the regular diet, but could also be useful for medicinal purposes. In this connection, *jambū* (*Xylia xylocarpa*, Roxb.) could be quoted as an example. The *Viṣṇu Purāṇa* highlights not only the nutritious value of this fruit but considers it quite useful for diabetes—

जम्बूद्वीपस्य सा जम्बूर्नामहेतुमहामुने।
महागज प्रमाणानि जम्ब्वास्तस्य फलानि वै।।
पतन्तिभूभृतः पृष्ठे शीर्य माणानि सर्वतः।
रसेन तेषां प्रख्याता तत्र जाम्बूनदीति वै।।

सरित प्रवर्तते चापि पीयते तन्निवासिभिः।
न स्वेदो न च दोर्गन्ध्यं न जरा नेन्द्रियक्षयः।।
तत्पानात् स्वच्छमनसां जनानां तत्र जायते।।

—*Viṣṇu Purāṇa*, II.2.18-21

The use of the juice of *jambū* fruit relieved people of several diseases including the excess of sweating, bad smell from body, old age, organic disability and mental depression. The symptoms (*Sveda*, etc.) are mostly met with in diabetes and it is interesting to learn that the *jambū* fruit and seeds have medicinal properties according to the Indian system of medicines. The *Garuḍa Purāṇa*, also goes on to testify the medicinal properties of a large number of fruits, pulses and vegetables, a glimpse of which is provided here—

केशरं मातुलुंगं च दीपनं कफवातनुत्।
वात पित्त हरो माषस्त्व किस्नग्धोष्णानिलापहः।।
सलमामलकं वृष्यं मधुरं हृद्यमम्लकृत्।
मुक्तप्ररोचका पुण्या हरीतकामृतोपमा।।
स्रंसनी कफवाताघ्नी ह्यक्षस्तद्वत्त्रिदोषजित।
वातश्लेष्महरं त्वम्लं स्रंसनं तिन्तिडी फलम्।।
दोषलं लकुचं स्वादु वकुलं कफवातजित्।
गुल्मवातकफश्वासकासघ्नं बीजपूरकम्।।
कपित्थं ग्राहि दोषघ्नं पक्वं गुरु विषापहम्।
कफपित्तकरं बालमापूर्णं पित्त वर्द्धनम्।।
पक्वाम्रं वातकृन्मांसशुक्रवर्णं बलप्रदम्।
वातघ्नं कफपित्तघ्नं ग्राहि विष्टम्भि जाम्बवम्।।

—*Garuḍa Purāṇa*, 1.169.21-26

The same text also spells out the medicinal properties of a number of fruits in Chapter-202 of Pūrva Khaṇḍa. The *Brahmavaivarta Purāṇa* (Gaṇeśakhaṇḍa, 13.28-30) mentions a number of fruits like *Dāḍima, Śrīphala*, Date-palm, *jambū, āmra, panāsa* and banana. Some seasonal fruits like orange and watermelon are also mentioned. The *Vāmana Purāṇa* has also an extensive list of fruits, some of which are noted below—

(i) Āmalaka—*Emblica officinalis*, Gaertn (17.55; 58.8; 68.28).

(ii) Kadalī—*Lagerstroemia speciosa*, Pers. or Plantain (7.11).

(iii) Cūta—Mango or *Mangifera Indica*, Linn (6.105; 12.51; 17.52; 58.8).

(iv) Dāḍima—*Punica grantum*, Linn, (64.97).

(v) Bilva—*Aegle marmelos*, C & Roxb. (1.22; 6.18; 18.8; 36.12; 58.8; 68.15).

The *Varāha Purāṇa* too provides an extensive list of fruits which were used for human consumption, offering to gods and to the departed souls of the dead—

(i) Kharjūra—Date-palm (80.70)

(ii) Cūta (Mango) *Mangifera indica*, Linn (39.44).

(iii) Jambira—*Citrus limonum*, Linn (168.24).

(iv) Jambū—*Xylia xylocarpa*, Roxb. (77.17; 140.73; 166.8).

(v) Dāḍima—*Punica grantum*, Linn. (170.36).

(vi) Drākṣa—*Vitis vinifera*, Linn.—grapes (80.70).

Some of the fruits mentioned in a few other Purāṇic texts are listed here—

(i) *Bhāgavata Purāṇa*—(X.12.43) Kapittha

(ii) *Vāyu Purāṇa*—Sugarcane (II.16.8); Patola (II.16.9); Dīrgha mūlaka (a kind of radish) (II.16.10); Garlic, Carrots, Onions, Karambha (II.16.11-14).

(iii) *Kūrma Purāṇa*—*Śyāmaka* grains, *Kasas*, *Nivaras*, *Priyaṅgu*, wheat, gingelly, grams (II.20.37). The Householder is required to offer the Brāhmaṇa after the meals at the *Śrāddha* ceremony, fruits of different kinds (II.22.62-64).

The *Padma Purāṇa* on the other hand considers the planting of fruit bearing trees as an act of great merit and to grow such trees in *āśramas* of ascetis or fruit orchards, were highly valued—

नामकेति स चूतस्य समारोप्य सहस्रकम्।
ततो द्वित्रिगुणनैव न्यूनेन वा प्रचुरेऽपि वा।।
आरोग्यं शौर्यसम्पन्नमारामादेव जायते।
फलानि यस्य खादन्ति जन्तवोऽथसहस्रशः।।

—*Padma Purāṇa*, I.60.32.34

In the *Kalpasūtra* a large number of fruits are found mentioned in various contexts, which include *āmalaka, āmra, ingudi, jambū, tinduka, dadhittha, dāḍima, palevata, priyāla, bilva, bhavya, mṛdvika, vikankata* etc. For religious sacrifices the powder (*saktu*) of various types of *badara* was prescribed. In the *Gṛhyasūtras* fruits are required to be offered in various ceremonies.

The *Aṣṭādhyāyī* of Pāṇini also does not neglect fruit and he mentions fruits like *aśvattha, āmra, jambū, tāla* and *bilva*.[5] Pātāñjali in his *Mahābhāṣya* has also mentioned fruits like *amalaba, āmra, udumbara, kadalī, kapittha, karkandu, kuvala, jambū, tintidika, dāḍima, drākṣā, nīpa, panāsa, badara* and *bilva*.[6]

Manu in his work prescribes for the hermits residing in the forests not only to live on fruits and roots of the trees but offer them in sacrifices—

मुन्यन्नैवर्विविधैर्मेध्यैः शाकमूलफलेन वा।
एतानेवमहायज्ञान्निर्वपेद् विविधपूर्वकम्।।

—(VI.5)

(Let him offer those great sacrifices according to rule, with various kinds of pure food fit for ascetics, or with herbs, roots and fruit.)

पुष्पमूलफलैर्वापि केवलैर्वर्त्तयेत् सदा।
कालपक्वैः स्वयं शीर्णैर्तैरवानस मते स्मृत स्थितः।।

—(VI.13)

(Let him eat vegetables and grow on dry land or in water, flower, roots and fruits, the production of pure trees, and oil extracted from forest fruits).

स्थल जौदक शाखानि पुष्पमूलफलानि च।
मैध्यवृक्षोद्भवानद्यात्स्नेहाश्च फल सम्मवान्।।

—(VI.21)

(or he may constantly subsist on flowers, roots and fruits alone which have been ripened by time and have fallen spontaneously).

Manu further prescribes the offering of fruits and other eatables to the ancestors in their *Śrāddha* ceremony—

तिलैव्रीहियवैमषैरद्भिर्मूल फलेन वा।
दत्तेन मासं तृप्यन्ति विधिवत्पितरो नृणाम्।।

—(III.267)

(The ancestors of men are satisfied for one month with sesamum grains, rice, barley, *māsa*, beans, water, roots and fruits, which have been given according to the prescribed rules.)

Kālidāsa was a great lover of nature and in his works he described besides trees, plants and creepers, also the fruits, a few illustrations of which are provided here—

(i) *Raghuvaṁśa* (4.69)—Akṣoṭa (Wall-nuts).

(ii) *Kumārasaṁbhava* (3.67)—Bimba, (3.32) Mango

(iii) *Meghadūta* (2.38)—Kadalī

(iv) *Mālavikāgnimitra* (Act-III—Bījapūraka (lemon).

Other poets of ancient Sanskrit literature like Daṇḍī, Māgha, Bhavabhūti and others have also mentioned the use of fruits in a variety of purposes. Varāhamihir has in his *Bṛhatsaṁhitā*[7] devoted an exclusive chapter on the growth of fruits and flowers. He has described the effect of flowering of different fruit trees on the crops of different species of grains. As for example he mentions (29.4), "that the growth of Jambū (the rose-apple) tree indicates the growth of gingelly and black grain; the growth of Śirīṣa indicates the growth of Kangu and the growth of Madhūka indicates the growth of wheat and the growth of Saptaparṇa indicates the growth of barley.

## Mentioned in Historical Mahākāvyas of Medieval Period

The fruit trees are abundantly mentioned in the historical Sanskrit literature of the medieval times (ranging from the 11th to the 13th century A.D.), some glimpses of which are provided here—

| | | |
|---|---|---|
| (i) | *Navasahasrāṅkacarita* by Parimal Padma Gupta | Mango (I.33; VI.77); *Gunjaphala* (I.70); *Jambū* (II.61); *Patala* (VI.81); *Bimba* (VII.2); *Nārikela* (XI.10); *Kadalī* (XI.11; XII.4); *Dāḍima* (XIII.24). |
| (ii) | *Vikramāṅkadevacaritam* by Bilhaṇa | *Gunjaphala* (I.25); *Bimbaphala* (I.35; III.17); *Nārikela* (V.22), *Kadalī* (V.22; VII.15); IX.17; XII.51)—*Tāḍa* (VII.71; XII.33). |
| (iii) | *Rājataraṅgiṇī* by Kalhaṇa | *Tamāla* (III.326); *Tāḍi* (III.326); *Bimba* (III.416); *Drākṣa* (IV.192); *Kapittha* (VI.219); *Āmalaka* (VII.526); *Kadalī* (VIII.843). |
| (iv) | *Pṛthvīrājavijaya* by Jayanka | *Khairavini* (V.19). |
| (v) | *Dvyasrayamahākāvya* by Hem Candra Sūri | *Badarī* (XV.80); *Plakṣa* fruit (XV.81); *Kapittha* (XV.83); *Cūta*-mango (XVI.73); *Kadalī* (VIII.79); *Khadīra* (II.30); *Kuṣmāṇḍa* (1.75); *Patali* (III.31); *Kharjūra* (IV.23); *Panāsa* (IV.23); *Śrīphala* (V.7); *Bimba* (V.10). |
| (vi) | *Kīrtikaumadī* by Someśvara | *Patali* (VI.8); *drākṣalatā* (VI.49). |
| (vii) | *Sukṛta Saṁkīrtana* by Arisimha | *Kadamba* (I.17; VI.4); *Patala* (IX.15); *Lavaṅga*. |
| (viii) | *Vasautavilāsa* by Bālacandra Sūri | *Patala* (VI.2); *Kadamba* (VI.23); *Tamāla* (III.45); *Lavaṅga* (VI.45); *Āmra* (VI.53); *Laknca* (VI.59); *Dāḍima* (IX.53). |
| (ix) | *Hammīr Kāvya* by Nayacandra Sūri | *Kadalī* (IV.8); *Kandaphala* (V.16); *Lavaṅga* (XII.27). |
| (x) | *Madhuravijaya* by Gaṅgādevī | *Kadalī* (I.48); *Patala* (V.16); *Tamāla* (VII.23); *Nārikela* (VIII.8). |
| (xi) | *Saluvabhyudaya* by Rājanātha Dāḍima | *Jambū* (V.31); *Āmra* (X); *Dāḍima* (XI). |

## Works on Indian System of Medicines

The treatises on the ancient Indian systems of medicine brought out the nourishing aspects of the different species of fruits, spelling out at the same time their favourable and unfavourable effects on their consumption in various ailments. Of these treatises (*i.e.*, Vāgbhaṭṭa, Śuśruta, etc.), *Aṣṭāṅgahṛdaya* (Sūtrasthāna) has defined the effects of the human consumption of some of the fruits as under—

### (i) Drākṣa (Grapes)

द्राक्षाफलोत्तमा वृष्या चक्षुष्या सृष्टमूत्रविद्।
स्वादु पाकरसास्निग्धा सकषाया हिमागुरु।।
निहंत्यनिलपित्तास्रतिक्तास्य त्वमदात्ययान्।
तृष्णाकासश्रमश्वास स्वरभेदक्षत क्षयान्।।

—(VI.116)

## (ii) Dāḍima (Pomegranate)

उद्रिक्तपित्तान् जयति त्रीन्दोषान्स्वादु दाडिमम्।
पित्तविरोधि नात्युष्मम्लं वात कफापहम।।
सर्वं हृद्यं लघु स्निग्धं ग्राहि रोचन दीपनम्।।

—(VI.117)

## (iii) Different kinds of fruits like Banana, Date-palm, Panāsa, Coconut, Tāla fruit, Khirnī, Madhūka, Wallnut, etc.

मोच खर्जूरपनस नारिकेलपरूषकम्।।
आभ्रात ताल काश्मर्यराजादनमधूकजम्।
सैवीरवदरांकोल्ल फल्गु श्लेष्मा तकोद्भवम्।।
वातामाभिषुकाक्षोडम कूलकनिकोचकम्।
उरुमाणं प्रियालं च बृहणं गुरु शीतलम्।।
दाहक्षतक्षयहरं रक्तपित्त प्रसादनम्।
स्वादु पाक रसं स्निग्धं विष्टंभि कफ शुक्रकृत्।।

—(VI.118-121)

## (iv) *Tāla* Fruits

फलं तु पित्तलं तालं सरं काश्मर्यजं हिमम्।
शकृन्मूत्रबिवन्धघ्नं केश्यं मेध्यं रसायनम्।।

—(VI.122)

## (v) Dry Fruits (Almond, *Cirougi*)

वातामाद्युष्णवीर्यं तु कफ पित्त करं सरम्।
परं वात हरं स्निग्धमनुष्णं तु प्रियालजम्।।
प्रियालमज्जा मधुरो वृष्यः पित्तनिलापहः।

## (vi) Ripe Bilva Fruit

पक्वं सुर्दुजरं बिल्वं दोषलं पूतिमारुतम्।
दीपनं कफवातघ्नं ग्राह्युभयं हि तत्।।

(VI.125)

## (vii) *Kapittha*

कपित्थमामं कठघ्न दोषलं दोषघातितु।
पक्वं हिध्मावमथुजित्सर्वं ग्राहि विषापहम्।।

—(VI.126)

### (viii) Jambū Fruit

जांबवगुरुविष्टंभि शीतलं भृशवातलम्।
संग्राहि मूत्रशकृतोरकंठ्यं कफ पित्तभुत्।।

—(VI.127)

### (ix) Mango

वातपित्तास्रकृद्वालं बद्धास्थि कफपित्त कृत्।
गुर्वाम्रं वातजित्प क्वं स्वाद्वम्लं कफशक्रकृत।।

—(VI.128)

### (x) Mātuluṅga

त्वक्तिक्तकटका स्निग्धा मातुलंगस्य वातजित्।
वृंहणं मधुरं मांसं वातपित्तं हरं गुरु।।
लघुतत्केसरं कास श्वासहिध्मामदात्ययान्।
आस्यशोषा निलेश्लेष्म विबन्धच्छर्द्यं रोचकाम्।।
गुलमोदर्राशः शूलानि मंदाग्नित्वं च नाशयेत्।।

—(VI.131-132)

### (xi) Different Species of Fruits

तथाम्लं कोलककेंधूल लकुचाम्रातमारुकम्।
ऐरावतं दंतशहं सतूदं मृगलिंडिकम्।।
नातिपित्तकरं पक्वं शुष्कं च करमदर्कम्।
दीपनं भेदनं शुष्कमम्लीकाकोलयोः फलम्।।
तृष्णा श्रम कलमच्छेदि लध्विष्टं कफवातयो।
फलानामवरं तत्र लकचं सर्व दोषकृत।।

—(VI.137-139)

## Notes and References

1. *Ṛgveda*, I.164.20.
2. *Śatapatha Brāhmaṇa*, XIV.7.41.
3. *Jaminīya Brāhmaṇa*, I.38.6.
4. *Chāndogya Upaniṣad*, VII.3.1.
5. McGrindle, *Ancient India, described by Meghasthanese and Arrian*, p. 31.
6. Sharma, P.V., "Plants in works of three Sages of Sanskrit Grammar", *Bulletin of the Indian Institute of History of Medicine*, Vol. VII, No. 1&2, (Jan.-April) 1977, pp. 5-19.
7. *Bṛhatsaṁhitā*, Chapter-29.

# 16
# The Flowers

Flowers have been associated with human beings from time immemorial and they are mentioned in the earliest literature of the country, *viz.*, the Vedas and the literature that followed. The earliest mention is that of the lotus flower, both of the white and blue varieties which were quite popular with the Vedic people—

(i) Puṇḍarīka (white lotus) (*Ṛgveda*, X.142.8; *Atharvaveda*, X.8.43; *Chāndogya Upaniṣad*, VIII.1.1)—This denotes the blossom of lotus flower. The *Pañcaviṁśa Brāhmaṇa* states that the lotus flower was born out of the light of *Nakṣatras*. The *Atharvaveda* and the *Chāndogya Upaniṣad* compare the lotus with the heart.

(ii) Puṣkara (Blue Lotus) (*Ṛgveda*, VI.16.3; VII.33.II; *Atharvaveda*, IX.3.8; XII.1.24; *Taittirīya Saṁhitā*, V.I.4.1; *Vājasneyī Saṁhitā*, IX.29; *Maitrāyaṇī Saṁhitā*, III.1.5)—This represents the blue lotus flower and its sweet perfume is mentioned in the *Atharvaveda*. the lotus grew in lakes which were then called as *Puṣkariṇī* (Lotus bearing). That the flower was used early for personal adornment is evidenced from the epithet of Aśvins known as *Puṣkara-Sraja* (lotus crowned). Possibly because of its resemblance with lotus in the bowl of the ladle it is called as Puṣkara.

(iii) Abja (lotus), (*Ṛgveda*, IV.40.5; VII.34.16; *Atharvaveda*, VII.16)—This is also a type of lotus, literally meaning the one which is born in water.

Besides the above, the *Atharvaveda* also speaks of *Sadmapuṣpa* (IV.20.9) which refers to an ever flowering plant and Naladi or Naluda (*Atharvaveda*, VI.102.3); Nardostachya Jaṭāmānasī, a plant, the flowers of which were used for the making of garlands. During the later literature, one comes across a good variety of flowers when we reach the Epic stage.

The flowers had not only become an essential part of human life but were used in all the religious practices, so much so that no religious function could be complete without them. In the Purāṇic literature, the early traces of offering flower (lotus) in oblation can be witnessed in the *Śatapatha Brāhmaṇa* (V.4.5.7). This aspect of presenting the lotus flowers is further stressed in the subsequent passages as well (I.4.5.8.12).

> अथ सारस्वतं चरुं बिर्वपति। वाग्वे सरस्वती वाचेन वचिव
> तद्वरुणोऽनुसमर्सपत्तथोऽश्वेषऽएवेष एतद्वचिवानुसंसर्पति तत्रेकं
> पुण्डरीकं प्रयछति।
>
> —*Śatapatha Brāhmaṇa*, V.4.5.7

(He then prepares a rice pap for Sarasvatī—for Sarasvatī is speech, and it was with speech that Varuṇa on that occasion stole along; and in like manner does this one now steal along with speech. At this (oblation), he presents one lotus flower).

The *Gṛhya Sūtras* too did not lag behing in presceibing the use of flowers on a variety of religious or social customs. As for example the *Gobhila Gṛhya Sūtra* (II.4.1), ordains that the chariot of the bride (at the time of her departure after wedding) should be adorned with Kiṁśuka flowers. The *Hiranyakeṣin Gṛhya Sūtra* (1.3.2.4), prescribes that a student at the time of his admission for studies had to pass through a series of religious ceremonies, in one of which, a wreath of flowers has to be placed on his head. (The offereing of the flowers are prescribed in other *Gṛhya Sūtras* in various ceremonies).

The use of the word *puṣpa* gained currency in the Upaniṣadic period as would be evident from the following tabulation—

*Bṛhadāraṇyaka* (VI. 4.1) — औषधीनां पुष्पाणि पुष्पाणां फलानि:।

*Chāndogya* (III.1.2) — ऋग्वेद एव पुष्पम्।।

(III.2.1) — यजुर्वेद एवं पुष्पम्।।

(III.3.1) — सामवेद एव पुष्पम्।।

(III.4.1) — इतिहास पुराणं पुष्पम्।।

(III.5.1) — ब्रह्मैव पुष्पम्।

*Kauśitaki* (I.3) — पुष्पान्यादाय।

Besides the *Chāndogya* (I.6.7), *Bṛhadāraṇyaka* (II.3.6); *Nārāyaṇiya*, (5); *Ātmaprabodha* (1) and *Kṣuriki* (10) Upaniṣads mention of Puṇḍarīka, *i.e.*, the lotus flower.

During the period of Epics the use of flowers became considerably widespread and even the human limbs were compared with flowers. While describing Citrakūṭa, Rāma tells Sītā that the fragrant flowers of the trees of the forest there, appear as if they were wearing floral *kavacas*—

कुर्वन्ति कुसुमापीडांशिर: सु सुरभी नमी।

—*Vālmīki Rāmāyaṇa*, (II.92.13)

The use of flowers for the purpose of personal bodies has been testified in the same text as follows—

मृदिताश्चापविद्धाश्च दृश्यन्ते कमलस्रज:।

—*Ibid.* (II.94.25)

Even the river Mandākinī is found to be full of fragrant flowers, beautifying the natural environments of the forest—

नानाबिद्यैस्ती रुरुहैव्रतां पुष्पफलद्रुमै।
राजन्ति राज राजस्य नलिनीमिवे सर्वत:।।

—*Ibid.*, II.95.4

The great sage Vālmīki has compared the eyes of Rāma to a lotus, thus—

अब्रवीच वरारोहां चन्द्र चारु निभाननाम्।
विदेहराजसुतां रामो राजीवलोचनः।।

—*Ibid.*, II.95.2

In the *Mahābhārata*, however, the practice of describing gardens and parks with fragrant flowers, which was initiated in the *Rāmāyaṇa*, was further developed and several such gardens and parks have been detailed elsewhere in this work. Besides, there is also a story of Bhīma's going to Kadalī forest in search of a unique flower, where he had an encounter with Hanumān (This episode has also been described elsewhere in this work).

In the Buddhist literature, some literary works were given the title of the flowers. the *Saddharma Puṇḍarīka* (or the Lotus of the True Law) would be mentioned in this connection. In the *FO-SHO-HING-TSAN-KING* (Life of the Buddha) by Aśvaghoṣa, the Buddha is said to have been born in the Lumbinī garden, which was a pleasant spot, a quiet forest retreat, with its trickling fountains, blooming flowers and fruits. The *Amitayur Dhyana-Sūtra* (verse-18), (verse-10) while discussing the subject of the meditation of the Buddha, says that "among the dense foliage (in the forest), there blossom various beautiful flowers, upon which are found miraculously fruits of seven jewels and there are many excellent flowers which have the colour of Jambūnada gold and appearance of fire-wheels in motion turning between the leaves in graceful fashion."

The *Dhammapada* (Chapter-XXVI.401), declares that "Him I call indeed a Brāhmaṇa who does not cling to sensual pleasures, like water on lotus leaf, like a mustard seed on the point of a needle.

The *Jātaka* stories of the Buddhists reflect the various aspects of the different strata of society emphasising on their social customs, traditions and the habits of the contemporary society at large. In these stories flowers are said to have been used for the purpose of laying wreath on corpses, offerings on graves (*Śyāma Jātaka*) No. 540. The Pillyakkha shot Sama an arrow on mistaken identity, which killed him. The king then paid homage to his body with all kinds of flowers and sprinkled it with water and thrice went round it. In the story of *Sarabhaṅga Jātaka* (No. 522), the Master, standing near the Elder, ordered his remains to be deposited, and for a space of a league all round about the spot where the body was burned flowers rained down on it and a festival was held for seven days. In the *Kimchanda-Jataka*, (No. 511), *Sarubhaṅga Jātaka* (No. 522), *Mahāthamsa Jātaka* (No. 534), *Sudhabhojana Jātaka* (No. 535), *Mahāsutasoma Jātaka* (No. 537), there is the mention of the fragrant garlands of flowers used on various occasions.

But in the Purāṇic literature the use of flowers was made almost obligatory in the worship of deities. As for example, in the *Vāmana Purāṇa*, the flowers which are required to be offered to lord Viṣṇu during his worship, and those which are prohibited have been listed—

विशेषतः प्रवक्ष्यामि पुष्पाणि तिथयस्तथा।
दानानि च प्रशस्तानि माधवप्रीणनाय तु।।

जाति शताह्वा सुमनाः कुन्दं बहुपुटं बहुपुटं तथा।
बाणञ्च चम्पकाशोकं करवीरं च यूथिका।।
पारिभद्रं पाटला च ब कुलं गिरि शालिनी।
तिलकं च जपाकुसुमं पीतकं नागरं त्वपि।।
एतानि हि प्रशस्तानि कुसुमान्य च्युतार्चने।
सुरभीणि तथान्यानि वर्जयित्वा तु केतकीम्।।
बिल्वपत्र शमीपत्रं पत्र भृङ्ग मृगाङ्कयोः।
तमालामलकीपत्रं शस्तं केशव पूजने।।

—*Vāmana Purāṇa*, 68.11-15

The same text in its Chapter-5, provides a lively picture of the appearance of Vasanta Lakṣmī, comparing all the limbs of her body with flowers—

रक्ताशोककरा तन्वी देवर्षे किंशु काङ्घ्रिका।
नीलाशोककचा श्यामा विकासिकमलानुजा।।
नीलेन्दीवरनेत्रा च ब्रह्मन् बिल्वफलस्तनी।
प्रफुल्लकुन्ददशना मञ्जरीकर शोभिता।।
बन्धुजीवाधरा शुभ्रा सिन्दुवारजखाद्भता।
पुंस्कोकिलस्वना दिव्या अङ्कोलवसनाशुभा।।
बर्हि वृन्द कलापा च सारस्वर नूपुरा।
प्राग्वंशरसना ब्रह्मन् मत्तहंस गतिस्तथा।।
पुत्र जीवांशु का मृङ्गरोम राजिविराजिता।
वसन्तलक्ष्मीः संप्राप्ता ब्रह्मन् बद्रिकाश्रमे।।

—*Vāmana Purāṇa*, 6.17-21

(O Devarṣi, O Brāhmaṇa, the young lady Vasanta Lakṣmī appeared in the Badrikāśrama, whose hands resembled the red *aśoka* flowers, the feet were like *palāśa* the hair like blue-*aśoka*, face like the blossomed lotus, the eyes like the blue-lotus, the breasts like the fruit of *bilva*, the teeth like the blossomed *kunda* flowers, the arms like creeper, the lips like bandhujiva (*Pentapetas phoenicea*, Linn), the nails like *Sindhuraraka* (*Vitex negundo*, Linn), the voice like cuckoo, the garments like *aṅkolā*, the ornaments like peacock, the *nupuras* resembling the *Sārasa* bird, etc.)

Gradually the flowers became an essential part of not only the human decoration but also an integral part of the worship of different deities, so much so, that no religious ceremony was complete without them. Even the practice of placing floral wreaths on the dead bodies came in vogue (which continues even during the modern times). Indeed the ancient literature of the country has dealt with the flowers variously, the glimpses of which are available elsewhere in this work.

# 17
# Poisonous Trees and Plants

The Indian traditions based on the ancient literature testifies that the Venom or poison has the same source of origin as the *amṛta*, since both of them emerged out of the churning of the ocean by the demons and gods. This venom Kālakūṭa, according to the Purāṇic account was consumed by Lord Śiva at the request of the *Devas* and *Asuras*. Since the deadly venom could cause instant death of the consumer, had it reached his stomach; it was retained by Śiva, at the instance of Pārvatī in the throat itself, which earned him the epithet of Nīlakaṇṭha or Viṣakaṇṭha, because the poison caused the colour of the neck of Śiva to turn into blue shade.

Of course, the venom or poison is found mentioned in the *Ṛgveda* as *Viṣa*,[1] followed by the *Atharvaveda*.[2] In the later texts, however, there appears a word *Viṣāṅka*[3] which stands for poisonous (?) plants which could be used as medicine for a certain disease. The *Āśvālayana Śrauta Sūtra* (X.75), however, testifies that there were people who were well versed in *Viṣavidyā* or the means to remove the ill effects of the poison. While *Viṣa-puṣpa* stood for blue lotus, the word *Viṣapuṣpaka* was believed to mean a disease caused by the consumption of poisonous flowers. Then *Viśavallī* has been mentioned the later literature,[4] which denotes a poisonous creeper. Interestingly enough, the literary genius of the ancient Sanskrit literature also introduced *Viṣa-Kanyā*[5] or a Venomous-girl, who was supposed to cause the death of the man who had a chance of having an intercourse with her, but in the *Kumārasaṁbhava*,[6] the *Viṣvṛkṣa* or the poisonous tree has been clearly mentioned—

विषवृक्षोऽपि सम्बर्ध्य स्वयं छेत्तुमसाम्प्रतम्।

Other texts do mention about *Viṣakanda*, which has been interpreted to mean a species of bulbous plant and *Viṣakāṣṭha* which has been identified to be *Thespesia populnea* L.

There are certain plants which have been termed in the medicinal treatises including *Suśruta* antidote to poison of *Viṣāghna*, some of which are noted hereunder—

(i) *Mimosa Sirisa*

(ii) *Hedy serum*, Alhagi

(iii) *Belleric myrobalan*

(iv) *Terminalia Bellerica*

The *Aṣṭāṅgahṛdaya* (Sūtrasthāna), however, in its Chapter-7 declares that the unripe fruits coming in contact with poison, ripen early and the ripe fruits in similar circumstances become gangrenous and sour—

कालीमध्याभसोः क्षौद्रे हरितैलेऽरुणोपमा।
पाकः फल नामानां पक्वानां परिकोथनम्।।

—(7.8)

Then there is a mention of the floral garlands having venomous effect. The same text therefore says that when floral garland come in contact with venom, their petals fade out and lose fragrance—

माल्यस्यस्फुटताग्रत्वं म्लानिर्गंधांतरोद्भवः।
ध्यामण्डलता वस्त्रे शदनं ततुं पक्ष्मणाम्।।

—*Ibid.*, (7.10)

George Watt,[7] however, provided a detailed list of Indian poisonous plants, which appear in appendix of this work.

## Notes and References

1. *Ṛgveda*, I.117.16; I.191.11; IV.61.3; X.87.18.
2. *Atharvaveda*, IV.6.2; V.19.10; VI.90.2.
3. *Ibid.*, VI.44.3.
4. *Kathāsaritsāgara*.
5. The works like *Mudrārākṣasa* and *Kathāsaritsāgara* could be mentioned in this connection.
6. *Kumārasaṁbhava*, 2.55.
7. Walt, George, *op. cit.*, Vol. VI, Part-I, pp. 310-11.

# 18
# Epilogue

Ever since the dawn of civilization in India, the Indian culture had been influenced by the environments of forests and plant kingdom in more than one ways. The early man lived in the forest surroundings, looked upon trees as the place for shelter, food and even for clothing himself. With the advancement of the society, the man came to know more about the trees and plant kingdom. His knowledge of the plant-life was actuallly based on his personal observations and practical experience he daily used to have with the nature in wilderness. With the passage of time there appeared on the Indian social and religious scene some ascetics who not only preached on the importance the plant-life but also adoring them, besides emphasising on the healing properties inherent in some of the trees and plants. This could be the reason, that the earliest specimens of Indian sculptural art of Bharhut, Sanchi, Amarāvatī, Bodhagayā and Mathura display such motifs.

The various aspects of modern botanical science like (i) Plant Morphology, (ii) Plant Physiology, (iii) Taxonomy and (iv) Ecology, were quite well known to the ancient seers, who always tried to keep the ecological balance. It is recorded in the *Rāmāyaṇa* that the Sage Viśvāmitra went to ask for the help of Rāma and Lakṣmaṇa from the king Daśaratha, in order to destroy the *Rākṣasas,* who were bent on destroying the ecological balance of his *āśrama* and the forests. Indeed the destruction of various *Rākṣasas* by Rāma in the forests of Daṇḍakāraṇya was aimed at ridding the forests of their unnatural habitants, which proved to be hazardous for the plant-life, besides the humans.

The various properties of plant-life were quite well known to the Vedic people. These early texts testify to the plants possessing healing properties and removal of ailments; besides, they could also be used for charms to allay jealousy, for arousing love and passion, in addition to several other usages.

The *Saṁhitās,* the *Brāhmaṇas,* the *Āraṇyakas* and *Upaniṣads,* necessarily speak of the plant-life in one way or the other. Some of these texts symbolise them with the universe or Oṁ, besides describing the seed; the birth of trees; trees serving as resting place for birds, and their usages as fodder for animals. These texts also deal extensively with the various species of grass, the worms, the insects and the remedies to get rid of the worms which are injurious for the plant-life, the humans and the animals. The *Gṛhya-Sūtra* literature also continues to emphasise the usages of trees and plants in many ways. The Epics and the Purāṇas have been more vocal about the plant-life and forests and the observations made therein are apparently based on the personal knowledge of the heroes and actors. As for example the king Daśaratha exiled Rāma for twelve years and whatever information on the subject is revealed in the Araṇyakāṇḍa, Sundarakāṇḍa and Laṅkākāṇḍa in the *Rāmāyaṇa,* is supposed to be based on the personal knowledge of Rāma and Lakṣmaṇa.

Another salient feature of the Epics has been that the great personalities have been equated with the huge trees. For instance, Yudhiṣṭhira and Duryodhana have been equated with huge trees in the *Mahābhārata*, while Rāma and Rāvaṇa have been symbolised with huge trees in the *Rāmāyaṇa*. The Ādiparvan of the *Mahābhārata* equates the epic with a huge tree.

The Buddhist and Jaina literature also did not lag behind in the treatment of plant life, wherein the Buddha and the Jaina Tīrthaṅkaras were closely associated with trees and groves.

The ancient texts also speak of the Damsel and tree motifs, wherein it is believed that some trees blossom with the touch of a female. This aspect is abundantly shown in the sculptural art of Bharhut and Sanchi. Though it may not be possible to provide a scientific support on the subject, but it goes without saying that the motif had its own importance in the contemporary society.

An astounding aspect of the Vedic literature has been the use of channels possibly of bamboo for the purpose of irrigating the fields, which is quite revealing in comparison to the present day system of irrigation. The Vedic texts also point to the use of cowdung manure, ploughing the fields and the farm labour, besides the implements used in the field work.

The ancient texts also speak of the forest deities, like Vanaspati, Śākambharī, Araṇyāṇī, Vanadurgā, Vana-Śaṅkarī, Vindhyavāsinī, Kātyāyanī and Pārvatī.

During the modern times one daily hears of the planting of trees by high dignitaries and organising *Vanamahotsavas* for augmenting the planting of trees by the Government agencies. But this aspect was duly projected in various Purāṇic texts, which pronounce earning of high merits by those who plant the trees—

भूदानेन ये लोका गोदानेन च कीर्तिता:।
ते लोका प्राप्यते पुंभि: पादपानां प्ररोपणे।।
अश्वत्थमेकं पिचुमन्दमेकं न्यग्रोधमेकं दशपुष्पनाति।
द्वे द्वे तथा दाडिममातुलिङ्गेपञ्चाम्रवापि नरकं न याति।।
यथा सुपुत्रः कुलमुद्धेद्धियाति कृत्स्नान्नियमप्रयत्नात्।
तथा प्रारोपाः फलपुष्पमूला स्वं स्वामिनं नरकादुद्धरन्ति।।

—*Varāha Purāṇa*, 170.35-37

(One who plants the tree, acquires the merit equivalent to that of donating land or a cow. One who plants an Aśvattha—*Ficus religiosa*, a Vaṭa—*Ficus benghalensis*, a Nīma tree—*Melia azadirachta* L., ten flower plants, two each of Dāḍima—*Pomegranate* and *Mātuliṅga*—Orange and ten mango trees; and constructs a step-well, never goes to hell. As the son liberates the dynasty, so does a tree planted with hard labour, and bearing fruits liberates its master from hell).

*Appendix-1*

# Plan of Mathura Showing Ancient Gardens as Drawn by Cunningham

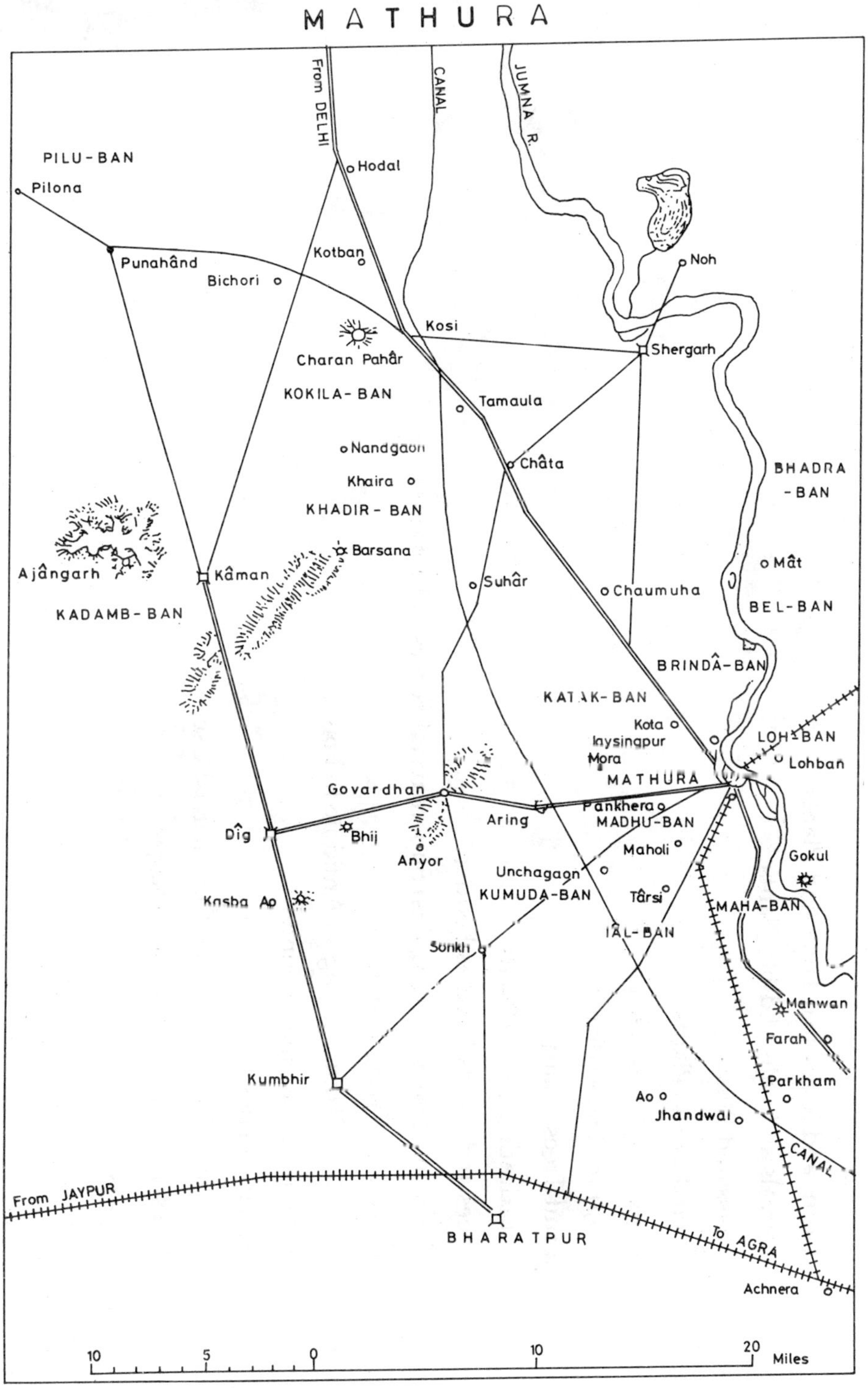


A. Cunningham. del.

*Appendix-2*

# Trees and Plants of Vedic and Post-Vedic Literature

(i) Herbs and Medicinal Trees and Plants
(ii) Cereals and Other Edible Plants
(iii) Grass and Reeds
(iv) Sacred Trees/Plants
(v) Flowering Trees/Plants
(vi) Fragrant Trees/Plants
(vii) Fruit Trees/Plants
(viii) Intoxicating Trees/Plants
(ix) Trees and Plants Used as Fodder
(x) Trees and Plants Used against Supernatural Forces or Demons
(xi) Poisonous Trees/Plants
(xii) Trees and Plants Serving as Antidote to Poison
(xiii) Trees of Cremation Grounds
(xiv) Aquatic Plants
(xv) Trees and Plants Meant for Commercial Use
(xvi) Trees and Plants of General Usages
(xvii) Forests
(xviii) Seeds

## (i) Herbs and Medicinal Trees and Plants

| Sl. No. | Saṁskṛt Name | Botanical Equivalent | Vedas | Saṁhitās | Brāhmaṇas | Āraṇyakas | Upaniṣads | Gṛhya-Sūtras Śrauta-Sūtras and Nirukta | Remarks |
|---|---|---|---|---|---|---|---|---|---|
| 1. | Adara | *Zingiber officinalis* | *A.V.* IV.35.5. | | | | | | Ginger has long been known both to Saṁskṛt and Mohammadan medicinal practitioners. The former describe it as an acrid, heating carminative and useful in several diseases. |
| 2. | Adhyaṇḍa | *Macaria pruriens* | | | *S.B.* XIII.8.1.16 | | | | |
| 3. | Ajaśṛṅgī | *Odina wodiar* | *A.V.* IV.37 | | | | | | The bark is very astringent, and, not although officinal, is described in the Pharmacopoeia of India. |
| 4. | Akṣa | *Terminalia bellerica* | | | | | *Chāndogya* 7.3.1 | | Seems to denote the nut of Vibhitaka. This is a large deciduous tree having medicinal properties. It produces gum which is soluble in water and burns away in the flames of a candle. |
| 5. | Āmūlā (without roots) | *Methonica Superba*, Linn. | *A.V.* V.31.4 | | | | | | The root is supposed by physicians to have valuable medicinal properties. It was used for poisoning the arrows. It constituted one of the seven poisons of Saṁskṛt |

| *Sl. No.* | *Saṁskṛt Name* | *Botanical Equivalent* | *Vedas* | *Saṁhitās* | *Brāhmaṇas* | *Āraṇyakas* | *Upaniṣads* | *Gṛhya-Sūtras Śrauta-Sūtras and Nirukta* | *Remarks* |
|---|---|---|---|---|---|---|---|---|---|
| | | | | | | | | | writers and could also cause abortion. |
| 6. | Apāmārgā | *Achyranthes aspera*, Linn. | *A.V.* IV. 7.16; IV. 18.7; IV. 19.4; VII. 6.2. | Vājasaneyī XXXV.2 | *S.B.* V.2.4.14; XIII.8.4.4; *Taittirīya*: I.7; I.8. | | | | In the Vedic literature its use is frequently mentioned in which craft practices and medicinal purposes especially *Kṣetriya*. It is described in the *A.V.* as having reverted leaves. In fact the whole plant has astringent and diuretic properties assigned to it. The leaves and the seeds relieve the pain of scorpion-bite. The juice of the leaves is used in snake bites. The dried leaves are smoked for relief in asthma. It is used for virility as well. |
| 7. | Aparājitā | *Clitoria ternatea*, Linn. | *A.V.* II.27.3. | | | | | | It is considered to be conducive to prosperity and long life. The root is a powerful cathartic like *jalap* and has been recommended to be used along with other laxatives diuretics in ascites and enlargements of abdominal viscera. |

| Sl. No. | Saṁskṛt Name | Botanical Equivalent | Vedas | Saṁhitās | Brāhmaṇas | Āraṇyakas | Upaniṣads | Gṛhya-Sūtras Śrauta-Sūtras and Nirukta | Remarks |
|---|---|---|---|---|---|---|---|---|---|
| 8. | Arātakī | *Accacia suma* | *A.V.* *IV. 37.6* | | | | | | This has also been identified with Ajaśṛṅgi. |
| 9. | Araka | *Gardenia enneandra* | *A.V.* VI.72.1 | | *S.B.* IX.1.14; IX. 1.1.9; IV. 61.14. | | | | Almost every part of this plant is used in medicines, but the bark of the root and dry milky juice are by far the best. |
| 10. | Arundhatī | *Sida cordifolia*, Linn. | *A.V.* IV.2.1; V.5.5; V.5.9; VI.59.1; VI.59.2; VII.7.16; XIX. 38.1. | | | | | | A medicinal climbing plant mentioned in several passages of the *A.V.* as possessing healing properties in case of wounds; as a fodder it induces to yield more milk. The creeper usually climbs Aśvattha, Nyagrodha and Parna trees. It is stated to be of golden colour (*hiraṇya-varṇa*) with a hairy stem. It is also called Silāca and the Lākṣā appears to be its product. |
| 11. | Aśvagandhā (Rock-smell) | *Withania coagulans*, Dunal. | | | *S.B.* XIII.8.1.16. | | | | This is also quite common in the Kātyāyana's *Śrauta Sūtra* as well as Śuśruta. The roots of the shrub have the medicinal properties. |

| *Sl. No.* | *Saṁskṛt Name* | *Botanical Equivalent* | *Vedas* | *Saṁhitās* | *Brāhmaṇas* | *Āraṇyakas* | *Upaniṣads* | *Gṛhya-Sūtras Śrauta-Sūtras and Nirukta* | *Remarks* |
|---|---|---|---|---|---|---|---|---|---|
| 12. | Aśvattha (Pīpal) | *Ficus religiosa* Linn. | *R.V.* I.164.20; X.97.5. *A.V.* III.6.1; IV.32.4; VI. 11.1; V. 4.3. | | *S.B.* XI.5.1.13. *Ait. Br.* VII.35.6. | | *Śvetāśvatāra:* IV.6.22; V.54.12. *Maṇḍūka:* III.1.1. *Bṛhadā-rṇyaka*: IV.1.41. | | The bark is astringent and is used in gonorrhoea. It has also nutritive properties. The root-bark is one of the five barks used by Saṁskṛt physicians. The fruit is laxative and helps digestion. |
| 13. | Aśvavaṭī | unidentified | *R.V.* X.97.7 | | | | | | An unidentified medicinal plant. |
| 14. | Baja | *Brassica Sp.* (A herb used as charm against evil spirits) | *A.V.* VIII.6.3; VIII.6.7; VIII.6.24. | | | | | | The seeds of this plant are used in medicines as poultice being a useful and simple rubefacient and vasicant. |
| 15. | Citraparṇī | unidentified | *A.V.* II. 25.3. | | | | | | The spotted leaves are used as protection of foetus. |
| 16. | Eraṇḍa or Tajaddhaṅga | *Ricinus communis* L. | *A.V.* VIII.3.8. | | | *Śānkhā-yama:* XII.8. | | | The plant is cultivated for both economical and medicinal purposes throughout the country. |
| 17. | Haridra | *Curcuma domestica,* Valeton. | *A.V.* I.24.2. | | | | | | This is used as a stimulant in medicines and is also externally applied in pains, bruises. |
| 18. | Jaṅgidā | *Terminalia arjuna* | *A.V.* II.9.1; | | | | | | This was used as remedy against all sorts |

| Sl. No. | Saṁskṛt Name | Botanical Equivalent | Vedas | Saṁhitās | Brāhmaṇas | Āraṇyakas | Upaniṣads | Gṛhya-Sūtras Śrauta-Sūtras and Nirukta | Remarks |
|---|---|---|---|---|---|---|---|---|---|
| | | | II.4.5; XIX.34.35; XIX.39.10. | | | | | | of diseases. This is the name of a healing plant and was also used for making amulets against diseases like fevers and rheumatic pains. Its bark was considered by ancient scholars to be tonic, astringent, cooling and was used in heart diseases, fractures and ulcers etc. |
| 19. | Karīra | *Capparis aphylla* or *Canthium decidua*, Edgew. | | *Taitterīya* II.4.9.2. *Kaṭhaka* XI.11; XXXVI.7. | | | | | The tender shoots and young leaves are made into powder which is used as a blister in indigenous system of medicines. Also relieves toothache when chewed. |
| 20. | Karakaṇḍa or Karakaṇḍū | *Zizyphus jujuba* | *R.V.* I.112.6. | *Kaṭhaka:* XII.10 *Maitrāyaṇī* III.11.2. *Vājasaneyī* XIX.23.91 XXI.32.1 XXIV.2.1 | *S.B.* V.5.4.10; XII.7.2.9; XII.9.1.5. *Jaiminīya :* II.152.5 | | | | In the *Bhāva-prakāśa* three different kinds of *ber* are said to be described viz. *Suvira*, *Kola* and *Karkaṇḍa*. The first is considered in the indigeneous medicines to be cooling aperient, astringent and nourishing and to be indicated in billious affections, fever, consumption and thirst. The second is descri- |

| *Sl. No.* | *Saṁskṛt Name* | *Botanical Equivalent* | *Vedas* | *Saṁhitās* | *Brāhmaṇas* | *Āraṇyakas* | *Upaniṣads* | *Gṛhya-Sūtras Śrauta-Sūtras and Nirukta* | *Remarks* |
|---|---|---|---|---|---|---|---|---|---|
| | | | | | | | | | bed as sweet and is good for constipation. The third form is inferior in quality than the first two. |
| 21. | Kārśmarya or Kasamarya-phala | *Gmelina arborea*, Roxb. | | *Taittirīya* V.2.7.3; VI.2.1.5. *Maitrāyaṇī* III. 2.6; III. 7.9. | *S.B.* III.4.1.6; III.8.2.17; IV. 3.3.6. | | | | The roots are used as medicine which are bitter, tonic, stomachic, laxative and useful in fever, indigestion and various other diseases. |
| 22. | Kuṣṭha | *Costus speciosus* | *A.V.* V.2.8; V.4.1; V.4.10; VI.102.3; XIX.39.1; XIX.39.9. | | | | | | This is a creeper which grows with Soma on the high peaks of the Himalayas where the eagles nest, and was then brought for men. As a remedy it held the highest place among herbs and was called by the auspicious names viz., Nāgamāra, Nagariṣa. It is useful to cure headache, bodily afflicting fever and eye ailments. It is called as *taktanāsa* i.e., destroyer of fever and also *Viśvabeṣaja* or all healing. Its aromatic qualities were apparently known as it is classed with *Āñjana* (Salve) Naladu. |

| *Sl. No.* | *Saṁskṛt Name* | *Botanical Equivalent* | *Vedas* | *Saṁhitās* | *Brāhmaṇas* | *Āraṇyakas* | *Upaniṣads* | *Gṛhya-Sūtras Śrauta-Sūtras and Nirukta* | *Remarks* |
|---|---|---|---|---|---|---|---|---|---|
| 23. | Lakṣmaṇa or Lakṣmā | *Ardea sibirica* | *A.V.* I.25.3. | | | | | | A plant having leaves with red spots and is used against abortion. |
| 24. | Madhugha or Madhuga | Honey plant | *R.V.* VI.70.1; VI.70.5. *A.V.* I.34.4. | | | | | | This is sweet-herbs plant. |
| 25. | Naktamālā | *Pongamia globra* | R.*V.* 42 | | | | | | The seeds are useful in skin deseases; the oil from seeds is useful in rheumatism and leaves are useful in ulcers. |
| 26. | Nyaṣṭikā | *Andropogan aciculatus* | *A.V.* VI.139.1. | | | | | | This has been identified as *Śaṁkhapuṣpikā.* |
| 27. | Oṣadhi | Medicinal Plants (in general) | *R.V.* I.32.5; X.97. *A.V.* XII.1.2; X.7.38. | *Taittirīya* II.5.3.2; VII. 3.19.1. *Vajasneyi* XXI.8. | | | *Chāndogya :* 1.1.2. *Bṛhadā-raṇyaka :* 1.1.1; 3.2.13; 6.3.6; 6.4.1; 6.4.5. *Mahānā-rāyaṇīya:* 9.8; 17.7; 1.4; 4.13. *Śvetāsvatāra:* 2.17. *Taittirīya :* 1.7.1; 2.1.1. *Maṇḍūkā:* 1.1.7; 2.1.5; 2.1 9. | | This term was used in general for medicinal plants. Roughly speaking, the vegetable world is divided in Vedic literature between Oṣadhi or Vi-rudha plants and Vana or Vṛkṣa-trees. Oṣadhi is employed in opposition to Virudha to denote plants as possessing healing power or some other quality, useful to men. |

| *Sl. No.* | *Saṁskṛt Name* | *Botanical Equivalent* | *Vedas* | *Saṁhitās* | *Brāhmaṇas* | *Āraṇyakas* | *Upaniṣads* | *Gṛhya-Sūtras Śrauta-Sūtras and Nirukta* | *Remarks* |
|---|---|---|---|---|---|---|---|---|---|
| | | | | | | | *Nṛsimha-tāpiṇī* : 1.4. *Nīlarudra* 21. *Siras*: 6 | | |
| 28. | *Oṣadhi-Vanaspati* | Green medicinal Plants or Herbs | | | *Aitreya:* 1.4; 2.4. | | *Chāndogya:* 5.10.6. *Mahānā-rāyaṇi:* 19.2; 23.1. *Āśrama:* 3. | | The terminology was possibly used for green plants or greenery having medicinal properties. |
| 29. | Pīlu | *Careya arborea* | *A.V.* XX.135. 12. | | | | | | The bark is used as an astringent medicine; when moistered, it gives much mucilage, and is on this account used in peparation of emolliant embrocation. The bark is applied to wounds in Snake-bite. |
| 30. | Pippalī-mūla | *Piper longinum* | *A.V.* VI.107.4. | | | | | | The dried unripe fruit and the root have been long used in medicines. They are considered treating, stimulant, carminative, alterative, laxative and useful in cough, asthama and other diseases. |
| 31. | Pītadāru or Putudru | *Pinus longifolia* | *A.V.* VIII.2.28 | *Kaṭhaka:* XXV.6 *Vājasneyī:* V.14. | *S.B.* III.5.2.15; III.4.4.5; III.4.4.17. | | | | The aromatic wood and the oliorasin are spoken as medicinal. The wood is considered as stimu- |

| Sl. No. | Saṁskṛt Name | Botanical Equivalent | Vedas | Saṁhitās | Brāhmaṇas | Āraṇyakas | Upaniṣads | Gṛhya-Sūtras Śrauta-Sūtras and Nirukta | Remarks |
|---|---|---|---|---|---|---|---|---|---|
| | | | | *Taittirīya:* IV.2 8.4. *Maitrāyaṇi* III.8 5. | *Pañcaviṁśa* XIV.5. *Aitareya:* I.28. | | | | lant, diaphroetic and "useful in burning of the body" cough, fainting and ulcer. |
| 32. | Putikā | *Caesalpina fonducella* | | *Kaṭhaka:* XXXIV.3. | *S.B.* XIV.1.2.12. | *P. Ar.* VIII.4.1; IX.5.3. | | | This is mentioned as substitute of Soma plant and means of making milk curdle. |
| 33. | Putrakoṇḍa | *Uraria longipoides* | *A.V.* II.25.3. | | | | | | The plant is an ingredient of *Daśamūla* and is often used in medicines. It is considered as alterative, tonic and anticatarrhal. It is also used as remedy against abortion. |
| 34. | Sahadevī or Arundhatī | *Vernonia cinerea* Less. | *A.V.* VI.59.2. | | | | | | The same as Arundhatī plant. |
| 35. | Śaṁkhapuṣpī | *Causcora decussata,* R&S. | *A.V.* VII.38.5 | | | | | | The plant is regarded as laxative, alterative and tonic and is used in diseases like insanity, epilepsy and nervous debility. |
| 36. | Sandadruva | *Cyperus rotundus* | *A.V.* XVIII.3.6. | | | | | | The roots are used medicinally as a diaphoratea and astringent. It contains diuratic and stimulant properties. |

| *Sl. No.* | *Saṁskṛt Name* | *Botanical Equivalent* | *Vedas* | *Saṁhitās* | *Brāhmaṇas* | *Āraṇyakas* | *Upaniṣads* | *Gṛhya-Sūtras Śrauta-Sūtras and Nirukta* | *Remarks* |
|---|---|---|---|---|---|---|---|---|---|
| 37. | Śatāvarī | *Asparagus racemosus,* Willd. | *A.V.* XIX.36. | | | | | | The root of this plant is useful in various ailments and prevents confluence of small-pox. |
| 38. | Silaci/Lākṣā | Unidentified | *A.V.* V.5.1.8. | | | | | | This is said to be a healing plant. |
| 39. | Siṁsapā | *Dalbergia sisu* | *R.V.* III.53.19. *A.V.* XX.129.7. | | | | | | This is a large deciduous tree, the wood of which yields medicinal oil and its leaves are used for recovery of various ailments. |
| 40. | Somavatī or Somavati | *Asclepias acida,* Roxb. | *R.V.* X.97.7. | | | | | | A major medicinal plant. |
| 41. | Talīśa | *Flacourtia cataphracta* | *A.V.* IV.15.3. | | | | | | Its leaves and tender shoots possess astringent and stomachic properties and are useful in diarrhoea, weakness and consumption. |
| 42. | Tilvaka | *Symplocos racemosus* | | *Maitrāyaṇī* III.1.9 | *S.B.* XIII.8.1.16. *Sadaviṁśa:* III.8; VI.61.3. | | | | Its bark is considered to be cooling, astringent and useful in bowel complaints, eye diseases and ulcer. |
| 43. | Upavaka | *Wrightia tinctoria* | | *Vājasneyi:* XIX.22; XIX.90; XXI.30. | *S.B.* XII.7.1.3. | | | | The root-bark of the plant and the seeds are used for medicinal purposes. |

| Sl. No. | Saṁskṛt Name | Botanical Equivalent | Vedas | Saṁhitās | Brāhmaṇas | Āraṇyakas | Upaniṣads | Gṛhya-Sūtras Śrauta-Sūtras and Nirukta | Remarks |
|---|---|---|---|---|---|---|---|---|---|
| 44. | Ūrjayauli | | *R.V.* X.97.7. | | | | | | An important medicinal plant. |
| 45. | Vanaspati (Lord of Forest) | Green trees and plants in general | *R.V.* I.35.5; 66.5; 188.16, II.3.9; III.4.10; 8.1; 8.3; 34.10; V.5.10; X.61.1; 61.12; 110.10. *A.V.* XII.1.8; X.7.38; IX.6.1; IX. 9.24. | *Taitteriya:* II.5 3.2; III.20 1; VII 3.19.1. *Vājasneyi:* XII 28. | *S.B.* IV.1.1.12. | | *Atharva-śira:* 6 *Bṛhadā-raṇyaka:* I.1.1.; III.2.13; 9.28; IV.3.6. *Chāndo-gya:* VII.2.1; VII.8.1. *Mahānā-rāyaṇi:* IX.10; XVII.7. *Maitrā-yaṇi:* 1.4 *Nīlarudra* 20. *Śvetāśva-tāra:* II.17. *Taittarīya* I.7.1. | | The king of wood or the forest tree, especially a large tree bearing fruit apparently without blossom, as several species of fig and the jack tree etc. The term applies to any tree and even the medicinal tree or plant. |
| 46. | Varuṇa | *Crateava roxburgii* | *A.V.* VI.85.1; X.3.1; XIX.32.9. | | *Pancaviṁśa* V.3.9.10. *S.B.:* XII.8.4.1. | | | | The bark of the stem and roots of this plant have medicinal properties. It is said to promote appetite, increase the secretion of bile, act as laxative and removes urinary disorders. |

| Sl. No. | Saṁskṛt Name | Botanical Equivalent | Vedas | Saṁhitās | Brāhmaṇas | Āraṇyakas | Upaniṣads | Gṛhya-Sūtras Śrauta-Sūtras and Nirukta | Remarks |
|---|---|---|---|---|---|---|---|---|---|
| 47. | Virāṇa (Viriṇa) | *Andropogon muricatus* | | | *S.B.* : XIII.8.1.15. | | | | An infusion of the roots is given as a fabrifuge and a powder in bilbous complaints. *Khaskhas* is regarded as stimulent diaphoratic, and stomachic. The essence is used as a tonic. |
| 48. | Virudha | A medicinal plant. | *R.V.* I.67.9; 141.4. | | | | | | It denotes an inferior species of plants but practically has the same sense as *Oṣadhi.* |
| 49. | Viṣṇaka | *Gymnema sylvestra* | *R.V.* VI.16.20; XI.96.8; *A.V.* VI.44.4. | | | | | | The root is considered as a remedy against the snake bite; the powdered root can be applied to the part bitten by snake and a decoction can be administered internally. |
| 50. | Yaṣthimadhu | *Glycyrrhiza glabra* | *A.V.* I.34.5. | | | | | | Liquoricea root is used for medicine. Which is known as sweet, demulcent, cooling and useful in inflammatory infections, cough, hoarseness and thirst. It is also used for virility and sexual pleasure. |

## (ii) Cereals and Other Edible Plants

| *Sl. No.* | *Saṁskṛt Name* | *Botanical Equivalent* | *Vedas* | *Saṁhitās* | *Brāhmaṇas* | *Āraṇyakas* | *Upaniṣads* | *Gṛhya-Sūtras Śrauta-Sūtras and Nirukta* | *Remarks* |
|---|---|---|---|---|---|---|---|---|---|
| 51. | Āla | Weed of corn *Fibraurea benghalensis,* Linn. | *A.V.* VI.16.3. | | | | | | Weed of corn. |
| 52. | Alaśāle | Green creeper | *A.V.* VI.16.4. | | | | | | |
| 53. | Aṇu | *Panicum miliaceum* | | *Vājasneyī:* XVIII 12. | | | *Bṛhadāraṇyaka :* VI.3.13. | | |
| 54. | Āmba | *Mangifera indica,* Linn. | | *Taittirīya:* I.8.10.1. *Kaṭhaka:* XV.5 | *S.B.* V.3.3.8. | | | | |
| 55. | Aṇḍikā | *Nymphaea alba* | *A.V.* IV.35.5; 35.17; 35.6. | | | | | | This is an edible plant leaf and its root is egg shaped. The root and seeds are eaten in times of scracity. |
| 56. | Avaka | *Blyxa octandra,* Rich. | *A.V.* VIII.7.9; 37. 8-10. | *Taittirīya:* VI.6.1.4. *Vājasneyī:* XXV.1. *Maitrāyaṇī* II.10.1. | *S.B.* VII.5.1.11; VIII.3.2.5; IX.1.2.20-22; XIII.8.3.13. | | | | This represents ordinary gram and is cultivated throughout the country. |
| 57. | Caṇaka | *Cicer arietinum,* Linn. | *A.V.* VI.139.3. | *Vājasneyī:* XVIII.1.2. | | | | | |
| 58. | Dhān | Grains of Corn. | *A.V.* IV.25.7; *R.V.* I.144.1; *II.3.2;* III.55 10. | *Vājasneyī:* IV 34. | | *Taittirīya:* II.7.4. | | | |

| Sl. No. | Saṁskṛt Name | Botanical Equivalent | Vedas | Saṁhitās | Brāhmaṇas | Āraṇyakas | Upaniṣads | Gṛhya-Sūtras Śrauta-Sūtras and Nirukta | Remarks |
|---|---|---|---|---|---|---|---|---|---|
| 59. | Gavadhuka | *Coix barbata* | | *Taittirīya:* V.4.3.2; I.8.7.1. *Maitrāyaṇī* II.6.5; IV.3.8. *Vājasneyī:* XV.5. | *S.B.* V.2.4.13; XIV.1.2.19. *Taittirīya :* I.7.3.6. | | | | |
| 60. | Garmut | Wild beans | | *Maitrāyaṇī* II.2.4. *Taittirīya:* II.4.4.12. | | | | | This also comes under pulses and beans. |
| 61. | Godhūma | *Tristania aestivum,* Linn. | | *Maitrāyaṇī* I.2.8. *Vājasneyī:* XVIII.12 | *S.B.* V.2.1.6. *Taittirīya:* I.3.7.2. | | *Bṛhadāraṇy-aka:* VI.3.22; XII.7.12-29. | | The wheat is commonly used in the entire length and breadth of India. |
| 62. | Khalkula | *Delichos biflorus* | | | | | *Bṛhadā-raṇyaka:* VI.3.22. | | This comes under pulses. |
| 63. | Khalva | *Phaseolus radiatus* | *A.V.* II.1; V.23.8. | *Vājasneyī:* XVIII.12. | | | *Bṛhadā-raṇyaka:* VI.3.32. | | This is one of the various varieties of pulses. |
| 64. | Kṣumpā | Mushroom | *R.V.* I.84.8. | | | | | *Nirukta:* 5.16. | This is used for eating purposes. |
| 65. | Kulmāśa | *Phaseolus mungo* | | | | | *Chāndo-gya:* I.10.2.1. | *Nirukta:* 1.4. | This is one of the varieties of pulses. |
| 66. | Māśaparṇī | *Teramonus labialis* | *A.V.* II.25.3. | | | | | | A species of pulses. |
| 67. | Masūra | *Lens esculenta* | | *Vājasneyī:* XVIII.12. | | | | | A species of pulses. |

| Sl. No. | Saṁskṛt Name | Botanical Equivalent | Vedas | Saṁhitās | Brāhmaṇas | Āraṇyakas | Upaniṣads | Gṛhya-Sūtras Śrauta-Sūtras and Nirukta | Remarks |
|---|---|---|---|---|---|---|---|---|---|
| 68. | Masuiya | | | | *Taittirīya:* III.8.14.6. | | | | This is a kind of grain. |
| 69. | Mudga | *Phaseolus mungo*, L. | | *Vājasaneyī:* XVIII.12. | | *Sankhayana:* XII.8. | | | It occurs in the list of vegetables. |
| 70. | Nāmba | A kind of grain. | | | *S.B.* V.3.3.8. | | | | |
| 71. | Nīlānjalaśala | Grain creeper | *A.V.* VI.16.4. | | | | | | |
| 72. | Nivara | A variety of paddy. | | *Kaṭhaka:* XII.4. *Maitrāyaṇī* II.4.10. *Vājasneyī:* XVIII.12. | | | | | This could represent the wild rice. |
| 73. | Nyaṣṭikā | *Andropogan aciculatus* | *A.V.* VI.139.1. | | | | | | |
| 74. | Pākadūrva | Edible millet. | *R.V.* X.16.3; *A.V.* XVIII.3.6. | | | *Taittirīya:* VI.4.1.2. | | | This is believed to be the plant usually found in the cremation grounds. It has egg-shaped roots. |
| 75. | Parṣu | Sheaf | *R.V.* X.48.7. | | *S.B.* VIII.4.2.5. | | | | |
| 76. | Plāśuka | Fast growing rice | | | *S.B.* V.3.3.2. | | | | |
| 77. | Priyaṅgu | *Setaria italica* | | *Taittirīya:* II.2.11, 4. *Maitrāyaṇī* II.1.8. | *Aitreya :* VIII.16. | | *Bṛhadāraṇyaka:* VI.3.22. | | |

| *Sl. No.* | *Saṁskṛt Name* | *Botanical Equivalent* | *Vedas* | *Saṁhitās* | *Brāhmaṇas* | *Āraṇyakas* | *Upaniṣads* | *Gṛhya-Sūtras Śrauta-Sūtras and Nirukta* | *Remarks* |
|---|---|---|---|---|---|---|---|---|---|
| | | | | *Vājasneyi :* XII.18. | | | | | |
| 78. | Saluṅjalā or Silanjalā | A grain creeper | *A.V.* VI.16.4. | | | | | *Kauśika Sūtra* II.16. | |
| 79. | Sophaka | Aquatic plant | *A.V.* IV.34.5. | | | | | *Āpastambha Śraut Sūtra* IX.14.14. | In the *A.S.S.* this has been mentioned as an edible plant. |
| 80. | Saphaka | *Trapa bispinosa* Roxb. | *A.V.* IV.39.5. | | | | | | |
| 81. | Silañjala | | | | | | | | Same as No. 78. |
| 82. | Śyāmaka | *Echinochloa colona* | *A.V.* XIX.50.2. | *Taittirīya:* I.8.1, 2; *Kaṭhaka:* X.2. *Maitrāyaṇī:* II.11.4; *Vājasneyi :* XVIII.12. | | *S.B.* X.6.3.2. *Kauśītaki:* IV.12. | | | Food for pigeons. |
| 83. | Saktu | Groats | | *Taittirīya:* VI.4.10.6. *Vājasneyī:* XIX.21. *Kaṭhaka :* XV.2. | *S.B.* I.6.13.16. | | | | |
| 84. | Sarṣapa | *Sarsopa brinicanapus* | | | *Śadaviṁśa:* V.2. | | *Chāndo-gya :* III.14.3. | | This represents oil seeds. |
| 85. | Taṇḍula | Rice grain | | *Maitrāyaṇī:* II.6.6. *Kaṭhaka:* X1. | *S.B.* I.1.4.3; *Aitareya:* I.1. | | *Chāndogya:* III.14.3. | | This represents oil seeds. |

| Sl. No. | Saṁskṛt Name | Botanical Equivalent | Vedas | Saṁhitās | Brāhmaṇas | Āraṇyakas | Upaniṣads | Gṛhya-Sūtras Śrauta-Sūtras and Nirukta | Remarks |
|---|---|---|---|---|---|---|---|---|---|
| 86. | Taraṣṭāgha | Mustard Plant | *A.V.* V.29.15. | | | | | *Kauśika Sūtra:* XXV.23. | This is a special tree mentioned in the *Kauśika Sūtra*, while the adjective formed from it Tāraṣṭāghī is derived from this tree in the *A.V.* |
| 87. | Tarastāgha | | | | | | | | Same as No. 86. |
| 88. | Tila | *Sesamum indicum* DC. | *A.V.* II.8.3; IV.7.3; VI.140.2; XVIII. 3.69. | *Taittirīya:* VI.3.2 *Vājasneyī:* XVIII.12. | | *Śankhāyana:* XII.8; XI.4. | *Bṛhadāraṇyaka:* VI.3.22; VI.4.16. *Chāndogya:* V.10.6. | | |
| 89. | Tokman | Shoot | *R.V.* X.62.8. | *Kaṭhaka:* XII.2. *Vājasneyī:* XIX.13, 81. | *Taittirīya:* II.6.4. *Aitreya:* VIII.5. | | | | |
| 90. | Upavaka | *Wrightia tinctoria* | | *Vājasneyī:* XIX.22,90; XXI.30. | *S.B.* XII.7.1.3. | | | | |
| 91. | Uravaru | *Cucumis mello C. Sativus* | *A.V.* VI.14.2; *R.V.* VII.59.12. | *Maitrāyaṇī* I.10.4. *Taittirīya :* I.8.62. *Vājasneyī:* III.60. | *Pañcaviṁśa* IX.2.19. | | | | |
| 92. | Usana | *Piper longium* | | | *S.B.* III.4.3.13; IV.2.5.15. | | | | This denotes the plant from which Soma was prepared. |

| *Sl. No.* | *Saṁskṛt Name* | *Botanical Equivalent* | *Vedas* | *Saṁhitās* | *Brāhmaṇas* | *Āraṇyakas* | *Upaniṣads* | *Gṛhya-Sūtras Śrauta-Sūtras and Nirukta* | *Remarks* |
|---|---|---|---|---|---|---|---|---|---|
| 93. | Vrihi | *Oryza sativa* | *A.V.* VI.140.2; VI.7.20; IX.6.14; IX.4.13. | *Taittirīya:* VIII.10.1 *Kaṭhaka:* X.6; IX.5. *Maitrāyaṇī* III.10.2. *Vājasneyi:* XVIII.12. | *Aiteraya:* II.8.7. *S.B.* V.5.5.9. *Jaiminiya:* I.43. | *Bṛhadā-raṇyaka:* VI.3.22. *Chāndogya:* III.14.3. | | | This is a type of paddy. |
| 94. | Yava(Ka) | *Hardeum vulgare*, L. | *R.V.* I.23.15; I.66.3. *A.V.* II.8.3; VI.30.1. | *Vājasneyī:* V.26 *Taittirīya:* VI.2.10.3; VII.2.10.2 *Kaṭhaka :* XV.10. *Maitrāyaṇī* IV.3.2. | *S.B.* 1.1.4.220; II.5.2.1. *Taittiriya:* I.8.4.1. | *Chāndogya:* III.14.3. | | | |

## (iii) Grass and Reeds

| | | | | | | | | | |
|---|---|---|---|---|---|---|---|---|---|
| 95. | Aśvavara or Aśvavala (Hair of horse's tail) | *Saccharum spontaneum* | | *Kaṭhaka:* XXIV.8; *Maitrāyaṇī* III.7.9; *Kapiṣṭhala:* XXXVIII.1. | *S.B.* III.4.1.7. | | | | The grass is used as thatched material for making ropes and mats, besides serving as animal fodder. |
| 96. | Balbaja | *Abutilon indicum*, Linn. | *R.V.* VIII.55.3. | *Taittirīya:* II.2.8.2. *Kaṭhaka:* X.10. *Maitrāyaṇī:* II.2.5. | | | | | |
| 97. | Darba or Darbha | *Desmostachya bipinata*, Staff. | *R.V.* I.191.3. | *Taittirīya:* I.5.1.4. | | | | | This is also used for calming the anger in a |

| *Sl. No.* | *Saṁskṛt Name* | *Botanical Equivalent* | *Vedas* | *Saṁhitās* | *Brāhmaṇas* | *Āraṇyakas* | *Upaniṣads* | *Gṛhya-Sūtras Śrauta-Sūtras and Nirukta* | *Remarks* |
|---|---|---|---|---|---|---|---|---|---|
| | | | *A.V.* VI.43.2; VII.7.20; X.4.13; XI.6.15. | | | | | | man according to the *A.V.* |
| 98. | Dūrbā, Dūrbā or Dūrvarā | *Desmostachya bipinnata*, Staff. | *R.V.* X.16.13; X.134.5; X.142.8. | *Taittirīya:* IV.2.9.2; V.2.8.3. *Vājasneyī:* XIII.20. | *Aiteraya:* VIII.5.8. *S.B.* IV.5.10.5. | | | | This is a species of grass having a special feature that its ears lie horizontal to the stem. |
| 99. | Ikṣu | Sugarcane; *Saccharum officinarum* Linn. | *A.V.* I.34.5. | *Maitrāyaṇī:* III.7.9. *Vājasneyī:* XXV.1. *Taittirīya:* VIII.3.16.1. | | | | | This is the generic name of sugarcane found in the *Atharvaveda* and the *Saṁhitās.* |
| 100. | Isika | Reed grass | *A.V.* VII.56.4; XII.2.54. | | *S.B.* IV.3.4.16. *Jaiminīya:* I.9; II.134. | | *Chāndogya:* V.24.3; *Kaṭhaka:* II.6.17. | *Nirukta:* IX.8. | |
| 101. | Kāsā | *Saccharum spontaneum* | *R.V.* X.100.10. | | | *Taittirīya:* VI.9.1. | | | This species of grass is used for mats. |
| 102. | Kuśa | *Poa cyonosuroides* | | *Maitrāyaṇī:* IV.5.7. | *S.B.* II.5.2.15; III.1.2.16; 6.2.9. *Taittirīya:* I.5.10; I.2.7. | | | | This is known as sacred grass which is used in all the religious sacrifices. |

| Sl. No. | Saṁskṛt Name | Botanical Equivalent | Vedas | Saṁhitās | Brāhmaṇas | Āraṇyakas | Upaniṣads | Gṛhya-Sūtras Śrauta-Sūtras and Nirukta | Remarks |
|---|---|---|---|---|---|---|---|---|---|
| 103. | Muñja | *Saccharum munja* | *R.V.* I.191.3; I.161.8. *A.V.* I.2.4. | *Taittirīya:* V.1.95. | *Kauśītaki:* II.8.7. *S.B.* IV.33.16; 6.1.23; 2.15.16; VI.3.1.26; XV.8.3.6. | | | | This is a species of grass which has the luxuriant growth attaining the height of over ten feet. |
| 104. | Nada | Reed | *R.V.* VIII.1.33; *A.V.* IV.19.1; VI.138.5; VI.137.2; XII.2.1. | *Kaṭhaka:* XXV.7. | *S.B.* I.14.19. | *Taittirīya:* VI.7.10. | | | It is said to be growing in lakes. The reeds were used for making mats etc. |
| 105. | Nadvala | Reed | | *Vājasneyī:* XXX.16. | *Taittirīya:* III.4.12.1. | | | | |
| 106. | Piñjula (Tejani) | Bundle of grass | | *Kaṭhaka:* XXIII.9; XXII.13. | *S.B.* XIII.8.3.12. *Aitareya:* I.11. | | | | |
| 107. | Sada | Grass | *R.V.* IX.15.6. | *Vājasneyī:* XXV.1, etc. | | | | | |
| 108. | Śairya | Grass | *R.V.* I.191.3. | | | | | | |
| 109. | Śara | *Saccharum arundinaceum* | *R.V.* I.191.3. *A.V.* IV.7.4; I.2.1; I.3.1. | *Taittirīya:* V.2.6.2. *Kaṭhaka:* IX.5; XXIII.4. | *S.B.* I.2.4.1; III.1.3.13. | | *Bṛhadāraṇyaka :* IV.4.11. | *Nirukta:* V.4. | |

| Sl. No. | Saṁskṛt Name | Botanical Equivalent | Vedas | Saṁhitās | Brāhmaṇas | Āraṇyakas | Upaniṣads | Gṛhya-Sūtras Śrauta-Sūtras and Nirukta | Remarks |
|---|---|---|---|---|---|---|---|---|---|
| 110. | Sasa | Herb or Grass | *R.V.* I.51.3; X.79.3. | | | | | | This word is also applied to Soma plant and the sacrificial straw. |
| 111. | Saspa | Tender grass | | *Vājasneyi:* XIX.13; XIX.81; XXI.29. | *S.B.* XII.7.2.8; XXII.9.1.2. *Aitareya :* VIII.5.3. | | | | |
| 112. | Stambha | Tuft of grass | *A.V.* VIII.6.14. | *Taittirīya:* V.6.4.1. | *Taittirīya:* II.7.17.3. *Aiteraya:* V.23.9. | | | | This is believed to be auspicious for building purposes. |
| 113. | Sugandhikā or Sugandhikatejanā | Fragrant grass | | *Taittirīya:* VI.2.8.4. *Kaṭhaka :* XXV.6. | *Pañcaviṁśa* XXIV.13.5. *Aiteraya:* I.28.28. *S.B.* III.5.2.17. | | | | |
| 114. | Sumbala | Straw | | | *S.B.* XII.5.2.3. | | | | |
| 115. | Tṛṇa | Straw | *R.V.* I.161.1; X.102.10; *A.V.* III.12.5; IX.4.3.7; II.30.1; VI.54.1. | | *Aiteraya:* III.22. | | | | This grass was possibly used for making roofs of huts. |
| 116. | Ulapa | *Imperata arundinaceae* | *R.V.* X.142.3; *A.V.* VII.66.1. | *Vājasneyi:* XIV.45 | | | | | This is the name of a species of grass. |

| Sl. No. | Saṁskṛt Name | Botanical Equivalent | Vedas | Saṁhitās | Brāhmaṇas | Āraṇyakas | Upaniṣads | Gṛhya-Sūtras Śrauta-Sūtras and Nirukta | Remarks |
|---|---|---|---|---|---|---|---|---|---|
| 117. | Vadala | Reed | *A.V.* VIII.8.3. | | *S.B.* II.4.5.14. | | | | This denotes a species of Reed (Sarkaṇḍā) |
| 118. | Veṇu | *Bambossa Sp.* | *A.V.* I.27.3. | *Taittiriya :* V.2.5.2; VII.4.19.2. *Kaṭhaka:* XIII.12. | *S.B.* I.1.4.19. *Kauśitaki:* IV.12. | | | | This denotes bamboo reed. In *Kauśitaki Br.* It is compared with Śasya, stating that they ripen in *Vasanta* (spring) season. |
| 119. | Viraṇa or Virina | *Andropogon muricatus* | | | *S.B.* XIII.8.1.15. | | | | |
| 120. | Yavasa | Grass | *R.V.* I.38.5; I.91.13. | *Vājasneyi:* XXI.43. | | | | | This species served as animal feed. |

## (iv) Sacred Trees/Plants

| Sl. No. | Saṁskṛt Name | Botanical Equivalent | Vedas | Saṁhitās | Brāhmaṇas | Āraṇyakas | Upaniṣads | Gṛhya-Sūtras Śrauta-Sūtras and Nirukta | Remarks |
|---|---|---|---|---|---|---|---|---|---|
| 121. | Aśvattha | *Ficus religiosa,* Linn. | *R.V.* I.135.8; I.164.20; X.97.5. *A.V.* III.6.1; IV.32.4; VI.11.1; V.4.3. | | *S.B.* XI.5.1.13; *Aiteraya:* VII.35.6. | | *Śvetāsvartāra:* IV.6.22; V.54.52. *Maṇḍūka:* III.1.1. *Bṛhadāraṇyaka:* IV.1.41. | | It is India's greatest tree and is also called Pīpala. If it is planted with its roots in other trees especially Khadira, it destroys them. Hence it is also called the destroyer. |
| 122. | Kadamba | *Anthocephalus cadamba,* Miq. | | | | | | | The flowers are offered to Śiva, are considered sacred to him. This tree is also associated with Kṛṣṇa. |
| 123. | Nyagrodha (Growing downwards) | *Ficus bengalensis* | *R.V.* I.24.7. *A.V.* IV.37.4; | *Taittiriya:* VII.4.12.1. *Vājasneyi:* XXIII.13. | *Aitareya:* VII.30.31; VII.35.5. *S.B.*: | | | | The staff of Kṣatriya and Brahmacārin is made of this tree. |

| Sl. No. | Saṁskṛt Name | Botanical Equivalent | Vedas | Saṁhitās | Brāhmaṇas | Āraṇyakas | Upaniṣads | Gṛhya-Sūtras Śrauta-Sūtras and Nirukta | Remarks |
|---|---|---|---|---|---|---|---|---|---|
| | | | V.5.5. | | V.3.5.1.3; VII.2.7.3; XIII.2.7.13. | | | | |
| 124. | Pārijāta | *Erythrina indica* | | | | | | | This is a coral tree and is believed to have emerged out of the churning of the ocean and taken possession of by Indra. It was then taken to Dvārkā by Kṛṣṇa. It blossoms large crimson flowers. |
| 125. | Pīpala | *Ficus religiosa,* L. | *R.V.* I.135.8; I.164.20; X 97.5; *A.V.* III.6.1; IV.32.4; VI.11.1; V.4.3. | | *S.B.* XI.5.1.13. *Aitareya:* VII.35.6. | | *Śvetāśvatāra:* IV.6.22; V.54.52. *Maṇḍūka:* III.1.1. *Bṛhadāraṇyaka :* IV.1.41. | | |
| 126. | Plākṣa or Prakṣa | *Ficus infactoria* | *A.V.* V.5.5. | *Taittirīya:* IV.12.1 *Maitrāyaṇī:* III.10.2. | *Aitareya:* VII.32; VIII.16. *Taittirīya:* III.8.19.2. *S.B.* III.8.3.12. | | | | This is a waved-leaf fig tree which is quite large and beautiful with small white fruit. It has several usages, besides serving as fodder for cattle. |
| 127. | Śyāmaka | *Echinochloa colona* | *A.V.* XIX.50.4. | *Taittirīya* I.8.[illegible].2; II.3 2.6; IV.[illegible].4.21. *Maitrāyaṇī:* II.11.4. | *S.B.* X.6.3.2. | | *Chāndogya:* III.14.3. | | |

| *Sl. No.* | *Saṁskṛt Name* | *Botanical Equivalent* | *Vedas* | *Saṁhitās* | *Brāhmaṇas* | *Āraṇyakas* | *Upaniṣads* | *Gṛhya-Sūtras Śrauta-Sūtras and Nirukta* | *Remarks* |
|---|---|---|---|---|---|---|---|---|---|
| | | | | *Vājasneyī:* XVIII.18. *Kaṭhaka:* X.2. | | | | | |
| 128. | Syandana or Spandana | | *R.V.* III.53.19 | | | | | *Kauśaka Sūtra:* VIII.15. | |
| 129. | Tilvaka | *Symplocos racemosus* | | *Maitrāyaṇi* III.19. | *S.B.* XIII.8.1.16 *Sadaviṁśa:* III.8. | | | | Yupa, the sacrificial part is believed to be made of the wood of this tree. |
| 130. | Tulasī | *Ocimum sanctum* L. or Holy basil. | | | | | | The plant is frequently mentioned in the Purāṇas. | A sacred plant which sanctifies the house and ensures the happiness of the house-holder in the other world. |
| 131. | Udumbara | *Ficus glomerata* | *A.V.* XIX.31.1. | *Taittirīya:* II.1; I.6; V.24. | *S.B.* III.2.1.33; VII.4.1.38. *Aitareya:* VII.15. *Pañcaviṁśa:* XXVI.6.4. | | | | The sacrificial part Yupa and ladle were made of the wood of this tree. The sweetness of its fruit is equated with honey. |
| 132. | Vikantaka (Vikantika) | *Flacourtia ramontchi* | *A.V.* XI.10.3. | *Taittirīya:* III.5.7.3; VI.4.10.5. *Kaṭhaka:* XIX.10. *Maitrāyaṇī:* III.1.9. | *S.B.* I.2.4.10; V.2.4.18. | | | | |

## (v) Flowering Trees/Plants

| Sl. No. | Saṁskṛt Name | Botanical Equivalent | Vedas | Saṁhitās | Brāhmaṇas | Āraṇyakas | Upaniṣads | Gṛhya-Sūtras Śrauta-Sūtras and Nirukta | Remarks |
|---|---|---|---|---|---|---|---|---|---|
| 133. | Abja | *Barringtonia acutangula* or Blue Lotus | *R.V.* IV.16.13; VII.33.11; X.134.2. *A.V.* VII.16. | | | | *Bṛhadāraṇyaka:* VI.4.1. *Dhyānabindu:* 7 & 9. *Āśrama:* 3; *Vāsudeva* 3. | | The plant is used in rituals against supernatural agencies. |
| 134. | Kāśa | *Saccharum spontaneum* L. | *R.V.* X.100.10. | *Taittirīya:* VI.9.1. | | | | | Its flowers are ultra white. |
| 135. | Kiṁśuka or Palāśa | *Butea frondosa* | *R.V.* X.85.20. | | | | | | It is mentioned in the wedding hymn of the *R.V.* wherein it is prescribed that the bridal car be decorated with the flowers of this tree. |
| 136. | Kumuda | *Nymphaea rubra* (white water lilly). | *A.V.* IV.34.5. | | | | | | This is mentioned with other flowering plants in *A.V.* |
| 137. | Naladi (Nalada) | *Nardostachys jatamanasi* | *A.V.* VI.10.2,3. | | *Aitareya:* III.24. *Sāṁkhāyana:* XI.4 | | | | Appears to have been used for the making of garlands. |
| 138. | Puṇḍarīka (white lotus) | *Artemesia indica* | *R.V.* X.142.8 *A.V.* X.8.43. IV.35.8. | *Taittirīya:* I.8.2.1. | *S.B.* V.5.5.6. *Taittirīya:* I.8.2.1. *Pañcaviṁśa* XVIII.9.6. | *Aitareya:* III.34. | *Chāndogya:* I.6.7. *Bṛhadāraṇyaka:* II.3.10; VI.3.14. | | This denotes the blossom of lotus flower. The *Pañcaviṁśa Brāhmaṇa* testifies that the lotus flower is born out of the light of the *Nakṣatras.* It is also compared with human heart. |

| Sl. No. | Saṁskṛt Name | Botanical Equivalent | Vedas | Saṁhitās | Brāhmaṇas | Āraṇyakas | Upaniṣads | Gṛhya-Sūtras Śrauta-Sūtras and Nirukta | Remarks |
|---|---|---|---|---|---|---|---|---|---|
| 139. | Puṣkara | Lotus flower (*Nymphaea nelumbo*) | *R.V.* VI.16.13; VII.33.11. *A.V.* IX.3.8; XII.1.24. | *Vājasneyi:* IX.29. *Taittiriya:* V.1.4.1. *Maitrāyaṇi:* III.1.5. | *S.B.* IV.5.1.16. *Taittirīya:* I.2.1.4. | | | | This represents blue lotus flower and its sweet perfume is mentioned in the *A.V.* The lakes in which the lotus grew were called as Puṣkarṇī (lotus bearing). |
| 140. | Puśpa | Flower | *A.V.* VIII.7.12. | *Vājasneyī:* XXII.28. *Taittiriya:* V.4.4.2. | *Pañcaviṁśa* VIII.4.1; XV.3.23. | | *Chāndogya:* III.1.2. *Bṛhadāraṇyaka:* VI.4.1. | | |
| 141. | Sadmapuśpa | A flowery Plant | *A.V.* IV.20.9. | | | | | | |
| 142. | Śālūka | *Nyṃphaea lotus* L. | *A.V.* IV.35.5. | | | | | | |

## (vi) Fragrant Trees/Plants

| | | | | | | | | | |
|---|---|---|---|---|---|---|---|---|---|
| 143. | Aśvagandhā (Rock smell) | *Withania coagulans*, Dunal. | | | *S.B.* XIII.1.16. | | | | |
| 144. | Auksagandhā | Having smell | *A.V.* IV.37.3. | | | | | | This word appears in the *Atharvaveda*, besides other names of Guggulu and Naladi. |
| 145. | Guggulu | *Commiphara mukul*, Hook | *A.V.* XIX.3.8. | | | | | | |
| 146. | Pila | *Careya arborea* A Fragrant plant. | *A.V.* IV.37.3. | | | | | | |

| Sl. No. | *Saṁskṛt Name* | *Botanical Equivalent* | *Vedas* | *Saṁhitās* | *Brāhmaṇas* | *Āraṇyakas* | *Upaniṣads* | *Gṛhya-Sūtras Śrauta-Sūtras and Nirukta* | *Remarks* |
|---|---|---|---|---|---|---|---|---|---|
| 147. | Pramandani | A sweet scented plant. | | | | | | *Kauśika Sūtra :* VIII.17; XXV.11; XXXII.22. | |
| 148. | Sugandhite-jana | *Andropogan squarrousus* | | *Taittirīya:* VI.2.3.4. *Kaṭhaka:* XXV.6. | *S.B.* III.2.5.17; III.5.3.17. *Pañcaviṁśa* XXIV.13.5. | | | | |

## (vii) Fruit Trees/Plants

| Sl. No. | *Saṁskṛt Name* | *Botanical Equivalent* | *Vedas* | *Saṁhitās* | *Brāhmaṇas* | *Āraṇyakas* | *Upaniṣads* | *Gṛhya-Sūtras Śrauta-Sūtras and Nirukta* | *Remarks* |
|---|---|---|---|---|---|---|---|---|---|
| 149. | Āmalaka | *Emblica officinalis,* Gaertn. | | | *Jaiminīya:* I.33.6. | | *Chāndogya:* VII.3.1. | | The juice of the fruit is considered as cooling, refrigerant, durative and laxative. It is also used in medicines. |
| 150. | Badarī or Badarā | *Zizania jujuba,* Mill or *Zizania mauritiana,* Lam. | | *Kaṭhaka:* III.11.2. *Vājasneyī:* XIX.22; XXI.30. | *Taittirīya:* I.8.5.1. *S.B.* V.5.4.10. *Jaiminīya:* II.156.5. | | | | The staff of Vaiśya Brahmacārin is required to be made of the wood of this tree. |
| 151. | Bilva | *Aegle marmelos,* C & Roxb. | *A.V.* XXI.136.3. | *Maitrāyaṇī:* III.9.3. *Taittirīya:* II.18.1, 2. | *Aitareya:* II.1. *S.B.* XIII.4.4.8. | *Śānkhyā-yana:* XII.20. | | | This is also a sacred tree and its leaves and fruits are offered to Śiva. |
| 152. | Bimba | *Momordica monadelpha* | | | *Jaiminīya:* III.5.6. | | | | A beautiful woman's lips are compared with its fruit. |
| 153. | Jambula | *Citrus aurantium;* C *Medica varacids* | | *Maitrāyaṇī* III 15.3. *Kaṭhaka:* V.12.1. | | | | | This represents the common type of orange. |

| *Sl. No.* | *Saṁskṛt Name* | *Botanical Equivalent* | *Vedas* | *Saṁhitās* | *Brāhmaṇas* | *Āraṇyakas* | *Upaniṣads* | *Gṛhya-Sūtras Śrauta-Sūtras and Nirukta* | *Remarks* |
|---|---|---|---|---|---|---|---|---|---|
| | | | | *Vājasneyi:* XXV.3. | | | | | |
| 154. | Kapittha | *Feronia limonia,* Linn. | *A.V.* IV.4.8. | | | | | | It also serves as an antidote against poison. |
| 155. | Karkaṇḍu | *Ziziphus mummularia* | *R.V.* I.112.6. | *Kaṭhaka:* XII.10. *Maitrāyaṇī:* III.11.2. *Vājasneyī:* XIX.23.91; XXI.32; XXIV.2. | *S.B.* V.5.4.10; XII.7.2.9; XII.9.1.5. *Jaiminīya:* II.15.2.5. | | | | |
| 156. | Kharjūra | *Phoenix sylvestris* | | *Taittirīya:* II.4.9.2. *Kaṭhaka :* XI.10; XXVI.7. | | | | | |

## (viii) Intoxicating Trees/Plants

| | | | | | | | | | |
|---|---|---|---|---|---|---|---|---|---|
| 157. | Bhāṅg | *Cannabis sativa,* Linn. | *R.V.* IX.61.13. *A.V.* XI.6.15. | *Śāṅkhāyana:* XII.19. | | | | | In *Ṛgveda* it is an epithet of Soma plant. |
| 158. | Madaratī or Madavatī | Intoxicating creeper. | *A.V.* VI.16.2. | | | | | | This is an intoxicating plant. |
| 159. | Mudgha | Honey plant or *Phaseolus mungo* L. | *A.V.* I.34.4. VI.120.3. | | | | | | This literally means probably yielding honey. |
| 160. | Madhūka | *Bassia latifolia* | *A.V.* I.34.5. | | | | | | An intoxicating drink is made from its flowers. |

| Sl. No. | Saṁskṛt Name | Botanical Equivalent | Vedas | Saṁhitās | Brāhmaṇas | Āraṇyakas | Upaniṣads | Gṛhya-Sūtras Śrauta-Sūtras and Nirukta | Remarks |
|---|---|---|---|---|---|---|---|---|---|
| 161. | Soma | *Asclepias acida*, Roxb. | *R.V.* X.34.1; X.124.1. | | | | | | The juice extracted from this plant was an intoxicating drink. It is personified as one of the Vedic gods to whose praise all the 114 hymns of the 9th book of the *R.V.* besides six in other books, are devoted. |

## (ix) Trees and Plants Used as Fodder

| Sl. No. | Saṁskṛt Name | Botanical Equivalent | Vedas | Saṁhitās | Brāhmaṇas | Āraṇyakas | Upaniṣads | Gṛhya-Sūtras Śrauta-Sūtras and Nirukta | Remarks |
|---|---|---|---|---|---|---|---|---|---|
| 162. | Anu | *Panicum milaceum* | | *Vājasneyī:* XVIII.12. | | | *Bṛhadāraṇyaka*: VI.3.13. | | |
| 163. | Aśvabalā | *Saccharum spontaneum* | | *Kapiṣṭhala:* XXXVIII.1. *Kaṭhaka:* XX.4, 8. *Maitrāyaṇī:* III.7.9. | *S.B.* III.4.1.17. | | | | |
| 164. | Gavadhūka | *Coix barbata* | | *Taittiriya:* I.8 7.1; I.8 9.2; V.4.3.2; *Maitrāyaṇī* II.6.5; IV 3.8. *Vājasneyī:* XV.5. | *S.B.* V.2.9.13; V.3.1.10; XIV.1.2.19. *Taittirīya:* I.7.3.6. | | | | |
| 165. | Manjiṣṭha | *Rubia tinctarium* L. | | | *Aitareya:* III.2.4. | | | | It is believed to be an antidote to poison. |

| *Sl. No.* | *Saṁskṛt Name* | *Botanical Equivalent* | *Vedas* | *Saṁhitās* | *Brāhmaṇas* | *Āraṇyakas* | *Upaniṣads* | *Gṛhya-Sūtras Śrauta-Sūtras and Nirukta* | *Remarks* |
|---|---|---|---|---|---|---|---|---|---|
| | | or *Rubia munijosta* | | | *Śāṅkhāyana* VII.7. | | | | |
| 166. | Śami | *Prosopis spicigera Minosa suma* | *A.V.* VI.11.1; XXX.2.3; VI.3.2-3. | *Taittirīya:* V.1.9.6; V.4.7.4. *Kaṭhaka:* XXVI.6. | *Taittirīya:* I.1.3.11. *S.B.* II.5.2.12; IX.2.3.37. | | | | The leaves are used as fodder. |
| 167. | Ulapa | *Imperata arundinaceae* | *R.V.* X.14.2. *A.V.* VIII.66.1. | *Vājasneyi:* XVI.45. *Maitrāyaṇī* I.7.2. | | | | | |
| 168. | Upavāka | *Wrightia tinctoria* | | *Vājasneyi:* XIX.22-90; XXI.30. | *S.B.* XII.7.1.3; XII.7.2.9. | | | | This is also believed to have healing properties. |
| 169. | Vaikanta | *Flacourtia sapide* | | *Taittirīya:* III.5.7.3. *Kaṭhaka:* XIX.10. *Maitrāyaṇī* III.1.9. | | | | | A small stiff spicy shrub found in dry forest. The fruits and leaves are used for human as well as animal consumption. |

## (x) Trees and Plants Used Against Supernatural Forces or Demons

| | | | | | | | | | |
|---|---|---|---|---|---|---|---|---|---|
| 170. | Abja | *Barringtonia acutangula* | Same as Puṣkara | | | | | | This is used against supernatural agencies. |
| 171. | Abaya/Abhya Abhayā | *Andropogon muricatus* | *A.V.* IV.35.5. | | | | | | This is meant for use against supernatural agencies. |
| 172. | Ajaśṛngi | *Odina wodiar* | *A.V.* IV.37.1. etc. | | | | | | It is supposed to destroy demons. |

| Sl. No. | Saṁskṛt Name | Botanical Equivalent | Vedas | Saṁhitās | Brāhmaṇas | Āraṇyakas | Upaniṣads | Gṛhya-Sūtras Śrauta-Sūtras and Nirukta | Remarks |
|---|---|---|---|---|---|---|---|---|---|
| 173. | Sadmapuśpā | Ever flowering plant. | *A.V.* IV.20.9. | | | | | | These are used for driving away the *pretas* and goblins. |

## (xi) Poisonous Trees/Plants

| Sl. No. | Saṁskṛt Name | Botanical Equivalent | Vedas | Saṁhitās | Brāhmaṇas | Āraṇyakas | Upaniṣads | Gṛhya-Sūtras Śrauta-Sūtras and Nirukta | Remarks |
|---|---|---|---|---|---|---|---|---|---|
| 174. | Kanaknaka or Kandaviṣa | Species of poison. | *A.V.* X.4.22. | | | | | | |
| 175. | Naraci | A poisonous plant. | *A.V.* V.3.4. | | | | | | |
| 176. | Śālmalī | *Salmalia malabarica* Schott. | *R.V.* VII.50.3; X 85.20. | *Taittirīya:* VII.4 12. *Vājasneyī:* XXIII.13. | *S.B.* XIII.2.7.4. *Pañcaviṁśa* IX.4.11. | | | | Its fruit is regarded as poisonous. |

## (xii) Trees and Plants Serving as Antidote to Poison

| Sl. No. | Saṁskṛt Name | Botanical Equivalent | Vedas | Saṁhitās | Brāhmaṇas | Āraṇyakas | Upaniṣads | Gṛhya-Sūtras Śrauta-Sūtras and Nirukta | Remarks |
|---|---|---|---|---|---|---|---|---|---|
| 177. | Amūla | *Methonica superba*, Linn. | *A.V.* V.32.5. | | | | | | This is believed to be a rootless plant. |
| 178. | Jivanti | *Desmotrichum fimbriatum*, Blume | *R.V.* V.78.9. *A.V.* III.14.5; VII.2.6. | | | | | | It is believed to serve as an antidote to poison. |
| 179. | Priyaṅgu | *Setaria italica* | | *Taittirīya:* II.2.11-14. *Kaṭhaka:* X.11. *Maitrāyaṇī* II.1.8. *Vājasneyī:* XVIII.12. | | | | | It is believed to blossom at the touch of a woman, besides being an antidote to poison. |

| *Sl. No.* | *Saṁskṛt Name* | *Botanical Equivalent* | *Vedas* | *Saṁhitās* | *Brāhmaṇas* | *Āraṇyakas* | *Upaniṣads* | *Gṛhya-Sūtras Śrauta-Sūtras and Nirukta* | *Remarks* |
|---|---|---|---|---|---|---|---|---|---|
| 180. | Pṛṣṇīparṇī | *Uraria logopioides,* L. | *A.V.* II.25.1. | | *S.B.* XIII.8.1.16. | | | | This is believed to serve as protection against wild creatures. |

## (xiii) Trees of Cremation Ground

| | | | | | | | | | |
|---|---|---|---|---|---|---|---|---|---|
| 181. | Kiyaṁbu | *V.L. Kayambu* (A kind of aquatic plant) | *R.V.* X.16.13. *A.V.* XVIII.3.6. | | | *Taittirīya:* VI.4.1.2. | | | Planted at places where dead bodies are burnt. |
| 182. | Pākadūrvā | Edible millet | *R.V.* X.16.3. *A.V.* XVII.3.6. | | | *Taittirīya:* VI.4.1.2. | | | This is said to grow at places where the dead bodies are burnt. Read as *Saṇḍadūrva* i.e., having egg shaped roots. |

## (xiv) Aquatic Plants

| | | | | | | | | | |
|---|---|---|---|---|---|---|---|---|---|
| 183. | Avaka, Avakin, Avakada or Avakalba | *Blyxa octandra,* Rich. otherwise called *Śaivāla* | *A.V.* VIII.7.9; IV.37.8-10 | *Vājasneyī:* XVII.4,25. *Maitrāyaṇī* II.101. *Taittirīya:* VI.6.1.1. | *S.B.* VII.5.1.11; VIII.3.2.5. | | | | A grassy plant growing in marshy lands. This has been identified with Siplā also. |
| 184. | Kiyaṁbu | A kind of aquatic plant. *V.L. Kayambu* | *R.V.* X.16.13; X.13.16; *A.V.* XVIII.3.6. | | | | | | This is said to grow at places where dead bodies were burnt. |

| *Sl. No.* | *Saṁskṛt Name* | *Botanical Equivalent* | *Vedas* | *Saṁhitās* | *Brāhmaṇas* | *Āraṇyakas* | *Upaniṣads* | *Gṛhya-Sūtras Śrauta-Sūtras and Nirukta* | *Remarks* |
|---|---|---|---|---|---|---|---|---|---|
| 185. | Kumuda (exciting with joy) | *Nymphaea rubra* (white water lilly) | *A.V.* IV.34.5. | | | | | | This plant is mentioned with other plants in the *A.V.* |
| 186. | Puṇḍarīka (white lotus) | *Artemesia indica* | *R.V.* X.142.8. *A.V.* IV.35.5. | *Taittirīya:* I.8.2.1. | *S.B.* V.5.5.6. | *Aitareya:* III.34. | *Bṛhadāraṇy-aka:* II.3.10; VI.3.14. *Chāndogya:* I.6.7. | | |
| 187. | Puṣkara (Blue lotus) | *Nymphaea nelumbo* | *R.V.* VI.16.13; VII.33;11; X.184.2. *A.V.* XI.3.8; XII.1.24. | | | | | | This is known for the sweet fragrance of its flowers. |
| 188. | Śalūka | *Nymphaea lotus*, L. | *A.V.* IV.34.5 | | | | | | |
| 189. | Sana | *Crotalaria juncea* | *A.V.* II.4.5. | | | | | | This is mainly used for commercial purposes, but it grows mainly in water. |
| 190. | Saphala | Water plant | *A.V.* IV.34.5. | | | | | *Āpastambha Śr. Sūtra :* IX.14.14. | It has been described as an edible plant as well. |
| 191. | Sipāla | *Blyxa octandra* | *R.V.* X.68.5. *A.V.* IV.12.31. | | *S.B.* III.1.1. | | | | |

| Sl. No. | Saṁskṛt Name | Botanical Equivalent | Vedas | Saṁhitās | Brāhmaṇas | Āraṇyakas | Upaniṣads | Gṛhya-Sūtras Śrauta-Sūtras and Nirukta | Remarks |
|---|---|---|---|---|---|---|---|---|---|
| 192. | Vanaparṇī | Water plant | *A.V.* III.18.1. | | | | | | |
| 193. | Vetasa | *Calamus rotang* | *A.V.* X.7.4. | | | | | | This is a water plant. In the *Taittirīya Saṁhitā* it is called golden and water born. |
| 194. | Vyaikasā | Water plant | *R.V.* X.16.3. | | | | | | |

## (xv) Trees and Plants for Commercial Use

| | | | | | | | | | |
|---|---|---|---|---|---|---|---|---|---|
| 195. | Alābu | *Lagenaria vulgaris*, Mol. | *R.V.* VIII.10.29, 30; XX.130.1. | *Maitrāyaṇī:* IV.2.13; VI.79.17. | | | | | This is also believed to be sacred and its seeds yield an oil and juice of the leaves is used for cure of ailments. |
| 196. | Aralu | *Ormocarpum indicum*, Vent. | *A.V.* XX.131.17 | | | | | | Axles of chariots were made of the wood of this tree. |
| 197. | Balbajā or Balbīja | *Abutilon indicum*, Linn. | *R.V.* VIII.53.3; *A.V.* XIV.2.22-23. | *Maitrāyaṇī* II.2.5; *Taittirīya:* II.2.8.2; *Kaṭhaka:* X.10. | | | | | |
| 198. | Dhava or Dava | *Anogeissus latifolia* or *Solanum jacquini* | *A.V.* V.5.5. | | | | | | It yields gum and bark is used for dye and tan. The wood is known for strength and toughness. |

| *Sl. No.* | *Saṁskṛt Name* | *Botanical Equivalent* | *Vedas* | *Saṁhitās* | *Brāhmaṇas* | *Āraṇyakas* | *Upaniṣads* | *Gṛhya-Sūtras Śrauta-Sūtras and Nirukta* | *Remarks* |
|---|---|---|---|---|---|---|---|---|---|
| 199. | Isika | *Polytoca barbata* | *A.V.* VIII.54.4; XII.2.54. | | *S.B.* I.1.4.19; IV.3.4.16. *Jaiminīya: Br.* I.9; II.134. | *Śānkhyāyana:* II.16. | *Chāndogya:* V.24.3. *Kaṭhaka:* II.6.17. | *Nirukta:* IX.8 | It was used for making baskets besides folds and pins of doors. |
| 200. | Khadira | *Abutilon catechu*, Willd | *R.V.* III.53.19. *A.V.* III.6.1; *A.V.* 5.5; VIII.8.3. | *Taittirīya:* III.5.7.1. | *Aitareya:* II.1. *S.B.* I.3.3.20; III.6.2.12. | | | | This tree is frequently mentioned in the *R.V.* In the *Atharvaveda* it is mentioned that the Aśvattha tree gets planted upon Khadira and destroys it. |
| 201. | Pītadārū | *Deodar* or *Cedrus deodara* Linn. | | *Kaṭhaka:* XV.6. | *S.B.* II.5.2.15; XIII.4.4.5, 17. *Pañcaviṁśa* XIV.13.5. | | | | This is a very large and tall tree found in the North-West Himalayas. An oil is extracted from its wood, which is dark coloured, thick and is used for medicinal purposes. |
| 202. | Sana | *Crotalaria juncea* | *A.V.* II.4.5. | | *S.B.* I.6.1.24. | | | | This is mostly used for commercial purposes. |
| 203. | Vibhītaka | *Terminalia bellerica* Roxb. | *R.V.* III.86.6; X.34.1. *A.V.* (Pippalala) XX.4.6. | | | | | | It is useful for making of gum, dye and tan. |

## (xvi) Trees and Plants of General Usages

| *Sl. No.* | *Saṁskṛt Name* | *Botanical Equivalent* | *Vedas* | *Saṁhitās* | *Brāhmaṇas* | *Āraṇyakas* | *Upaniṣads* | *Gṛhya-Sūtras Śrauta-Sūtras and Nirukta* | *Remarks* |
|---|---|---|---|---|---|---|---|---|---|
| 204. | Aratu | *Calosanthus indica* | *R.V.* VIII.46.27. *A.V.* XX.131.17 | | | | | | Chariots were sometimes made of this wood. |
| 205. | Caṇḍā | *Macadamia peltata*, M&A | *A.V.* VI.139.3. | | | | | | |
| 206. | Daśavṛkṣa | | *A.V.* II.9.1. | | | | | | |
| 207. | Kakambara | | *R.V.* VI.48.17. | | | | | | This represents a useful plant. |
| 208. | Kṛṣṇalā | *Arbus precatorius* | | *Taittirīya:* II.3.2.1. *Kaṭhaka:* IX.4. | *Taittiriya :* I.3.67. | | | | The berry of this tree is used as weight. In the sense of weight it occurs in several texts. |
| 209. | Nyaṣṭika | *Andropogan aciculatus* | *A.V.* VI.139.1. | | | | | | |
| 210. | Pītarajju | | | *Kaṭhaka:* XXIV.3. | | | | | This is an unknown species of plant. |
| 211. | Pramanda | | *R.V.* III.53.14. | | | | | | This is a species of plant. |
| 212. | Putrīrajjis | | | *Kaṭhaka:* XXIV.3. | | | | | This is a species of plant. |
| 213. | Putadru | *Pinus longifolia* | | *Kaṭhaka:* XXV.6. | *S.B.* III.5.2.15; III.4.4.5,17. *Pañcaviṁśa* XXIV.5. | | | | |

| Sl. No. | Saṁskṛt Name | Botanical Equivalent | Vedas | Saṁhitās | Brāhmaṇas | Āraṇyakas | Upaniṣads | Gṛhya-Sūtras Śrauta-Sūtras and Nirukta | Remarks |
|---|---|---|---|---|---|---|---|---|---|
| 214. | Rajanī | *Curcuma longa.* The tree which causes fear. | *A.V.* I.23.1. | | | | | | It is so called because of its power of colouring. |
| 215. | Saha | *Aloe perfolita* | *A V.* XI.6.15. | | *Sāmvidhāna* II.6.10. | | | | |
| 216. | Sahadeva | *Sida cardifolia* | *A.V.* IV.59.2. | | | | | | |
| 217. | Sahamana | | *A.V.* II.25.2; IV.17.2; VIII.2.6; VIII.7.5. | | | | | | |
| 218. | Sidhvaka | | | | *Saḍaviṁśa:* III.6.9. *Taittirīya:* III.4.10. | | *Bandhāyana dharma Sūtra :* II.1.15. | | |
| 219. | Simbala (Sillaka) | *Alingia excelsa,* N | *R.V.* III.53.22. | | | | | | |
| 220. | Spandana | | *R.V.* III.53.19. | | | | | | |
| 221. | Sphurjaka | *Diospyros perrigeri* | | | *S.B.* XIII.8.1.16. | | | | |
| 222. | Sraktya | *Clerodendran phlomidis* | *A.V.* VIII. 5.4.7, 8. | | | | | | The wood was used for making amulets. |
| 223. | Śrekparṇa | *Nerium odcrum* | | | *Taittirīya :* III.6.6.3. *Aitareya:* II.6.15. | | | | |

| *Sl. No.* | *Saṁskṛt Name* | *Botanical Equivalent* | *Vedas* | *Saṁhitās* | *Brāhmaṇas* | *Āraṇyakas* | *Upaniṣads* | *Gṛhya-Sūtras Śrauta-Sūtras and Nirukta* | *Remarks* |
|---|---|---|---|---|---|---|---|---|---|
| 224. | Svadhā | | *A.V.* VI.96.3. | | | | | | |
| 225. | Svadhiti | | *R.V.* II.32.10; IX.96.6. | | | | | | This is a tree with hard wood. |
| 226. | Trayamana | *Gentiana kurroo* Royle. | *A.V.* VIII.2.6. | | | | | | |
| 227. | Triplākṣa | Three Fig tree. | | | *Pañcaviṁśa* XXV.13.4. | | | | This is also represents a place where Dṛṣadavatī disappeared near the Yamunā. |
| 228. | Varaṇa | *Cratoosva nurvalla* | | | *Aitareya:* VI.85.1; X.3.1; XIX.32.9. *Pañcaviṁśa* V.3.9.10. *S.B.* XII.8.4.1. | | | | |
| 229. | Vihalha | | *A.V.* VI.10.2. | | | | | | This denotes a plant. |
| 230. | Vṛṣa | *Gendarussa vulgaris* | | *Kaṭhaka :* XXX.1. | | | | | |
| 231. | Vyalkṣa | | *R.V.* X.16.13. | | | | | | This is also a species of plant. |

## (xvii) Forests

| | | | | | | | | | |
|---|---|---|---|---|---|---|---|---|---|
| 232. | Āvaṇya | Forest | *R.V.* VI.24.10; X.97.1-23; X.146.1-6. | | *S.B.* V.2.3.5. | | *Bṛhadāraṇyaka*: V.11. | | The term denotes the uncultivated land and not always the forest, beyond the village. It is |

| Sl. No. | Saṁskṛt Name | Botanical Equivalent | Vedas | Saṁhitās | Brāhmaṇas | Āraṇyakas | Upaniṣads | Gṛhya-Sūtras Śrauta-Sūtras and Nirukta | Remarks |
|---|---|---|---|---|---|---|---|---|---|
| | | | *A.V.* II.4.5. | | | | | | contrasted with home (*amā*) and with plough land (*kṛṣi*). It is also contrasted with *grama* (*A.V.* II.1.54) and according to the *S.B.* (V.2.3.5.) it is the place where the thieves reside. The character of the forest is described in Sukta-146 of the tenth Mandala of the *R.V.* to the forest spirit Araṇyāṇī. The dead were carried there for burial. |
| 233. | Udumbara | *Ficus glomerata* | | | *Pañcaviṁśa* XVI.6.4. | | | | A forest of Udumbara trees is mentioned in this text. |
| 234. | Khāṇḍava | Forest | | | | *Taittirīya:* V.1.1. | | | This forest is stated to be located on the boundary of Kurukṣetra. It was subsequently identified in the *Mahābhārata.* |

## (xviii) Seeds

| Sl. No. | Saṁskṛt Name | Botanical Equivalent | Vedas | Saṁhitās | Brāhmaṇas | Āraṇyakas | Upaniṣads | Gṛhya-Sūtras Śrauta-Sūtras and Nirukta | Remarks |
|---|---|---|---|---|---|---|---|---|---|
| 235. | Bīja | Seeds | *R.V.* X.94.13; X.85.37; V.5.3, 13. *A.V.* X.6.33. | | *S.B.* VII.2.2.4. | *Aitareya:* 235. | *Chāndogya:* VI.3.1. | | *Dhānya-bīja* means the seed which produces corn according to the *R.V.* |

*Appendix-3(a)*

# Trees and Plants Described in Epics, Purāṇas etc.

| S. No. | Name | Botanical equivalent | Epic | | Purāṇas | | | | Other Sources | Remarks |
|---|---|---|---|---|---|---|---|---|---|---|
| | | | *Rāmāyaṇa* | *Mahā-bhārata* | *Matsya* | *Vāyu* | *Varāha* | *Vāmana* | | |
| 1. | Abdaka | *Cyperus rotandus* | | | 118.6 | | | | | |
| 2. | Abhayā | *Andropogon muricatus* | | | | | | | | |
| 3. | Abja (Born in water or lotus) | *Barringtonia acutangula* | | | 118.6 | | 9.32; 45.3; 60.7; 73.28. | | | |
| 4. | Abjini (Multitrade of lotus flowers) | (Lotus) | | | | | | 38.57. | | |
| 5. | Āḍhākya | *Cajanus indicus* Spreng. | | | | 8.152 | | | | |
| 6. | Agaru | *Aporusa agallocha,* Roxb. | II.76.16 | Dron: 34.16. Santi: 170.18 | 161.63 | 75.32 | 101.3; 103.3; 104.4; 178.8; 179.11; 194.31. | 17.60; 36.13; 58.8; 68.20. | Kangra Paintings. | A large evergreen tree with fragrant heart wood. |
| 7. | Agstiphala | *Sesbania grandiflora,* Pers. | | | 96.9 | | | | | |
| 8. | Aila or Ela | *Elettaria cardamomum,* Maton. | | | 217.66 | | | | | |
| 9. | Ailāka | A species of plant | | | 96.7 | | | | | |

| S. No. | Name | Botanical equivalent | Epic | | Purāṇas | | | | Other Sources | Remarks |
|---|---|---|---|---|---|---|---|---|---|---|
| | | | *Rāmāyaṇa* | *Mahā-bhārata* | *Matsya* | *Vāyu* | *Varāha* | *Vāmana* | | |
| 10. | Airaṇḍaka | A species of plant | | | 217.66 | | | | | |
| 11. | Āka | *Calotropis gigantea*, Linn. | | Udyoga: 100.10 Vana: 3.42; Sabha: 10.7. | | | | | | |
| 12. | Ākalla | *Anthemis pyrethrum* | | | 118.11. | | | | | |
| 13. | Aksoṭā | Wall nut. *Jatropa regia*, Linn. | | | 118.6; 217.53. | 38.66. | | | | |
| 14. | Alābu | *Lagenaria vulgaris*, Mol. | | | 118.27. | | | | | |
| 15. | Ambuja | Lotus | | | 62.24. | 94.37. | 127.45. | 36.25. | | |
| 16. | Āmalaka or Āmalakī | *Emblica officinalis*, Gaertn. | | | | | 117.29; 146.66. | 17.55; 58.8; 68.28; 64.49; 68.15. | | |
| 17. | Amalavetsa or Amalaveta | *Emblica officinalis*, Gaertn. | | Yes | 217.55. | | | | | A small short tree with pendolous large bright flowers which bloom in April-May. |
| 18. | Amalatāsa | *Cassia fistula* | | | | | | | Kangra Paintings | |

| S. No. | Name | Botanical equivalent | Epic | | Purāṇas | | | | Other Sources | Remarks |
|---|---|---|---|---|---|---|---|---|---|---|
| | | | *Rāmāyaṇa* | *Mahā-bhārata* | *Matsya* | *Vāyu* | *Varāha* | *Vāmana* | | |
| 19. | Amamsi | *Amallasa Cassia fistula* | | | | 118.38 | | | | |
| 20. | Āmra | *Mangifera indica* | 1,51.23; 4.1.60; 1.34.16. | Vana: 158.61. | 96.11; 187.29; 257.8. | 69.307; 69.308. | 55.42; 146.64; 168.24. | | | This represents the mango tree. |
| 21. | Amaratika | *Spondias pinnata* Kurz. | | | 96.5; 118.8; 217.55-56. | | | | | |
| 22. | Amṛta | *Eulophia campestris*, Wall. | | | 217.65. | | | | | |
| 23. | Andira | | | | 217.65. | | | | | |
| 24. | Anjana | *Cenchrus ciliaris*, Linn. | | | 118.12; 217.77. | | | | | |
| 25. | Anjira | *Ficus carica*, Linn. | | Yes | | | | | | |
| 26. | Ankada | *Alangium salvifolium* | | Yes | | | | 6.19. | | |
| 27. | Ankurasphuta | | | | 217.73. | | | | | |
| 27a. | Anu | *Panicum miliaceum* | | | | | | 8.50 | | |
| 28. | Apāmārgā | *Achyranthes aspera*, Linn. | | | 93.27; 215.31. | | | | | |
| 29. | Arani | *Premna obtusi-folia, R.Br.* | | Yes | | | | | | |
| 30. | Arjuna | *Terminalia arjuna*, Roxb. | Yes | Yes | 118.6; 161.56; | | 17.12; 123.45; | 1.18; 58.9. | | A large shady tree. |

| S. No. | Name | Botanical equivalent | Epic | | Purāṇas | | | | Other Sources | Remarks |
|---|---|---|---|---|---|---|---|---|---|---|
| | | | *Rāmāyaṇa* | *Mahābhārata* | *Matsya* | *Vāyu* | *Varāha* | *Vāmana* | | |
| | | | | | 186.40; 217.76; 257.11. | | 138.70; 154.4; 155.21. | | | |
| 31. | Araka | *Gardenia enneandra* | | | | | 155.12; 158.20. | 17.55; 44.86. | | |
| 32. | Aravinda | *Nelumbium speciosum* | | | | | 184.31; | 58.42. | | |
| 33. | Asanā | *Bridelia stepularis*, Blume | Yes | | | | | | | |
| 34. | Aśoka | *Saraca asoca*, Roxb. | Sundara Kāṇḍa | Ādi: 262.48 Droṇ: 131.51 Anu: 32.7 Vana: 64.10. | | | 11.33; 91.14; 125.45. 150.26. | 12.51; 62.18; 68.12. | Kangra Paintings | A herald of spring, has scarlet crimson bunches of flowers. It is believed to blossom with the touch of a lady. |
| 35. | Aśvaparṇa | *Vatica robusta* | | | | | 117.26. | | | |
| 36. | Aśvattha | *Ficus religiosa*, Linn. | 2.10.13; 4.1.30; 4.10.11. | | | | 77.23; 125.78; 126.34; 127.4; 147.94; 170.36. | 14.37; 18.8; 15.32, 38; 58.69. | | |
| 37. | Atasi | *Linum usitatissimum* Linn. | | | | | | 44.34. | Kangra Paintings. | |
| 38. | Atimuktā (Creeper) | *Hiptage madablota*, Gaertn. | | Yes | | | | 36.13. | | |

| S. No. | Name | Botanical equivalent | Epic | | Purāṇas | | | | Other Sources | Remarks |
|---|---|---|---|---|---|---|---|---|---|---|
| | | | *Rāmāyaṇa* | *Mahā-bhārata* | *Matsya* | *Vāyu* | *Varāha* | *Vāmana* | | |
| 39. | Badarī/Badara | *Zizania jujuba*, Mill. or *Zizania mauritiana*, Lam. | | Vana: 145.17. | 96.6; 118.17 187.33. | 38.66-70. | 48.7; 144.60; 16.24. | | | |
| 40. | Bākula | *Manilkara hescandra*, Roxb. | | | 118.16; 180.25; 161.56; 255.23. | | | 6.19; 68.13. | | |
| 41. | Bāṇa | *Latipes nudi-caulis*, Hook. | | | 60.38; 81.28; 118.4. | | | 68.12. | | |
| 42. | Bandujihvā | *Pentapetas phoenicea*, L. | | | 62.22. | | | 6.19; 18.8; 39.44. | | |
| 43. | Banduka | *Polygonum aviculare*, Linn. | | | 118.18. | | | | | |
| 44. | Barhis | *Celosia cristata* | | | | | | 68.17. | | |
| 45. | Betal | A climber | | | | | | | Kangra Paintings. | |
| 46. | Bhadra | *Tithyamalus antiquorum*, L. | | | | | | 17.38. | | |
| 47. | Bhadradārū | *Pinus deodora* | | | 161.61. | | | | | |
| 48. | Bhallātaka(ki) | *Samecarpus anacardium*, Linn. | | Yes | 118.23; 217.62. | | | | | |
| 49. | Bhaṇḍi | *Rubia munijosta*, L. | | | | 69.307-08. | | | | |
| 50. | Bhandira | *Acacia sirissa* | | | 118.30. | | | | | |

| S. No. | Name | Botanical equivalent | Epic | | Purāṇas | | | | Other Sources | Remarks |
|---|---|---|---|---|---|---|---|---|---|---|
| | | | *Rāmāyaṇa* | *Mahābhārata* | *Matsya* | *Vāyu* | *Varāha* | *Vāmana* | | |
| 51. | Bhavya | *Dillenia indica*, L. | | | | | | | Kangra Paintings. | |
| 52. | Bhṛṅga/Bhṛṅga-rāja | *Vigna aconitifolia*, Jacq. | | | 217.71. | | | 6.21; 68.15. | | |
| 53. | Bhūmipuṣpa | A species of flower. | | | 118.4; 217.39; 217.77. | | | | | |
| 54. | Bhurja or Bhurjapatra | *Betula arloides*, Buah-Ham. | Yes | | 118.4; 217.64. | | | | | |
| 55. | Bhūstṛṇa | *Cymbopogon citratus*, Stapff. | | | 118.39; 217.64. | | | | | |
| 56. | Bhūtumbi | A species of plant. | | | 217.69. | | | | | |
| 57. | Bibhitakā | *Terminalia anacardium* | | | 118.4; 217.75. | | 177.8. | | | |
| 58. | Bījapinaka | *Semicarpus anacardium* | | | 118.20; 161.64. | | | | | |
| 59. | Bījapūraka | *Citrus limonum*, Linn. | | | 217.56. | | 168.29. | | | |
| 60. | Bilva | *Aegle marmelos*, C & Roxb. | 2.14.34. | | 56.16; 60.34. | | 27.39; 63.6; 79.8,12; 80.25, 26-70; and many others. | 1.22; 6.18; 18.8; 36.12; 58.8; 27.39; 68.15. | | |
| 61. | Bilvataki | *Aegle mermelus* | | | 218.2. | | | | | |

| S. No. | Name | Botanical equivalent | Epic | | Purāṇas | | | | Other Sources | Remarks |
|---|---|---|---|---|---|---|---|---|---|---|
| | | | *Rāmāyaṇa* | *Mahā-bhārata* | *Matsya* | *Vāyu* | *Varāha* | *Vāmana* | | |
| 62. | Bimba | *Momordica monadelpha*, L. | | Asva: 46.6. | 118.20 | | 91.3 | | Kangra paintings. | A creeper with bright red fruit. |
| 63. | Cakramardana | A species of plant | | | 118.21 | | | | | |
| 64. | Campaka | *Michelia champaca* Linn. | 2.10.13 | Anu : 42.16. Virat: 55-10. | 118.14; 118.18; 161.57; 255.23. | | 133.33; 91.14. | 6.98; 58.8; 68.12. | Kangra Paintings | |
| 65. | Canaka | *Cicer arietinum*, Linn. | | | | 8.152. | | | | |
| 66. | Candana | *Santalum album*, Linn | | Droṇa : 34.16. Virāṭa: 20.24. Anu: 200.24. | 64.16; 118.6 | | 50.90: 103.3; 104.4; 150.26; 174.7; 178.108; 179.12; 180.27; 194.31. | 12.7; 17.47; 25.6; 36.12, 13; 41.37; 42.8; 45.5; 58.8; 68.19. | | |
| 67. | Caturaṅgala | *Cathartocarpus fistula* | | | 217.72. | | | | | |
| 68. | Cavya | *Scindapsus officinalis*, Schott | | | 217.62; 118.15. | | | | | |
| 69. | Citraka | *Racenus communis* | | | 161.59; 217.62. | | | | | |
| 70. | Citravalliphala | *Cucumis coloquintha*, L. | | | 96.10. | | | | | |

| S. No. | Name | Botanical equivalent | Epic | | Purāṇas | | | | Other Sources | Remarks |
|---|---|---|---|---|---|---|---|---|---|---|
| | | | *Rāmāyaṇa* | *Mahābhārata* | *Matsya* | *Vāyu* | *Varāha* | *Vāmana* | | |
| 71. | Cukraphala | Tamarind tree | | | 217.57 | | | | | |
| 72. | Cūta | *Mangifera indica*, Linn | | | | | 39.44 | 6.105; 12.51; 17.52; 58.8. | | |
| 73. | Dāḍima | *Punica granatum*, Linn. (Pomegranate) | Yes | Karna: 21.30 | 96.6; 118.18; 154.520; 161.64; 217.55; 255.23. | | 170.36. | 64.97 | | |
| 74. | Dantaśṭha | *Citrus aurantium* | | | 217.57. | | | | | |
| 75. | Dantikā | *Croton polyandrum* | | | 217.71. | | | | | |
| 76. | Dantitvāka | A species of plant. | | | 217.69. | | | | | |
| 77. | Darbha | *Desmostachya bipinata*, Staff | | | | 75.17-18; 75.79; 75.83. | 5.47; 14.16; 99.8. 104.6; 106.5; 197.5; 108.3; 186.43. 188.56; 81, 114. | 10.80 | | |
| 78. | Darbharenukā | *Saccharum cylindricum*, W. | | | 217.75. | | | | | |

| S. No. | Name | Botanical equivalent | Epic | | Purāṇas | | | | Other Sources | Remarks |
|---|---|---|---|---|---|---|---|---|---|---|
| | | | *Rāmāyaṇa* | *Mahābhārata* | *Matsya* | *Vāyu* | *Varāha* | *Vāmana* | | |
| 79. | Darvika | *Curcuma aromatica* | | | 217.49. | | | | | |
| 80. | Devadārū | *Plyalthia longifolia*, Thw. | | | 274.32. | | | 68.48; 68.20. | Kangra Paintings | |
| 81. | Dhanikā | *Panicum italium* | | | 118.7. | | | | | |
| 82. | Dharvana | *Grewia tiliaefolia* Vohl. | | | 217.46. | | | | | |
| 83. | Dhānya | *Coriandrum sativam*, Linn. | | | 62.9; 217.63; 217.80. | | 71.12; 76.15; 103.6. | | | |
| 84. | Dhanyaka | *Coriandrum sativum* Linn. | | | 217.50 | | | | | |
| 85. | Dhātakī | *Woodfordia fruitcosa*, Kurz. | | | 118.22; 217.16. | | 211.70. | | | |
| 86. | Dhattūra | *Datura tatula*, Linn. | | | 95.24 | | | 16.32; 17.32; 18.4; 36.12. | | |
| 87. | Dhava | *Anogeissus latifolia*, Wall. | Yes | | 118.8; 161.56. | | | | | |
| 88. | Drākṣa | Grapes *Vitis vinifera*, Linn. | | | 96.8; 118.24. | | 80.70 | | | |
| 89. | Drukāla | *Pinus deodara* | | | 161.62. | | | | | |
| 90. | Duḥasparsa | A species of plant. | | | 217.72. | | | | | |

| S. No. | Name | Botanical equivalent | Epic | | Purāṇas | | | | Other Sources | Remarks |
|---|---|---|---|---|---|---|---|---|---|---|
| | | | *Rāmāyaṇa* | *Mahābhārata* | *Matsya* | *Vāyu* | *Varāha* | *Vāmana* | | |
| 91. | Dūrvā or Durvara | *Desmostachya bipinnata*, Staff. | | | 93.27; 217.69. | 75.39. | 99.33 | 14.36; 18.9; 68.18. | | |
| 92. | Elāvatī | *Elettaria cardamum*, Maton | | | 118.9. | | | | | |
| 93. | Eraṇḍa | *Ricinus communis*, Linn. | | | 217.65 | | | | | |
| 94. | Gairika (Jala madhūka) | Name of a tree | | | 217.77. | | | | | |
| 95. | Gandhapalatā | *Hyperanthera moringa* | | | 62.24. | | | 122.35. | | |
| 96. | Gandhapatra | Turmeric *Aegle marmelos* | | | | 122.35. | | | | |
| 97. | Ghana | Bulbous root or *Cyperus hexastachyos communis* | | | 217.79. | | | | | |
| 98. | Gandharāja | *Gardenia jasminoides*, Ellis | | | | | | | Kangra Paintings. | A shrub with highly fragrant flowers. |
| 99. | Girikarṇa | *Clitoria ternatea*, Linn. | Āraṇya kāṇḍam | | | | | | | |
| 100. | Giriśālinī | *Clitoria ternatea*, Linn. | | | | | | 68.13. | | |
| 101. | Godhūma | *Tristania aestivum*, Linn | | | 266.12 | 8.150. | 8.82; 33.29; 38.11; | 68.21 | | |

| S. No. | Name | Botanical equivalent | Epic | | Purāṇas | | | | Other Sources | Remarks |
|---|---|---|---|---|---|---|---|---|---|---|
| | | | *Rāmāyaṇa* | *Mahābhārata* | *Matsya* | *Vāyu* | *Varāha* | *Vāmana* | | |
| | | | | | | | 58.10; 114.21. | | | |
| 102. | Govalli | A species of tree. | | | 217.50. | | | | | |
| 103. | Gṛnjana | *Allium ascalonicum* Linn. | | | | 78.12-14. | | | | |
| 104. | Guḍūci | *Cocculus cordifolius* | | | 217.74. | | | | | |
| 105. | Guggulu | *Commiphara mukul*, Hook. | | | 118.21. | | | 17.49 | | |
| 106. | Gulma | A climber creeper. | | | | | 148.29; 162.2; 206.50. | | | |
| 107. | Gulmohar | | | | | | | | Kangra Paintings. | An umbrella like tree with pinnate feathery leaves. |
| 108. | Gunja | *Abrus precatorius*, Linn. | | | | | | | Kangra Paintings. | |
| 109. | Gunjalātakā | *Abrus precatorius*, Linn. (berry) | | | 118.44; 217.51. | | | | | |
| 110. | Hanjikā | *Clemodendrum siphonantus* | | | 217.65. | | | | | |
| 111. | Harenuka | | | | 217.73. | | | | | |
| 112. | Harenumamsi | | | | 218.11. | | | | | |

| S. No. | Name | Botanical equivalent | Epic | | Purāṇas | | | | Other Sources | Remarks |
|---|---|---|---|---|---|---|---|---|---|---|
| | | | *Rāmāyaṇa* | *Mahābhārata* | *Matsya* | *Vāyu* | *Varāha* | *Vāmana* | | |
| 113. | Haridra | *Curcuma domestica*, Valeton | | | 118.16; 161.57; 217.68. | | | | | |
| 114. | Haridrava | *Mesua roxburgii* | | | 257.9. | | | | | |
| 115. | Haridruma | *Chloroxylon suietenia* | | | 118.5. | | | | | |
| 116. | Haritaki | *Terminalia citirina*, Roxb. | | | 217.66. | | | | | |
| 116a | Harilatā | *Panicum dactrylcn* | | | 217.64. | | | | | |
| 116b | Hingu | *Ferula foetida* Regd. | | | 161.62; 217.63. | | | | | |
| 117. | Hingupatrikā | *Gardenia gummifera* L. | | | 217.66. | | | | | |
| 118. | Harivara | *Mesua roxburgii* | | | 217.68. | | | | | |
| 119. | Honey Apple | | | | | | | | Kangra Paintings | A large round fruit. |
| 120. | Ikṣu | *Saccharum officinarum*, Linn. | | | 60.8-9; 62.28; 118.36. | 49.29-30. | 54.17; 99.9; 100.3; 101.4; 102.7; 103.4; 104.6; 106.5; 107.2; 109.8. | 62.17 64.43. | | |

| S. No. | Name | Botanical equivalent | Epic | | Purāṇas | | | | Other Sources | Remarks |
|---|---|---|---|---|---|---|---|---|---|---|
| | | | *Rāmāyaṇa* | *Mahābhārata* | *Matsya* | *Vāyu* | *Varāha* | *Vāmana* | | |
| 121. | Indivara | *Nymphaea stellata* | | | | | | 22.32; 68.17. | | |
| 122. | Indrayava | *Wrightia antidysenterica* | | | 118.23. | | | | | |
| 123. | Inguda or Ingudi | *Terminalia catappa* or *Balanites aegyptiaca*, Linn. | 2.103.20; 2.103.29. | Anu : 142.7 | 274.32, 118.8. | | 154.4. | | | |
| 124. | Indrayava | *Wrightia antidysenterica* | | Yes | 118.33. | | 54.17; 99.9; 100.3; 101.4; 102.7; 103.4; 106.5; 107.2. | | | |
| 125. | Irapuṣpa | Reed grass | | | 118.38. | | | | | |
| 126. | Jambava | *Eruca sativa*, Mill. | | | 217.58. | | | | | |
| 127. | Jambira | *Citrus lemonum*, Linn. | | | 96.6; 118.23; 255.24. | | 168.24. | | | |
| 128. | Jambu | *Xylia xylocarpa*, Roxb. | 4.228.30. | Bhīṣma: 7.20. | 96.7; 161.61 | | 77.17 140.73; 166.8. | | | |
| 129. | Japā (Kusum) | *Hibiscus rosasinensis*, Linn. | | | 60.39. | 75.37. | | 68.13. | | |

| S. No. | Name | Botanical equivalent | Epic | | Purāṇas | | | | Other Sources | Remarks |
|---|---|---|---|---|---|---|---|---|---|---|
| | | | *Rāmāyaṇa* | *Mahā-bhārata* | *Matsya* | *Vāyu* | *Varāha* | *Vāmana* | | |
| 130. | Jaṭāmānasī | *Nardostachys jatamansi* | | | | | 99.4 | | | |
| 131. | Jāti | *Jasminum grandiflorum*, Linn. | | | 118.14; 118.36; 255.24. | | 6.101, 12.50, 68.12. | 6.101, 12.50; 68.12, 20. | | |
| 132. | Jātiphala | *Myrica fragrans* | | | 217.65. | | | 68.20. | | |
| 133. | Jīvaka | *Terminalia tomentosa* | | | 217.43. | | | | | |
| 134. | Jīvitā | *Lepta demaretriculata* | | | 217.74. | | | | | |
| 135. | Kachanāra | *Bauhinia tomentosa* Linn. | | | | | | | Kangra Paintings | A medium sized ornamental tree with drooping branches. |
| 136. | Kadalī | *Lagerstroemia speciosa*, Pers. | | | 96.6; 118.37; 180.36; 255.24. | | 201.26; 213.17. | 7.11. | | |
| 137. | Kadamba | *Anthocephalus cadamba*, Miq. | | | 62.24; 95.24; 118.16; 161.56; 180.27; 217.80. | | 17.12; 77.11; 123.45; 138.70; 147.63; 158.26; 162.26. | 1.18; 17.9; 18.2; 26.71; 58.8. | Kangra Paintings | It has ball-like flowers in rains. Believed to be the favourite of Kṛṣṇa. |
| 138. | Kākajihvā | *Arbus precatorius* | | | 217.51. | | | | | |

| S. No. | Name | Botanical equivalent | Epic | | Purāṇas | | | | Other Sources | Remarks |
|---|---|---|---|---|---|---|---|---|---|---|
| | | | *Rāmāyaṇa* | *Mahā-bhārata* | *Matsya* | *Vāyu* | *Varāha* | *Vāmana* | | |
| 139. | Kākamacī | *Solanum nigrum*, Linn. | | | | | | 12.53. | | |
| 140. | Kākodumbara (barikā) | | | | 217.70. | | | | | |
| 141. | Kākola | *Piper cubeba*, Linn. | | | 217.43. | | | | | |
| 142. | Kālābhra | *Andrographis peneiculata* | | Bhīṣma : 7.14. | | | | | | |
| 143. | Kalahāra | *Nymphaea lotus* | | | | | 78.8; 122.8; 138.76; 211.21. | 18.17; 22.32; 8.17. | | |
| 144. | Kalama | A type of Śali. | | | | | | 27.46; 58.17. | | |
| 145. | Kalamālikā | *Rubia munijosla*, L. | | | 217.63. | | | | | |
| 146. | Kālāmra | Black Mango | | | | | 82.20. | | | |
| 147. | Kālīyaka | | | | 161.62. | | 180.27. | 68.19; 36.13. | | |
| 148. | Kalpavṛkṣa | A divine tree. | | Karṇa : 93.44. | 11.44; 92.5; 121.1. | 45.40. | | | | |
| 149. | Kālayakādruma | A kind of aloe wood. | | | 118.5. | | 180.27. | | | |
| 150. | Kāmā | *Cycas rumphii*, Miq. | | | 217.72. | | | | | |
| 151. | Kamalā | *Nelumbo nucifera*, Gaertn. | | | | | 1.22; 24.11; | 6.17; 22.32; | | |

| S. No. | Name | Botanical equivalent | Epic | | Purāṇas | | | | Other Sources | Remarks |
|---|---|---|---|---|---|---|---|---|---|---|
| | | | *Rāmāyaṇa* | *Mahā-bhārata* | *Matsya* | *Vāyu* | *Varāha* | *Vāmana* | | |
| | | | | | | | 55.41; 73.27; 78.8; 89.37; 149.4; 172.70; 188.76; 209.67. | 31.20; 58.17; 36.12; 62.14. | | |
| 152. | Kancana | *Toricellia asiatica*, Linn | | | | | 58.17. | | | |
| 153. | Kanakabhadrā | *Butea frondosa* | | | 83.20. | | | | | |
| 154. | Kandalī | *Amorpho-phallus campanulatus* or mushroom | | | | | | | Kangra Paintings. | An annual herb with blue purple flowers. |
| 155. | Kāṇḍaka | *Amorpho-phallus campanulatus* | | | 118.31. | | | | | |
| 156. | Kundūra | *Mucuna pruriens* | | | 217.66. | | | | | |
| 157. | Kankota | *Piper cubeba*, Linn. | | | 96.7; 118.25; 217.80. | | 80.70. | | | |
| 158. | Kangu | *Selaria italica* Beaux | | | | | 118.11; 114.21. | | | |
| 159. | Kapittha | *Feronia limonia* Linn. | | | 96.5; 118.6; 217.5; 218.5. | 75.75-76. | | | | |

| S. No. | Name | Botanical equivalent | Epic | | Purāṇas | | | | Other Sources | Remarks |
|---|---|---|---|---|---|---|---|---|---|---|
| | | | *Rāmāyaṇa* | *Mahā-bhārata* | *Matsya* | *Vāyu* | *Varāha* | *Vāmana* | | |
| 160. | Karamarda | *Carissa carandas*, L. | | | 96.7; 118.24; 217.55. | | | | | |
| 161. | Karakoti/ Karkata | *Centropus pyropterus* | | | 187.33. | | 117.26. | | | |
| 162. | Karakotakī | *Momordia dioica*, Roxb. | | | 118.28. | | | | | |
| 163. | Karakṛṣṇa | | | | 217.74. | | | | | |
| 164. | Karṇikāra | *Pterospermum acerifolium*, Willd. | | Vana: 158.66 Udyoga: 22.29. | 118.3; 180.25. | | | 6.12; 58.8; 26.135. | Kangra paintings. | Rich in golden colour but is devoid of fragrance. |
| 165. | Karakandu | *Zizyphus mummularia* | | | 217.55. | | | | | |
| 166. | Karanja | *Pongamia pinnata*, Pierre | | Vana: 230.55. | | | | | | |
| 167. | Kārpāsi/ Karpasa | *Gyossipum hertacaum* L. | | | 118.38; 217.77. | | | 12.52; 15.6. | | |
| 168. | Karata | *Carthanus tinctorius* | | | 118.44. | | | | | |
| 169. | Kāravī | *Carvia callosa*, Nees. | | | 217.63. | | | | | |
| 170. | Karavīrā | *Nerium indicum*, Mill. | 4.1.21; 2.30.12. | Droṇa: 23.62. | 60.38; 75.3; 118.19, 21. | | 125.52. | 17.36; 50.36; 68.12. | | |
| 171. | Karīra | *Capparis aphylla* | | | 96.7; 217.80. | | 126.37; 154.4. | | | |

| S. No. | Name | Botanical equivalent | Epic | | Purāṇas | | | | Other Sources | Remarks |
|---|---|---|---|---|---|---|---|---|---|---|
| | | | *Rāmāyaṇa* | *Mahābhārata* | *Matsya* | *Vāyu* | *Varāha* | *Vāmana* | | |
| 172. | Karusa | A species of fruit. *Terminalia bellarica* | | | 217.55, 58. | 8.152. | | | | |
| 173. | Kasa | *Saccharum spontaneum*, L. | | | 61.50; 118.35; 154.300. | 75.41. | | | Kangra Paintings. | |
| 174. | Kasari | *Scirpus mysoor* Roxb. | | | 217.79. | | | | | |
| 175. | Kasamarda | *Cassia occidentalis* Linn. | | | 118.24, 31. | | | | | |
| 176. | Kaseru | *Cyperus esculantus*, Linn. | | | 217.52. | | | | | |
| 177. | Kāśmārī | *Gmelina arborea*, Roxb. | | | 96.6; 271.47. | 75.2 | | | | The vessels were made of the wood of this tree, when used, drives away demons. |
| 178. | Kasamaryaphala | *Gmelina arborea*, Roxb. | | | 96.6. | | | | | |
| 179. | Kāśmīrī | *Ficus elastica*, L. | | | 118.8; 217.52. | | | | | |
| 180. | Kaṭaphala | *Solanum melongena* | | | 118.9, 17; 217.40, 76. | | | | | |

| S. No. | Name | Botanical equivalent | Epic | | Purāṇas | | | | Other Sources | Remarks |
|---|---|---|---|---|---|---|---|---|---|---|
| | | | *Rāmāyaṇa* | *Mahā-bhārata* | *Matsya* | *Vāyu* | *Varāha* | *Vāmana* | | |
| 181. | Katu(ka) | *Hypuanthala moringa*, L. | | | 217.46. | | | | | |
| 182. | Kayastha | *Myrobalanus chebula*, L. | | | 217.64. | | | | | |
| 183. | Kesara | Saffron *Rottleria tinctoria* | | | 57.16; 118.12; 217.52. | | 183.5. | 6.99. | Kangra Paintings | Girdles made of the flowers of this tree are used by ladies. |
| 184. | Ketaka | *Pandanus odoratissimus* | | | 57.16; 81.28; 118.12; 255.24. | | | | | The plant when planted in a house increases its prosperity. |
| 185. | Ketakī | *Pandanus odoratissimus* | | | | | 211.17. | 1.18; . .68.14 | Kangra Paintings. | This is a highly fragrant plant with sword-like pointed leaves. |
| 186. | Khadira | *Abutilon catachu*, Willd. | 3.15.18; . 1.14.22 | Śānti: 38.40. | 69.29; 93.27; 217.80. | 74.9. | 127.54; . 151.42 | 18.5. | | |
| 187. | Kharjura (Date Palm) | *Phoenix dactylifera*, Linn. | | | 118.8; 161.62; 217.46. | 65.75. | 80.70. | | | |
| 188. | Khasa | *Vetveria zizanoides*, L. | | Udyoga: 191.21. | | | | | | |
| 189. | Kīcaka | *Arundo bamboo* | | | 118.35. | | | 58.18, 68. | | Produces charming sound with the blowing of the wind. |
| 190. | Kaṁjalka | A type of lotus. | | | 154.30. | | | | | |

| S. No. | Name | Botanical equivalent | Epic | | Purāṇas | | | | Other Sources | Remarks |
|---|---|---|---|---|---|---|---|---|---|---|
| | | | *Rāmāyaṇa* | *Mahā-bhārata* | *Matsya* | *Vāyu* | *Varāha* | *Vāmana* | | |
| 191. | Kiṁśuka | *Butea frondosa* | | | 118.9; 217.78. | | 80.26. 4.29; 6.9; 16.46. | | Kangra Paintings. | |
| 192. | Kinki | A species of plant. | | | 217.63. | | | | | |
| 193. | Kinkiriṭa | *Flacourtia sapide* | | | 118.12. | | | | | |
| 194. | Kirita | *Swertia chirata* Buch. Ham. | | | 118.12; 217.69. | | | | | |
| 195. | Kakamnada | *Nelumbium speciosum,* Willd | | | | | | 22.32; 62.14. | | |
| 196. | Kolaka (aṇḍa) | A sort of bulbous plant (used as remedy for worms) *Piper cubeba* | | | 217.56. | | | | | |
| 197. | Kovidāra | *Bauhinia variegata* | 2.96.18. | Vana Parva. | 118.7,9. | | | | Kangra Paintings. | A small tree yielding flowers in November. |
| 198. | Kṛṣṇadruma-barka | *Ficus hispida* | | | | | | 18.7. | | |
| 199. | Kṛtamālaka | *Cassia fistula* | | | 217.68. | | | | | |
| 200. | Kṣīrakasāka | *Ficus arnotiana* | | Bhisma: 8.4.5. | 118.32. | | 84.5; 98.18. | | | |
| 201. | Kṣīravṛksa | The tree which emits milk like white fluid. | | | 118.22. | | | | | |

| S. No. | Name | Botanical equivalent | Epic | | Purāṇas | | | | Other Sources | Remarks |
|---|---|---|---|---|---|---|---|---|---|---|
| | | | *Rāmāyaṇa* | *Mahā-bhārata* | *Matsya* | *Vāyu* | *Varāha* | *Vāmana* | | |
| 202. | Kṣīrika (-druma) | A divine tree. | | | 118.22. | | | | | By the planting of this tree one is believed to ensure happiness for seven generations. |
| 203. | Kuberaka | *Cedrela toona* | | | 118.40; 217.62. | | | | | |
| 204. | Kubjaka or (Apāmārgā) | *Achyranthes aspera* | | | 60.38; 81.28. | | | | | |
| 205. | Kultha | *Dolichos biflarus*, Linn. | | | | | 118.11. | | | |
| 206. | Kumkuma | *Crocus sativus* Linn. | | | 60.38; 81.28; 118.38; 154.230; 217.65. | | | 68.19. | | |
| 207. | Kumuda | *Nymphaea rubra* | | | | | 78.9; 80.50; 82.24; 122.8; 148.30; 184.26. | 22.32; 58.17. | | |
| 208. | Kunda | *Jasminum multiforum* | | | 62.23; 118.11; 161.63. | | 151.35; 184.26; 194.23; 211.17; 213.18. | 6.11; 17.47; 18.6; 27.12; 68.12, 19. | Kangra Paintings. | A water lilly having white flowers opening at night. |

| S. No. | Name | Botanical equivalent | Epic | | Purāṇas | | | | Other Sources | Remarks |
|---|---|---|---|---|---|---|---|---|---|---|
| | | | *Rāmāyaṇa* | *Mahā-bhārata* | *Matsya* | *Vāyu* | *Varāha* | *Vāmana* | | |
| 209. | Kuraṇḍaka or Kuraṇṭaka | *Barleria prionitis*, Linn. | | | | 75.34. | | | | |
| 210. | Kurbaka | *Barleria cristala* | | | 118.19. | | 137.37. | | Kangra Paintings. | |
| 211. | Kuraṇṭa | *Barleria prionitis*. Linn. | | | 62.23; 161.63. | | | | | |
| 212. | Kuśa | *Poa cyonosurloides* | 2.4.23 | Śānti : 196.14 | 93.27; 118.36. | | 34.14; 47.10; 126.13; 127.53; 136.48; 155.17; 158.20; 178.64; 185.95; 196.3; 209.21. | 17.42; 26.17; 25.42; 64.38; 46.45; 26.17. | | |
| 213. | Kuśeśya | A king of Lotus. | | | 118.15. | | | | | |
| 214. | Kuṣṭha | *Costus speciosus* L. | | | 217.63; 218.9. | | 99.4. | | | |
| 215. | Kusmāṇḍa | *Benincasa cerifera*, Savi | | | 96.5; 118.27. | | | | | |
| 216. | Kusumbha | *Carthamus tinctorius*, Linn. | | | 60.9,39; 96.7; 118.16; 161.63. | | | | | |
| 217. | Kuṭaja | *Holarrhena antidysenterica*, Wall | | | | | 123.45; 138.70. | | Kangra Paintings | |

| S. No. | Name | Botanical equivalent | Epic | | Purāṇas | | | | Other Sources | Remarks |
|---|---|---|---|---|---|---|---|---|---|---|
| | | | *Rāmāyaṇa* | *Mahā-bhārata* | *Matsya* | *Vāyu* | *Varāha* | *Vāmana* | | |
| 218. | Kutaphala | *Wrightia antidysenterica* | | | 96.10. | | | | | |
| 219. | Lakṣmaṇa | *Ardea sibirica* A plant with red spotted leaves. | | | 266.14. | | | | | |
| 220. | Lakta | A red variety of the *Lodha* tree, bark of which is used in dyeing. | | | 161.63. | | | | | |
| 221. | Lakucha | *Artocarpus lacoocha* | Araṇya-kāṇḍa | Vana Parva | 118.15; 161.61; 217.55. | | 84.6; 94.7. | | | |
| 222. | Lasuna | *Alluim sativum*, Linn | | | 217.64. | | | | | |
| 223. | Lasoḍha | *Cardia oblique* | | Vana : 134.38. | | | | | | |
| 224. | Lāva | *Perdix chinensis* | | | 118.53. | 75.31. | | | | |
| 225. | Lavanga | *Spzygium aro-maticum*, Linn. | | | 118.25. | | | | Kangra Paintings. | |
| 226. | Loca | *Celosis cristala* | | | 118.15. | | | | | |
| 227. | Lodhara | *Symplocos racemosa* | Araṇya-kāṇḍa | Vana Parva | 118.22; 161.61. | | | | Kangra Paintings. | |
| 228. | Lodhara Puśpanī | *Bassia latifolia* | | | 217.3. | | | | | |
| 229. | Lodhrapṛṣṭhā | *Symplocos racemosus* | | | 118.52. | | | | | |

| S. No. | Name | Botanical equivalent | Epic | | Purāṇas | | | | Other Sources | Remarks |
|---|---|---|---|---|---|---|---|---|---|---|
| | | | *Rāmāyaṇa* | *Mahā-bhārata* | *Matsya* | *Vāyu* | *Varāha* | *Vāmana* | | |
| 230. | Lomaka | A species of plant. | | | 217.77. | | | | | |
| 231. | Mādhavī | *Hiptage madalota*, Gaertn. | | | | | | 45.5. | Kangra Paintings. | It blossoms in spring. |
| 232. | Madhucchiṣṭha | *Flacourtia sapide* | | | 217.78. | | | | | |
| 233. | Madhuka or Mahuā | *Bassia latifolia* | Araṇya-kāṇḍa | | | | 117.26; 118.6; 179.10. | 17.40. | Kangra Paintings. | |
| 234. | Madhūlikā | *Aletris hyacinthoides* | | | 217.48. | | | | | |
| 235. | Madhupargu | *Gmelina arborea* | | | 217.45. | | | | | |
| 236. | Madhuvallikā | A kind of grape. | | | 217.50. | | | | | |
| 237. | Mahākṣīra | Sugarcane | | | 217.45. | | | | | |
| 238. | Mahādantāśattha | | | | 218.5. | | | | | |
| 239. | Mahānimba | *Ailantus excelsa*, Roxb. | | | 118.5. | | | | | |
| 240. | Mahāsaṁbha | | | Śalya: 37.21. | | | | | | |
| 241. | Mahātapasa | | | | 217.45. | | | | | |
| 242. | Mahuā | *Bassia latifolia* | | | | | | | Kangra Paintings | A common shade tree in central India. Alcohal is made from its flowers. |

| S. No. | Name | Botanical equivalent | Epic | | Purāṇas | | | | Other Sources | Remarks |
|---|---|---|---|---|---|---|---|---|---|---|
| | | | *Rāmāyaṇa* | *Mahābhārata* | *Matsya* | *Vāyu* | *Varāha* | *Vāmana* | | |
| 243. | Mālatī | *Jasminum grandiflorum* Linn. | | | 60.39; 62.24; 70.4; 95.24; 217.56; 56.74 | | | | Kangra Paintings. | |
| 244. | Mallata | *Jasminum sambac* | Araṇya-kāṇḍa | | | | | | | |
| 245. | Malli | *Jasminum sambac* | | | | | 158.20. | 6.102. | | |
| 246. | Manḍāra or Mandāraka | *Erythrina indica*, Lamk | | | 60.34; 62.24; 118.9; 154.381; 161.62. | | 143.9. | 17.49; 36.13. | Kangra Paintings. | A small tree having red flowers. |
| 247. | Manjiṣṭha | *Rubia munjista* | | | 217.72; 218.6. | | | | | |
| 248. | Manju | *Rubia munijosta* | Vana: 11.18. | | | | | | | |
| 249. | Mārica | A species of plant (Pepper) | | | 118.26. | | | | | |
| 250. | Marika | *Piper nigrum*, Linn. | | | 118.26; 217.62. | | | | | |
| 251. | Māṣa | *Phaseolus radiatus* | | | | 8.151. | 118.11 | 17.61; 68.21. | | |
| 252. | Māśaparṇī | *Teramonus labialis* | | | 217.44. | | | | | |
| 253. | Masūra | *Lens esculenta* | | | | 8.151. | | | | |

| S. No. | Name | Botanical equivalent | Epic | | Purāṇas | | | | Other Sources | Remarks |
|---|---|---|---|---|---|---|---|---|---|---|
| | | | *Rāmāyaṇa* | *Mahā-bhārata* | *Matsya* | *Vāyu* | *Varāha* | *Vāmana* | | |
| 254. | Mātāmra | | | | 118.6. | | | | | |
| 255. | Mātuliṅga or Mātulunga | *Citrus medica* | | | 217.44. | | 170.36; 80.45. | | | |
| 256. | Moca (ka), Mocakhya | *Moringa pterygoperma* or *Musa sapientum.* Linn. | | | 217.76. | | | | | |
| 257. | Mṛdvika | *Vitis vinifera* | | | 118.26; 217.46. | | | | | |
| 258. | Mṛtasanjīvanī | A medicinal plant | Yuddha: 74.33. | | 122.56. | | | | | |
| 259. | Mucukunda | *Pterospermum suberifolium* | | | 118.11. | | | | | |
| 260. | Mudga | *Phaseolus mungo* L. | Araṇya-kāṇḍam | | 96.11. | 8.151. | 118.11. | 16.41; 26.122; 54.17; 68.21, 22. | | |
| 261. | Mudgaparṇī | *Phaseolus trilolus* | | | 217.43. | | | | | |
| 262. | Mukustha | *Phaseolus aconitifolius,* Jacqus | | | | | 118.11. | | | |
| 263. | Mūlaka | *Raphanus sativus,* L. | | | 96.7; 217.63; 217.77. | | 136.114. | | | |
| 264. | Munja | *Saccharum munja* | | Vana: 11.18. | 118.4. | | | | | |

| S. No. | Name | Botanical equivalent | Epic | | Purāṇas | | | | Other Sources | Remarks |
|---|---|---|---|---|---|---|---|---|---|---|
| | | | *Rāmāyaṇa* | *Mahā-bhārata* | *Matsya* | *Vāyu* | *Varāha* | *Vāmana* | | |
| 265. | Mūrva | *Sanseviera roahuburghiana* | | | 218.4. | | | | | |
| 266. | Muṣṭa | *Cyperus rotundus* | | | 217.68. | | | | | |
| 267. | Nagakeśara | *Mesuaferrea*, Roxb. | Kiṣkindhā-kāṇḍa | | 118.24. | | 11.33; 91.14. | | | A fruit tree of eastern India. |
| 268. | Nagara | *Euphorbia antiquorum* | | | 217.62. | | | 68.13. | | |
| 269. | Nagaraka | Dry ginger | | | 217.57. | | | | | |
| 270. | Nagavalli | *Piper betle* | | | 118.25. | | | | | |
| 271. | Naktamālā | *Pongamia globra* | | | 217.68. | | | | | |
| 272. | Nalada | *Naradostachys jatamanasi* | | | 217.68. | | | | | |
| 273. | Nalikera or Narikela | *Cocos mucifera* | | | 118.8;. 161.62; 255.24. | | 53.15. | | | |
| 274. | Nalina | *Nelumbium speciosum* | | | 161.52. | | | | | |
| 275. | Nalinī or Padminī | *Nelumbium speciosum* Water lillies in general. | | | | 17.3. | 194.5. | 12.54; 37.39. | | |
| 276. | Nāranga | (Orange) *Citrus raticulata*, Blane. | | | | | 161.24, 39. | | | |
| 277. | Narakula | *Phragenites karka* | | Vana: 107.3. | | | | | | |

| S. No. | Name | Botanical equivalent | Epic | | Purāṇas | | | | Other Sources | Remarks |
|---|---|---|---|---|---|---|---|---|---|---|
| | | | *Rāmāyaṇa* | *Mahā-bhārata* | *Matsya* | *Vāyu* | *Varāha* | *Vāmana* | | |
| 278. | Navamālikā | *Jasminum sambac* | | | 118.26. | | | | Kangra Paintings. | |
| 279. | Nicula | *Barringtonia acutangula* | | | 180.31. | | | | | |
| 280. | Nikotaka | *Pinus gerardiana*, Wall. | | | 118.12. | | | | | |
| 281. | Nilandivara | *Nymphaea stellata*, Willd. | | | | | | 25.4; 58.6, 18. | Kangra Paintings. | A medicinal shady tree. |
| 282. | Nilasoka | *Amtherstia nobilis*, Wall | | | | | | 6.17. | | |
| 283. | Nilotpala | *Nymphaea cyanea* | | | | | 89.19; 155.84. | 5.18; 17-13, 25.4. | | |
| 284. | Nīma | | | | | | | | Kangra Paintings. | |
| 285. | Nimba | *Melia azadirachta*, L. | | | | | | | Kangra Paintings | Same as No. 284. |
| 286. | Nipa | *Nauclea cadamba* | | | 118.17; 161.56; 217.53; 257.7. | | 17.12. | 1.22; 6.13; 58.8. | | |
| 287. | Nispana | *Dolichos lablab*, Linn. | | | 62.28; 96.11. | | | | | |
| 288. | Nirgundi | *Vitese negundo*, L. | | | 118.5. | | | | | |
| 289. | Nispana | *Dolichos leblab* Linn. | | | 62.28; 96.11. | 8.151. | | | | |

| S. No. | Name | Botanical equivalent | Epic | | Purāṇas | | | | Other Sources | Remarks |
|---|---|---|---|---|---|---|---|---|---|---|
| | | | *Rāmāyaṇa* | *Mahābhārata* | *Matsya* | *Vāyu* | *Varāha* | *Vāmana* | | |
| 290. | Nivāra | Rice grown without cultivation. | | | 118.45; 266.16. | | 114.21. | | | A variety of paddy. |
| 291. | Niyagrodha or Nyagrodha | *Ficus bengalensis* | 6.87.3. | 69.29. | 77.27; 81.55; 84.3; 118.3; 217.78. | 38.68; 35.40, 2; 60.24; 69.150; 72.8-9; 45.3-5. | | | | |
| 292. | Nyagrodha | *Ficus bengalensis* | | | 118.33; 217.78. | 35.40-42; 69.50; 72.8-9; 45.3-5. | 77.27; 81.55; 84.3; 97.6; 170.36. | 38.68; 60.24. | | Same as Niyagrodha. |
| 293. | Auṣadhi or Auuṣadhī | Small creeper | | | | | 34.14; 75.57; 8.29; 11.17; 35.7; 137.1. | | | |
| 294. | Padma | *Clerodendrum siphonanthus* | | | | | 2.12; 9.10; 17.10; 21.50; and many others. | 1.4, 22, 25; 2.24; 3.47; 12.45; 18.1; 26.3; 25.3; 28.23; 44.32. | Kangra Paintings. | Wild cherry found in the temprate Himalayas at an altitude of 600-1000 feet. |

| S. No. | Name | Botanical equivalent | Epic | | Purāṇas | | | | Other Sources | Remarks |
|---|---|---|---|---|---|---|---|---|---|---|
| | | | *Rāmāyaṇa* | *Mahā-bhārata* | *Matsya* | *Vāyu* | *Varāha* | *Vāmana* | | |
| 295. | Padmaka | *Cerasus puddum* | 5.56.26. | | 118.6; 217.79; 257.11. | | | 68.19. | | |
| 296. | Palāṣa | *Butea frondosa* | | | 93.27; 265.31. | 37.18-19; 62.191; 75.1. | 39.46; 126.13; 127.3. | 6.10, 100; 18.7; 62.17. | | Same as Kiṁsuka. |
| 297. | Panāsa | *Artocarpus heterophyillus*, Lam. | | | 96.5; 118.7; 161.60; 180.38; 255.24; 257.11. | 43.4. | 82.16. | | | |
| 298. | Parābata | A species of plant. | | | 217.57. | | | | | |
| 299. | Pañcāmra | | | | | | 170.36. | | | These denote *Āmra*, *Bargada*, *Pipala*, *Pakara* and *Gūlar*. |
| 300. | Paṅkaja | *Nelumbium speciosum* | | | | | 79.14, 147.5; 211.21. | 2.3; 17.34; 18.31; 51.57; 58.22. | | |
| 301. | Paraṇāsa (Reed) | *Grawia asestrea* or *Xylocarpus gromatum* | | | 118.10. | | | | | |
| 302. | Pāribhadra | *Erythrina indica* | | | 118.16; 277.8. | | | 68.13. | | |

| S. No. | Name | Botanical equivalent | Epic | | Purāṇas | | | | Other Sources | Remarks |
|---|---|---|---|---|---|---|---|---|---|---|
| | | | *Rāmāyaṇa* | *Mahābhārata* | *Matsya* | *Vāyu* | *Varāha* | *Vāmana* | | |
| 303. | Pārijāta | *Erythrina indica* | | Ādi : 18.37; Stri: 12.19. | 118.25; 161.61. | 39.11 ff. | 69.28. | 36.13. | | Drops flowers in the morning. |
| 304. | Pariplva | *Azadirachta indica* | | | 217.71. | | | | | |
| 305. | Parpaṭa | *Gardenia latifolia* Ait. | | | 217.71. | | 58.9. | | | |
| 306. | Paruṣka | | | | 118.11. | 38.63 ff. | | | | |
| 307. | Parvata | A species of pot-herb. | | | 217.51. | | | | | |
| 308. | Pāṭala | *Bignonia suavealens* | | | | 78.9. | | | | |
| 309. | Pāṭalā | *Bignonia suavealens* | | Droṇ : 23.16. | 95.24; 118.19; 154.309, 161.57; 218.2; 255.24. | | | 6, 100; 58.8; 68.13. | Kangra Paintings | Trumpet flower, herald of spring. |
| 310. | Pāṭalapatra | A species of tree. | | | 217.77. | | | | | |
| 311. | Paṭola | *Trichosanthes dioica*, Roxb. | | | | 78.9. | | 54.19. | | |
| 312. | Paṭollaka | *Trichosanthes dioica* Roxb. | | | 96.11. | | | | | |
| 313. | Pathya | *Terminalia chebula* Retz. | | | | | | 112.51. | | |

| S. No. | Name | Botanical equivalent | Epic | | Purāṇas | | | | Other Sources | Remarks |
|---|---|---|---|---|---|---|---|---|---|---|
| | | | *Rāmāyaṇa* | *Mahābhārata* | *Matsya* | *Vāyu* | *Varāha* | *Vāmana* | | |
| 314. | Patolikāra | *Trichosanthes dioica*, Roxb. | | | 118.28. | | | | | |
| 315. | Patangi | | | | 217.71. | | | | | |
| 316. | Patrajihvā | *Cinnamomum zeylanicum* Blume. | | | | | 127.70. | 62.21. | | |
| 317. | Phaṅijihvā | *Rivea ornato*, Chois | | | 217.64. | | | | | |
| 318. | Picumaṇḍa | *Azadirachta indica* | | | 187.31. | | 170.36. | | | |
| 319. | Pilu | *Careya arborea* | | | 118.22; 217.79. | | | | | |
| 320. | Piluka | *Salvadora persica* | | | | | 154.4. | | | |
| 321. | Piṇḍāra | A species of bulbous plant. | | | 96.9. | | | | | |
| 322. | Pīppala or Pīpala | *Ficus religiosa* L. | | | 93.27; 217.49; 255.20. | | 117.26; 158.64. | | Kangra Paintings. | A large sacred tree. |
| 323. | Pippalī | *Piper longinum* | 3.1 38. | | | | | | | |
| 324. | Pippalimūla | *Piper longinum* | | | 118.25; 217.62; 255.23. | | | | | |
| 325. | Pitaka | *Odiana pannata* | | | | | | 68.13. | | |
| 326. | Plākṣa | *Ficus infactoria* | 1.4.39. | Vana: 93.10. | 8.8; 217.79; 255.21. | 75.1. | 147.33. | 11.35. | | Its bark is used in religious rites. It symbolises the lordship of the |

| S. No. | Name | Botanical equivalent | Epic | | Purāṇas | | | | Other Sources | Remarks |
|---|---|---|---|---|---|---|---|---|---|---|
| | | | *Rāmāyaṇa* | *Mahābhārata* | *Matsya* | *Vāyu* | *Varāha* | *Vāmana* | | |
| | | | | | | | | | | world and is believed to be auspicious. |
| 327. | Prasthapuṣpā | Tulasī or holy basil. | | | 161.56. | | | | | |
| 328. | Pracinaruka | | | | 217.57. | | | | | |
| 329. | Pratibimba | *Momordica monodelphea* | | | 118.20; 217.80. | | | | | |
| 330. | Priyaka | | | | 217.80. | | | | | |
| 331. | Priyakancana | *Terminalia tomentosa* | | | 274.32. | | | | | |
| 332. | Priyala(ka) | *Buchanania latifolia* | | | 118.8; 118.40; 217.79. | | | | | |
| 333. | Priyaṅgu | *Setaria italica* | | | 161.57; 180.24, 39; 217.76; 218.6; 267.16. | 8.151. | 183.5. | 54.23. | | Blossoms at the touch of a woman. Serves as an antidote to poison. |
| 334. | Pṛṣniparṇī | *Uraria logopioides*, L. | | | 217.43. | | | | | |
| 335. | Pugā | *Araca catechu*, Linn. | | | 118.9; 255.24. | | | | | It grows on sea side. |
| 336. | Punarnava | *Boerhaavia procumbens* | | | 217.51. | | | | | |
| 337. | Puṇḍarīka (Lotus) | *Artemesia indica* | | | | | 6.10; 79.3; 149.30. | 58.17; 62.14. | | |

| S. No. | Name | Botanical equivalent | Epic | | Purāṇas | | | | Other Sources | Remarks |
|---|---|---|---|---|---|---|---|---|---|---|
| | | | Rāmāyaṇa | Mahā-bhārata | Matsya | Vāyu | Varāha | Vāmana | | |
| 338. | Punnāga | *Colcphyllum inophyllum* Linn | | | | 161.56; 180.25; 255.22. | 11.33; 91.14; | 58.8. | Kangra Paintings. | A tree with glabrous leaves and fragrant flowers. |
| 339. | Puṣkara | *Nymphaea nelumbo* | | | | | 21.72; 49.2. | 41.40, 58.53. | | |
| 340. | Puṣpahaṁsa | *Ficres glomerato* | | | 217.48. | | 127.70. | 6.21. | | |
| 341. | Putrajīva | *Putranjiva roxburgii* | | | | | 127.70. | 6.21. | | |
| 342. | Rājādana | *Buchanania latifolia* | | | 217.80. | | | | | |
| 343. | Rājadanti | | | | 67.7. | | | | | |
| 344. | Rājajambu | A species of Jambu tree. | | | 118.30. | | | | | |
| 345. | Rājasarṣapa (black mustard) | *Sanipis racemosa* | | | 217.50. | | | | | |
| 346. | Rajaśirṣaki | | | | 217.49. | | | | | |
| 347. | Raktālukaṇḍika | Red lotus | | | 96.10. | | | | | |
| 348. | Raktacandana | *Caesalpina sappan* | | | | | 154.1. | 50.36. | | |
| 349. | Raktaśālī | A type or rice. | | | | | 118.9. | 17.39; 54.23. | | |
| 350. | Raktāśoka (Red Aśoka) | *Jovesia asoka* | | | | | | 6.17. | | |
| 351. | Rambha | Same as Plantain *Musa sapientum*, L. | | | 217.43. | | | 29.6; 62.18; 64.5. | | |

| S. No. | Name | Botanical equivalent | Epic | | Purāṇas | | | | Other Sources | Remarks |
|---|---|---|---|---|---|---|---|---|---|---|
| | | | *Rāmāyaṇa* | *Mahā-bhārata* | *Matsya* | *Vāyu* | *Varāha* | *Vāmana* | | |
| 352. | Rasānjana | *Amomum anthorrhza* | | | 217.71. | | | | | |
| 353. | Reṇḍī | *Ricinus communis* | | Anu : 147.7. | | | | | | |
| 354. | Risāṭāka | *Sapindus detergens* | | | 118.6. | | | | | |
| 355. | Rodhara | | | | | | 117.26. | | | |
| 356. | Rohisa | A fragrant tree. | | | 217.65. | | | | | |
| 357. | Rohita | *Aphanamixis polystachia* | | | | | 84.3. | | | |
| 358. | Ṛṣabha | *Coropogan pruriens* | | | 217.43. | | | | | |
| 359. | Ṛṣyaprokta | *Glycine debibi* | | | 217.50. | | | | | |
| 360. | Rudanti | A species of a fragrant tree. | | | 217.65. | | | | | |
| 361. | Rudrākṣa | *Elaeocarpus ganitrus*, Roxb. | | | 118.24. | | 127.70. | | | |
| 362. | Rurulā | A fragrant tree. | | | 118.29. | | | | | |
| 363. | Sadagrantha | A flowery plant. | | | 217.71. | | | | | |
| 364. | Sahadeva | *Sida cardifolia* | | | 217.46. | | | | | |
| 365. | Sahadevī | *Vernonia cinerea*, Less. | | | 217.48. | | | | | |
| 366. | Sahakara | *Aloe perfolita* | | | 118.3; 180.28. | | | | | |

| S. No. | Name | Botanical equivalent | Epic | | Purāṇas | | | | Other Sources | Remarks |
|---|---|---|---|---|---|---|---|---|---|---|
| | | | *Rāmāyaṇa* | *Mahābhārata* | *Matsya* | *Vāyu* | *Varāha* | *Vāmana* | | |
| 367. | Saivala | *Ceratophyllum demersum*, Linn. | | | | | | 9.37. | | |
| 368. | Sāla | *Artocarpus locuca* | | Bhīṣma: 7.14; Virāṭa : 17.12. | 154.576; 161.57; 257.8. | | 17.12; 123.9, 43; 146.34; 144.18, 20; 154.4. | 7.43; 58.9. | Kangra Paintings. | A tall tree associated with the birth of Buddha. |
| 369. | Sallakī | *Artocarpus locuca* | | | 217.65; 218.2. | | | | | |
| 370. | Śāli | *Oryza sativa*, Linn. | | | | | 38.11; 55.6, 57.7, 8; 71.14; 114.21; 116.3. | 2.50; 54.18; 56.6,7; 68.21. | | |
| 371. | Śālmalī | *Salmalia malabarica*, Schott | Sundara-kāṇḍam | Śānti : 134.5. Udyoga : 75.19. | 161.57. | | 196.47 and several others. | 12.10. | Kangra Paintings. | The fruit is poisonous. |
| 372. | Saluka | *Nymphaea lotus*, L. | | | 217.52. | | | | | |
| 373. | Samala | | | | 118.3. | | | | | |
| 374. | Samanga | | | | 217.77; 218.2. | | | | | |
| 375. | Śami | *Prosopis spicigera* | Araṇya-kāṇḍaṁ | Vana Parvan | 93.27; 118.10; | | | 18.8; 53.17; | | |

| S. No. | Name | Botanical equivalent | Epic | | Purāṇas | | | | Other Sources | Remarks |
|---|---|---|---|---|---|---|---|---|---|---|
| | | | *Rāmāyaṇa* | *Mahābhārata* | *Matsya* | *Vāyu* | *Varāha* | *Vāmana* | | |
| | | | | | 255.23. | | | 18, 21, 41, 59; 68.15, 31. | | |
| 376. | Samidhā | Fire wood | | | 93.27. | | | | | |
| 377. | Samidhānya | A kind of grain | | | 217.52. | | | | | |
| 378. | Sana | *Crotalaria juncea* | | | 217.39. | | | | | |
| 379. | Śaṁkhapuṣpī | *Causcara decussota,* R&S | | | 267.15. | | | | | |
| 380. | Santana | | | | 92.6 etc. | | | | | |
| 381. | Saptaparṇa | *Alstonia scholaris,* R.Br. | | | 118.4; 161.64-65; 217.70. | | | | | |
| 382. | Sara | *Saccharum arundinaceum* | Araṇya-kāṇḍaṁ | Vana: 225.10. | | | | 18.9. | | |
| 383. | Saraja | *Vateia robusta* | | | 118.30; 217.62; 266.12. | 78.11. | | 1.18, 22; 17.34. 53; 26.71. | | |
| 384. | Sarala | *Pinus roxburghii* | 5.2.6; 4.1.8. | Vana parvam | 161.56; 257.11. | | | 58.9. | | |
| 385. | Sarasapa | *Sarsopa brinicanapus* | | | | | 99.4. | | | |
| 386. | Sariva | *Ichnocarpus frutescens,* R. | | | 217.44. | | | | | |

| S. No. | Name | Botanical equivalent | Epic | | Purāṇas | | | | Other Sources | Remarks |
|---|---|---|---|---|---|---|---|---|---|---|
| | | | *Rāmāyaṇa* | *Mahā-bhārata* | *Matsya* | *Vāyu* | *Varāha* | *Vāmana* | | |
| 387. | Śataparṇi or Satapatra | *Rosa centifolia*, L. | | | 95.25; 217.46. | | | 58.17. | | |
| 388. | Śatapuṣpa | *Peucedanum soiva*, Kurz. | | | 217.48. | | | | | |
| 389. | Śatāvarī | *Asparagus racemosus*, Willd. | | | 217.49. | | | | | |
| 390. | Sauvira | | | | 217.58. | | | | | |
| 391. | Savi(ti)naka | | | | | 8.151. | | | | |
| 392. | Siddharthaka | *Brassica pirta* Moench | | | 243.18. | | | 18.17. | | |
| 393. | Sigru | *Moringa pterygosperma* Gaertn. | | | 217.62. | | | | | |
| 394. | Sillaka | *Alingia excelsa* Noronha | | | | | 179.12. | 68.20. | | |
| 395. | Silapatra | | | | 217.77. | | | | | |
| 396. | Simbi | *Phaseolus trilolus* | | | 118.32. | | | | | |
| 397. | Siṁsapa | *Dalbergia sisu* | Sundara-kāṇḍaṁ | Ādi: 58.26. | 118.3; 217.78; 257.10. | | | | | |
| 398. | Sindhuvara | *Vitex negundo* Linn. | | Anu : 98.17. | 60.38. | | | 6.19; 18.6. | | |
| 399. | Siriśa | *Albizzia lebbeck* Benth | Sundara-kāṇḍaṁ | Bhīṣma: 79.21. | 56.6; 118.3. | | | | | A tree with fragrant flowers. |

| S. No. | Name | Botanical equivalent | Epic | | Purāṇas | | | | Other Sources | Remarks |
|---|---|---|---|---|---|---|---|---|---|---|
| | | | *Rāmāyaṇa* | *Mahā-bhārata* | *Matsya* | *Vāyu* | *Varāha* | *Vāmana* | | |
| 400. | Sisam | *Dalbergia sisu* | | | | | | | Kangra Paintings. | A tree usually found in the Himalayan ranges. |
| 401. | Sitapaki | *Sida cordifolia* | | | 217.51. | | | | | |
| 402. | Sleṣmantikā | *Cordia myxa* Roxb. | | | 118.4; 257.8. | | | | | |
| 403. | Sobha | *Maringa pterygosperma* | | | 118.12. | | | | | |
| 404. | Somabanaka | | | | 217.53. | | | | | |
| 405. | Śrīparṇī | *Premna spinosa* | | | 217.10; 274.32. | | | | | |
| 406. | Śrīphala | *Aegle marmelos* | | | | | | 17.58. | | |
| 407. | Śṛṅgātaka | *Trapa bispinosa* Roxb. | | | 217.45. | | | | | |
| 408. | Śrīvatsa | *The Obeoresin of Penus* Roxb. | | | | | | 17.36. | | |
| 409. | Śṛṅgavera | *Typha angustifolia* | | | 118.40. | | | | | |
| 410. | Śṛṅgī | *Ficus infectoria* A kind of medicinal or poisonous plant. | | | 217.45. | | | | | |
| 411. | Stri | Priyaṅgu plant | | | 217.76. | | | 68.27. | | |
| 412. | Sucandus | *Santalum album*, Linn | | | | | | | | |

| S. No. | Name | Botanical equivalent | Epic | | Purāṇas | | | | Other Sources | Remarks |
|---|---|---|---|---|---|---|---|---|---|---|
| | | | *Rāmāyaṇa* | *Mahā-bhārata* | *Matsya* | *Vāyu* | *Varāha* | *Vāmana* | | |
| 413. | Sukadhanya | | | | 217.52. | | | | | |
| 414. | Śukra | | | | 217.47. | | | | | |
| 415. | Sumana | *Rosa centifolia* Linn. | | | | | | 68.12. | | |
| 416. | Sumukha | A species of plant. | | | 217.63. | | | | | |
| 417. | Surgodha | | | | 118.38. | | 144.71. | | | |
| 418. | Surmandaka | A bulbous plant. | | | 96.9. | | 123.43. | | | |
| 419. | Suvarcala | *Catea fistula*, L. | | | 118.30, 267.15. | | | | | |
| 420. | Svadukantaka | *Flacourtia sapide* | | | 118.29. | | | | | |
| 421. | Śvasantī | | | | 217.44. | | | | | |
| 422. | Svetakadruma | *Crataeva roxburgii* | | | 118.23. | | | | | |
| 423. | Śyāma | *Artemisia indica* | | | 217.72, 79; 266.12. | 78.6-9. | | | | |
| 424. | Śyāmaka | *Echinochioa colona* | | | 217.72, 79; 266.12. | 78.6-9. | 58.11, 114.21; 118.12. | | | |
| 425. | Syonaka | *Panicum frumentacec* | | | 217.76. | | | | | |
| 426. | Tagara | *Tubernae-montana cornata*, Willd. | | | 118.38; 217.70. | | | 70.40. | | |

| S. No. | Name | Botanical equivalent | Epic | | Purāṇas | | | | Other Sources | Remarks |
|---|---|---|---|---|---|---|---|---|---|---|
| | | | *Rāmāyaṇa* | *Mahābhārata* | *Matsya* | *Vāyu* | *Varāha* | *Vāmana* | | |
| 427. | Tala | *Borassus flabelliformis,* Linn. | | Virāṭa: 14.15; Ādi: 39.86. | 96.9; 118.3; 161.57; 217.49. | | 79.21; 149.37; 151.49; 154.4; 155.37; 197.26; 198.1. | 2.49, 12.54; 16.47; 42.48; 47.48; 47.49; 58.9; 68.27. | Kangra paintings | Palmyra having round purple fruits, symbol of female charm. |
| 428. | Talasphoṭa | *Flacourtia cataphracta* | | | 217.73. | | | | | |
| 429. | Tamāla | *Crataeva roxburgii* | | | 118.3; 161.57; 161.65; 180.25. | | | 58.9. | Kangra Paintings | It has straight stem and dark fragrant leaves; symbol of Kṛṣṇa. |
| 430. | Tanduliya | *Amaranthus polygonoides* | | | | | 177.8. | | | |
| 431. | Tantumāla | Palmyra tree | | | 118.8. | | | | | |
| 432. | Tejapāta | *Laurus cassia* | | Śānti: 170.18. | | | | | | |
| 433. | Tila | *Sesamum indicum* D.C. | | Vana: 279.37. | | 8.150. | 13.56; 14.25; 50.10; 53.16, 17, 20; 34.15; 39.44; 57.8; 59.2, 4; 60.6; 100.5; 103.7; | 15.6; 17.35; 15.5, 60; 18.13, 17; 24.27; 24.9; 50.38; 53.49; 50.20-22; | | |

| S. No. | Name | Botanical equivalent | Epic | | Purāṇas | | | | Other Sources | Remarks |
|---|---|---|---|---|---|---|---|---|---|---|
| | | | *Rāmāyaṇa* | *Mahābhārata* | *Matsya* | *Vāyu* | *Varāha* | *Vāmana* | | |
| | | | | | | | 104.8; 106.6; 108.3; 118.11; 126.15, 52; 135.49, 56; 178.64, 108; 186.64; 188.40, 52, 110; 196.41; 205.24; 209.23, 27; 213.104. | 68.21. | | |
| 434. | Tilaka | *Clerodendrum phlomoides* | | | 161.56; 180.35; 255.23. | | | 68.13. | | |
| 435. | Tilapuṣpa | *Terminalia bellerica* | | | 118.15. | | | | | |
| 436. | Tinduka | *Diospyros embryopteris* | 1.4.22; 2.76.16. | Udyoga: 133.14. | 161.60; 257.10. | | | 26.122. | | |
| 437. | Tinisa | *Dalbergia ujjeinensis* | 4.228.3. | | | | | | | |
| 438. | Titindi | *Tamarindus indica* L. | | | 96.7; 118.22; 217.55. | | | | | |

| S. No. | Name | Botanical equivalent | Epic | | Purāṇas | | | | Other Sources | Remarks |
|---|---|---|---|---|---|---|---|---|---|---|
| | | | *Rāmāyaṇa* | *Mahābhārata* | *Matsya* | *Vāyu* | *Varāha* | *Vāmana* | | |
| 439. | Toṣī | *Linum usitatissimum* | | Droṇa: 29.20. Vana: 188.95. | | | | | | |
| 440. | Trapuṣa | *Cucumis sativus,* Linn. | | | 118.27; 217.51; 227.111. | | | | | |
| 441. | Trivata | *Operculina turpethum,* Linn. | | | 217.77. | | | | | |
| 442. | Tṛṇa | Hay | | | | | 11.20; 52.11; 213.6. | | | |
| 443. | Tundira | *Cucumis utilissimus* | | | 96.7. | | | | | |
| 444. | Turukṣa | *Styrax officinale,* Linn. | | | | | 181.23. | | | |
| 445. | Tuśaparṇī | | | | 217.72. | | | | | |
| 446. | Tuvarika | | | | 217.76. | | | | | |
| 447. | Tvagodruma | | | | 118.25. | | | | | |
| 448. | Udara or Udaraka | *Vanguiers spinosa* | | | | 8.151. | | | | |
| 449. | Udbhija | Sprouting | | | | | 112.44; 121.9. | | | |
| 450. | Udumbara | *Ficus glomerata* | 5.58.56; 3.62.4. | Śānti: 194.39. Anu: 4.27. | 217.78; 255.20; 265.31. | 75.67. | 80.1; 118.6; 126.35; 127.16; | 15.13; 17.19. | | |

| S. No. | Name | Botanical equivalent | Epic | | Purāṇas | | | | Other Sources | Remarks |
|---|---|---|---|---|---|---|---|---|---|---|
| | | | *Rāmāyaṇa* | *Mahā-bhārata* | *Matsya* | *Vāyu* | *Varāha* | *Vāmana* | | |
| | | | | | | | 186.67; 187.10; 188.42; 190.18; 195.8; 209.16. | | | |
| 451. | Urupuṣpika | A species of plant. | | | 217.51. | | | | | |
| 452. | Uśira | *Andropogon squarrosus,* Hack | | | 217.68; 266.13. | 75.32. | 179.2. | 12.7; 68.19. | | |
| 453. | Uśarika | *Vetveria zizanoides* Linn. | | | | | 179.12. | | | |
| 454. | Utkaṭa | *Laurus cassia* | | | 217.50. | 80.1; 118.6; 126.35; 127.16. | | | | |
| 455. | Utpala | *Nymphaea coerulea* | | | | | 69.18; 78.89; 82.16; 122.8, 28; 138.101; 147.80; 148.30; 150.26; 155.15; 170.30; 194.14; 198.18. | 3.47; 58.17. | | |

| S. No. | Name | Botanical equivalent | Epic | | Purāṇas | | | | Other Sources | Remarks |
|---|---|---|---|---|---|---|---|---|---|---|
| | | | *Rāmāyaṇa* | *Mahā-bhārata* | *Matsya* | *Vāyu* | *Varāha* | *Vāmana* | | |
| 456. | Valukā | | | | 96.5. | | | | Kangra Paintings. | A dwarf tree with highly fragrant leaves. |
| 457. | Vaṁśa | *Shorea robusta* | Araṇya-kāṇḍa | Ādi: 62.17; Virāṭa: 21.3. | | | | 64.93. | | |
| 458. | Vaṁśikarīra | | | | | | | 78.11. | | |
| 459. | Vaca | *Acorus calamus* | | | 267.14. | | | | | |
| 460. | Vaiṣṇavī | *Asparagus racemosus* | | | 266.14. | | | | | |
| 461. | Variṣṭhaka | | | | 118.24. | | | | | |
| 462. | Varaṣābhu | *Boerhaavia diffusa*, Linn. | | | 217.45. | | | | | |
| 463. | Vārtakī | *Solanum melongana* | | | 118.28. | | | | | |
| 464. | Varuna | *Crataeva roxburgii* | | | 161.60. | | | | | |
| 465. | Vaṭa | *Ficus benghalensis*, L. | | Vana: 72.41. | 96.11. | 30.3.50. | 51,211 77.27; 126.34; 140.53; 145.28; 185.15; 198.34. | 12.54; 18.3; 22.4, 8, 38; 24.25, 31; 25.1, 2.8, 9.11.12. 25; 38.20, | | |

| S. No. | Name | Botanical equivalent | Epic | | Purāṇas | | | | Other Sources | Remarks |
|---|---|---|---|---|---|---|---|---|---|---|
| | | | *Rāmāyaṇa* | *Mahā-bhārata* | *Matsya* | *Vāyu* | *Varāha* | *Vāmana* | | |
| | | | | | | | | 22, 26, 36.69; 72.75; 39.95. | | |
| 466. | Vātāma | *Prunus amygdalus* Batsch | | | 217.353. | | | | | |
| 467. | Vatsanābha | *Calamus rotang* | | | 161.60. | | | | | |
| 468. | Vayaḥsthā | | | | 217.64. | | | | | |
| 469. | Venu | *Bambcssa sp.* | | | 118.35. | 7.5. | | 44.25. | | |
| 470. | Vetasa | *Calamus rotang* | Sundara-kāṇḍaṁ | Vana: 106.28,29. | 118.10; 217.73. | | | 6.16. | | |
| 471. | Vibhitaka | *Terminalia bellerica*, Roxb. | Sundara-kāṇḍaṁ | | 161.62; 257.7. | | | 6.16. | | |
| 472. | Vidāṅga | *Erycibe paniculata* | | | 118.22. | | | | | |
| 473. | Vidari | *Ipomcea paniculata*, R.Br. | | | 118.29. | | | | | |
| 474. | Vidārya | | | | 217.45. | | | | | |
| 475. | Vidruma | | | | 217.78. | | 1.23. | | | |
| 476. | Vidusara | | | | 118.30. | | | | | |
| 477. | Viraṇa | *Andropogon muricatus* | | | 217.47. | | | | | |
| 478. | Viraṁksa | | | | 217.44. | | | | | |
| 479. | Viṣa | *Aconitum napelluis* L. | | | 217.69. | | | | | |

| S. No. | Name | Botanical equivalent | Epic | | Purāṇas | | | | Other Sources | Remarks |
|---|---|---|---|---|---|---|---|---|---|---|
| | | | *Rāmāyaṇa* | *Mahā-bhārata* | *Matsya* | *Vāyu* | *Varāha* | *Vāmana* | | |
| 480. | Viśalyakarṇī | *Croton polyandram* | | | 122.56. | | | | | |
| 481. | Vrihi | *Oryza sativa*, Linn. | | | | | 8.22, 32; 100.5; 109.3; and several others. | 15.2; 18.13; 68.21, 24. | | |
| 482. | Vṛṣyā | *Gendarussa vulgaris* | | | 217.44. | | | | | |
| 483. | Vyāgtoranakha | *Martyaia diandra*, Roxb. | | | 217.72. | | | | | |
| 484. | Vyāghrī | *Solanum jacquini* | | | 267.14. | | | | | |
| 485. | Yajiya | A sacred plant. | | | 217.63. | | | | | |
| 486. | Yavaka | *Hardeum vulgare* L. | | | | | | 8.150. | | |
| 487. | Yava | Barley | | | 118.10. | | 14.15; 38.11; 63.5 and several others. | 17.59; 18.13; 68.21, 58. | | |
| 488. | Yavas | Grass | | | | | 174.3; 178.36, 37. | | | |
| 489. | Yuthika | *Jasminum auriculatum* Vahl. | | | 118.14. | | | 68.12. | | Jasmine with fragrant white flowers. |

# List of Trees and Plants Mentioned in the Brahmavaivarta Purāṇa

| | *Name* | *Botanical Name* |
|---|---|---|
| 1. | Alābu | *Lagenaria vulgaris*, Mol. |
| 2. | Āmalaka | *Emblica officinalis*, Gaertn. |
| 3. | Āmra | *Mangifera indica*, Linn. |
| 4. | Aparājita | *Clitoria ternatea*, Linn. |
| 5. | Ārdraka | *Zingiber officinale*, Rosc. |
| 6. | Arjuna | *Terminalia Arjuna*, Roxb. |
| 7. | Aśoka | *Saraca asoca*, Roxb. |
| 8. | Aśvattha | *Ficus religiosa*, Linn. |
| 9. | Badarī | *Zizania mauritiana*, Lam. |
| 10. | Bākula | *Manilkara hescandra*, Roxb. |
| 11. | Campaka | *Michelia champaca*, Linn. |
| 12. | Canaka | *Cicer arietinum*, Linn. |
| 13. | Dāḍim | *Punica grantum*, Linn. (Pomegranate) |
| 14. | Dhattūra | *Datura tatula*, Linn. |
| 15. | Dūrva | *Desmostachya bipinnata*, Staff. |
| 16. | Haritaki | *Terminalia citirina*, Roxb. |
| 17. | Eraṇḍa | *Ricinus communis*, Linn. |
| 18. | Ikṣu | *Saccharum officinurum*, Linn. |
| 19. | Jambīra | *Citrus lemorum*, Linn. |
| 20. | Jambu | *Xylia xylocarpa*, Roxb. |
| 21. | Jātiphala | *Myricaria fragrans*, Houtt. |
| 22. | Kadalī | Banana (*Lagerstroemia speeicsa*, Pers.) |
| 23. | Kadamba | *Anthocephalus cadamba*, Miq. |
| 24. | Karkaṭi or Madhu-Karakaṭi | *Centropus pyropterus* |
| 25. | Keśara | *Rottleria tinctoria* |
| 26. | Ketakī | *Pandanus odoratissimus* |
| 27. | Khadira | *Abutilon catechu*, Willd. |
| 28. | Kharjūra | *Phoenix dactylifera*, Linn. |
| 29. | Kiṁśuka | *Butea frondosa* |
| 30. | Kuśā | *Poa cyonosurloides* |
| 31. | Kūṣmaṇḍa | *Benincasa cerifera*, Savi. |
| 32. | Mādhavī | *Hiptage madalota*, Gaertn. |
| 33. | Mālatī | *Jasminum grandiflorum*, Linn. |
| 34. | Mallikā | *Gaertnera racemosa* |
| 35. | Mandāra | *Erythrina indica*, Lamk. |
| 36. | Mudga | *Phaseolus mungo*, L. |
| 37. | Nageśvara | A kind of plant. |
| 38. | Nagranga | Orange |
| 39. | Nārikela | *Cocos mucifera* Linn. |
| 40. | Nimba | *Melia azadirachta*, L. |
| 41. | Palāśa | *Butea frondosa* |
| 42. | Panāsa | *Artocarpus heterophyillus* |
| 43. | Paribhadra | *Erythrina indica.* |
| 44. | Paṭolā | *Tricosamtheo dioica*, Roxb. |
| 45. | Pingraka | *Ferula assafoetida* L. |
| 46. | Pippala or Pipate | *Ficus religiosa*, L. |
| 47. | Priyāla | *Buchanania latifolia* |
| 48. | Puṣkara | *Nymphaea nelumbo* |
| 49. | Rambhā | *Musa sapientum*, L. |
| 50. | Sāla | *Artocarpus locuca* |
| 51. | Saptaparṇaka | *Alstonia scholaris*, R.Br. |
| 52. | Sasarpa | *Sarsopa brinicanapus* |
| 53. | Simbī | *Phaseolus trilolus* |
| 54. | Sindhuvara | *Vitex negundo*, Linn. |
| 55. | Tāla | *Borassus flabelliformis*, Linn. |
| 56. | Tāmbūla | *Piper betle*, Linn. |
| 57. | Tila | *Sesamum indicum*, D.C. |
| 58. | Titaṅḍika | *Diospyros embroyoptaris* |
| 59. | Tulasī | *Ocimum sanctum*, L. |
| 60. | Vaṁśa | *Shorea robusta* |
| 61. | Vartakī | *Solanum melongana* |
| 62. | Vaṭa | *Ficus benghalensis*, L. |
| 63. | Yuthikā | *Jasminum auriculatum*, Vahl. |

# Trees and Plants Mentioned in Classical Sanskrit Works Works of Kālidāsa and Others

| S. No. | *Name of Plant* | *Botanical equivalent* | *Pāṇini's Aṣṭādhyāyī* | *Abhijñāna Śākuntalaṁ* | *Meghadūta* | *Kumāra Saṁbhava* | *Mālavikāgnimitra* | *Ritusaṁhāra* | *Raghuvaṁśa* | *Vikramorvaśiam* | *Kāmasūtra* | *Harṣacarita* | *Amarkoṣa* |
|---|---|---|---|---|---|---|---|---|---|---|---|---|---|
| 1. | Abhaya | *Andropogon muricatum* A species of tree. | | | | | | | | | | | 55.59, 122.65. |
| 2. | Abhīru | *Asparagus racemosus* A species of plant. | | | | | | | | | | | 106.101. |
| 3. | Abja or Lotus | *Barringtonia acutangula* | | | | | | 6.33 | 4.61. | | | | |
| 4. | Agada | A medicinal plant. | | | | | | | | | | | 114.50. |
| 5. | Agamā | A tree. | | | | | | | | | | | 80.5. |
| 6. | Agnijvala | *Grislea tomentosa* | | | | | | | | | | | 112.125. |
| 7. | Agnimantha | *Premna spinosa* | | | | 4.1.29. | | | | | | | 97.66. |
| 8. | Agnimukhi | *Semicarpus anacardium* | | | | | | | | | | | 91.43. |
| 9. | Agniśikhā | *Gloriosa superba* Saffron plant. | | | | | | | | | | | 110.119; 115.137. |
| 10. | Aguru | *Aporusa agallocha*, Roxb. | | | | | | 5.12. | 8.71; 19.41. | | | | 96.63. |
| 11. | Aheru | A species of plant. | | | | | | | | | | | 106.102. |
| 12. | Airāvata | A species of plant. | | | | | | | | | | | 90.38. |
| 13. | Ajagandhikā | *Ageratum conyzoides*, Linn. | | | | | | | | | | | 116.40. |

| *S. No.* | *Name of Plant* | *Botanical equivalent* | *Pāṇini's Aṣṭādhyāyī* | *Abhijñāna Śākuntalaṁ* | *Meghadūta* | *Kumāra Saṁbhava* | *Mālavikāgnimitra* | *Ritusaṁhāra* | *Raghuvaṁśa* | *Vikramorvaṣiam* | *Kāmasūtra* | *Harṣacarita* | *Amarkoṣa* |
|---|---|---|---|---|---|---|---|---|---|---|---|---|---|
| 14. | Ajahā | A species of plant. | | | | | | | | | | | 102.87. |
| 15. | Ajāmodā | *Apium graveolens*, Linn. | | | | | | | | | 6.1.6. | | 117.45. |
| 16. | Ajarūṣā | | | | | | | | | | | | 107.104 |
| 17. | Ajaśṛṅgī | *Odina wodier* | | | | | | | | | | | 111.20. |
| 18. | Ajjhaṭā | | | | | | | | | | | | 113.27. |
| 19. | Ākṛḍā | | | | | | | | | | | | 80.3. |
| 20. | Akṣiva | *Hyperanthera moringa* | | | | | | | | | | | 87.31. |
| 21. | Akṣoṭā | *Jatropa regia*, Linn. | | | | | | | | | | | 87.29. |
| 22. | Alarka | A king of Alce plant. | | | | | | | | | | | 101.81. |
| 23. | Alābu | *Lagenaria vulgaris*, Mol. | | | | | | | | | | | 120.56. |
| 24. | Āli | *Aram campanulatum* | | | | | | | | | | | 80.14. |
| 25. | Ālūka | *Elytracia officinalis*, Gaertn. | | | | 4.1.29. | | | | | | | |
| 26. | Āmalaka | *Emblica officinalis*, Gaertn. | | | | 4.1.7; 7.1.6; 7.2.36. | | | | | | | 95.98. |
| 27. | Amaṇḍa | *Ricinus communis* | | | | | | | | | | | 93.52. |
| 28. | Ambuṣṭha | *Jasminum auriculatum* | | | | | | | | | | | 98.72; 102.85; 116.41. |

| S. No. | Name of Plant | Botanical equivalent | *Pāṇini's Aṣṭādhyāyī* | *Abhijñāna Śākuntalaṁ* | *Meghadūta* | *Kumāra Saṁbhava* | *Mālavikāgnimitra* | *Ritusaṁhāra* | *Raghuvaṁśa* | *Vikramorvaṣiam* | *Kāmasūtra* | *Harṣacarita* | *Amarkoṣa* |
|---|---|---|---|---|---|---|---|---|---|---|---|---|---|
| 29. | Ambuja | *Barringtonia acutangula* Gaertn. | | | | 11.25; 14.13. | | 3.20, 4.4, 5; 6.14; 6.7. | | | | | |
| 30. | Amuruha | Lotus | | | | | | | | | | | |
| 31. | Amalvetasā | *Emblica officinalis,* Gaertn. | | | | | | | | | | | 116.41. |
| 32. | Ambalikā | *Wrightia antidysenterica* | | | | | | | | | | | 91.44. |
| 33. | Āmiṣi | A plant. | | | | | | | | | | | 114.134. |
| 34. | Amlāna | *Emblica officinalis,* Gaertn. | | | | | | | | | | | 99.47. |
| 35. | Amogha | *Bignonia suavealens,* Roxb. | | | | | | | | | | | 94.55; 107.106. |
| 36. | Āmra | *Mangifera indica,* Linn. | | | | 8.78. | | 6.23. | | | | | 88.33. |
| 37. | Amaratika | *Spondias pinnatta,* Kurz. | | | | 4.1.29. | | | | | | | 86.27. |
| 38. | Amṛta | *Eulophia campestris,* Wall. | | | | | | | | | | | 95.58; 59; 101.83. |
| 39. | Ananta | *Ervatamia divaricata,* Linn. and Aru. | | | | | | | | | | | 103.92; 115.137; 121.159. |
| 40. | Anāryatika | A creeper | | | | | | | | | | | 117.143. |
| 41. | Angīatṛṇa | *Vulgaris indica* | | | | 7.2.45. | | | | | | | |
| 42. | Angaravallarī | *Clerodendrum serratum,* Linn. | | | | | | | | | | | 93.49; 103.90. |

| S. No. | Name of Plant | Botanical equivalent | Pāṇini's Aṣṭā-dhyāyī | Abhijñā-na Śāku-ntalaṁ | Megha-dūta | Kumā-ra Saṁ-bhava | Mālavi-kāgni-mitra | Ritusa-ṁhāra | Raghu-vaṁśa | Vikramo-rvaṣiam | Kāma-sūtra | Harṣa-carita | Amar-koṣa |
|---|---|---|---|---|---|---|---|---|---|---|---|---|---|
| 43. | Anghriparṇikā | *Uraria lagopodioides*, Linn. | | | | | | | | | | | 104.93. |
| 44. | Anjnakaśī | *Ammonium xanghoarhiza* | | | | | | | | | | | 113.30. |
| 45. | Ankolā | *Alangium salvifolium*, Linn. | | | | | | | | | | | 87.29. |
| 46. | Apāmārgā | *Achyranthes aspera*, Linn. | | | | | | | | | | | 103.89. |
| 47. | Aparājita | *Clitoria ternatea*, Linn. | | | | | | | | | | | 107.104; 119.150. |
| 48. | Āra | A tree. | 8.4 5. | | | | | | | | | | |
| 49. | Aragavedha | | | | | | | | | | | | 85.23. |
| 50. | Aralu | *Ormocarpum indicum*, Vent. | | | | | | | | | | | 95.58. |
| 51. | Aravinda | Lotus *Nelumbium speciosum* | | 3.4. | 1.32; 3.56; 9.1.20; 13.12.7; 17.49. | | 4.13. | 1.43; 6.13; 13.23; 14.50, 5. | | | | | |
| 52. | Ārevalā | *Ipomoea pescaprai*, Sw. | | | | | | | | | | | 85.24. |
| 53. | Arimeda | *Vachellia farnesiana*, Linn. | | | | | | | | | | | 93.50. |
| 54. | Ariṣṭa or Ariṣṭaka | *Xanthium spinosum*, Linn. | | | | | | | | | | | 87.31; 96.62; 118.49. |
| 55. | Arjaka | *Ocimum gratisimum*, Linn. | | | | | | | | | | | 100.80. |

| S. No. | Name of Plant | Botanical equivalent | *Pāṇini's Aṣṭādhyāyī* | *Abhijñāna Śākuṁtalaṁ* | *Meghadūta* | *Kumāra Saṁbhava* | *Mālavikāgnimitra* | *Ritusaṁhāra* | *Raghuvaṁśa* | *Vikramorvaṣīam* | *Kāmasūtra* | *Harṣacarita* | *Amarkoṣa* |
|---|---|---|---|---|---|---|---|---|---|---|---|---|---|
| 56. | Arjuna | *Terminalia arjuna,* Roxb. | | | | | | | | | | | 92.45; 123.68. |
| 57. | Arka | *Calotropis gigantea,* R.Br. | | | | | | | | | 7.2.36. | | |
| 58. | Askālana | A species of plant. | | | | | | | | | | | 101.81. |
| 59. | Arkaparṇi | *Calotropis gigantea,* R.Br. | | | | | | | | | | | 101.81. |
| 60. | Ārtgala | A species of plant. | | | | | | | | | | | 99.75. |
| 61. | Aruṇā | *Rottleria tinctoria,* Linn. | | | | | | | | | | | 106.100. |
| 62. | Aruṣkara | *Semecarpus anacardium* | | | | | | | | | | | 91.43. |
| 63. | Āsana | *Bridelia sepularis,* Blume. | | | | | | | | | | | 91.44. |
| 64. | Aśmapuṣpa | *Coleus senttelariodes* | | | | | | | | | | | 111.123. |
| 65. | Aśoka | *Saraca asoca,* Roxb. | | Act-VII | | 3.26, 53. | 3 Verses, 5, 12 and several others. | 6.5.16. | 7.21; 8.62, 63; 9.28; 13.32. | 2.7; 4.62; 5.3. | | | 97.65. |
| 66. | Aśokarohinī | A medicinal plant. | | | | | | | | | | | 102.86. |
| 67. | Āsphota | A species of plant. | | | | | | | | | | | 101.81; 98.70, 107.104. |
| 68. | Aśvakarna (ka) | *Vatica robusta* | | | | | | | | | | | 92.45. |

| S. No. | Name of Plant | Botanical equivalent | Pāṇini's Aṣṭādhyāyi | Abhijñāna Śākuntalaṁ | Meghadūta | Kumāra Saṁbhava | Mālavikāgnimitra | Ritusaṁhāra | Raghuvaṁśa | Vikramorvaṣiam | Kāmasūtra | Harṣacarita | Amarkoṣa |
|---|---|---|---|---|---|---|---|---|---|---|---|---|---|
| 69. | Aśvattha | *Ficus religiosa*, Linn. | 4.3.43. | | | | | | | | | | 83.18; 85.21. |
| 70. | Aṭarūṣa | *Justicia adhatoda* | | | | | | | | | | | 107.104. |
| 71. | Aticarā | *Hibiscus mutabilis* | | | | | | | | | | | 116.146. |
| 72. | Atichatrā (Mushroom) | *Anethum sowa* | | | | | | | | | | | 119.152; 123.167. |
| 73. | Atimukta | *Hiptage madablota*, Gaertn. | | Act-III | | | 4.13. | 6.17. | | 7.7.1. | | | |
| 74. | Atimuktaka | *Dalbergia ujjeinensis* | | | | | | | | | | | 86.26; 98.72. |
| 75. | Atisaurabha | A species of plant. | | | | | | | | | | | 88.33. |
| 76. | Ativiṣa (Poisonous plant) | *Aconitum heterophyllum*, Wall. | | | | | | | | | | | 106.100. |
| 77. | Avadāha | A fragrant plant. *Andropogon muricatus*, Linn. | | | | | | | | | | | 122.165. |
| 78. | Avakeśin | An unfruitful or a barren plant. | | | | | | | | | | | 81.7. |
| 79. | Avākapuṣpi | *Blyxa octandra*, Rich. | | | | | | | | | | | 119.153. |
| 80. | Avalguja (ka) | *Vernonia anthelmithica* | | | | 7.2.31. | | | | | | | 105.196. |
| 81. | Āvali | A kind of bulbous plant. | | | | | | | | | | | 80.4. |
| 82. | Avandhyā | | | | | | | | | | | | 81.6. |
| 83. | Avegi | | | | | | | | | | | | 115.137. |

| S. No. | Name of Plant | Botanical equivalent | Pāṇini's Aṣṭādhyāyī | Abhijñāna Śākuṅtalaṁ | Meghadūta | Kumāra Saṁbhava | Mālavikāgnimitra | Ritusaṁhāra | Raghuvaṁśa | Vikramorvaṣiam | Kāmasūtra | Harṣacarita | Amarkoṣa |
|---|---|---|---|---|---|---|---|---|---|---|---|---|---|
| 84. | Avigna | *Carrissa carandas*, Linn. | | | | | | | | | | | 97.68. |
| 85. | Avyaṅgā | *Carpopogan prupriens*, Roxb. | | | | | | | | | | | 102.87. |
| 86. | Avyathā | *Terminalia citrina*, Roxb. | | | | | | | | | | | 95.59; 117.146. |
| 87. | Badara/Badarī | *Zizania jujuba*, Mill. | 5.2.24. | | | | | | | | | 2.7.18. | 119.151. |
| 88. | Bahupadika | Indian fig tree. | | | | | | | | | | 7.2.41. | |
| 89. | Bākula | *Manilkara hescandra*, Roxb. | | | | | | 2.24. | 7.64; 9.30; 12.2. | | | 57. | 97.65. |
| 90. | Bāla (Balā) | *Valeriana hardwickii*, Wall. | | | | | | | | | | | 111.22. |
| 91. | Balbaja | *Abutilon indicum* | | | | | | | | | | | 122.163. |
| 92. | Bālakṣira | | | | | 4.1, 7. | | | | | | | |
| 93. | Bāṇa | *Latipes nudicaulis*, Hook. | | | | | | | | | | | 99.75. |
| 94. | Bandujihvā | *Pentapetas phoenicea*, L. | | | | 8.40. | | 3.24. | 9.25. | | | | |
| 95. | Bandhuka | *Polygonum aviculare*, Linn. | | | | | | 3.5, 25. | | | | | 99.73. |
| 96. | Barbara | *Clerodendrum siphonantus* | | | | | | | | | | | 116.40. |
| 97. | Bhanṭākī | *Solanum indicum*, Linn. | | | | | | | | | | | 109.114. |
| 98. | Bhardvāji | The wild cotton shrub. | | | | | | | | | | | 110.117. |
| 99. | Bhṛṅga | Indian fig tree. | | | | 8.2.31. | | | | | | | |
| 100. | Bhṛṅgarāja | *Eclipta alba*, Linn. | | | | | | | | | | | 119.152. |
| 101. | Bhunimba | *Sutherlandia bimaculat*, Hook. | | | | | | | | | | | 117.143. |

| S. No. | Name of Plant | Botanical equivalent | Pāṇini's Aṣṭādhyāyī | Abhijñāna Śākuntalam | Meghadūta | Kumāra Saṁbhava | Mālavikāgnimitra | Ritusaṁhāra | Raghuvaṁśa | Vikramorvaṣīam | Kāmasūtra | Harṣacarita | Amarkoṣa |
|---|---|---|---|---|---|---|---|---|---|---|---|---|---|
| 102. | Bhūrja | *Betula anlodes*, Buah-Ham | | | | 1.7,55; 7.11.2; 17.10; 18.1.4; 9.9; 1.3. | | | 4.78. | | | | |
| 103. | Bhūruṇḍi | *Heliotropium indicum*, Linn. | | | | | | | | | | | 98.70. |
| 104. | Bibhitaka | *Terminalia anacardium* | | | | | | | | | | | 95.58. |
| 105. | Bījapūraka | *Citrus limonum*, Linn. | | | | | | | | | | | 100.79. |
| 106. | Bilva | *Aegle marmelos*, C & Roxb. | 4.3.136. | | | | | | | | | | 88.32. |
| 107. | Bimbikā | *Momordica monadelpha* | | | | | | | | | | | 116.139. |
| 108. | Bisa | *Aconitum napellus* L. | | | | 4.29; 8.35. | | | | 4.31. | | | |
| 109. | Bisini | Lotus plant | | | | | Act-V | | | | | | |
| 110. | Brāhmaṇi | | | | | | | | | | | | 103.90. |
| 111. | Brāhmī | *Centella asiatica*, Linn. | | | | | | | | | | | 115.138. |
| 112. | Bṛhati | | | | | 7.2.28. | | | | | | | 104.94. |
| 113. | Buka | *Agati grandiflora* | | | | | | | | | | | 101.82. |
| 114. | Campaka | *Michelia champaca*, Linn. | | | | | 6.29. | | | | | | 96.64. |
| 115. | Caṇḍā | *Macadamia peltata*, M & A | | | | | | | | | | | 113.129. |
| 116. | Candana | *Satinum album*, Linn. | | 4.144; 7.18. | | 7.25. | | 1.4,6,8; 2.21; 3.20; 5.3; 6.6; 6.12, 32. | 4.48, 51; 6.64; 8.71; 10.42; 11.64. | | | | |

| S. No. | Name of Plant | Botanical equivalent | *Pāṇini's Aṣṭā-dhyāyī* | *Abhijñā-na Śāku-ntalaṁ* | *Megha-dūta* | *Kumā-ra Saṁ-bhava* | *Mālavi-kāgni-mitra* | *Ritusa-ṁhāra* | *Raghu-vaṁśa* | *Vikramo-rvaṣiam* | *Kāma-sūtra* | *Harṣa-carita* | *Amar-koṣa* |
|---|---|---|---|---|---|---|---|---|---|---|---|---|---|
| 117. | Cāṅgerī | *Oxalis corniculata*, Linn. | | | | | | | | | | | 116.41. |
| 118. | Carmakaṣā | *Mimosa abstergens* | | | | | | | | | | | 117.44. |
| 119. | Chatra | *Andropogon schoenanthus*, L. | | | | | | | | | | | 123.167. |
| 120. | Corakatumbī | *Trigonella corniculata* | | | | | | | | | 5.6.25. | | |
| 121. | Cūta (Āmra)-Mango | *Mangifera indica*, Linn. | | 4.12; 5.1, 6. verse 4 and 8. | | 1.27; 2.64; 3.27,30, 32; 4.14, 38. | 3.4; 4.2. | 6.1.3,14, 15, 30. | 8.21; 13.34, 19.40. | 2.7. | 2.8.10. | | |
| 122. | Dāḍima | Pomegranate *Punica grantum*, Linn. | | | | | | | | | 7.2.28. | | 97.65. |
| 123. | Damanaka | *Artemisia indica* | | | | | | | | | 4.1.29. | | |
| 124. | Dantikā | *Croton polyandrum* | | | | | | | | | | | 117.145. |
| 125. | Darbha | *Desmostachya bipinata*, Staff. | | Act-2,3,4. | | | | | 13.25, 49. 14.89. | | | | |
| 126. | Dāru | *Pinus deodora* | 4.3.152. | | | | | | | | | | |
| 127. | Daruharidra | Wooden spoon or ladle *Berberis aristate*, D.C. | | | | | | | | | | | 106.102. |
| 128. | Devadāru | *Plyalthia longifolia*, Thw. | | | | 1.15, 54; 3.44; 6.51. | | | 2.36, 56, 4.76. | | 7.2.41. | | 94.54. |

| S. No. | Name of Plant | Botanical equivalent | Pāṇini's Aṣṭādhyāyī | Abhijñāna Śakuntalam | Meghadūta | Kumāra Sambhava | Mālavikāgnimitra | Ritusaṁhāra | Raghuvaṁśa | Vikramorvaṣiam | Kāmasūtra | Harṣacarita | Amarkoṣa |
|---|---|---|---|---|---|---|---|---|---|---|---|---|---|
| 129. | Devatāḍa | *Erythropsis colorala*, Roxb. | | | | | | | | | | | 98.69. |
| 130. | Dhātakī | *Woodfordia fruitcosa*, Kurz. | | | | | | | | | | | 112.125. |
| 131. | Dhattūra | *Datura tatula*, Linn. | | | | | | | | | | | 100.78. |
| 132. | Drākṣā (Mṛdvika) | *Vitis vinifera*, Linn. | | | | | | | 4.65. | | | 161. | 108.108. |
| 133. | Dugdhika | *Brassica nigra*, Linn. | | | | | | | | | | | 106.100. |
| 134. | Dūrvā | *Desmostachya bipinnata*, Staff | | Act-IV | | 7.7,14. | | | 6.25; 17.12. | | | | 121.158. |
| 135. | Ela | *Elettaria cardamum*, Maton. | | | | | | | | | | | 112.125. |
| 136. | Elālatā | *Eletteria cardamum*, Maton. | | | | | | | 4.47, 6.64. | | | | |
| 137. | Elāparṇi | *Alpinia cardamum*, Maton. | | | | | | | | | | | 116.40. |
| 138. | Elavalukā | | | | | | | | | | | | 111.121. |
| 139. | Eraṇḍa | *Ricinus communis*, Linn. | | | | | | | | | | | 93.51. |
| 140. | Gambhiri | *Ocimum americanum*, Linn. | | | | | | | | | | | 89.36. |
| 141. | Gandhamūlī | *Alpinia galanga*, L. | | | | | | | | | | | 119.154. |

| S. No. | Name of Plant | Botanical equivalent | Pāṇini's Aṣṭādhyāyī | Abhijñāna Śākuṁtalaṁ | Meghadūta | Kumāra Saṁbhava | Mālavikāgnimitra | Ritusaṁhāra | Raghuvaṁśa | Vikramorvaṣiam | Kāmasūtra | Harṣacarita | Amarkoṣa |
|---|---|---|---|---|---|---|---|---|---|---|---|---|---|
| 142. | Gaṇḍira | *Ranunculus sceleratus*, Linn. | | | | | | | | | | | 120.157. |
| 143. | Gāṇikārikā | *Premną spinosa* | | | | | | | | | | | 97.66. |
| 144. | Gardābhanda | *Thespesia populaea*, Soland | | | | | | | | | | | 91.43. |
| 145. | Ghaṇṭāpātali | *Schreberia swictenoides* | | | | | | | | | | | 90.40. |
| 146. | Girikarmika | A variety of Achyranthes with white blossoms. | | | | | | | | | 7.2.37. | | |
| 147. | Girimallikā | *Wrightia antidysenterica* | | | | | | | | | | | 97.67. |
| 148. | Goḍumbā | | | | | | | | | | | | 120.157. |
| 149. | Gokṣura | *Tribulus alatus*, Delile | | | | | | | | | | | 105.99. |
| 150. | Granthiparṇa | A species of plant or plant Jatikā | | | | | | | | | | 410. | 114.133. |
| 151. | Gṛñjana | *Alluim ascalonicum*, Linn. | | | | | | | | | | | 118.149. |
| 152. | Gaḍuci | *Cocculus cordifolius* | | | | | | | | | | | 101.83. |
| 153. | Guggula | *Commiphara mukul*, Hook. | | | | | | | | | | 170. | 88.34. |
| 154. | Gunja | *Abrus precatorius*, Linn. | | | | | | | | | | | 105.98. |

| S. No. | Name of Plant | Botanical equivalent | Pāṇini's Aṣṭādhyāyi | Abhijñāna Śākuntalam | Meghadūta | Kumāra Saṁbhava | Mālavikāgnimitra | Ritusaṁhāra | Raghuvaṁśa | Vikramorvaṣiam | Kāmasūtra | Harṣacarita | Amarkoṣa |
|---|---|---|---|---|---|---|---|---|---|---|---|---|---|
| 155. | Haridrā | *Curcuma domestica*, Valenton | | | | | | | | | | 420. | |
| 156. | Haritaki | *Terminalia citrina*, Roxb. | 4.3.167. | | | | | | | | 7.2.460. | | 95.60. |
| 157. | Hijjula | | | | | | | | | | | | 96.81. |
| 158. | Hilamocaka | *Enhydra hingcha*, L. | | | | | | | | | | | 121.158. |
| 159. | Hintala | | | | | | | | | | | | 124.170. |
| 160. | Idhuma | | | | | | | | | | | | 82.13. |
| 161. | Ikṣu | *Saccharum officinarum*, Linn. | | | | | | 5.1.16. | 4.20. | | 4.1.6. | 410. | 122.164. |
| 162. | Ikṣugandhā | *Tribulus terrestris*, Linn. | | | | | | | | | | | 105.99; 107.105; 108.10; 122.163. |
| 163. | Ikṣura | Sugarcane | | | | | | | | | | | 107.106. |
| 164. | Indivara | Lotus, *Nymphaea stellata*, | | | | | | 2.12. | 4.65. | | | | 106.101. |
| 165. | Indradārū | *Terminalia arjuna*, Roxb. | | | | | | | | | | | 92.45. |
| 166. | Indrānikā | | | | | | | | | | | | 98.69 |
| 167. | Indra Surasā | *Vitex negundo*, L. | | | | | | | | | | | 98.69. |
| 168. | Indravāruṇī | *Citrullus colocynthes*, Shoad | | | | | | | | | | | 120.157. |

| S. No. | Name of Plant | Botanical equivalent | *Pāṇini's Aṣṭādhyāyī* | *Abhijñāna Śākuṁtalaṁ* | *Meghadūta* | *Kumāra Saṁbhava* | *Mālavikāgnimitra* | *Ritusaṁhāra* | *Raghuvaṁśa* | *Vikramorvaṣīam* | *Kāmasūtra* | *Harṣacarita* | *Amarkoṣa* |
|---|---|---|---|---|---|---|---|---|---|---|---|---|---|
| 169. | Indrayava | *Wrightia antidysenterica* | | | | | | | | | | | 97.67. |
| 170. | Ingudī | *Balanites aegyptiaca,* Linn. | 4.3.164. | | | | | | 14.81. | | | | 92.46. |
| 171. | Jambira | *Citrus limonum,* Linn. | | | | | | | | | | | 85.25; 100.80. |
| 172. | Jambū | *Xylia xylocarpa,* Roxb. | 4.3.165. | | | | | | | | 7.2.31, 33. | | 84.19. |
| 173. | Japa | *Hibiscus rosa-sinensis,* Linn. | | | | | | | | | 4.1.7. | | 99.76. |
| 174. | Jaṭāmānasī | *Nardostachys jatamansi* | | | | | | | | | | | 114.134. |
| 175. | Jāti | *Jasminum grandiflorum,* Linn. | | | | | | | | | 4.1.7. | | 99.73. |
| 176. | Jayā | *Premna spinosa* | | | | | | | | | | | 97.67. |
| 177. | Jayantī | *Sesbania cannabina,* Retz. | | | | | | | | | | | 97.66. |
| 178. | Jīraka | *Ipomoea nil,* Linn. | | | | | | | | | 6.1.6. | | |
| 179. | Jīvaka | *Terminalia tomentosa* | | | | | | | | | | | 117.143. |
| 180. | Jīvantī | *Desmotrichum fimbriatum,* Blume | | | | | | | | | | | 116.142. |
| 181. | Kadalī | *Lagerstroemia speciosa,* Pers. | | | | 1.36. | | | 12.96. | | | | 109.113. |
| 182. | Kadamba (Nipa) | *Anthocephalus cadamba* Miq. | | | | 3.68. | | 2.17,20, 23, 24; 3.13. | 13.27; 14.99. | 4.61. | | 39. | 91.42. |

| S. No. | Name of Plant | Botanical equivalent | Pāṇini's Aṣṭādhyāyī | Abhijñāna Śākuntalam | Meghadūta | Kumāra Saṁbhava | Mālavikāgnimitra | Ritusaṁhāra | Raghuvaṁśa | Vikramorvaṣīam | Kāmasūtra | Harṣacarita | Amarkoṣa |
|---|---|---|---|---|---|---|---|---|---|---|---|---|---|
| 183. | Kairavaruka | | | | | | | | | | 4.1.29. | | |
| 184. | Kakubha | *Terminalia arjuna* A wreath of *Campaka* flowers. | | | | | | 2.20. | | 1.61.14. | | | |
| 185. | Kālāgoru | *Stereospermum suaveolens*, D.C. | | | | | | 2.21; 4.5; 5.5; 6.13. | 4.81; 13.55; 14.12. | | | | |
| 186. | Kalama | A sort of rice. | | | | 5.47. | | | | | | | |
| 187. | Kālambī | *Ipomoea aquatica*, Forsk. | | | | | | | | | | | 121.158. |
| 188. | Kalhāra | *Nymphaea lotus* | | | | | | 3.15. | | | | | |
| 189. | Kaliyaka | | | | | | | 4.5; 6.12. | | | | | |
| 190. | Kamala | *Nelumbo nucifera*, Gaertn. | | | | 4.21; 7.89; 8.19, 85. | | 1.28; 3.5, 8, 11; 5.13; 6.32; 17.40. | 3.36; 9.16; 9.27; 11.12; 19.1,11. | 2.12; 2; 4.40, 1.54, 73. | | | |
| 191. | Kandalī | *Amorphophallus campanulatus* | | | | | | 2.5. | 13.29. | 4.15. | | | |
| 192. | Kankelī | | | | | | | 3.18. | | | | | |
| 193. | Kaṇṭakāra | *Solanum surattense*, Burm | 4.3.152. | | | | | | | | | | 104.94. |
| 194. | Kapittha | *Feronia limonia*, Linn. | | | | | | | | | 2.1.29. | | 85.21. |
| 195. | Karanjaka | *Pongamia pinnata*, Pierre | | | | | | | | | | | 92.48. |

| S. No. | Name of Plant | Botanical equivalent | Pāṇini's Aṣṭādhyāyī | Abhijñāna Śākuṁtalaṁ | Meghadūta | Kumāra Saṁbhava | Mālavikāgnimitra | Ritusaṁhāra | Raghuvaṁśa | Vikramorvaṣiam | Kāmasūtra | Harṣacarita | Amarkoṣa |
|---|---|---|---|---|---|---|---|---|---|---|---|---|---|
| 196. | Kāraskara | *Strychnos nux-vomica*, Linn. | 6.1.156. | | | | | | | | | | |
| 197. | Karavi | *Carvia callosa*, Nees. | | | | | | | | | | | 119.112, 153. |
| 198. | Karavīra | *Nerium indicum*, Mill. | | | | | | | | | | | 99.77. |
| 199. | Karacūra | *Curcuma zedoaria*, Rose. | | | | | | | | | | | 120.155. |
| 200. | Karīra | *Canthium decidua*, Edgew. | 6.3.141. | | | | | | | | | | |
| 201. | Karkaṇḍu | *Ziziphus mummularia* | 5.2.24. | | | | | | | | | | 89.37. |
| 202. | Karakoṭi | *Centropus pyropterus* | | | | | | | | | | | 120.156. |
| 203. | Karṇikāra | *Pterospermum acerifolium*, Willd. | | | | 3.28, 5, 3, 62. | | 6.5, 20, 27. | | 2.22; 3.3. | | | 96.61. |
| 204. | Karpasi | *Gyossipum hertaceum* L. | | | | | | | | | | | 110.116. |
| 205. | Kārśya | *Artocarpus lacucha*, L. | 8.4.5. | | | | | | | | | | |
| 206. | Kāśa | *Sacciolepis spontaneum*, Linn. | 6.2.80; 6.2.82. | | | 7.11. | | 3.1.2, 26. | 6.17. | | | | 122.168. |
| 207. | Kaseru (-kā) | *Cyperus esculantus*, Linn. | | | | | | | | | 8.1.39, 43. | | |
| 208. | Kaṭaphala | *Solanum melongena* | | | | | | | | | | | 90.41. |
| 209. | Katambarā | *Helleborusniger* | | | | | | | | | | | 119.153. |

| S. No. | *Name of Plant* | *Botanical equivalent* | *Pāṇini's Aṣṭādhyāyī* | *Abhijñāna Śākuntalam* | *Meghadūta* | *Kumāra Saṁbhava* | *Mālavikāgnimitra* | *Ritusaṁhāra* | *Raghuvaṁśa* | *Vikramorvaṣiam* | *Kāmasūtra* | *Harṣacarita* | *Amarkoṣa* |
|---|---|---|---|---|---|---|---|---|---|---|---|---|---|
| 210. | Kattṛṇa | | 6.3.103. | | | | | | | | | | |
| 211. | Kaṭuka | | | | | | | | | | | | 102.86. |
| 212. | Kavijjaka | | | | | | | | | | 8.2.37. | | |
| 213. | Kaśara (Bākula) | *Manilkara hescandra,* Roxb. | | Acts-1, 4. | 86. | 2.55; 4.8, 8.76. | | 2.20. | 9.36.40. | | | | 96.64. |
| 214. | Keśarī | *Rottleria tinctoria* | | | | | | | | | | | 134.70. |
| 215. | Ketaka | *Pandamus odoratissimus* | | | 23. | 8.58. | Act-4. | 2.17, 20, 23, 26. | 4.55; 6.17. | | | 450. | |
| 216. | Khadira | *Abutilon catechu,* Willd. | 8.4.5. | | | | | | | | 8.1.31. | | 93.50; 116.42. |
| 217. | Kharjūra | *Phoenix sylvestris* | | | | | | | 4.57. | | | | 124.170. |
| 218. | Kicaka | *Arundo bamboo* | | | 58. | 1.8. | | | 2.12; 6.73. | | | | 122.162. |
| 219. | Kimśuka | *Butea frondosa* | 4.3 141. | | | 3.29. | | | | | | | 87.29; 120.155. |
| 220. | Kokamada | *Nilumbium specisburm,* Willd. | | | | 17.16. | | | | | | | |
| 221. | Kokilākṣa | *Hygrophila spinosa,* T. | | | | | | | | | 7.2.34. | | |
| 222. | Kovidāra | *Bauhinia variegata* | | | | | | 3.6. | | | | | 85.22. |
| 223. | Kṣīrakokali (ka) | Name of a root from the Himalaya yielding milky juice. | | | | | | | | | 7.1.39. | | |

| S. No. | Name of Plant | Botanical equivalent | Pāṇini's Aṣṭādhyāyī | Abhijñāna Śākuntalam | Meghadūta | Kumāra Saṁbhava | Mālavikāgnimitra | Ritusaṁhāra | Raghuvaṁśa | Vikramorvaṣiam | Kāmasūtra | Harṣacarita | Amarkoṣa |
|---|---|---|---|---|---|---|---|---|---|---|---|---|---|
| 224. | Kuberākṣī | *Bignonia suaveolens* | | | | | | | | | | | 94.55. |
| 225. | Kubjaka | *Achyranthes aspera* | | | | | | | | | 4.1.27. | | |
| 226. | Kumkuma | *Rottleria tinctoria* | | | | | | 5.4, 9, 12. | 4.67. | | | 368. | |
| 227. | Kumuda | *Nymphaea rubra* | 4.2.10, 87. | 5.28. | 42, 60. | 7.74; 8.70. | | 3.2, 15, 21, 23, 26. | 4.19; 6.86; 16.76; 81, 86; 17.6.75; 18.3; 19.34. | 3.16. | | | |
| 228. | Kumudavatī (Same as Kumuda) | —do— | | 3.14; 4.2. | | | | | 6.36; 16.85; 18.1, 6. | | | | |
| 229. | Kunda | *Jasminum multiforum* | | | | 17.27. | 3.8. | 4.2; 6.23, 31. | | 2.4. | | | 99.73. |
| 230. | Kuranthaka | *Barleria prionitis*, Linn. | | | | | | | | | | | 99.74. |
| 231. | Kurbaka | *Barleria cristala* | | 6.4. | 71, 86. | | Acts-3, 5. | 3.10; 6.18, 33. | 9.29. | 2.7. | | | 99.74. |
| 232. | Kuraṇṭa | *Barleria prioniti*, Linn. | | | | 4.1.7. | | | | | | | |
| 233. | Kuruvinda | | | | | | | | | | | | 121.160. |
| 234. | Kuśā | *Poa cyonosurloides* | 5.3.105; 5.2.63. | Act-3 | | | | | 1.95; 5.4, 7; 8.18; 13.43; 14.28; 19.2. | | | | 123.166. |

| S. No. | Name of Plant | Botanical equivalent | Pāṇini's Aṣṭādhyāyī | Abhijñāna Śākuntalaṁ | Meghadūta | Kumāra Saṁbhava | Mālavikāgnimitra | Ritusaṁhāra | Raghuvaṁśa | Vikramorvaṣiam | Kāmasūtra | Harṣacarita | Amarkoṣa |
|---|---|---|---|---|---|---|---|---|---|---|---|---|---|
| 235. | Kūśmāṇḍa | *Benincasa cerifera* Savi. | | | | | | | | | 4.1.29. | | 120.156. |
| 236. | Kuṣṭha | *Costus speciosus* | | | | | | | | | 7.1.4; 7.2.4. | | 112.126. |
| 237. | Kusumbha | *Carthamus tinctorius* L. | 5.1.50. | | 4. | | | 1.24; 4.4; 3.13. | | | 7.2, 19, 37. | 285. | 97.67. |
| 238. | Kutamata | | | | | | | | | | | | 114.132. |
| 239. | Kuvala | *Zizyphus jijube* | 5.2.24. | | | | | | | | | | |
| 240. | Kuvalaya or Kubalaya | Water Lilly | | | 46, 100. | | | 2.22. | | | | | |
| 241. | Lasuna | *Allium sativum* | | | | | | | | | 4.1.29. | | 118.149. |
| 242. | Lavaṅga | *Syzgium aromaticum* | | | | 8.25. | | | 6.57. | | | | |
| 243. | Lodhara | *Symplocus racemosa* | | | | 7.9.16. | | 4.1; 6.23. | 2.29; 3.2. | | | | 88.33 |
| 244. | Madayantika | *Sanseviera roxburghiana* | | | | | | | | | 7.2.37. | | |
| 245. | Mādhavī | *Hiptage madalota*, Gaertn | | Acts-3, 4. | 86. | | | | | 2.4. | | | 98.72. |
| 246. | Madhūka | *Bassia latifolia* | | | | 7.14. | | | 6.25. | | 7.1.39. | | 86.27; 108.110. |
| 247. | Māgadhī | | | | | | | | | | | | 105.97. |
| 248. | Makanda | | | | | | | | | | | | 88.34. |
| 248a. | Makūlaka | *Croton polyandrum* | | | | | | | | | | | 117.144. |

| S. No. | Name of Plant | Botanical equivalent | Pāṇini's Aṣṭā-dhyāyī | Abhijñā-na Śāku-ntalaṁ | Megha-dūta | Kumā-ra Saṁ-bhava | Mālavi-kāgni-mitra | Ritusa-ṁhāra | Raghu-vaṁśa | Vikramo-rvaṣiam | Kāma-sūtra | Harṣa-carita | Amar-koṣa |
|---|---|---|---|---|---|---|---|---|---|---|---|---|---|
| 249. | Malāpū | | | | | | | | | | | | 96.62. |
| 250. | Mālatī | *Jasminum grandiflorum* | | | 103. | | | 2.24; 3.2, 19. | | | | | 99.73. |
| 251. | Mallikā | *Gaertnera racemosa* | | | | | | | 16.47, 50. | | 4.1.7. | | 98.70. |
| 252. | Malura | | | | | | | | | | | | 88.32. |
| 253. | Mañjiṣṭhā | *Rubia tinctorium*, L. | | | | | | | | | | 313. | 103.91. |
| 254. | Mārica | A species of plant. | | | | | | | 4.46. | | 7.1.24. | | |
| 255. | Maruvaka | *Majarana hortensis*, Moench | | | | | | | | | | | 94.53; 100.79. |
| 256. | Māṣaparṇi | *Teramonus labialis* | | | | | | | | | | | 116.39. |
| 257. | Māṣiṇa | *Linum usitatissionum*, L. A type of bean or bean field. | 5.2.4. | | | | | | | | | 375. | |
| 258. | Maskara | Bamboo | 6.1.154. | | | | | | | | | | |
| 259. | Mātulunga or Mātuliṅga | *Citrus medica*, Linn. | | | | | | | | | | | 100.79. |
| 260. | Moca or Mocakhya | *Moringa pterygoperma* or *Musa sapientum* | | | | | | | | | | | 87.31. |
| 261. | Moraṭa | *Sanseviera roxburghiana* | | | | | | | | | 7.1.38. | | |
| 262. | Mṛṇāla | *Andropogon muricatus* Lotus stalk | | Acts-3 and 6, 18. | | 1.40; 3.49; 5.29; 6.68. | | 1.19. | 16.7,16. | 3.13. | | | |

| S. No. | Name of Plant | Botanical equivalent | Pāṇini's Aṣṭādhyāyī | Abhijñāna Śākuntalaṁ | Meghadūta | Kumāra Saṁbhava | Mālavikāgnimitra | Ritusaṁhāra | Raghuvaṁśa | Vikramorvaṣiam | Kāmasūtra | Harṣacarita | Amarkoṣa |
|---|---|---|---|---|---|---|---|---|---|---|---|---|---|
| 263. | Mudgaparṇi | *Phaseolus trilolus* | | | | | | | | | | | 109.114. |
| 264. | Mūlaka | *Raphanus sativus*, L. | | | | | | | | | 4.1.29. | | 121.158. |
| 265. | Munja | *Saccharum munja* | 3.1.117. | | | | | | | | | | |
| 266. | Mūra | | | | | | | | | | | | 111.123. |
| 267. | Mūrva | *Sanseviera roxburghiana* | | | | | | | | | | | 101.84. |
| 268. | Musakaparṇi | *Anthericum tuberosum* | | | | | | | | | | | 102.88. |
| 269. | Muṣṭhā | *Cyperus rotundus* | | | | | | | 1.17. | 9.59; 15.19. | | | 121.160. |
| 270. | Naḍa | *Phragnites maxima* | 4.2.87. | | | | | | | | | | 122.163. |
| 271. | Nāgabalā | *Sida spinosa* | | | | | | | | | | | 110.117. |
| 272. | Nāgakesara | *Mesua ferrea*, L. | | | | | | | | | 7.1.8, 32. | | 87.65. |
| 273. | Nakha | | | | | | | | | | | | 113.131. |
| 274. | Naktamālā | *Pongamia globra* | | | | | | | | | | | 92.48. |
| 275. | Naḷi | | | | | | | | | | | | 113.130. |
| 276. | Nalikera/ Narikela | *Cocos mucifera* | | | | | | | 4.42. | | | | 123.169. |
| 277. | Nalinī | Lotus | Act-3 | 41. | 6.6. | | 3.10. | | | | | | |
| 278. | Nameru | *Elaeocarpus ganitrus* | | | 1.55; 3.43. | | | | 4.74. | | | | |

| S. No. | Name of Plant | Botanical equivalent | Pāṇini's Aṣṭādhyāyī | Abhijñāna Śākuṅtalaṁ | Meghadūta | Kumāra Saṁbhava | Mālavikāgnimitra | Ritusaṁhāra | Raghuvaṁśa | Vikramorvaśīam | Kāmasūtra | Harṣacarita | Amarkoṣa |
|---|---|---|---|---|---|---|---|---|---|---|---|---|---|
| 279. | Nandyāvrata | | | | | | | | | | 6.1.7. | | |
| 280. | Navamālikā | *Jasminum sambac* | | | | 18.28. | | 3.18; 4.5. | 9.42. | | 4.1.7. | | 99.73. |
| 281. | Niculā | *Barringtonia acutangula* | | | 14. | | | | | 4.13. | | | 96.61. |
| 282. | Nīlī | | | | | | | | | | | 57. | 104.95. |
| 283. | Nīlotpala (Blue lotus) | *Nymphaea stellata,* Willd. | | | | 1.46. | | 2.2; 3.17, 19, 26; 4.9. | | | | | |
| 284. | Nimba | *Melia azadirachta,* L. | | | | | | | | | | | 86.26; 96.63. |
| 285. | Nipa (Kadamba) | *Nauclea cadamba* | 4.3.152. | | 21, 71. | | | 2.17; 3.13. | 19.37. | 4.42, 2. | 7.2.33. | | 91.42. |
| 285a. | Nivāra | A kind of paddy. | | 1.13; 4.20. | | | | | | 9.53. | | | 88.32. |
| 286. | Nyagrodha | *Ficus bengalensis* | 7.3.5. | | | | | | | | | | |
| 287. | Padma | *Clerodendrum siphonanthus,* Lotus | | | 70. | 1.16, 43; 5.19, 27; 7.16, 38; 8.23, 30; 9.25.46. | | 2.14; 3.1; 3.15; 4.1, 14; 6.2. | 3.17; 6.26, 36, 86; 9.80; 10.8, 62; 13.51; 16.16, 48, 68; 17.75. | | | 18. | |
| 288. | Padmacārṇī | *Hibiscus mutabilis* | | | | | | | | | | | 117.46. |
| 289. | Padminī | *Nelumbium speciosum,* Lotus | | | 84. | 3.76. | | | | 14.33. | 4.1.29. | | 118.148. |

| S. No. | Name of Plant | Botanical equivalent | Pāṇini's Aṣṭādhyāyī | Abhijñāna Śākuntalam | Meghadūta | Kumāra Saṁbhava | Mālavikāgnimitra | Ritusaṁhāra | Raghuvaṁśa | Vikramorvaṣīam | Kāmasūtra | Harṣacarita | Amarkoṣa |
|---|---|---|---|---|---|---|---|---|---|---|---|---|---|
| 290. | Palāndu | Onion | | | | | | | | | | | |
| 291. | Pālaṅki or Plankya | *Spinacea oleracea,* Linn. | | | | | | | | | 4.1.29. | | 111.122. |
| 292. | Palāśa (Kiṁśuka) | *Butea frondosa* | 4.3.141. | | | 3.29. | | | | | | | 87.29, 120, 155. |
| 293. | Panāsa | *Artocarpus heterophyllus* | | | | | | | | | | | 96.61. |
| 294. | Paṅkaja | *Nelumbium speciosum,* Lotus | | | | 5.9; 13.39. | | 3.10; 3.23; 5.5. | 3.8; 7.64; 8.55; 13.54; 15.52; 18.20; 19.8. | | | | |
| 295. | Paribhadra | *Erythrina indica* | | | | | | | | | | | 88.26; 94.53. |
| 296. | Parkaṭi | *Ficus infectoria* | | | | | | | | | | | 88.32. |
| 297. | Pātala | *Bignonia suaveolens* | | 1.3. | | | | 1.28. | 16.52; 19.46. | | | | |
| 298. | Paṭola | *Trichosanthes dioica,* Roxb. | | | | | | | | | | | 120.155. |
| 299. | Patali (Rice) | *Bignonia suaveolens* | 4.3.136. | | | | | | | | | | 94.55. |
| 300. | Pāthā | | | | | | | | | | | | 102.85. |
| 301. | Pāthi | | | | | | | | | | | | 100.80. |
| 302. | Pauṇḍrya | | | | | | | | | | | | 113.128. |
| 303. | Phalini | A fruit tree. | | | | | | | 8.61. | | | | |
| 304. | Picula | Cotton plant | | | | | | | | | | | 90.40. |

| S. No. | Name of Plant | Botanical equivalent | Pāṇini's Aṣṭādhyāyi | Abhijñāna Śākuntalaṁ | Meghadūta | Kumāra Saṁbhava | Mālavikāgnimitra | Ritusaṁhāra | Raghuvaṁśa | Vikramorvaṣiam | Kāmasūtra | Harṣacarita | Amarkoṣa |
|---|---|---|---|---|---|---|---|---|---|---|---|---|---|
| 305. | Picumanda | Nimba tree. *Azadirachta indica* | | | | | | | | | | | 96.63. |
| 306. | Pīlu | *Careya arborea* | 5.2.24; 6.3.21. | | | | | | | | | | 86.88. |
| 307. | Pīpala (Aśvattha) | *Ficus religiosa,* L. | | | | | | | | | | | 84.20. |
| 308. | Pippalī-mūla | *Piper longinum* | | | | | | | | | 7.1.24, 45. | | 105.97. |
| 309. | Pitala | | | | | | | | | | 7.1.38. | | |
| 310. | Pitanaka | | | | | | | | | | | | 86.27. |
| 311. | Pujala | *Buchanania latifolia* | | | | 3.31. | | | | | | | |
| 312. | Piyūkṣa | | 8.4.5. | | | | | | | | | | |
| 313. | Plākṣa | *Ficus infactoria* | 4.3.164; 8.4.5. | | | | | | 8.93; 13.71; 17.12. | | | | 88.32, 91.44. |
| 314. | Potagala | *Basella lucida* | | | | | | | | | | | 122.163. |
| 315. | Prapannada | | | | | | | | | | | | 118.147. |
| 316. | Piryala (ka) | *Buchanania latifolia* | | | | | | | | | | | 89.35. |
| 317. | Priyaṅgu | *Setaria italica* | | | | | Act-3 | 4.10; 6.12. | | | 7.1.32. | | 94.56. |
| 318. | Pṛṣṇīparṇī | *Uraria lagopoioides* | | | | | | | | | | | 104.93. |
| 319. | Puga | *Araca catechu* | | | | | | | 4.44; 6.64; 13.17. | | 3.3.16; 3.5.29. | | 123.169. |

| S. No. | Name of Plant | Botanical equivalent | Pāṇini's Aṣṭādhyāyi | Abhijñāna Śākuntalaṁ | Meghadūta | Kumāra Saṁbhava | Mālavikāgnimitra | Ritusaṁhāra | Raghuvaṁśa | Vikramorvaṣiam | Kāmasūtra | Harṣacarita | Amarkoṣa |
|---|---|---|---|---|---|---|---|---|---|---|---|---|---|
| 320. | Punarnavā | *Boerhaavia procumbens* | | | | | | | | | | | 118.149. |
| 321. | Puṇḍarīka | *Artemesia indica* | | | | 8.58. | 4.7. | | | 4.42. | | | |
| 322. | Punnāga | *Colophyllum inophyllum* | | | | | | | | | | | 86.25. |
| 323. | Puṣkara (Blue lotus) | *Nymphaea nelumbo* | 5.2.135. | | | 3.65. | | | 18.30. 37. | | | | |
| 324. | Putikā | *Caesalpinia borducella* | | | | | | | | | | | 92.48. |
| 325. | Rājadānā | *Buchanania latifolia* | | | | | | | | | | | 92.46. |
| 326. | Rājajambu | | | | | | | | | 4.27. | | | |
| 327. | Rājatali | Areca nut tree | | | | | | | 6.56. | | | | |
| 328. | Rājiva | *Trichosanthes dicica* | | | | 3.45. | | | | | | | |
| 329. | Rambhā | *Musa sapientum, L.* | | | | | 3.10. | | | | | | |
| 330. | Rāsnā | *Vanda roxburgii* | | | | | | | | | | | 109.115, 116.140. |
| 331. | Rasonā | *Allium ascalonicum* | | | | | | | | | | | 118.140. |
| 332. | Renukā | | | | | | | | | | | | 111.121. |
| 333. | Rocanī | *Croton polyandrum* | | | | | | | | | | | 118.147. |
| 334. | Rohataka | *Aphanamixis polystachia* | 4.3.152. | | | | | | | | | | 93.49. |
| 335. | Ṛṣyagandhikā | | | | | | | | | | | | 108.111. |

| S. No. | Name of Plant | Botanical equivalent | Pāṇini's Aṣṭā-dhyāyī | Abhijñā-na Śāku-ntalaṁ | Megha-dūta | Kumā-ra Saṁ-bhava | Mālavi-kāgni-mitra | Ritusa-ṁhāra | Raghu-vaṁśa | Vikramo-rvaṣiam | Kāma-sūtra | Harṣa-carita | Amar-koṣa |
|---|---|---|---|---|---|---|---|---|---|---|---|---|---|
| 336. | Sada | | 4.2.88. | | | | | | | | | | |
| 337. | Sahā | *Aloe perfolita* | | | | | | | | | | | 99.74. |
| 338. | Sahakāra (Āmra) | Mango *Aloe perfolita* | | Act-1, 3. | | 4.38; 8.78. | 4.13; 5.4. | 7.1, 30, 32. | 6.69; 8.61; 9.33; 16.52; 19.46. | 2.6. | | | |
| 339. | Śaileya | | | | | | | | | | | | 111.125. |
| 340. | Sairajaka | | | | | | | | | | | | 99.75. |
| 341. | Sāla | *Artocarpus locuca* | | | | 13.33. | | | 1.38. | | | | 92.45. |
| 342. | Sallakī | *Anrtocarpus locuca* | | | | 8.33. | | | | 4.44. | | | 112.124. |
| 343. | Sālmalī | *Salmalia malabarica,* Schott. | 4.2.82. | | | | | 1.26. | | | | | 92.47. |
| 344. | Śami | *Prosopis spicigera minosa suma* | 5.3.88. | 4.3. | | | | | 7.26. | | | | 93.52. |
| 345. | Śaṇapuṣpikā | *Causcora decussafa* | | | | | | | | | | | 107.107. |
| 346. | Saṁkhinī | *Xanthium stramonium,* L. | | | | | | | | | | | 113.127. |
| 347. | Saptacchaḍa | *Alstonia scholaris,* R.Br. | | | | | | 3:2.13. | 5.48. | | | | |
| 348. | Sataparṇa | *Alstonia scholaris,* R.Br. | | | | | | | 4.23. | | 2.2.5. | | 85.23. |
| 349. | Śara | *Saccharum arundinaceum* | 8.4.5. | | 47. | 8.74; 11.13. | | | 13.26. | | | | 122.162. |

| S. No. | Name of Plant | Botanical equivalent | Pāṇini's Aṣṭādhyāyī | Abhijñāna Śākuntalam | Meghadūta | Kumāra Sambhava | Mālavikāgnimitra | Ritusamhāra | Raghuvaṁśa | Vikramorvaṣiam | Kāmasūtra | Harṣacarita | Amarkoṣa |
|---|---|---|---|---|---|---|---|---|---|---|---|---|---|
| 350. | Sarala | *Pinus roxburgii* | | | 55. | 1.9. | | | 4.75. | | | | 95.60. |
| 351. | Saraṇa | | | | | | | | | | | | 119.153. |
| 352. | Sarañja | | | 1.17. | | | | | 5.69. | | | | |
| 353. | Saraja or Sarajaka | | | | | | | 2.17; 3.13. | | 1.61.14. | | | |
| 354. | Saroruha | | | | | 1.16;9. 27; 11.21. | | | | | | | |
| 355. | Sarṣopa | *Brassicanapus*, L. | | | | | | | | 1.79.17. | 4.1.6. | | |
| 356. | Saspa | Tender grass | | | | | | 1.22. | | 4.57. | | | 123.168. |
| 357. | Satamūlī | *Asparagus racemosus*, Willd. | | | | | | | | | | | 106.101. |
| 358. | Sataprasa | | | | | | | | | | | | 99.77. |
| 359. | Satapuṣpā | *Pencedanum sowa*, Kurz. | | | | | | | | | 6.1.6. | | 119.152. |
| 360. | Śatāvarī | *Asparagus racemosus* Willd. | | | | | | | | | | | 106.101. |
| 361. | Śelu | | | | | | | | | | | | 89.35. |
| 362. | Śaphalika | *Nyctanthes arbortristist* | | | | | | 3.14. | | | | | 98.71. |
| 363. | Sidhraka | | 8.4.4. | | | | | | | | | | |
| 364. | Simhi | | | | | | | | | | | | 106.103. |
| 365. | Simsapā | *Dalbergia sisu* | 7.3.1. | | | | | | | | 7.1.30. | | |
| 366. | Sindhuvarā | *Vitex negundo* | | | | 3.53. | | | | | | | 98.69. |
| 367. | Śiriṣa | *Albizzia lebbeck*, Benth. | 4.2.8. | 1.4, 26; 6.13. | 71. | 1.41, 5.4. | | | 16.48; 61; 18.45. | | | | 96.63. |

| S. No. | Name of Plant | Botanical equivalent | *Pāṇini's Aṣṭā-dhyāyī* | *Abhijñā-na Śāku-ntalaṁ* | *Megha-dūta* | *Kumā-ra Saṁ-bhava* | *Mālavi-kāgni-mitra* | *Ritusa-ṁhāra* | *Raghu-vaṁśa* | *Vikramo-rvaṣiam* | *Kāma-sūtra* | *Harṣa-carita* | *Amar-koṣa* |
|---|---|---|---|---|---|---|---|---|---|---|---|---|---|
| 368. | Sitapatra | | | | | | | | | | 7.2.21. | | |
| 369. | Slākṣaparṇi | | | | | | | | | | 7.2.37. | | |
| 370. | Śleṣamāntaka | | | | | | | | | 1.61.2. | | | 89.35. |
| 371. | Snūhi | *Euphorbia nerifolia*, L. | | | | | | | | | | 410. | |
| 372. | Snuk | | | | | | | | | | | | 107.106. |
| 373. | Synthistaka | | | | | | | | | | 7.2.30, 36. | | |
| 374. | Sobhāñjana | | | | | | | | | | | | 87.31. |
| 375. | Soma | *Asclepias acida*, Roxb. | | | | | | | | | 7.2.31, 36. | | |
| 376. | Somarāji | *Vernonia anthelmentica*, Willd. | | | | | | | | | | | 105.96. |
| 377. | Sonaka | *Crotolaria juncea* | | | | | | | | | | | 95.58. |
| 378. | Spandana | | 4.3.141. | | | | | | | | | | |
| 379. | Śravanapiṅgaka | | | | | | | | | | 7.2.37. | | |
| 380. | Śriparṇi | *Premna spinosa* | | | | | | | | | 7.1.46. | | |
| 381. | Śṛṅgāṭaka | *Trapa bispinosa*, Roxb. | | | | | | | | | 7.1.39, 42. | | |
| 382. | Śṛṅgī | *Ficus infectoria* | | | | | | | | | | | 106.100; 110.117. |
| 383. | Stambha | Tuft of grass | | | | | | | | | 6.1.6. | | |
| 383a. | Sthalakamala | | | | 97. | | | | | | | | |

| S. No. | Name of Plant | Botanical equivalent | Pāṇini's Aṣṭādhyāyi | Abhijñāna Śākuntlaṁ | Meghadūta | Kumāra Saṁbhava | Mālavikāgnimitra | Ritusaṁhāra | Raghuvaṁśa | Vikramorvaṣiam | Kāmasūtra | Harṣacarita | Amarkoṣa |
|---|---|---|---|---|---|---|---|---|---|---|---|---|---|
| 384. | Sthalacarvinda | | | | | 1.33 | | | | | | | |
| 385. | Śukanāsa | | | | | | | | | | 4.1.29. | | |
| 386. | Śukasinhi | | | | | | | | | | | | 102.87. |
| 387. | Surana | | | | | | | | | | 4.1.29. | | |
| 388. | Surasa | | | | | | | | | | | | 109.115. |
| 389. | Susira | | | | | | | | | | | | 118.130. |
| 390. | Suvāhā | | | | | | | | | | | | 110.119; 116.140. |
| 391. | Svayamgupta | | | | | | | | | | 4.1.29; 7.1.41, 43. | | |
| 392. | Śyāma | *Artemisia indica* | | | 109. | | | 3.18. | | | | | 94.55; 109.112. |
| 393. | Śyāmaka | *Echinochloa colona* | | 4.13 | | | | | | | | | 123.166. |
| 394. | Tagara | *Tabernaemontana coronata*, Willd. | | | | | | | | | 4.1.7; 8.1.4, 9, 32; 7.2.41. | | |
| 395. | Tāla | *Borassus flabelliformis* | 4.3.152. | | 32. | | | | 4.34; 6.57; 15.23. | 5.10. | | | 113.127; 113.123, 169, 124; 124.170. |
| 396. | Talīsa | *Flacourtia cataphracta* | | | | | | | | | 7.1.4,9; 7.2.41. | | |
| 397. | Tamāla | *Crataeva roxburgii* | | | | 3.53. | | | 6.64; 13.49. | | 4.1.6; 7.1.9. | | 97.68. |

| S. No. | Name of Plant | Botanical equivalent | Pāṇini's Aṣṭā-dhyāyī | Abhijñā-na Śāku-ntalaṁ | Megha-dūta | Kumā-ra Saṁ-bhava | Mālavi-kāgni-mitra | Ritusa-ṁhāra | Raghu-vaṁśa | Vikramo-rvaṣiam | Kāma-sūtra | Harṣa-carita | Amar-koṣa |
|---|---|---|---|---|---|---|---|---|---|---|---|---|---|
| 398. | Tamalaki | | | | | | | | | | | | 113.127. |
| 399. | Tamasā | | | | | 8.26, 32. | | | 4.37; 9.12, 37. | | | | |
| 400. | Tāmbūlī or Tāmbūla | *Piper betle* | | | | | | 5.5 | 4.42; 6.64 | | | | 111.120 |
| 401. | Tāpiccha | | | | | | | | | | | | 97.68 |
| 402. | Tilaka | *Clerodendrum phlomoides* | | | | 3.30; 8.40 | | | 1.41, 44. | | | | 90.40 |
| 403. | Tilaparṇikā | *Pterocarpus santalimus* | | | | | | | | | 4.1.29 | | |
| 404. | Tinduka | *Diospyros embryopteris* | | | | | | | | 1.61.3 | | | 90.39 |
| 405. | Titiṇḍī | *Tamarindus indica* L. | | | | | | | | | | | 91.44 |
| 406. | Trapuṣā | *Cucumis sativis,* L. | | | | | | | | | 4.1.29; 7.2.28. | | |
| 407. | Trayamanā | *Gentiana kurroo,* Royle | | | | | | | | | | | 119.151 |
| 408. | Tūdā | | | | | | | | | | | | 91.42 |
| 409. | Tumbī | | | | | | | | | | | | 120.156 |
| 410. | Tanna | *Cedrela toona* Roxb. | | | | | | | | | | | 113.128 |
| 411. | Tūvarikā | | | | | | | | | | | | 113.131 |
| 412. | Uccarakaṇḍa | | | | | | | | | | 7.1.35 | | |
| 413. | Uccaṭā | | | | | | | | | | | | 121.160 |
| 414. | Udumbara | *Ficus glomerata* | 4.3.152 | | 44. | | | | | 11.54; 27.12 | | | 85.22. |

| S. No. | *Name of Plant* | *Botanical equivalent* | *Pāṇini's Aṣṭādhyāyi* | *Abhijñāna Śākuntalam* | *Meghadūta* | *Kumāra Saṁbhava* | *Mālavikāgnimitra* | *Ritusaṁhāra* | *Raghuvaṁśa* | *Vikramorvaṣiam* | *Kāmasūtra* | *Harṣacarita* | *Amarkoṣa* |
|---|---|---|---|---|---|---|---|---|---|---|---|---|---|
| 415. | Udumbaraparṇī | | | | | | | | | | | | 117.45 |
| 416. | Ugragandhā | | | | | | | | | | | | 106.103 |
| 417. | Ulapā | *Imperata arundinaceae* | | | | | | | | | | | 81.9 |
| 418. | Upodakā | *Basella cardifolia* | | | | | | | | | | | 121.158 |
| 419. | Uraṇākṣā | | | | | | | | | | | | 118.148 |
| 420. | Uruvarka | | | | | | | | | | | | 93.51 |
| 421. | Usaṇā | *Piper longium* | | | | | | | | | | | 105.97 |
| 422. | Usira | *Andropogon squarrosus* Hack. | | 1.84 | | | | | | | | | 122.164 |
| 423. | Utapala | *Nymphaea coerulea* | | Act-1 | 26. | 4.8; 7.20,81. 9.26; 17.35. | | 2.9, 14,15; 3.14, 24; 5.10. | 3.36; 4.42; 7.26; 12.86; 13.31. | | 7.1.6,8; 7.2.35. | | 68.37; 122.126 |
| 424. | Utapala Śarīvā | | | | | | | | | | | | 109.112 |
| 425. | Vacā | *Acorus calamus* | | | | | | | | | | | 106.103 |
| 426. | Vajrakandā | | | | | | | | | | 7.2.41. | | |
| 427. | Valgujā | *Amaranttus polygamus* | | | | | | | | | 7.2.36. | | |
| 428. | Vaṁśa | *Shorea robusta* | 5.1.50. | | | | | 1.25. | | | | 4.10. | 121.161. |
| 429. | Vanajā | *Cyperus rotundus* | | | | | | | 5.73. | | | | |
| 430. | Vanadhuka | | | | | | | | | 1.61. | | | |
| 431. | Vānīra | | | | 43. | | | | 13.30, 35; 16.21. | | | | 87.30. |

| S. No. | Name of Plant | Botanical equivalent | Pāṇini's Aṣṭādhyāyi | Abhijñāna Śākuntalaṁ | Meghadūta | Kumāra Saṁbhava | Mālavikāgnimitra | Ritusaṁhāra | Raghuvaṁśa | Vikramorvaṣīam | Kāmasūtra | Harṣacarita | Amarkoṣa |
|---|---|---|---|---|---|---|---|---|---|---|---|---|---|
| 432. | Vaṅkoṭa | | | | | | | | | | | | 87.29. |
| 433. | Vārāhī | *Physalis flexuosa* | | | | | | | | | | | 119.151. |
| 434. | Varaṇā | *Cratoosva murvalla* | 4.2.82. | | | | | | | | | | 86.25. |
| 435. | Varhis | | | | | 1.60 | | | | | | | |
| 436. | Varijā | | | | | 15.12. | | | | | | | |
| 437. | Vāriruha | | | | | 8.33. | | | | | | | |
| 438. | Vartābhrānta patra | | | | | | | | | | 7.1.25. | | |
| 439. | Vartakā | | | | | | | | | | 4.1.29. | | |
| 440. | Vārtākī | *Solamum melongana* | | | | | | | | | | | 109.114. |
| 441. | Vastukā | | | | | | | | | | | | 121.158. |
| 442. | Vaṭa or Nyagrodha | *Ficus bengalensis* | 6.2.82. | | | | | | 13.53. | | | | 88.32. |
| 443. | Vaṭaphala | | | | | | | | | | | | 87.29. |
| 444. | Vāṭyālaka | *Sida cordifolia* L. | | | | | | | | | | | 107.107. |
| 445. | Venu | *Bambossa sp.* | | | | | | | | | | | 121.161. |
| 446. | Vatsa | *Calamus rotang* | 4.2.17. | Acts-2.3. | | | | | 9.75. | | 7.2.19. | 387. | 87.29. |
| 447. | Vibhītaka | *Terminalia bellerica,* Roxb. | 6.3.152. | | | | | | | | | | |
| 448. | Viḍaṅga | *Erycibe paniculata* | | | | | | | | | | | 107.107. |
| 449. | Vidari | *Ipomoea paniculata,* R.Br. | | | | | | | | | 7.1.37, 38, 41. | | |

| S. No. | Name of Plant | Botanical equivalent | Pāṇini's Aṣṭādhyāyī | Abhijñāna Śākuntalam | Meghadūta | Kumāra Saṁbhava | Mālavikāgnimitra | Ritusaṁhāra | Raghuvaṁśa | Vikramorvaśiam | Kāmasūtra | Harṣacarita | Amarkoṣa |
|---|---|---|---|---|---|---|---|---|---|---|---|---|---|
| 450. | Vikankaṭa | *Flacourtia ramontchi* | 4.3.141. | | | | | | 11.25. | | | | 90.38. |
| 451. | Viraṇa | *Andropogon muricatus* | | | | | | | | | | | 122.164. |
| 452. | Vīravṛkṣa or Vīrā | | | | | | | | | | | | 91.43. |
| 453. | Viṣtara | | 8.3.93. | | | | | | | | | | |
| 454. | Vitunna | | | | | | | | | | | | 118.149. |
| 455. | Vyāghranakhā | *Martiyaia diandra*, Roxb. | | | | | | | | | | | 113.129. |
| 456. | Yaṣṭhimadhu | *Glycyrrhiza glabra* | | | | | | | | | | | 108.110. |
| 457. | Yava (Ka) | *Hordeum vulgare*, L. | | | | 7.17, 82; 8.62. | | | 9.43; 13.49; 17.12. | | | | |
| 458. | Yavaṅika | *Carum copticum* Benth. | | | | | | | | | | | 117.145. |
| 459. | Yavasa | Grass | | | | | | | | | | | 103.92 |
| 460. | Yuthikā | *Jasminum auriculatum*, Vahl. | | | 26. | | | 2.24. | | 4.46. | | 455. | 98.72. |

# Medicinal Trees and Plants Mentioned in the Indian Medical Treatises

| *S. No.* | *Name in the list* | *Botanical Equivalent* | *Hindi version* | *Caraka Saṁhitā (100 to 500 BC)* | *Śuśruta Saṁhitā (500 BC)* | *Vāgbhaṭṭa Aṣṭāṅgahṛdaya (4th Century A.D.)* | *Rasārṇava (12th Century A.D.)* |
|---|---|---|---|---|---|---|---|
| 1. | Akṣa | *Terminalia bellerica* | बहेड़ा | | Sū: 46.281, 282. | Sū: 5.61 | 12.364. |
| 2. | Agastyapuṣpā | *Sesbania grandiflora,* Pers. | हथिया | | | | 6.10. |
| 3. | Aguru | *Aporusa agallocha,* Roxb. | अगर | | | | 1.39 |
| 4. | Agnikā | *Plumbago zeylanisa* | चिता | | | | 11.86 |
| 5. | Agnijarā | *Methonica superba* | | | | | 6.19 |
| 6. | Agnidhamnī | *Solanum jacquini* | | | | | 5.8 |
| 7. | Ankoṭa | *Alangium salvifolium,* Linn. | | Sū: 27.159 | Sū: 46.197 | | 7.129 |
| 8. | Aṅganāyikā/Ajanāyikā | Black Cotton | काली कपास | | | | 5.10 |
| 9. | Ajaśṛngikā | *Odina wodiar* | मेढ़ाशिंगी | | | | 19.89 |
| 10. | Ajamāri | *Cawateh, coropogonpruriens* | | | | | 5.12 |
| 11. | Atasi | *Linum usitatissimum,* Linn. | तीसी | | | | 18.124 |
| 12. | Atibalā | *Sida rhombifolia,* Linn. | गंगेरन | | | | 6.104. |
| 13. | Ananta | *Ervatamia divaricata,* Linn. | अनन्तमूल | | | | 5.3 |
| 14. | Apāmārga | *Achyranthes aspera,* Linn. | चिचड़ा | | | | 6.25 |
| 15. | Abhayā | *Andropogon muricatus* | हरड़ | | | | 12.34 |
| 16. | Amṛtā | *Eulophia campestris,* Wall. | गिलोय | | | | 6.98 |

| S. No. | Name in the list | Botanical Equivalent | Hindi version | Caraka Saṁhitā (100 to 500 BC) | Śuśruta Saṁhitā (500 BC) | Vāgbhaṭṭa Aṣṭāṅgahṛdaya (4th Century A.D.) | Rasārṇava (12th Century A.D.) |
|---|---|---|---|---|---|---|---|
| 17. | Amalavetasā | *Emblica officinalis* | अमलतास (?) | Vi: to 8.137 | Sū: 27.152-153 | | 5.31 |
| 18. | Aruṇotpala | Red lotus | लाल कमल | | | | 6.11 |
| 19. | Arka | *Calotropis gigantea*, R.Br. | मदार | | Sū: 46.284 | | 6.34 |
| 20. | Arjūna | *Terminalia arjuna*, Roxb. | अर्जुन वृक्ष | | | | 8.76 |
| 21. | Alaka | *Andropogon serratus*, Rotz. | बंदाल | | | | 7.116 |
| 22. | Alambuṣā | | | | | | 11.26 |
| 23. | Alārka | *Calotropis gigantia*, Linn. | | | | | 11.178 |
| 24. | Aśvagandhā | *Withenia coagulans*, Dunal | | | | | 7.139 |
| 25. | Aśvattha | *Ficus religiosa*, Linn | पीपल | | | | 6.99 |
| 26. | Aśvavāra | *Saccharum spontaneum* | सफेद कनेर | | | Sū: 4.37.7. | 15.193 |
| 27. | Asana | *Bridelia stepularis*, Blume | विजयसार | | Sū: 46.284 | | 5.39 |
| 28. | Asthiśṛṇkhalā | *Vitis quarangularia*, Wall | हड़जोड़ | | | | |
| 29. | Ahimāra | *Vachellia farnesiana* | | | | | 8.77 |
| 30. | Ahivallī or Nāgavallī | *Piper betle* | पान | | | | 15.138 |
| 31. | Ākhukarṇī or Ākhuparṇī | *Salvinia cucullata*, Roxb. | मूसाकर्णी | | Sū: 27.113 | | 6.79; 5.6 |
| 32. | Ātarukṣā | *Justicia adhatoda* | | | | | 14.141 |
| 33. | Āmalaka/Āmalakī | *Emblica officinalis*, Gaertn. | | Sū: 4.13, 24, 36, 39 | Sū: 38.53, 55, 59 | | 12.370 |
| 34. | Āraktaka | | लाल मदार | | | | 6.92 |
| 35. | Āraktavallī | | मंजिष्ठा | | | | 8.43 |

| S. No. | Name in the list | Botanical Equivalent | Hindi version | Caraka Saṁhitā (100 to 500 BC) | Śuśruta Saṁhitā (500 BC) | Vāgbhaṭṭa Aṣṭāṅgahṛdaya (4th Century A.D.) | Rasārṇava (12th Century A.D.) |
|---|---|---|---|---|---|---|---|
| 36. | Āragvedha | *Cassia fistula*, Linn. | अमलतास | | | | 10.42 |
| 37. | Ārdaka | *Zingiber officinalis*, Rosc | अदरक | Sū: 27.160 | Sū: 46.227 | Sū: 6.108 | 5.30 |
| 38. | Āśupuṅkha or Sarapaṅkha | | सरफोंका | | | | 11.178 |
| 39. | Āsurī | *Sinapis ramosa*, Linn. | राई | Sū: 27.100 | Sū: 46.221 | | 11.88 |
| 40. | Āsphoṭa | *Jalan regia* | | | | | 11.26 |
| 41. | Ingudā or Ingudī | *Terminalia catappa* or *Balanites aegyptiaca*, Linn. | | Sū: 27.146 | | | 7.117 |
| 42. | Indurī or Indurekhā or Vacuki | *Vernonia anthelmenthica* | | | | | 6.101 |
| 43. | Indravāruṇī | *Citrullus colocynttes*, Shoad | | | | | 5.14 |
| 44. | Īśvarī | *Aristolochia indica*, Linn. | बांझ खेकसा | | | | 5.20 |
| 45. | Uccaṭa | | सफेद गुँजा | | | | 5.10 |
| 46. | Uccatti | | | | | | 12.112 |
| 47. | Utakata | | सहलीपिपली | | | | 5.19 |
| 48. | Uttaravāruṇī or Indravāruṇī | *Citrullus colocynthes*, Sehrad | | | | | 6.94 |
| 49. | Udakakarṇā | *Comnelina salicifolia*, Roxb. | जलपिपली | | | | 7.89 |
| 50. | Udumbara | *Ficus glomerata*, Roxb. | गूलर | Sū: 4.33; 27.105 | | | 6.103 |
| 51. | Unmatta | *Dhatura fastuosa* | धतूरा | | | | 6.79 |
| 52. | Umāphala | *Linum usitatissimum* | वीसी | | | | 6.18 |
| 53. | Uragā | | नागी | | | | 10.39 |

| S. No. | Name in the list | Botanical Equivalent | Hindi version | Caraka Saṁhitā (100 to 500 BC) | Śuśruta Saṁhitā (500 BC) | Vāgbhaṭṭa Aṣṭāṅgahṛdaya (4th Century A.D.) | Rasārṇava (12th Century A.D.) |
|---|---|---|---|---|---|---|---|
| 54. | Uṣaṇa | Dry ginger<br>*Piper longium* | सोंठ | | | | 10.59 |
| 55. | Ūṣaṇa | Black pepper | काली मिरच | | | | 10.59 |
| 56. | Ekaparṇikā | *Ocimum grantissimum* | कचूर | | | | 5.6 |
| 57. | Ekavīrā or Vandhyā Karkotaki | | | | | Sū: 5.58, 59 | 5.10 |
| 58. | Eraṇḍa | *Ricinus communis*, Linn | अरण्डी | | | | 9.10 |
| 59. | Kukubha or Arjūna | *Terminalia arjuna*, Roxb. | | | | | 12.328 |
| 60. | Kankala Khecarī | | | | | | 12.53 |
| 61. | Kankoṭa | *Piper cubeba*, Linn. | शीतल चीनी | | | | 18.112 |
| 62. | Kaṅguṇī | *Calastrus paniculatus* | | | | | 6.34 |
| 63. | Kañcukī or Kṣira Kañcukī | *Lipeocercis serrata* | | | | | 5.11 |
| 64. | Kaṭuka Rohiṇī | Ginger black or long pepper.<br>*Pieroliza kurroa*, Royle | | | | | 18.5 |
| 65. | Kaṭukalābu | *Trichosanthes dioica* | कड़वी तुम्बी | | | | 7.30 |
| 66. | Kaṭu Kosatakī | | कड़वी तुरई | | | | 15.94 |
| 67. | Kaṭutumbara or Kaṭutumbī | *Lagenaria vulgaris*, Sering | कड़वी तुम्बी | | | | 5.14 |
| 68. | Kaṭutrayaya | | सोंठ, मिरच, पीपल | | | | 7.91 |
| 69. | Kaṇa or Kaṇā | *Obligochaeta cuspidata*, Royle | पीपल | | | | 18.7 |
| 70. | Kaṇṭakārī | *Solanum surattense*, Burm. | भरकटैया | | Sū: 46.262, 267 | | 14.149 |

| S. No. | Name in the list | Botanical Equivalent | Hindi version | Caraka Saṁhitā (100 to 500 BC) | Śuśruta Saṁhitā (500 BC) | Vāgbhaṭṭa Aṣṭāṅgahṛdaya (4th Century A.D.) | Rasārṇava (12th Century A.D.) |
|---|---|---|---|---|---|---|---|
| 71. | Kandulsūrmā | *Amorphophallus campanulatus*, Blume | | | | | 6.105 |
| 72. | Kaṭaka | *Strychnos potatorum*, Linn | निर्मली | | | | 11.101 |
| 73. | Kadambaka or Kadamka | *Anthocephalus Indicus*, A. Rich | कदम्ब | Sū: 27.113 | | | 14.192 |
| 74. | Kadali | *Lagerstroemia speciosa*, Pers. | केला | | | | 5.22 |
| 75. | Kaṇaka | *Daucus stramonium*, Linn. | धतूरा | | | | 15.81 |
| 76. | Kanda-Sūraṇa | | | | | | 6.98 |
| 77. | Kandapadminī | | | | | | 14.168 |
| 78. | Kanyā-Cūtakumārī | *Aloe perfolita* | | | | | 16.89 |
| 79. | Kapali | *cucuonis utilissimus* | वायविड़ंग | | | | 6.26 |
| 80. | Kapi or Kapikacchu | *Emblica officinalis* or *Mucuna pruriens*, DC. | केवाच | | | | 6.11 |
| 81. | Kapittha | *Feronia limonia*, Linn. | कैथा | | | Sū: 6.126 | 8.29 |
| 82. | Karaka (Palāśa) | *Butea frondosa*, Roxb. | पलाश | | | | 15.182 |
| 83. | Karañja | *Pongamia pinnata*, Pierre | | | | | 7.116 |
| 84. | Karañjaphala | *Holoptelea integrifolia*, Plauch | कैथाफल | Sū: 26.84 | Sū: 42.11; 46. 147-148 | | 12.374 |
| 85. | Karavira | *Nerium indicum*, Mill | कनेर | | | | 5.34 |
| 86. | Karakaṭī | *Cucumis mello*, Linn | खीरा | | | | 18.120 |
| 87. | Karkoṭa or Karkoṭi | | खेकसा | | | Sū: 6.76 | 18.120 |
| 88. | Karṇākhya | *Barleria cristata* | सफेद सहचर | | | | 8.36 |

| S. No. | Name in the list | Botanical Equivalent | Hindi version | Caraka Saṁhitā (100 to 500 BC) | Śuśruta Saṁhitā (500 BC) | Vāgbhaṭṭa Aṣṭāṅgahṛdaya (4th Century A.D.) | Rasārṇava (12th Century A.D.) |
|---|---|---|---|---|---|---|---|
| 89. | Karpūra | Camphor | | | | | 1.39 |
| 90. | Kākajañghā | *Leea hirta*, Roxb | | | | | 5.3 |
| 91. | Kākatuṇḍi | *Asclepias carassavica*, Linn. | | | | | 9.83 |
| 92. | Kākamācī | *Sclanum nigrum*, Linn. | मकोय | Sū: 27-89 | Sū: 46.266 | Sū: 6.74 | 5.3 |
| 93. | Kākamuṇḍī | | | | | | 5.23 |
| 94. | Kākāṇḍī or Kolaśimbī | *Diospyrcs tomentosa* | | | | | 15.148 |
| 95. | Kākinī or Kamaci | *Arbus precatorius* | | | | | 6.34 |
| 96. | Kāñcana | *Toricellia asiatica*, Linn. | कचनार | | | | 9.10 |
| 97. | Kāñcanāra | *Bauhinia tomentosa*, Linn | लाल कचनार | | | | 14.144 |
| 98. | Kāṇcanī or Harids | *Curcuma longa*, Roxb | हलदी | | | | 5.12 |
| 99. | Kāntā or Pṛyaṅgu | *Barringtonia acutangula* or *Setaria italica* | | | | | 10.39 |
| 100. | Kāniā or Madana Vṛksa | *Artemisia vulgaris*, Linn | | | | | 8.43 |
| 101. | Kāravallikā or Karavallā | *Momordica charantia*, Linn. | करेला | | Sū: 46.262, 269 | Sū: 6.76 | 5.15; 5.25 |
| 102. | Kārpāsī | *Gyossypum hertacaum*, Linn. | | | | | 5.14 |
| 103. | Kāśamarda | *Cassia occidentalis*, Linn. | कसौंदी | Sū: 27.88 | Sū: 46.236 | Su : 6.100 | 5.24 |
| 104. | Kiṁśuka | *Butea frondosa*, Roxb. | पलाश | | Sū: 46.288 | | 5.39 |
| 105. | Koṭemāria or Rakta Lajjalukā | | | | | | 12.42 |
| 106. | Kilala or Śallaki | *Boswellia serrata*, Roxb | | | | | 6.81 |
| 107. | Kukkuṭī | *Bombax malabaricum*, DC. | सेमर | | Sū: 46.287 | | 5.6 |

| S. No. | Name in the list | Botanical Equivalent | Hindi version | Caraka Saṁhitā (100 to 500 BC) | Śuśruta Saṁhitā (500 BC) | Vāgbhaṭṭa Aṣṭāṅgahṛdaya (4th Century A.D.) | Rasārṇava (12th Century A.D.) |
|---|---|---|---|---|---|---|---|
| 108. | Kumkuma | Saffron<br>*Crocus sativus*, Linn. | केसर | | | | 1.39 |
| 109. | Kuberākṣī or Vallikaranja | *Bignonia sauvealens* | | | | | 7.113 |
| 110. | Kumārī or Ghṛta-kumārī | *Aloe vera* Linn,<br>*Var. officinalis* | | | | | 8.77 |
| 111. | Kumuda | *Nymphaea rubra* | | | | | 6.10 |
| 112. | Kumudini | | कुंई | | | | 12.129 |
| 113. | Kumudi | | कायफल | | | | 15.2 |
| 114. | Kurbaka | *Cidrela toona* | | | | | 5.25 |
| 115. | Kuringaṇī or Mudgaparṇī | *Phaseolus trilolus*, Ait. | | | | | 5.4 |
| 116. | Kultha | *Dolichos biflarus*, Linn. | | | | Sū: 6.18 | 6.79 |
| 117. | Kuliśa | *Heliotropium indicum* | होहुण्ड, स्नूही | | | | 11.131 |
| 118. | Kuṣṭha | *Costus speciosus* | सुगन्ध कूट | | | | 11.136 |
| 119. | Kuṣṭhacchali | | | | | | 14.98 |
| 120. | Kūṣmāṇḍa | *Benincasa cerifera*, Savi | | Sū: 27.101 | Sū: 46.213, 214 | Sū: 6.87, 88 | 7.74 |
| 121. | Kusumbha | *Carthamus tinctorius*, Linn. | | Sū: 27.100, 108 | Sū: 46.127 | Sū: 6.23, 101 | 5.39 |
| 122. | Kuhukambika or Kuraṇḍa | | कुरण्ड या कुरुण्डिका | | | | 5.13 |
| 123. | Kṛtāñjali or Lajjāvatī | *Mimosa pudica* | छुई मुई | | | | 5.24 |
| 124. | Kṛṣṇa-jīraka | *Nigella sativa*, Linn | काला जीरा | | | | 5.14 |
| 125. | Kṛṣṇatilā | *Xanthochymus pictorius* | काला तिल | | | | 6.132 |

| S. No. | Name in the list | Botanical Equivalent | Hindi version | Caraka Saṁhitā (100 to 500 BC) | Śuśruta Saṁhitā (500 BC) | Vāgbhaṭṭa Aṣṭāṅgahṛdaya (4th Century A.D.) | Rasārṇava (12th Century A.D.) |
|---|---|---|---|---|---|---|---|
| 126. | Kṛṣṇaparṇī | Black Tulasī *Ocimum pilosum* | काली तुलसी | | | | 5.7 |
| 127. | Kṛṣṇa-māñjarī | | कृष्ण मञ्जरी | | | | 5.18 |
| 128. | Kṛṣṇa-raji | | काली सरसों | | | | 12.153 |
| 129. | Kṛṣṇā Pīpalī | *Ficus religiosa*, L. | | | | | 5.4 |
| 130. | Kṛṣṇāgorū | Aloe wood | काला अगर | | | | 14.17 |
| 131. | Kṛṣṇonamatta | | काला धतूरा | | | | 12.223 |
| 132. | Ketakī | *Pandanus odoratissimus* Roxb. | केवड़ा | | | | 7.137 |
| 133. | Keśanī | *Chrysopogon aciculatus* | जटामानसी | | | | 5.22 |
| 134. | Kokilā | *Calculus indicus* | काकोली | | | | 15.175 |
| 135. | Kokilākṣa | *Hygrophila spinosa*, T. | ताल मखाना | | | | 5.5 |
| 136. | Kotarākṣī | *Alangium decapetalum* | वघारा | | | | 5.12 |
| 137. | Kodrava | *Paspalum scrobiculatum*, Linn | कोंदों | | | | 6.79 |
| 138. | Kola | *Piper longum* | बेर | | | | 10.26 |
| 139. | Kosaṭkī | *Trichosanthes dioica* | कड़वी तोरई | | Sū: 46.262 | Sū: 6.81 | 5.22 |
| 140. | Kaumārī or Kumārī | *Aloe vera*, Linn. | | | | | 5.27 |
| 141. | Kuraṇḍaka | *Randia dumetorum*, Lamk | मदन वृक्ष | | | | 9.76 |
| 142. | Kauberimūlā | *Costus speciosus* | उत्तर वारुणी की जड़ | | | | 11.25 |
| 143. | Krāntā or Bṛhatī | *Solanum indicum* | बड़ी कटेरी | | | | 12.211 |
| 144. | Kṣamāpāli | | | | | | 2.28 |

| S. No. | Name in the list | Botanical Equivalent | Hindi version | Caraka Saṁhitā (100 to 500 BC) | Śuśruta Saṁhitā (500 BC) | Vāgbhaṭṭa Aṣṭāṅgahṛdaya (4th Century A.D.) | Rasārṇava (12th Century A.D.) |
|---|---|---|---|---|---|---|---|
| 145. | Kṣirakuñcuka or Kañcukī | *Lipeocercis serrata* | | | | | 10.54 |
| 146. | Kṣirakanda | *Batatas paniculata* | भूमि कूष्माण्ड | | | | 6.13 |
| 147. | Kṣirakuṭi | | | | | | 5.18 |
| 148. | Kṣiramālī | | | | | | 5.25 |
| 149. | Kṣirinī | *Mimosa kauki* | खन्नी | | | | 5.6 |
| 150. | Kṣudrāmala | *Oxalis corniculata,* Linn. | चांगेरी | | | | 10.59 |
| 151. | Kṣuraka | | गोखरू | | | | 7.113 |
| 152. | Kṣmāpāla | | | | | | 12.85 |
| 153. | Khadgārī | *Seirpus maximus* | | | | | 5.8 |
| 154. | Khadira | *Abutilon catechu,* Willd. | कत्था | | | | 5.39 |
| 155. | Kharmañjarī or Apāmārga | *Achyranthes aspera,* Linn | चिचड़ा | | | | 11.26 |
| 156. | Gajārī or Sallakī | *Artocarpus locuca* | | | | | 12.90 |
| 157. | Gadā | | पाटल वृक्ष | | | | 11.24 |
| 158. | Gāngerī or Nāgabalā | *Sida spinosa,* Linn | | | | | 10.52 |
| 159. | Girikarmikā or Aparājitā | *Clitoria ternatea,* Linn. | | | | | 5.7 |
| 160. | Guggulu | *Commiphara mukul,* Hook. | गुगल | | | | 7.91 |
| 161. | Guñjā | *Abrus precatorius,* Linn. | घुंघची | | | | 4.47 |
| 162. | Gudūcī | *Cocculus cordifolius* | गिलोच | | Sū: 46.262, 263 | | 7.108 |
| 163. | Gṛdhrakarṇī | A dark coloured plant. | | | | | 17.10 |
| 164. | Gṛhakanyā or Ghṛtakumārī | *Aloe vera,* Linn | | | | | 10.45 |

| S. No. | Name in the list | Botanical Equivalent | Hindi version | Caraka Saṁhitā (100 to 500 BC) | Śuśruta Saṁhitā (500 BC) | Vāgbhaṭṭa Aṣṭāṅgahṛdaya (4th Century A.D.) | Rasārṇava (12th Century A.D.) |
|---|---|---|---|---|---|---|---|
| 165. | Gokarṇa or Aśvagandhā | *Sanseviera zeylanica*, L | पलाश (?) | | | | 10.50 |
| 166. | Gokṣura | *Tribulus alatus*, Delile | गोखरू | | | | 7.151 |
| 167. | Gojihvā | *Elephantopus scaber*, Linn. | गोभी | Sū: 27.96 | Sū: 46.264 | Sū: 6.76 | 5.5 |
| 168. | Godhūma | *Tristania aestivum*, Linn. | गेहूं | | | Sū: 6.14, 15, 16 | 6.93 |
| 169. | Godhavatī | *Vitex pedeta* ,Vahl | | | | | 17.95 |
| 170. | Gopalaki or Nāgabalā | | | | | | 17.109 |
| 171. | Gorambhā or Gorakṣā | | | | | | 15.136 |
| 172. | Gosandhī | | | | | | 5.14 |
| 173. | Gorabhā or Haridrī | *Curcuma aromatica* | | | | | 6.107 |
| 174. | Gauriphala or Akhā | | | | | | 7.113 |
| 175. | Grīṣma Sundara | *Mollugo cerviana*, Seringe | | | | | 10.53 |
| 176. | Ghānadhavara | *Cyperus rotundus*, Linn | नागर मोथा | | | | 5.5 |
| 177. | Cakramarda | *Cassia tora*, Linn | चकबड़ | | | | 5.4 |
| 178. | Cākraka or Cakramardā | A species of plant. | | | | | 5.18 |
| 179. | Cakrī or Cakremardā | *Dalbergia ujjeinensis* | | | | | 5.19 |
| 180. | Caṭurparṇā | | | | | | 12.97 |
| 181. | Caṇaka | *Cicer arictinum*, Linn. | चना | | | | 5.31 |
| 182. | Cạndalikhaṇḍa | *Amcporgon aciculatus* | | | | | 15.94 |
| 183. | Candana | *Satinum album*, Linn. | | | | | 1.39 |
| 184. | Candrāvallī | | गोमलता | | | | 15.148 |

| S. No. | Name in the list | Botanical Equivalent | Hindi version | *Caraka Saṁhitā (100 to 500 BC)* | *Śuśruta Saṁhitā (500 BC)* | *Vāgbhaṭṭa Aṣṭāṅgahṛdaya (4th Century A.D.)* | *Rasārṇava (12th Century A.D.)* |
|---|---|---|---|---|---|---|---|
| 185. | Cāvi | *Piper retrofractum*, Vahl. | चविका | | | | 5.25 |
| 186. | Cāngeri | *Oxalis corniculata*, Linn. | | Sū: 27.92 | Sū: 46.273 | | 5.31 |
| 187. | Cāṇḍālī | | | | | | 15.133 |
| 188. | Cāmīkara | | शाक वृक्ष अथवा धतूरा | | | | 17.85 |
| 189. | Ciñcā | *Rumex vescarious* | इमली | | | | 5.30 |
| 190. | Citraka | *Racenus communis* | चीता | | | | 7.116 |
| 191. | Cirbilva or Karanja | *Pongamia pinnata*, Pierre | | | | | 18.126 |
| 192. | Jambīra | *Citrus limonum*, Linn | जम्बीरी नींबू | Sū: 27.167 | Sū: 46.139, 232 | | 5.31 |
| 193. | Jayā or Vijayā | *Premna spinosa* | भांग | | | | 10.52 |
| 194. | Jalabindujā | | | | | | 5.2 |
| 195. | Jati | *Jasminum grandiflorum*, Linn. | जातिकाफल | | | | 18.112; 18.200 |
| 196. | Jiraka | *Ipomoea nil*, Linn. | जीरा | | | | 15.195 |
| 197. | Jīvaka | *Terminalia tomentosa* | | | Sū: 46.274 | | 14.51 |
| 198. | Jyotiṣmatī | *Cardiospermum helicacabum*, Linn | मालकंगनी | | | | 8.83 |
| 199. | Jvālāmukhī or Lālitraks | | | | | | 6.118 |
| 200. | Jvālini or Mūrvā | *Sanseviera roahuburghiana* | | | | | 5.5 |
| 201. | Tagara | *Tubernaemontana cornata*, Willd.. | | | | | 15.158 |
| 202. | Tandulīyaka | *Emblia ribes*, Burm | विडंग | Sū: 27.94; | Sū: 46.258 | Sū: 6.82 | 18.12 |
| 203. | Tamā | *Xanthochymus pictorius* | इरफरेवड़ी | | | | 15.153 |

| S. No. | Name in the list | Botanical Equivalent | Hindi version | Caraka Saṁhitā (100 to 500 BC) | Śuśruta Saṁhitā (500 BC) | Vāgbhaṭṭa Aṣṭāṅgahṛdaya (4th Century A.D.) | Rasārṇava (12th Century A.D.) |
|---|---|---|---|---|---|---|---|
| 204. | Tāmbūla | *Piper betle*, Linn | पान | | | | 18.200 |
| 205. | Tāmraparṇī | | जीठ | | | | 5.20 |
| 206. | Tālamūla | *Borassus flabellifermis* | | | | | 17.93 |
| 207. | Tiktaśoka | *Terminalia catappa* | | | | | 11.26 |
| 208. | Titiṇḍi | *Tamarindus indica*, L. | इमली | | | | 5.6 |
| 209. | Tiniśa | *Dalbergia ujjeinensis* | | | | | 7.142 |
| 210. | Tintiṇi | *Tamarindus indica*, L. | इमली | Vi: 8.40 | Sū: 26.84 | | 12.170 |
| 211. | Tinduka | *Diophyros embryopteris*, Pers. | | Sū: 27.144 | Sū: 46.168 | | 6.11 |
| 212. | Timira or Tamāla | *Crataeva roxburgii* | | | | | 14.150 |
| 213. | Tila | *Sesamum indicum*, DC | तिल | | | Sū: 5.56; 6.22 | 5.30 |
| 214. | Tilaparṇī | *Gynandropsis pentaphylla* DC | तिलोनी | Sū: 27.96 | Sū: 86.239 | Sū: 6.75 | 11.41 |
| 215. | Tumbī | *Long gourd* | | | | Sū: 6.89 | 15.122 |
| 216. | Tumburu | *Xanthoxylumalatum*, Roxb | नेपाली धनिया | Sū: 27.171, 307 | | | 7.133 |
| 217. | Tursimhanī | | | | | | 12.66 |
| 218. | Tulasī | *Ocimum sanctum*, Linn | तुलसी | | | | 5.7 |
| 219. | Tṛṇajyoti | A kind of grass | | | | | 5.19 |
| 220. | Tṛṇauṣadhi | A kind of grass | लवा | | | | 12.75 |
| 221. | Toyavallī | *Momordica charantia* | रुदन्ती | | | | 5.13 |
| 222. | Tridāṇḍī | | | | | | 5.19 |
| 223. | Triparṇī | *Desmodium gangeticum*, DC | | Sū: 27.102 | Sū: 46.247, 275 | | 5.6 |

| S. No. | Name in the list | Botanical Equivalent | Hindi version | Caraka Saṁhitā (100 to 500 BC) | Śuśruta Saṁhitā (500 BC) | Vāgbhaṭṭa Aṣṭāṅgahṛdaya (4th Century A.D.) | Rasārṇava (12th Century A.D.) |
|---|---|---|---|---|---|---|---|
| 224. | Trailokyajananī | | अजनायका | | | | 12.28 |
| 225. | Tvakkṣira | *Bamboo mana* | वंश लोचन | | | | 6.98 |
| 226. | Dagdharuhā | | | | | | 5.17; 12.149; |
| 227. | Daniyāmalā | | गमलवेतस | | | | 17.121 |
| 228. | Dantinī or Dantī | *Gymnosporia montana,* Roth. | जमालगोटा | | | | 17.95 |
| 229. | Dārvi or Dāruharidra | *Berberis aristate,* DC | | | | | 6.28 |
| 230. | Dali or Devadālī | *Andropogon serratus,* Retz. | | | | | 7.133 |
| 231. | Devadāru | *Plyalthia longifolia,* Thw. | देवदारु | | | | 8.80 |
| 232. | Devadālī | *Andropogon serratus,* Retz. | बन्दाल | | | | 5.14 |
| 233. | Drākṣā | Grapes *Vitis vinifera,* Linn. | अंगूर | Sū: 27.125, 126 | Sū: 46.182 | Sū: 6.116 | 18.125 |
| 234. | Dvipadi | *Mesua ferra* Linn | नागकेसर | | | | 12.8 |
| 235. | Dhattura | *Datura tatula,* Linn. | धतूरा | | | | 12.8 |
| 236. | Dhāka | *Anogeissus latifolia,* Wall | | | | | 12.169 |
| 237. | Dhatāki | *Woodfordia fruitcosa,* Kurz. | धाय का फल | | | | 8.37 |
| 238. | Dhātrī | *Emblica officinalis,* Gaertn. | आमताकी (धात्री) | | | | 12.16 |
| 239. | Naktamāla | *Pongamia globra* | करञ्ज | | | | 7.108 |
| 240. | Narakasā | | नीलवृक्ष (?) | | | | 5.10 |
| 241. | Narajihvā | | | | | | 12.86; 5.13 |

| S. No. | Name in the list | Botanical Equivalent | Hindi version | Caraka Saṁhitā (100 to 500 BC) | Śuśruta Saṁhitā (500 BC) | Vāgbhaṭṭa Aṣṭāṅgahṛdaya (4th Century A.D.) | Rasārṇava (12th Century A.D.) |
|---|---|---|---|---|---|---|---|
| 242. | Nāgakanyā | *Mesua roxburgii* | गड्ची | | | | 8.77 |
| 243. | Nāgakarṇī | Red Ricinus *Salvinia cuculata*, Roxb | | | | | 5.20 |
| 244. | Nagajihvā | *Tiardium indicum* | अनन्ता | | | | 5.20 |
| 245. | Nāgabalā | *Sida spinosa*, Linn. | | | | | 5.4 |
| 246. | Nāgar | *Euphorbia antiquorum* | नागर मोथा अथवा सोंठ | | | Sū: 6.163 | 18.7 |
| 247. | Nāgavalli | *Piper betle* | पान | | | | 5.27 |
| 248. | Nāgaśuṇḍi | A kind of cucumber *Heliotropium indicum*, Linn | | | | | 11.44 |
| 249. | Nāginī or Nāgadantī | | | | | | 5.25 |
| 250. | Nāginikanda | | | | | | 12.141 |
| 251. | Nāranga | *Citrus aurantium*, Linn | नारंगी | Sū: 27.154 | Sū: 46.161 | | 5.31 |
| 252. | Nārī or Aśvabalā | *Saccharum spontaneum* | | | | | 7.139 |
| 253. | Nicula | *Barringtonia acutangula*, Gaertn. | | | | | 6.17 |
| 254. | Nimba | *Melia azadirachta*, L. | नीम | | | Sū: 5.61; 6.75 | 11.34 |
| 255. | Nimbuka | *Citrus medica*, Linn. | नींबू | Sū: 27.97 | Sū: 46.70 | | 7.73 |
| 256. | Nirguṇḍi | *Vitex negundo*, Linn. | | | | | 10.53 |
| 257. | Niśā or Haridrā | *Curcuma domestica*, Valeton | | | | | 7.35 |
| 258. | Niśācara | | सेमलता | | | | 12.4 |
| 259. | Nīlacitraka | *Black Plumbago zeylanica*, Linn. | काला चीता | | | | 5.9 |

| S. No. | Name in the list | Botanical Equivalent | Hindi version | Caraka Saṁhitā (100 to 500 BC) | Śuśruta Saṁhitā (500 BC) | Vāgbhaṭṭa Aṣṭāṅgahṛdaya (4th Century A.D.) | Rasārṇava (12th Century A.D.) |
|---|---|---|---|---|---|---|---|
| 260. | Nīlajyoti | | | | | | 5.19 |
| 261. | Nīlasindūri | *Blue negundo* | | | | | 14.168 |
| 262. | Nīla or Nālā | *Indigofera tinctoria,* Linn | | | | | 5.5 |
| 263. | Paṭoli | *Trichosanthes dioica,* Roxb. | मीठा परवर | | | Sū: 6.75; 78 | 5.16 |
| 264. | Patanga | *Caesalpinia sappan,* Linn | | | | | 5.39 |
| 265. | Pattya | *Terminalia chebula,* Retz. | हर्र | | | | 7.35 |
| 266. | Padma | *Clerodendrum siphonanthus* | कमल | | | | 12.377 |
| 267. | Padmacāriṇī | *Hibiscus mutabilis,* Linn. | स्थाल पद्म | | | | 5.5 |
| 268. | Panāsa | *Artocarpus heterophyillus* | कटहल | Sū: 27.143 | Sū: 46.178, 181 | | 12.376 |
| 269. | Purūṣka | *Grewia asiatica,* Linn. | फालसा | | | | 18.125 |
| 270. | Palāśa | *Butea frondosa,* Roxb | | | | | 5.30 |
| 271. | Pāṭalī | *Stereospermum suaveolens,* DC | लोध | | | | 8.83 |
| 272. | Paṭhā | *Stephania hernandifolia,* Walp | पाढी | | | Sū: 6.71, 72 | 5.5 |
| 273. | Pippalī | *Piper longinum,* Linn. | | | Sū: 46.221, 262 | Sū: 6.161, 162 | 8.83 |
| 274. | Piluka | *Salvadora persica,* Linn. | दिठोरी | Sū: 4.24, 39; 25.49; 27.142 | Sū: 42.18; 46.184 | | 11.25 |
| 275. | Punarnavā | *Boerhaavia procumbens* | | | Sū: 46.255 | Sū: 6.71, 72, 97 | 5.2 |
| 276. | Pūgaphala | *Areca catechu* | सुपारी | | | | 18.156 |
| 277. | Peṭārī | *Abutilon indicum* | | | | | 6.99 |

| S. No. | Name in the list | Botanical Equivalent | Hindi version | Caraka Saṁhitā (100 to 500 BC) | Śuśruta Saṁhitā (500 BC) | Vāgbhaṭṭa Aṣṭāṅgahṛdaya (4th Century A.D.) | Rasārṇava (12th Century A.D.) |
|---|---|---|---|---|---|---|---|
| 278. | Potakī | | पोई | | | | 7.133 |
| 279. | Phanijihvā | *Rivea ornato* | शतावरी | | | | 5.5 |
| 280. | Bākula | *Manilkara hescandra*, Roxb. | मौलश्री | | | 46.169 | 8.76 |
| 281. | Badara | *Ziziana jujuba*, Mill. | बेर | Sū: 4.10, 24; 6.8.137 | Sū: 27.142 | Sū: 6.124 | 18.126 |
| 282. | Badarī | *Zizania mauritiana*, Lam | बेर | | | | 6.102 |
| 283. | Bandhyā | *Pentapetas phoenicea*, L | | | | | 5.22; 14.168 |
| 284. | Balā | *Valeriana hardwickii*, Wall. | | Sū: 27.107 | | | 5.4 |
| 285. | Bākucī | *Cymopsis tetragonoloba*, Linn. | बावची | | | | 5.14 |
| 286. | Bālamocanā | *Bombax malabaricum*, DC | छोटी सेमर | | | | 5.8 |
| 287. | Bālā | *Valeriana hardwickii*, Wall. | खरा | | | | 6.104 |
| 288. | Bibhītaka | *Terminalia bellerica*, Roxb | | | | | 8.76; 15.126 |
| 289. | Bimbi | *Momordica monadelpha* | कदरूं | Sū: 27.142 | Sū: 46.176 | | 10.57 |
| 290. | Bilva | *Aegle marmelos*, C & Roxb. | बेल | Sū: 25.39; 27.138; | Sū: 38.45 | Sū: 6.125 | 5.16 |
| 291. | Brāhmī | *Centella asiatica*, Linn. | | | | | 5.10 |
| 292. | Brahmadaṇḍa | *Centella asiatica*, Linn. | | | | | 5.26 |
| 293. | Brahmadaṇḍī | *Lampracheenium microcephalum* | | | | | 5.3 |
| 294. | Brahmavṛkṣa | A divine tree *Butea frondosa* | पलाश | | | | 15.126 |

| S. No. | Name in the list | Botanical Equivalent | Hindi version | Caraka Saṁhitā (100 to 500 BC) | Śuśruta Saṁhitā (500 BC) | Vāgbhaṭṭa Aṣṭāṅgahṛdaya (4th Century A.D.) | Rasārṇava (12th Century A.D.) |
|---|---|---|---|---|---|---|---|
| 295. | Brahmasomā | *Centella asiatica,* Linn. | ब्राह्मी | | | | 17.10 |
| 296. | Bhadrā | *Tithyamalus antiquorum,* L. | | | | | 10.98 |
| 297. | Bhallata | *Semecarpus anacardium,* Linn. | भिलावा | | Sū: 46.249 | Sū: 6.133 | 17.129 |
| 298. | Bhiṇḍi | *Abelmoschus esculentus,* Linn. | रामतोरई | Sū: 27.100 | Sū: 38.9 | | 6.102 |
| 299. | Bhūpatali | A kind of plant | | | | | 5.27 |
| 300. | Bhūmyāmalaka | *Phyllagathis fraternus,* Webster | | | | | 15.9. |
| 301. | Bhūrja or Bhūrjapatra | *Betula anloides,* Buah-Ham | | | | | 11.115; 14.164 |
| 302. | Bhṛṅgarāja | *Eclipta alba,* Linn. | भांगरा | | | | 5.2: 5.27: 18.12. |
| 303. | Majjiṣṭhā | *Rubia cordifolia,* Linn | मजीठ | | | | 5.39 |
| 304. | Maṇḍūkaparṇī | *Calosanthus indica* | | Sū: 27.95, 97 | Sū: 46.264 | Sū: 6.76 | 5.3; 10.52 |
| 305. | Matsyākṣī | *Hincha repens* | | | | | 5.3 |
| 306. | Madhyantikā | Jasmine *Sanseviera roxburghiana* | मोगरा | | | | 5.39 |
| 307. | Madhūka | *Bassia latifolia,* Roxb. | मुलेठी | Sū: 27.125; 13.9; 25.49 | Sū: 38.39; 46.182, 186 | | 12.328 |
| 308. | Madhūkūṣmāṇḍa | *Benincasa cerifera,* Savi | मीठा कोंहड़ा | | | | 7.76 |
| 309. | Mārīca | *Piper nigrum,* Linn. | गोल मिरच | | | | 6.13 |
| 310. | Markaṭī or Apāmārgā | *Achyranthes aspera,* Linn. | चिंचिटा | | | | 15.84 |
| 311. | Mallikā | *Gaertnera recemosa* | मोगरा | | Sū: 46.286 | | 7.113 |

| S. No. | *Name in the list* | *Botanical Equivalent* | *Hindi version* | *Caraka Saṁhitā (100 to 500 BC)* | *Śuśruta Saṁhitā (500 BC)* | *Vāgbhaṭṭa Aṣṭāṅgahṛdaya (4th Century A.D.)* | *Rasārṇava (12th Century A.D.)* |
|---|---|---|---|---|---|---|---|
| 312. | Masūraka | *Ichnocarpus frustescens* | मसूर | | | | 18.124 |
| 313. | Mahākālī | *Tragia involucrata*, Linn | बिच्छी | | | | 5.16 |
| 314. | Mahābalā | *Sida cordifolia* | | | | | 5.4 |
| 315. | Mahābalā | *Sida cordifolia* | | | | | 15.138 |
| 316. | Mātulunga | *Citrus medica* Linn. | बिजोरा नींबू | Sū: 4.10; vi: 8.137 | | Sū: 6.131-132 | 11.183 |
| 317. | Māninī | | लक्षणा | | | | 5.10 |
| 318. | Mārjārī | | | | | | 14.150 |
| 319. | Māṣa | *Phaseolus radiatus* | उड़द | | | Sū: 6.20 | 7.93 |
| 320. | Mīnākṣī | Species of some plants | | | | | 10.39 |
| 321. | Muṇḍī | *Sphaeranthus hirtus* | | | | | 11.42 |
| 322. | Mudga | *Phaseolus mungo*, Linn. | मूँग | | | Sū: 6.17; 31, 40 | 18.131 |
| 323. | Mudgaparṇī | *Phaseolus trilolus*, Ait. | | | | | 10.52 |
| 324. | Muni | *Agati grandiflora* | | | | | 11.31 |
| 325. | Muṣalī | *Curculigo orchioides*, Gaertn. | तालमूली | | | | 7.133 |
| 326. | Mūlaka | *Raphanus sativus*, Linn. | मूली | Sū: 27.168 | Sū: 46.240-242 | Sū: 6.102; 104 | 5.30 |
| 327. | Mṛgadūrvā | *Cynodon dactyton*, Pers. | सफेद दूब | | | | 15.148 |
| 328. | Mṛdu-dūrvā | *Cynadon dactylon*, Pers. | | | | | 7.141 |
| 329. | Meghanadā | *Amaranthus polygonoides* | चौराई | | | | 6.11 |
| 330. | Meṣavallī | *Odina pannata* | | | | | 5.6; 15.83 |
| 331. | Mokṣaka | *Stereospermum sucveolens*, DC | पाटला | | | | 18.242 |

| S. No. | Name in the list | Botanical Equivalent | Hindi version | *Caraka Saṁhitā (100 to 500 BC)* | *Śuśruta Saṁhitā (500 BC)* | *Vāgbhaṭṭa Aṣṭāṅgahṛdaya (4th Century A.D.)* | *Rasārṇava (12th Century A.D.)* |
|---|---|---|---|---|---|---|---|
| 332. | Modā | *Engenia jambulana* | अजमोदा | | | | 5.11 |
| 333. | Morāta | *Senseviera roxburghiana* | ढाक | | | | 6.134 |
| 334. | Yavaciñcā | Tamarind | इमली | | | | 5.15; 15; 136 |
| 335. | Raktakiñcukī | *Baubinia variegata* | | | | | 12.99 |
| 336. | Raktacandana | *Pterocarpus santalinus,* Linn | लाल चन्दन | | | | 8.80 |
| 337. | Raktacitraka | *Plumbago tinctoria* | लाल चीता एरण्ड | | | | 5.10 |
| 338. | Raktaśṛṅgī | *Rhus saccedanea,* Linn. | | | | | 5.9 |
| 339. | Raktasnūī | *Red Euphorbia nerifolia,* Linn | | | | | 5.9 |
| 340. | Raktikā | | लाल गुञ्जा | | | | 5.9 |
| 341. | Rajanī (Haridra) | *Curcuma domestica,* Valeton. | | | | | 7.113 |
| 342. | Rambhā (Kadalī) | Plantain *Lagerstroemia speciosa,* Pers. | | | | | 11.24 |
| 343. | Rasona | *Allium ascalonicum* | लशुन | | | | 13.17 |
| 344. | Rakṣasā | | सोम | | | | 12.3 |
| 345. | Rāgini (Aśoka) | *Jovesia asoka* | अशोक वृक्ष | | | | 11.86 |
| 346. | Rājakośātākī | *Luffa foetida* | पीली तोरई | | | | 7.82 |
| 347. | Rājikā or Rājī | *Sinapis ramosa* | राई | | Sū: 46.221 | | 5.9; 11.67 |
| 348. | Rāmaṭha | (Plantain) *Asafoctida* | हिंगु | | | | 6.118 |
| 349. | Rāsanā | *Vanda roxburgii,* R.Br. | | | | | 5.6 |
| 350. | Rudantī or Rudantikā | Fragrant tree | | | | | 5.10 |

| S. No. | Name in the list | Botanical Equivalent | Hindi version | Caraka Saṁhitā (100 to 500 BC) | Śuśruta Saṁhitā (500 BC) | Vāgbhaṭṭa Aṣṭāṅgahṛdaya (4th Century A.D.) | Rasārṇava (12th Century A.D.) |
|---|---|---|---|---|---|---|---|
| 351. | Rudhira | | कुमकुम, केशर | | | | 17.7 |
| 352. | Lakṣanā | *Solanum xanthocarpum*, Sehand | | | | | 5.25 |
| 353. | Lakṣmī | *Hibiscus Mutabilia* | श्वेत तुलसी | | | | 5.5 |
| 354. | Laghuparṇi or Mūrvā | *Sanseviera zeylanica*, Willd | | | | | 10.39 |
| 355. | Laśūna | *Allium sativum*, Linn. | लहसुन | Sū: 27.176; | Sū: 46.245 | Sū: 6.109 | 6.105; 16.6 |
| 356. | Lāṅgalī | *Gloriosa superba*, Linn | कलिहारी | Sū: 27.108 | | | 65.34 |
| 357. | Lunga or Mātulunga | *Citrus medica* | | Sū: 4.10; vi: 8.137 | Sū: 38-39; 46.182, 186 | | 5.31 |
| 358. | Vaṁśapatri | *Bambus arundinacea*, Willd. | | | | | 5.20 |
| 359. | Vacā | *Acorus colamus*, Linn. | वच | | | | 5.34 |
| 360. | Vajrakandā | A species of bulbous plant | शकरकन्द | | | | 5.22 |
| 361. | Vajralatā | *Heliotropium indicum* | | | | | 6.13; 6.115 |
| 362. | Vajrī or Vakrī | *Euphorbia antiquorum*, Linn. | सेहुँड़ | | | | 5.9 |
| 363. | Vaḍavāmukhī | | | | | | 9.5 |
| 364. | Vadara or Badara | *Ziziana jujuba*, Mill. | बेर | | | | 18.126 |
| 365. | Vanarāji | A kind of plant | लसोड़ा | | | | 5.12 |
| 366. | Vanaśigruka | | जंगली सहजन | | | | 8.28 |
| 367. | Vanasūrana | | जंगली सूरन | | | | 6.12 |
| 368. | Varaṇa or Varuṇa | *Crataeva roxburgii* | बरना | | Sū: 46.254 | | 12.371 |

| S. No. | Name in the list | Botanical Equivalent | Hindi version | Caraka Saṁhitā (100 to 500 BC) | Śuśruta Saṁhitā (500 BC) | Vāgbhaṭṭa Aṣṭāṅgahṛdaya (4th Century A.D.) | Rasārṇava (12th Century A.D.) |
|---|---|---|---|---|---|---|---|
| 369. | Varāhakarnī or Aśvagandhā | *Physalis flexousa* | | | | | 5.24 |
| 370. | Varaṇīkā (Haridra) | *Curcuma domestica*, Valeton | | | | | 11.178 |
| 371. | Vartulaparṇikā | *Scindapus officcinalis* | | | | | 5.20 |
| 372. | Varaṣābhū (Punarnava) | *Boerhaavia diffusa*, Linn. | | | Sū: 46.239 | Sū: 6.97 | 6.79 |
| 373. | Vasuhalata (Agasta) | *Sesbenia grandiflora*, Pers | | | | | 10.23 |
| 374. | Vanhi or Citraka | *Racenus communis* | | | | | 7.89 |
| 375. | Vanhikarkaṭī | | | | | | 5.12 |
| 376. | Vanhiśikhā | *Gloriosa superba*, Linn. | | | | | 11.197 |
| 377. | Vakucī | | | | | | 5.14 |
| 378. | Vārāhī | *Physalis flexuosa* (yava) | | | | | 15.180 |
| 379. | Vāruṇī | Durva grass or similar species | | | | | 15.175 |
| 380. | Vāsaka | *Adhatoda vasica*, Nees | अड़ूसा | | | | 8.76; 9.90 |
| 381. | Vāstuka | *Chenopodium album*, Linn | वथुआ | Sū: 26.88 | Sū: 46.260 | | 9.7 |
| 382. | Viḍaṅga | *Erycribe paniculata* | | | | | 12.355 |
| 383. | Vibhuta | *Terminalia bellerica*, Roxb. | बहेड़ा | | | | 8.76; 15.126 |
| 384. | Viśalyā | *Croton polyandram* | गड़ूची | | | | 15.45 |
| 385. | Viṣṇubheṣaja | | शुण्ठी | | | | 12.158 |
| 386. | Viśvāmitra-kalpa | Coconut | नारिकेल | | | | 14.188 |
| 387. | Viṣṭṛṇa | | | | | | 12.229 |

| S. No. | Name in the list | Botanical Equivalent | Hindi version | *Caraka Saṁhitā (100 to 500 BC)* | *Śuśruta Saṁhitā (500 BC)* | *Vāgbhaṭṭa Aṣṭāṅgahṛdaya (4th Century A.D.)* | *Rasārṇava (12th Century A.D.)* |
|---|---|---|---|---|---|---|---|
| 388. | Viṣamuṣṭikā | *Strychnos nuxvomica*, Linn. | | | | | 15.141 |
| 389. | Viṣṇukāntā | *Clitoria ternatea*, Linn. | | | | | 15.138 |
| 390. | Vījapura | Citron | नींबू | | | | 11.18 |
| 391. | Vikā | *Croton tinglium*, Linn. | जमालगोटा | | | | 6.98 |
| 392. | Viravallī | | विदारी | | | | 5.20 |
| 393. | Vṛddha-dāraka | *Argyreia speciosa*, Siot | | | | | 5.17 |
| 394. | Vṛhati | *Solanum indicum*, Linn. | भरकटैया | | | Sū: 6.75 | 5.9 |
| 395. | Vṛścaka | *Fragaria involucrata*, Linn. | वईण्टा | | | | 5.9 |
| 396. | Vegā (Jyotiṣmatī) | | | | | | 6.30 |
| 397. | Vetasā | *Calamus rotang* | अमलवेतस | Sū: 27.108 | | Sū: 6.75 | 6.87 |
| 398. | Vyaghri (Karañja) | *Pongamia pinnata*, Pierre | | | | | 12.248 |
| 399. | Vyāghrī (Kaṇṭakārī) | *Solanum surattense*, Burm. | | | | | 5.8 |
| 400. | Vyadhighata | | अमलतास | | | | 7.76 |
| 401. | Vyomavallī | *Cassytha filiformis*, Linn. | | | | | 13.5 |
| 402. | Śakralatā | | | | | | 16.89; 7.117 |
| 403. | Śakodaya | | | | | | 14.42 |
| 404. | Śaṅkara | | मदार | | | | 15.198 |
| 405. | Śaṁkhapuṣpī | *Causcora decussata*, R & S | शंखाहुल | | | | 5.8 |
| 406. | Śatāvarī | *Asparagus racemosus* Willd. | | Sū: 27.107 | Sū: 46.298, 301, 302 | | 5.2 |

| S. No. | Name in the list | Botanical Equivalent | Hindi version | Caraka Saṁhitā (100 to 500 BC) | Śuśruta Saṁhitā (500 BC) | Vāgbhaṭṭa Aṣṭāṅgahṛdaya (4th Century A.D.) | Rasārṇava (12th Century A.D.) |
|---|---|---|---|---|---|---|---|
| 407. | Śamī | *Prosopis spicigera minosa suma* | सागवान | | | | 5.11 |
| 408. | Śambarī | | | | | | 5.16 |
| 409. | Śaraphankha | *Tephrosia purpurea*, Pers. | | | | | 5.2; 6.91 |
| 410. | Śāka | *Shorea robusta*, Gaertn. | सागवान | | | | 8.76 |
| 411. | Śākhoṭaka | *Streblus asper*, Lour | सागवान | | | | 5.11 |
| 412. | Śiṁśapā | *Dalbergia sisu*, Roxb. | शीशम | | | | 17.86 |
| 413. | Śikhaṇḍinī | *Arbus prectorius* | गुञ्जा | | | | 5.3 |
| 414. | Śigru | *Moringa pterygosperma*, Gaertn. | सहजन | | Sū: 46.221, 225, 239, 289 | | 5.22; 5.30 |
| 415. | Śiphā | | पदनकन्द | | | | 6.134 |
| 416. | Śiriṣa | *Albizzia lebbeck*, Benth | | | | | 17.126 |
| 417. | Śivā or Śyama-latā | *Ichnocarpus frutescens*, R.Br | | | | | 16.89 |
| 418. | Śunṭhi | *Ginger* (dry) | सोंठ | | | | 16.6 |
| 419. | Śuṣka (Dhava) | *Anogeissus latefolia*, Wall. | | | | | 7.114 |
| 420. | Śukari or Vārāhī | *Physalis flexuosa* | | | | | 5.27 |
| 421. | Śūraṇa | *Amorphophallus campanulatus*, Blume | | | | | 7.6 |
| 422. | Śūrpaparṇī | | | | | | 10.54 |
| 423. | Śūlinī | | | | | | 10.54 |
| 424. | Śṛgālajihvā | *Asteracantha longifolia* | | | | | 5.9 |
| 425. | Śṛṅgāra | | | Sū: 27.171 | | | 7.72 |

| S. No. | Name in the list | Botanical Equivalent | Hindi version | Caraka Saṁhitā (100 to 500 BC) | Śuśruta Saṁhitā (500 BC) | Vāgbhaṭṭa Aṣṭāṅgahṛdaya (4th Century A.D.) | Rasārṇava (12th Century A.D.) |
|---|---|---|---|---|---|---|---|
| 426. | Śṛṅgī | *Ficus infectoria* | | | | | 12.223 |
| 427. | Śobhañjanā | *Moringa pterygosperma* Gaertn. | | | | | 18.126 |
| 428. | Śyāmā | *Artemisia indica* | | | | | 6.79 |
| 429. | Śrīkhaṇḍa | | पीत चन्दन | | | | 17.13 |
| 430. | Śvetā | | श्वेत अर्क | | | | 6.101 |
| 431. | Śveta-guñjā or Śvetā | *Arbrus precatarius* | | | | | 5.16; 6.101 |
| 432. | Śvetāka | | सफेद मदार | | | | 5.15 |
| 433. | Ṣaṣṭikā | | सांठी चावल | | | | 6.18 |
| 434. | Saṭīrī | *Cercuma zerumbet.* Roxb | | | | | 5.24 |
| 435. | Saralā | *Pinus roxburgii* | देवदारु | | | | 8.80 |
| 436. | Sarpākṣī | *Ophiorrhiza mungos,* Linn. | | | | | 5.2 |
| 437. | Sarṣapa | Mustard<br>*Scropa brinicanapus* | सरसों | Sū: 27.122; 290 | Sū: 2.31, 15, 20 | Sū: 5.60 | 7.75 |
| 438. | Sahakāra | Mango<br>*Aloe perfolita* | आम | | | | 18.126 |
| 439. | Sahacarā | *Barleria cristala,* Linn | पीली झिंटी | | | | 5.4 |
| 440. | Sahadevī | *Sonchus arvensis* | | | | | 10.39; 5.4 |
| 441. | Simhavallī | | | | | | 5.27 |
| 442. | Sitta | | श्वेत चंदन | | | | 10.14 |
| 443. | Sitaguñjā | | | | | | 12.185 |

| S. No. | Name in the list | Botanical Equivalent | Hindi version | Caraka Saṁhitā (100 to 500 BC) | Śuśruta Saṁhitā (500 BC) | Vāgbhaṭṭa Aṣṭāṅgahṛdaya (4th Century A.D.) | Rasārṇava (12th Century A.D.) |
|---|---|---|---|---|---|---|---|
| 444. | Sitamuṣṭā | | सफेद मोथा | | | | 5.33 |
| 445. | Sitarāmalha | | सफेद हींग | | | | 13.17 |
| 446. | Sitasarṣapa | | सफेद सरसों | | | | 10.46 |
| 447. | Sitāṅkola | | सफेद अंकोल | | | | 5.16 |
| 448. | Sitaraka | | सफेद मदार | | | | 17.130 |
| 449. | Sitaṣupuṅkhā | | सफेद सरपुंखा | | | | 11.39 |
| 450. | Siddhārtha | | सरसों | | | | 11.188 |
| 451. | Sitanāḍikā | | | | | | 6.61 |
| 452. | Sudaṇḍa | | गोरक्षी | | | | 5.26 |
| 453. | Sumukha | | राई | | | | 6.13 |
| 454. | Suraghoṣa | | | | | | 7.120 |
| 455. | Surataru | | देवदारू | | | | 18.9 |
| 456. | Suradāli | | देवदाली | | | | 7.120 |
| 457. | Surasā | | लाल अथवा काली तुलसी | | | Sū: 6.107 | 8.35 |
| 458. | Surendragopa | | | | | | 17.12 |
| 459. | Suvarṇa | | हरिद्रा | | | | 17.74 |
| 460. | Sūraṇa | *Amorphophallus campanulatus,* Blume | जिमीकन्द | | | Sū: 6.133 | 6.99 |
| 461. | Sūryabhaktā | *Cleome viscosa,* Linn | हुरहुर | | | | 5.13 |
| 462. | Sūryavarta | *Gynandropsis pentaphylla* DC | | | | | 5.12 |
| 463. | Somalatā | *Sarcostemma brevistigna* W.A | | | | | 5.11 |

| S. No. | Name in the list | Botanical Equivalent | Hindi version | Caraka Saṁhitā (100 to 500 BC) | Śuśruta Saṁhitā (500 BC) | Vāgbhaṭṭa Aṣṭāṅgahṛdaya (4th Century A.D.) | Rasārṇava (12th Century A.D.) |
|---|---|---|---|---|---|---|---|
| 464. | Saubhāñjana | | सहजन | | | | 11.185 |
| 465. | Sthalakumbhī | | | | | | 6.118 |
| 466. | Sthalapadmini | *Hibiscus mutabilis*, Linn. | | | | | 9.19 |
| 467. | Snuha | *Ocimum sanctum*, Linn | सेहुंड़ | | | Sū: 6.107 | 6.1; 5.34 |
| 468. | Hansadāhī | | अगरू चन्दन | | | | 5.24 |
| 469. | Hamsāgri | *Red mimosa pudica*, Linn | लाल लज्जावती | | | | 12.107 |
| 470. | Hanumati | | | | | | 5.12 |
| 471. | Hayagna | | | | | | 17.95 |
| 472. | Haridri | *Curcuma aromatica* | | | | | 12.68 |
| 473. | Haridrā | *Curcuma domestica*, Valeton | हलदी | | | | 12.322 |
| 474. | Halinī | *Tactona grandis* | | | | | 7.116 |
| 475. | Himsrā | *Nardostachys jatamansi*, DC | जटामांसी | | | | 17.17 |
| 476. | Hingu | *Ferula assafoetida*, Linn. | | | | | 9.14 |
| 477. | Himavati | | | | | | 5.11 |
| 478. | Hilamucā | *Enhydra fluctuans*, Lour | | | | | 11.25 |

# Poisonous Plants

**(Poisonous Plants as Brought Out by George Watt, in the Dictionary of Economic Products of India, Vol. VI, Part-I, pp. 310-311.)**

| | |
|---|---|
| 1. *Aconitum ferox*, Wall; Ranunculaceae. | 23. *Crinum asiaticum* Linn, *var - toxicarum*, Herbert, Amaryllideae. |
| 2. *A. napellus* Linn. | 24. *Crocus sativus* Linn; Isideae. |
| 3. *A. palmatum* Don. | 25. *Croton tinglium* Linn; Euphorbiaceae. |
| 4. *Actaea spicata* Linn; Ranunculaceae. | 26. *Daphne oleoides* Sohrab; Thymalacaceae. |
| 5. *Ammania baccifera* Linn; Lythraceae. | 27. *Datura fastuosa* Linn; Solanceae. |
| 6. *Anamirta cocculus*, W&A; Henispermaceae. | 28. *D. stramonium* Linn. |
| 7. *Andrachne cordifolia* Muell, Arg; Euphorbiaceae. | 29. *Dioscores ? deltoides*, Wall; Dioscoreaceae. |
| 8. *Andropogan sp.*, Stewart; Possibly *A. Scandens.* Roxb.; Gramineve. | 30. *Diospyros montana* Roxb.; Ebenaceae. |
| 9. *Arisaesma curvatum*, Kunth; Aroideae. | 31. *Eleodendron glaucum* Pers., Celastrineae. |
| 10. *A. speciosum*, Mart. | 32. *Euphorbia tirucalli* Linn., Euphorbiaceae. |
| 11. *Atropa belladona* Linn; Solanceae. | 33. *Excoecaria agallocha* Linn.; Euphorbiaceae. |
| 12. *Avena fetua* Linn; Gramminease. | 34. *Flueggea microcarpa* Blume, (Fl. Br. Ind., V., 328); Euphorbiaceae. The bark is said to be poisonous. |
| 13. *Beliospermum axillare* Blume; Euphobiaceae. | 35. *Gloriosa superba* Linn.; Euphorbiaceae. |
| 14. *B. monanum* Huell, Arg. | 36. *Hera crepitans* Linn.; Euphorbiaceae. |
| 15. *Buxus sempervirens* Linn; Euphorbiaceae. | 37. *Hyoscyamus niger* Linn., Solanaeae. |
| 16. *Calotropis procera* R.Br.; Asclepiadeae. | 38. *Jatropha curcas* Linn., Euphorbiaceae. |
| 17. *Caltha palustris* Linn; Ranunculaceae. | 39. *Jatropha multifida* Linn., (Fl. Br. Ind., V. 383), is cultivated and naturalised in India; the seeds are a powerful acid poison, similar to those of the preceding. |
| 18. *Cannabis sativa* Linn; Urticaceae. | 40. *Kalanchoe spathulata* D.C., Crassulaceae. |
| 19. *Casearea graveoleus* Dalz; Samydaceae. | 41. *Lacturia tatarica* C.A., Meyer; Compositae |
| 20. *Cerbera odollam* Gaertn; Apocynaceae. | 42. *Lagenaria vulgaris* Seringe; Cucarbitaceae. |
| 21. *Chrozophora plicata* A. fuss; Fl. Bk. Ind. V. 209; Euphorbiaceae. The whole plant has acid poisonous properties. | 43. *Lasiosiphon ericephalus* Dene; Thymelaeaceae. |
| 22. *Coriaria nepalensis* Wall; Coriareae. | |

| | |
|---|---|
| 44. *Lathyrus sativus* Linn.; Loguminosae. | 59. *Rhododendron arboreum* Sm.; Ericaceae. |
| 45. *Lobelia nicoteanaefolia* Hayne, Campanulaceae. | 60. *R. canpanulatum* Don. |
| 46. *Manihot utilissima* Pohl., Euphorbiaceae. | 61. *Ricinus communis* Linn.; Euphorbiaceae. |
| 47. *Heconopsis aculeata* Royle; Papaveraceae. | 62. *Sapium insigne* Benth; Euphorbiaceae. |
| 48. *Heliotus alba* Lamk (Fl.Br.Ind. II.89); Leguminose; said to produce piosonous symptoms in cattle, which browse it. | 63. *Semecarpus anacardium* Linn. |
| 49. *Syn. Nerium odorum* Saland; Apocynaeceae. | 64. *Sophora mollis* Grah, *var hydaopis*; Leguminosae. |
| 50. *Papaver somniforum* Linn.; Papaveraceae. | 65. *Sorghum halepense* Pers., Gramineae. |
| 51. *Paspalum scrobiculatum* Linn.; Gramineae. | 66. *Strychnos colubrina* Linn, Lognaisceae. |
| 52. *Phoenix sylvestris* Roxb., Palmae (Fruit are said to be occassionally poisonous). | 67. *S. nux-vomica* Linn. |
| 53. *Physochlaina praealta* Hook, F., Solanceae. | 68. *Taxus baccata* Linn, Coniferae. |
| 54. *Phytolacea acinosa* Roxb., Phytolaccaceae. | 69. *Syn. Thevetia neriifolia* DC.; Apocynaceae. |
| 55. *Plumbago indica* Linn., Plumbagineae. | 70. *Trianthema peutandra* Linn, Ficoideae. |
| 56. *P. zeylanica* Linn. | 71. *Trigonella foenum-graecum* Linn; Leguminosae. |
| 57. *Ranunculus arvensis* Linn. Ranunculaceae. | 72. *T. polycerata* Linn. |
| 58. *Rauwolfia serpentina* Benth; Apocynaceae. | 73. *Tylophora fasciculata* Ham.; Ascesfiodeae. |
| | 74. *Withania somnifera* Dunal; Solanaceae. |

The list comprises most of the plants, parts of which are poisonous when eaten, or most commonly cause deleterious effects, when browsed for fodder by cattle, sheeps, goats or camels.

# Bibliography

## (A) Original Sources

### (1) Vedic Literature


*Atharvaveda* — English Translation by Whitney, William Dwight, Motilal Banarasi Das, Delhi, 1962, Svadhyaya Maṇḍala Aundh, V.S. 1995.

*Ṛgveda* — with comments of Sayana, Bombay, 1889-90.

*Sāmaveda* — Ed. by Rāmacharan Acharya, Mathura, 1960.

*Śukla Yajurveda* — Moradabad, 1940.


### (2) Post Vedic Literature

#### (a) Saṁhitās


*Taittirīya Saṁhitā* — With commentary of Sayaṇa, Asiatic Society of Bengal, Calcutta, 1860-1899.

*Vājasneyī Saṁhitā* — With commentary of Mahidhara, Calcutta, 1874.

*Maitrāyaṇīya Saṁhitā* — Leipzig, 1881-86, 1923.


#### (b) Brāhmaṇas


*Aitareya* — Anandāśrama Sanskrit Series, Poona, 1896.

*Jaiminīya* — Sarasvati Vihar Sanskrit Series, Lahore, 1937.

*Kauśītaki* — Ed. by E.B. Cowell, Calcutta, 1861.

*Pañcaviṁśa* — Ed. by A. Vedantavagisa, Calcutta, 1869-74.

*Śatapatha* — Srivenkatesvara Press, Bombay, 1942.

*Tāṇḍya* — Chaukhamba Sanskrit Series, Vārāṇasī, 1941.

*Ṣaḍaviṁśa* — Calcutta, 1911.


#### (c) Āraṇyakas


*Aitereya* — A.S.B., Calcutta, 1876.

*Sāṁkhyāyana* — Ed. A.B. Keith, Oxford, 1909.

*Taittirīya* — A.S.B., Calcutta, 1872.


#### (d) Upaniṣads


*Bṛhadāraṇyaka* — A.S.S., Poona, 1915.

*Chāndogya* — A.S.S., Poona, 1915.

*Īśavasyopaniṣad* — Astadasa Upaniṣads, Poona, 1958.

*Kaṭha* — —do—

*Kena* — —do—

*Kauśītaki* — —do—

| | | |
|---|---|---|
| | *Maṇḍūka* | Astadasa Upaniṣad, Poona, 1958. |
| | *Praśna* | —do— |
| | *Śvetāsvatāra* | —do— |
| | *Taittirīya* | —do— |
| (e) | **Gṛhya Sūtras** | |
| | *Āśvālayana* | Trivandrum Sanskrit Series, 1915. |
| | *Gobhila* | Translated by Oldenberg, S.B.E., Vol. XXX, Part-II. |
| | *Paraskara* | Kashi Sanskrit Series, Vārāṇasī, 1920. |
| **(3)** | **The Epics** | |
| | *Vālmīki Rāmāyaṇa* | Translated by Ralph and T.H. Griffith, Vārāṇasī, 1895. |
| | | —Geeta Press, Gorakhpur |
| | *Ādhyātma Rāmāyaṇa* | Ed., Vārāṇasī Sanskrit Sansthan, Vārāṇasī, V.S. 2041. |
| | *Mahābhārata* | Translated P.C. Roy, (12 Vols.), Calcutta. |
| **(4)** | **Purāṇas** | |
| | *Agni* | A.S.I., Poona, 1900. |
| | *Bhāgavata* | Gita Press, Gorakhpur. |
| | *Bhaviṣya* | S.V. Press, Bombay. |
| | *Brahma* | Anandasrama Press, Poona, 1895. |
| | *Brahmāṇḍa* | S.V. Press, Bombay. |
| | *Brahmavaivarta* | A.S.S., Poona, 1895. |
| | *Devī Bhāgavata* | S.V. Press, Bombay. |
| | *Garuḍa* | Translated M.N. Dutt, Calcutta, 1908. |
| | *Harivaṁśa* | S.V. Press, Bombay, 1911. |
| | *Kūrma* | Ed., N. Mukhopadhyaya, Calcutta, 1960. |
| | *Liṅga* | Ed., J. Vidyasagara, B.I. Calcutta, 1885. |
| | *Mārkaṇḍeya* | A.S.B., Calcutta, 1862. |
| | *Matsya* | A.S.S., Poona, 1907. |
| | *Padma* | A.S.S., Poona, 1893-94. |
| | *Śiva* | S.V. Press, Bombay. |
| | *Skanda* | S.V. Press, Bombay, 1910. |
| | *Vāmana* | All India Kasiraja Trust, Vārāṇasī, 1968. |
| | *Varāha* | —do—, 1963. |
| | *Vāyu* | A.S.S., Poona, 1905. |

| | | |
|---|---|---|
| | *Viṣṇu* | S.V. Press, Bombay, 1921. |
| | *Viṣṇudharmottara* | Gaekwad Oriental Series, No. 137, Baroda, 1961. |
| **(5)** | **Commentaries and Early Sanskrit Works** | |
| | *Nirkuta* by Yash | Bhandarkar Oriental Institute, Poona, 1918. |
| | *Aṣṭādhyāyī* by Pāṇini | Motilal Banarasi Das, Delhi, 1962. |
| | *Arthaśāstra* by Kauṭilya | Mehar Chand Laksman Das, New Delhi, 1988. |
| | *Bṛhatsaṁhitā* by Varāhamihir | Ed., Rāma Krishna Bhatta, Delhi, 1981-82. |
| | *Mahābhāṣya* by Pātañjali | Ed., B.J. Sastracarya, Bombay, 1911. |
| | *Manusmṛti* | Asiatic Society of Bengal, Calcutta, 1934. |
| **(6)** | **Buddhist Literature** | |
| | *Bandhacaryāvatāra* | Trs. Sharma, Parmananda, New Delhi, 1990. |
| | *FO-SHO-HING-TSAN-KING* (Life of the Buddha) | Samnal Beal and F. Max Mullers, Rep. Delhi, 1990. |
| | *Lalita Vistāra Mahāvastu* | Ed. R. Mitra, Calcutta, 1817. |
| | *Majjhima Nikaya* | S.B.B., London, 1926-27. |
| | *Milindapanho* | Trs. L.B. Horner, London, 1969. |
| | *Niṣpannayogāvalī* | Gaekwad Oriental Series, No. 53, Baroda, 1958. |
| | *Prākritadhammapada* | With Hindi Trs. Rahul Sankrityayana, Lucknow, 1957. |
| | *Sādhanamālā* | Gaekwad Oriental Series, Baroda, 1925-28. |
| | *Saddharmapuṇḍarīka* | Atlantic Publishers and Distributors, New Delhi, 1990. |
| | *Sumaṅgala-vilāsinī* | Ceylonese Text. |
| **(7)** | **Jaina Literature** | |
| | *Acarāṅga Sūtra* | *Jaina Sūtras*, Oxford, 1884. |
| | *Anuttaropapatikadasa-Sūtraṁ* | Lahore, Ed. 1936. |
| | *Bhagavatī Sūtra* | K.S. Lalwani, ed. |
| | *Jaina Jatakas* | Lahore, ed., 1925. |
| | *Triṣaṣṭiślākāpuruṣacarita* | Ed. M.M. Johu, Book-II. |
| | *Uttaradhyānakalpasūtra* | K.C. Lalwani, edn., Calcutta, 1977. |
| | *Uttara Purāṇa* | |
| **(8)** | **Works on Indian Medicines** | |
| | *Aṣṭāṅgahṛdaya* (Sūtra Sthāna) | Lakṣmī Venkatesvara Steam Press, Bombay, 1928. |
| | *Aṣṭāṅgahṛdaya* | Ed., Vijaya Ratna Sen, Calcutta, 1886. |

| | |
|---|---|
| *Caraka Cintan* | Ed., Priyavrata Sharma, Vārāṇasī, 1970. |
| *Sarangdhara Saṁhitā* | Nirnaya Sagara Press, Bombay, 1931. |
| *Śuśruta Saṁhitā* | Moti Lal Banarasi Das, Lahore, 1939. |
| *Vāgbhaṭṭa-Vivecana* | Ed., Priyavrata Sharma, Vārāṇasī, 1968. |

**(9) Classical Sanskrit Literature**

**4th Century B.C.**

| | |
|---|---|
| *Arthaśāstra* by Kauṭilya | Mehar Chand Lachhman Das Publishors, New Delhi, 1988. |
| *Pratijñāyaugandharayana* by Bhāṣa | Chaukhamba Sanskrit Series, II, Vārāṇasī, 1958. |
| *Svapnavāsavadattā* by Bhāṣa | Asiatic Society of Bengal. |
| *Manusmṛti* | Calcutta, 1934. |

**4th-5th Century A.D.**

**Kālidāsa's works**

| | |
|---|---|
| *Abhijñānaśākuntalaṁ* | Ed., M.R. Kale, Bombay, 1957. |
| *Kumārasaṁbhava* | Ed., M.R. Kale, Delhi, 1967. |
| *Mālavikāgnimitra* | Ed., T.H. Tawney, 2nd ed., Calcutta, 1891. |
| *Meghadūtam* | |
| *Raghuvaṁśa* | Chaukhamba Sanskrit Series, Vārāṇasī, 1961. |
| *Vikramorvaśīyam* | Ed., Pandit S.P., Bombay, 1879. |

**5th Century A.D.**

| | |
|---|---|
| *Kirātārjuniyaṁ* by Bharavi | Chaukhamba Sanskrit Series, Vārāṇasī, 1958. |

**7th Century A.D.**

| | |
|---|---|
| *Mālatīmādhava* by Bhavabhūti | Ed., Sahitya Prakashan, Vārāṇasī. |
| *Uttararāmacaritam* by Bhavabhūti | *Six Sanskrit Plays*, by H.H. Wells, Bombay, 1964. |
| *Nāgananda* by Śrī Harṣa | *Sri Harsa's Plays*, by Bak Kum Bae, (I.C.C.R.), Bombay, 1961. |
| *Ratnāvalī* by Śrī Harṣa | —do— |
| *Kādaṁbarī* by Bāṇa | |

**8th Century A.D.**

| | |
|---|---|
| *Mudrārākṣasa* by Viśākhadatta | Chaukhamba Sanskrit Series, Vārāṇasī, 1990. |
| *Veṇīsaṁhāra* by Bhaṭṭanārāyaṇa | |

**10th Century A.D.**

*Naladamayantī* or *Nalacampū* by Trivikramabhaṭṭa

*Navasahasrāṅkacarita* by Parimalapadmagupta

**11th Century A.D.**

*Vikramāṅkadevacaritaṁ* by Bilhaṇa — B.H.U., Vārāṇasī, 1964.

*Kāvyamīmānsā*

**12th Century A.D.**

*Abhilāṣitārthacintāmaṇi* (*Mānasollāsa*) by Someśvaradeva — Trs. by R. Shama Sastry, Mysore, 1926.

*Dvgosrayamahākāvya* by Hema Chandra Sūri

*Gītagovinda* by Jayadeva

*Pṛthvīrājavijaya* by Jayanka

*Rājataraṅgiṇī* by Kalhaṇa — Ed., Raghunatha Singh, Vārāṇasī, 1970.

**(10) Historical Mahākāvyas**

*Dvasvaya Mahākāvya* by Hema Chandra Sūri — Historical Mahakavyas by Chandra Prabha, New Delhi, 1976.

*Hammīrmahākāvya* by Nayachandra Sūri — —do—

*Kīrtikaumadī* by Someśvaradeva — —do—

*Madhurvijaya* by Gaṅgā Devī — —do—

*Navasahasrāṅkacarita* by Parimal Padmagupta — —do—

*Pṛthvīrājavijaya* by Jayanka — —do—

*Rājataraṅgiṇī* by Kalhaṇa — —do—

*Saluvabhyudaya*, by Rajanātha Dadema — —do—

*Sukṛta Saṅkırtana* by Arisimha — —do—

*Vasantavilāsa* by Bālacandra Sūri — —do—

*Vikramāṅkadevacaritam* by Bilhaṇa — —do—

## (B) Modern Works

Abul Fazl — *Aın-e-Akbari*, Trs. by Blackmen and Javel, Calcutta, 1873-74.

Agrawala, V.S., — *Indian Art*, Vārāṇasī, 1965.

—*Gupta Art*, Vārāṇasī.

—*Matsya Purāṇa—A Study*, Vārāṇasī, 1963.

Allan, J., — *Catalogue of Coins of Ancient India in British Museum*, London, 1936.

Archer, W.G., — *Indian Paintings*, London, 1936.

Banerjea, J.N., *Development of Hindu Iconography*, Calcutta, 1956.

Banerjea, R.D., *History of Orissa*, Calcutta, 1930.

Banerjea, S.C., *Flora and Fauna in Ancient Sanskrit Literature*, Calcutta, 1980.

Barrett Douglas and Basil Grey *Paintings of India*, Ohio, 1963.

Barua, B.M., *Bharhut.*

Beal, Samual, *Chinese Account of India*, Calcutta, 1969 (Rep.).

Beal, Samual and Max Muller, *The FO-SHO-HING-TSAN-KING* (Life of the Buddha) Rep. Delhi, 1990.

Bhattacharya, B., *Indian Buddhist Iconography*, Calcutta, 1925.

Bloomfield and Maurice *Hymns of the Atharvaveda*, Moti Lal Banarasi Das, Rep. Vārāṇasī, 1964.

Brown, Norman, *Chanhudaro Excavations*, New Heavens, 1947.

Bussagli Mario, *Indian Miniatures*, London, 1969.

Candolle, A.D., *Origin of Cultivated Plants*, New York, 1959 (Rep.).

Chandra, Pramod, *Bundi Paintings*, New Delhi, 1949.

Coomaraswamy *Yakṣa*, 2 Vols., Washington, 1928 and 1931.

—*History of Indian and Indonesian Art*, New York, 1965.

Crowesylvia and others *The Gardens of Mughal India*, London, 1972.

Devi Chandra *The Yajurveda*, Munshi Ram Manohar Lal, 1980.

Das, A.C., *Ṛgvedic India*, I, Calcutta, 1921.

Elliot and Dawson, *History of India as told by its own Historians*, London, 1857-77.

Folk Toby and Archer Mildrel *Indian Miniatures in India Office Library*, London, 1991.

George, Watt, *History of Economic Products of India*, New Delhi, 1972.

Ghosh, A., *Jaina Art and Architecture*, New Delhi, 1975.

Gode, P.K., *Studies in Indian Cultural History*, Bombay, 1939.

Griffith, R.T.H., *Hymns of the Atharvaveda*, Munshi Ram Manohar Lal, Delhi, 1985.

Ivan Stchoukine, *La Picture Indienne a l' Epoque des Grand Mughals*, Paris, 1929.

James Fergusson, *Tree and Serpent Worship in India*, Rep. Delhi, 1971.

John, H.M., *Triṣaṣṭiślākāpuraṣacarita*, Oriental Institute, Baroda, 1937.

Khandelawala, Kail, *Miniatures and Sculptures*, Bombay, 1965.

Keith, A.B., *Aitareya Āraṇyaka*, Oxford, 1909.

—*Religion and Philosophy of the Vedas and Upaniṣads*, Rep. Delhi, 1970.

—*Veda of the Black Yajurveda School,*

*Taittirīya Saṁhitā*, Part-2, Khaṇḍa-IV-VIII, Cambridge, 1914.

Kern H, and Max Muller *The Saddharama Puṇḍarīka*, Rep. Delhi, 1990.

Lal, Mukundi, *Garhwal Paintings*, New Delhi, 1968.

Limaya, V.P. and Vadekar, R.D., *Eighteen Principal Upaniṣads*, Vol.-I, Poona, 1958.

Mackey, E.J.H., *Further Excavations at Mohenjodaro*, New Delhi, 1937.

Marshall, John, *Mohenjodaro and Indus Civilization*, London, 1931.

Max Muller, E., *The Upaniṣads*, 2 Vols., Reprint, Delhi, 1990.

Mehta, Rustam J., *The Handicrafts and Industrial Art of India*, Bombay, 1960.

Moti Chandra *Mewar Paintings*, Bombay, 1967.

Nagar, S.L., *Jātakas in Indian Art*, New Delhi.

—*Protection, Conservation and Preservation of Indian Monuments*, New Delhi.

—*Temples of Himachal Pradesh*, New Delhi.

—*The Universal Mother*, Delhi.

Oldenberg, H. and F. Max Muller *The Gṛhya Sūtras*, Reprint Delhi, 1990.

Prabha Chandra *Historical Mahākāvyas*, New Delhi, 1976.

Randhawa, M.S., *Basohli Paintings*, New Delhi, 1959.

—*Garden Through the Ages*, New Delhi.

—*Kangra Paintings on Love*, New Delhi, 1962.

—*Kangra Valley Paintings*, New Delhi.

—*Paintings from the Bhāgavata Purāṇa*, New Delhi, 1960.

Randhawa, M.S. and Bhambri, *Basohli Paintings on Rasamañjarī*, Delhi, 1981.

Rao, S.R., *Lothal—A Harappan Township*, New Delhi, 1979.

Rhys, Davids, T.W., and F. Max Muller *The Questions of King Malinda*, Reprint, Delhi, 1990.

Sharma, Acharya Priyavrata *Āyurveda Kā Vaijñānika Itihāsa* (Hindi), Vārāṇasī, 1981.

Sharma, Parmananda, *Bodhicaryāvatāra*, New Delhi, 1990.

Sharma, P.V., *Fruits and Vegetables in Ancient India*, Delhi, 1979.

Singh Govardhan, Mian, *Art and Architecture of Himachal Pradesh*, Delhi, 1983.

Singh, Satya Prakash — *Upaniṣadic Symbolism*, Delhi, 1981.

Sinha, R.N.P., — *Gīta Govinda in Basohli Paintings*, Calcutta, 1958.

Smith, V.A., — *History of Fine Arts in India and Ceylone*, Oxford, 1930.

Vats, M.S., — *Excavations at Harappa*, Delhi, 1940.

Vātsyāyana, — *Kāmasūtra*, Trs. by Richard Burton and F.F. Arbuthnot, London, 1963.

Voghel, J. Ph., — *Women and Tree or Śālabhañjikā*.

White Head R.B., — *Punjab Museum Catalogue of Coins*, Oxford, 1914.

Williams Moneer, — *Sanskrit-English Dictionary*, Oxford, 1960.

Yazdani, G., — *Ajanta*, Oxford, 1955.

## (C) Memoirs, Periodicals and Journals

*Memoirs of Jehangir*, — Trs. by Henry Biveridge, II, Ed., Delhi, 1969.

*Memoirs of Zahiruddin Muhammad Babur*, — Trs. by John Leyden and William, Erskina, London, 1926, Vol. II.

*A.S.I.A.R.*, — 1911-12; 1927-28.

*A.S.R.*, — III, X.

*Journal of Indian Antiquary*, — VIII, 1900.

*Bulletin of Indian Institute of History of Medicine*, — Vol. VIII, No. 1 & 2 (January and April), 1977.

# Index of Indigenous Terms

## A

## B

## J



## K

## L



## M

## N



## O



## P

## R

## S

## T

## U

## V

## W



## Y

# Index of Botanical Terms

## A

## B



## C

# D

# E



# F



# G

## H



## I



## J



## K



## L

# Q



# R

# General Index

## A

## B

## C

# D

## E



## F

## G

## H



## I

## J



## K

## L



## M

# N

# O



# P

## Q



## R

## S

## V

## W



## Y



## Z